PLATE 1

1 2

3 4

5 6

Plate 1. 1. Maize (*Zea mays*) showing external signs of stem-borer damage. 2. Pearl millet (*Pennisetum glaucum*), Niger. 3. Rice showing deadheart symptoms of stem-borer damage (Nigeria). 4. Rice *(Oryza sativa)* showing whitehead symptoms of stem-borer damage (Nigeria). 5. Sorghum (*Sorghum bicolor*), Niger. 6. Papyrus (*Cyperus papyrus*), a wild host of the stem borer (*Eldana saccharina*), South Africa.

PLATE 2

Plate 2. 1. *Chilo partellus* larva diapausing in winter in the lower part of a sorghum stalk. 2. *Busseola fusca* larvae diapause in winter on the lower part of a maize stalk. 3. A *Chilo partellus* larva in a sorghum stalk. 4. Damage to sorghum by *Busseola fusca*. 5. Damage to a maize cob by *Busseola fusca*. 6. An adult of the parasitoid *Cotesia sesamiae* (Braconidae).

PLATE 3

Plate 3. 1. *Cotesia sesamiae* larvae emerging from their host, a *Chilo partellus* larva. 2. Cocoons of *Cotesia sesamiae* next to the cadaver of their host, a *Chilo partellus* larva. 3. *Bracon sesamiae* larvae feeding as gregarious ectoparasitoids on a *Busseola fusca* larva. 4. Gregarious pupal endoparasitoids, *Pediobius furvus* (Eulophidae). 5. Omni-directional pheromone traps for monitoring *Busseola fusca* flight. 6. Moth emergence traps set in a grain sorghum field to determine the effects of various mechanical treatments of stalks on the survival of diapausing *Busseola fusca* and *Chilo partellus* in winter.

PLATE 4

1 2 3 4 5 6

Plate 4. Noctuidae: 1–4. *Busseola fusca*. 1. Eggs. 2. Larva. 3. Pupa. 4. Adult. 5, 6. *Sesamia calamistis*. 5. Larva. 6. Pupa.

PLATE 5

1 2 3 4 5 6

Plate 5. 1. Noctuidae: *Sesamia calamistis*, adult. 2–3. *Sesamia cretica*, adult. 4–6. Cossidae: *Phragmataecia boisduvalii* in *Phragmites* sp. 4. Larva. 5. Pupa. 6. Adult.

PLATE 6

Plate 6. 1, 2. Crambidae: *Scirpophaga* sp. 1. Egg mass on rice. 2. Adult. 3–6. *Chilo zacconius*. 3. Egg mass. 4. Larva. 5. Pupa. 6. Adults.

PLATE 7

Plate 7. 1–3. Crambidae: *Chilo partellus*. 1. Larva. 2. Diapause larva. 3. Adult. 4. *Chilo orichalcociliellus*, adult. 5, 6. Differing forms of *Chilo sacchariphagus* larvae.

PLATE 8

Plate 8. 1–4. Crambidae: *Coniesta ignefusalis*. 1. Egg mass in the sheath of millet. 2. Larva. 3. Pupa. 4. Adult. 5, 6. Pyralidae: *Eldana saccharina*. 5. Eggs. 6. Larva.

PLATE 9

Plate 9. 1. Pyralidae: *Eldana saccharina*, adult. 2–6. Pyralidae: *Maliarpha separatella*. 2. Exposed eggs. 3. 'Pinched' rice leaf concealing egg mass. 4. Larva. 5. Pupa. 6. Adult.

PLATE 10

1 2 3 4 5 6

Plate 10. 1. Chalcididae: *Psilochalcis soudanensis*. 2. Braconidae: *Cotesia sesamiae*, larvae. 3. Braconidae: *Parallorhogas pyralophagus*. 4. Braconidae: *Stenobracon rufus*. 5. Ichneumonidae: *Dentichasmias busseolae*. 6. Ichneumonidae: *Xanthopimpla stemmator*.

PLATE 11

1

4

3

4

5

6

Plate 11. 1. Tachinidae: *Chromatophania picta*, adult. 2. Pupa of *Schembria eldana* in *Cyperus* sp., Naivasha, Kenya. 3. A trichogrammatid ovipositing in eggs of *Eldana saccharina*. 4. *Telenomus applanatus* ovipositing in eggs of *Eldana saccharina*. 5. Eurytomidae: *Eurytoma braconidis* a hyperparasitoid of *Stenobracon unifasciatus* in Casamance, Senegal. 6. Pteromalidae: *Norbanus* sp., a gregarious ectoparasitoid of *Sesamia cretica* in Sudan.

PLATE 12

1

2

3

4

5

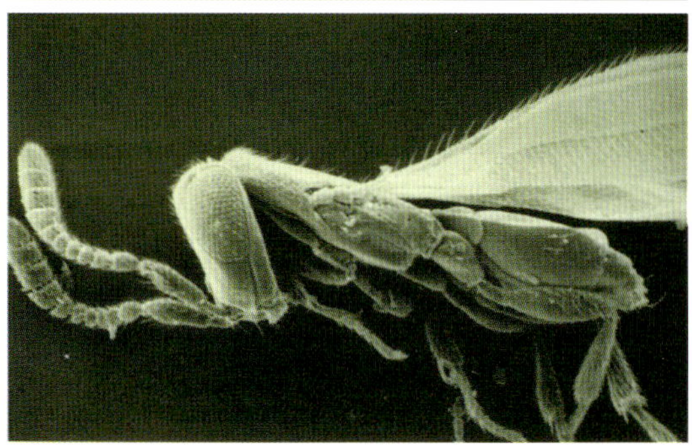
6

Plate 12. 1. Tettigoniidae: *Metioche* sp. feeding on *Chilo* eggs. 2. Formicidae: ants predating on *Chilo partellus*. 3–5. Bethylidae: *Goniozus indicus*. 3. Eggs on *Eldana saccharina* larva. 4. Developing larvae (gregarious ectoparasitoids). 5. Adult female. 6. Scelionidae: *Telenomus isis*, a dorso-ventrally depressed species adapted for seeking eggs concealed in leaf sheaths.

African Cereal Stem Borers: Economic Importance, Taxonomy, Natural Enemies and Control

African Cereal Stem Borers

Economic Importance, Taxonomy, Natural Enemies and Control

Edited by

ANDREW POLASZEK

CABI Bioscience UK Centre, c/o Entomology Department, The Natural History Museum, London SW7 5BD, UK

1998

CAB INTERNATIONAL

in association with the

ACP-EU Technical Centre for Agricultural and Rural Co-operation (CTA)

CAB INTERNATIONAL
Wallingford
Oxon OX10 8DE
UK

Tel: +44 (0)1491 832111
Fax: +44 (0)1491 833508
E-mail: cabi@cabi.org

CAB INTERNATIONAL
198 Madison Avenue
New York, NY 10016-4314
USA

Tel: +1 212 726 6490
Fax: +1 212 686 7993
E-mail: cabi-nao@cabi.org

A catalogue record for this book is available from the British Library, London, UK.

Library of Congress Cataloging-in-Publication Data

African cereal stem borers : economic importance, taxonomy, natural
enemies and control / edited by Andrew Polaszek.
p. cm.
Includes bibliographical references (p.) and index.
ISBN 0-85199-175-0 (alk. paper)
1. Stem borers–Integrated control–Africa. 2. Grain–Diseases
and pests–Integrated control–Africa. I. Polaszek, Andrew.
II. ACP-EU Technical Centre for Agricultural and Rural Co-operation
(Wageningen, Netherlands)
SB945.S74A47 1998
633.1′04978′096–dc21 97-46738
CIP

ISBN 0 85199 175 0

Published in association with:
The ACP-EU Technical Centre for Agricultural and Rural Co-operation
(CTA) Postbus 380
6700 AJ Wageningen
The Netherlands

Typeset in 10 on 12pt Garamond by Columns Design Ltd, Reading
Printed and bound in the UK at the University Press, Cambridge

Contents

Preface

Stem borers are preeminent among Africa's most serious cereal pests, and have therefore attracted a great deal of attention from economic entomologists for most of this century. The aims of these entomologists working earlier this century were no different from ours today, i.e. to prevent or reduce damage to cereal crops, whether on large-scale farms or under subsistence conditions. However, the methods and philosophies underlying these aims have changed drastically over the course of the century. A decreasing reliance on chemical control, a realization of the unsuitability of chemical methods for subsistence farmers in particular, and an increasing awareness of the uses of, and effects upon, arthropod biodiversity, have characterized the evolution of our ideas on stem-borer control. The catastrophic introduction into Africa of the Asian borer, *Chilo partellus*, has naturally led to many investigations into classical biological control of stem borers, which finally appear to be succeeding. It is the increasing awareness of the effects of various control methods, especially chemical methods, upon Africa's arthropod biodiversity, coupled with a holistic approach to understanding and utilizing agrobiodiversity, that have led to the current study. A number of previous investigations, many otherwise of great merit, have been marred by a lack of adequate characterization of the pest and, particularly, the beneficial organisms involved. It is one of the main aims of this book to enable that accurate characterization. Previous notable surveys of African borers and their natural enemies include those by Harris (1962) for Nigeria, Jordan (1966) on rice borers in West Africa, Milner's (1967) memorable report for East Africa, and several recent papers by D. Conlong, and R. Kfir for South Africa, and G. Phiri for Malawi. Useful contributions from francophone Africa have been produced by J. Appert, J. Brenière and J. Etienne, although much of this important work remains obscured in unpublished reports. Among previous attempts to synthesize available information those by H.E. Box and W.F. Jepson, published by the International Institute of Entomology in subsequent years (1953 and 1954, respectively) are still very useful. However, the cereal stem borer situation in Africa is changing rapidly from one year to the next, and a contemporary synthesis is much needed.

Although our understanding of the agroecology of borers has progressed well in recent years, particularly for Kenya and South Africa, a great deal of work is still required, especially for central Africa (the Congos, Central African Republic) and

West-Central Africa (Angola, Cameroon, Gabon). This book should go some way towards assisting the characterization of stem borer-associated agrobiodiversity resulting from future surveys in the continent.

This book has evolved from several earlier drafts, written specifically for training courses held at ICIPE (The International Centre of Insect Physiology and Ecology), Nairobi, Kenya, 8–9 August 1994 and 15–28 October 1995 and the Département de Formation en Protection de Végétaux in Niamey, Niger, 19–30 September, 1994. Its purpose is to give a comprehensive account of the economic importance of lepidopterous cereal stem boring pests in Africa, with particular emphasis on the correct recognition of both the borers and their natural enemies, and brief accounts of control principles. It is arranged in four main sections, dealing with economic importance, taxonomy of borers, natural enemies and control.

The authors' intention is not for the book to be read from front to back cover or in any particular order, rather that various chapters can be used to provide some of the basic information required for region-specific IPM development, for the major African cereals. The introductory sections to Lepidoptera, Diptera and Hymenoptera should be thoroughly studied before attempting the identification of either borers or their natural enemies, particularly where the reader may be unfamiliar with the terminology used.

ACKNOWLEDGEMENTS AND ABBREVIATIONS

Without the productive collaboration of many research institutes both in Africa and Europe, including various National Programme staff, the project leading ultimately to this book would not have been possible. That project was largely financed by the Netherlands Government's Directorate General for International Cooperation (DGIS), following a request from the Phytosanitary Council of the Organization for African Unity (OAU) through the Food and Agriculture Organization of the United Nations (FAO). Without the initial support given to the project proposal by Arnold van Huis (Wageningen) and Gerard Schulten (FAO) the project would not have gone forward. Publication was financed by the ACP-EU Technical Centre for Agricultural and Rural Co-operation (CTA), Wageningen, The Netherlands, and the cooperation of Alan Jackson and Chantal Guiot is particularly acknowledged.

As well as the various authors contributing to this book, whom I thank above all for their patience, the following institutes and individuals deserve particular mention.

Field collection in Africa was facilitated by the cooperation of the following organizations: Centre National de Recherches Appliquées au Dévéloppement Rural (CENRADERU), Madagascar; Département des Cultures Vivrières de l'Institut des Savanes (DCV-IDESSA), Côte d'Ivoire; Institut Sénégalais de Recherches Agricoles (ISRA), Senegal; International Institute of Tropical Agriculture (IITA), Nigeria; National Irrigation Board, Kenya; Plant Protection Division, Ministry of Agriculture & Livestock Development, Tanzania; Sous-Direction de la Protection des Végétaux – Ministère de l'Agriculture (SDPV-MINAGRI), Cameroon.

Specimens were kindly made available by the following institutes and/or individuals: O. Ajayi, International Crops Research Institute for the Semi-Arid Tropics

(ICRISAT), N.D. Bafokuzara, Kawanda Perennial Crops Research Station (KPCRS) Uganda; B. Bassa and V. Caleca (Palermo); Philippe Bousses, Projet 'Projection intégrée en riziculture', Lac Alaotra, Madagascar; J. Casevitz-Weulersse, M. LaChaise and C. Villemant, Muséum National d'Histoire Naturelle, Paris, France (MNHN); Centre de Coopération Internationale en Recherche Agronomique pour le Développement, Montpellier, France (CIRAD: Gérard Delvare, Henri-Pierre Aberlenc); Des Conlong, South African Sugar Association Experiment Station (SASA), Mount Edgecombe; Eliane De Coninck, Musée Royale de l'Afrique Centrale, (MRAC, Tervuren, Belgium); Domingos Cugala, Universidade Eduardo Mondlane, Maputo, Moçambique; John Deeming (National Museum of Wales, Cardiff, UK); Paul Dessart, Institute Royal des Sciences Naturelles de Belgique (IRSNB), Brussels, Belgium; S. Djiba, Institute Senegalais de la Recherche Agronomique, Ziguinchor, Senegal (ISRA); E. Diller, Zoologische Staatssammlung, Munich, Germany; M. Fischer, Naturhistorisches Museum, Vienna (NMW); International Centre of Insect Physiology and Ecology, Nairobi, Kenya (ICIPE: Susan Wangari Kimani-Njogu, Raphael Odhiambo, Bill Overholt); International Institute of Tropical Agriculture, Cotonou, Benin (IITA); F. Koch and A. Kleine-Möllhoff, Zoological Museum of the Humboldt-University, Berlin, Germany (ZMB); Karl Krombein and Paul Marsh, National Museum of Natural History, Smithsonian Institution (USNM), Washington, USA; K. Maetô, Ibaraki, Japan; National Cereals Research Institute, Moor Plantation substation, Nigeria (NCRI/MP); National Museums of Kenya, Nairobi (NMK); The Natural History Museum, London, UK (NHM: Z. Bouček, Mike Fitton, Ian Gauld, Martin Honey, Tom Huddleston, Suzanne Lewis, John Noyes, Laraine Tarel); J. Papp, Magyar Természettudományi Museum, Budapest, Hungary; George Phiri, Makoka Research Station, Malawi; Gerhard Prinsloo, Plant Protection Reserach Institute (PPRI), Pretoria, South Africa; Donald Quicke, Imperial College, London, UK; H. Robertson and M.A. Cochrane, South African Museum, Cape Town, South Africa; G. Simbolotti, L'Aquila, Italy; Bob Wharton, Texas A & M University, College Station, USA.

The following helped greatly with the retrieval of literature: Paul Cochereau, Marie-Claude Deboine, A. Diallo, J. Etienne, S. Quartey and the library staff at the Natural History Museum, London, Leiden Museum and the entomology library at Wageningen Agricultural University. Several people assisted the editor in various ways: Ineke Kok, Linda Millar, Mo Mookherjee and Lisa Rolaston. Henri-Pierre Aberlenc, Roger Crosskey, Ian Gauld, and David Wahl very generously provided either original illustrations or excellent copies of their own or others' figures. Several authors have also requested that the following be acknowledged for their help: B. Bolton, R.G. Booth, J.D. Möhr, P. Hillyard, A. van Huis and N.J. van Rensburg.

I was very ably assisted during the final stages of manuscript preparation by Lynn Meijerman who put together most of the plates and prepared illustrated camera ready copy for the key to Ichneumonidae genera, and Geoffrey Kibby, who designed and prepared the cover and colour plates, using computerized image enhancement, as well as preparing the glossary.

Finally, I will always be grateful to Tom Huddleston (NHM) for providing the initial stimulus and encouragement that led to the realization of this project.

PHOTO CREDITS

CIRAD: Plate 4:1,6. Plate 6:1–6; Plate 7:6; Plate 8:5; Plate 9:2–6.
D. Conlong: Plate 11:2,3,4; Plate 12:3,4.
Z. Dabrowski; Plate 4:2,3; Plate 8:2.
K.M. Harris: Plate 1:1; Plate 11:1.
IIBC: Plate 10:6. Plate 12:1.
G. du Heaume: Plate 4:4.
R.C. Joshi: Plate 1:3,4; Plate 5:1.
R. Kfir: Plates 2 and 3.
R. Markham: Plate 10:4.
M. Matthews: Plate 8:1,3.
A. Ngi-Song: Plate 10:2.
W.A. Overholt: Plate 7:2; Plate 12:2.
W.W. Page: Plate 10:1,5.
P. Päts: Plate 7:4.
A. Polaszek: Plate 1:2,5,6; Plate 4:5. Plate 5:2,3; Plate 7:3; Plate 8:4,6; Plate 9:1; Plate 11:5,6; Plate 12:5,6.
R. Potting: Plate 7:5.
M. Rose: Plate 10:3.
The Natural History Museum: Plate 5: 4,5,6; Plate 7:1.

I Economic Importance

1 Host Plants

A. Polaszek[1] and Z.R. Khan[2]
[1]*CABI Bioscience UK Centre, c/o Entomology Department, The Natural History Museum, London SW7 5BD, UK;*
[2]*International Centre of Insect Physiology and Ecology, PO Box 30772, Nairobi, Kenya*

INTRODUCTION

In order to effectively prevent or reduce stem-borer damage to our cereal crops, a detailed knowledge and proper understanding of the potential host plants of stem borers are essential. We need to know, if possible, their original or natural hosts, their possible host ranges and the hosts on which they develop most successfully. We can assume that the original hosts of all cereal stem borers were wild grasses, but which grasses in particular and whether the original hosts are still present in the areas where particular borer species are pests are often not known and merit further study.

In Africa, the most important cereals or graminaceous crops are maize, millets, rice and sorghum, cereals whose history and biology are extremely diverse, but an understanding of which is essential in the development of control strategies for borers. Within the cereal/borer complex in Africa are both indigenous and introduced cereals, as well as both indigenous and introduced stem borers. The following account is a summary of the history, distribution and importance of the main cereal crops grown in sub-Saharan Africa and some of the associated islands, in relation to African cereal stem borers. Alternative hosts of cereal stem borers are also summarized.

MAIZE

Cultivated maize (*Zea mays*, Plate 1) as we know it today originated in Central America, probably within the area equivalent to modern Mexico. When Columbus arrived in America at the end of the 15th century, maize was already an important food crop among the South Americans from Mexico to Peru. By the 16th century, maize was already being cultivated in southern Europe, parts of West and southern Africa, and Asia. By the 17th century, maize was present in virtually all areas of Africa in which it could be cultivated. Despite the existence of African maize for over 300 years, it must still be considered a relatively recent introduction in comparison with the more traditional African cereals, such as millet and sorghum, in terms of both agroecology and insect–plant coevolution. In Africa, maize is used as both human and animal food, eaten directly as grilled cobs or as various products of maize flour. It is easily stored after drying and/or milling.

In general, maize in Africa is grown by small-scale farmers for local consumption, and yields tend to be very low, averaging less than half of Asian and Latin American yields (FAO, 1993). In recent years, however, there have been attempts to cultivate more extensive maize monocultures with the aid of irrigation, for example in northern Senegal. On small farms, maize is very often intercropped with legumes, such as groundnuts, cowpeas or haricot beans. Principal producers of maize in sub-Saharan Africa are Kenya, South Africa and Tanzania, followed by Ethiopia and Nigeria, although South Africa is the only one of these exporting maize in any quantity.

The stem-borer species most often associated with maize are the noctuids *Busseola fusca* and various *Sesamia* species, particularly *S. calamistis* and *S. cretica.* Throughout Africa, *B. fusca* is known as the maize stalk borer. Important crambid and pyralid borers of maize in Africa are *Eldana saccharina* (Pyralidae) and several *Chilo* species (Crambidae). *Chilo partellus,* a species introduced accidentally from Asia earlier this century is one of the major pests of maize in East and Southern Africa. At the time of writing, it has not yet become a pest in West Africa. Other *Chilo* species that can be pests of maize include *C. aleniellus* and *C. diffusilineus* in West Africa and *C. orichalcociliellus* in East Africa and Madagascar. *Coniesta ignefusalis,* the principal borer of millet, also attacks maize in some parts of West Africa.

Major references: Acland (1971); Harrison (1981); Rouanet (1984); Ibekwe (1986).

MILLETS

Africa produces almost half of the world's millet, and almost a quarter is produced by Niger and Nigeria alone (FAO, 1993). Four main kinds of millet are cultivated in Africa, summarized in Table 1.1. Among these, pearl millet, *Pennisetum glaucum* (Plate 1:2), is by far the most important. Available evidence suggests that pearl millet, also known as bulrush millet, or *mil à chandelles* or *petit-mil* in West Africa, originated in Africa and was subsequently spread throughout the world. There is still some debate over whether pearl millet originated in West Africa or in the Abyssinian region (present-day Ethiopia) (Rachie and Majmudar, 1980).

Pearl millet is a fast-growing, robust cereal that uses water more efficiently

Table 1.1. Millets in Africa.

Species	Common names	Regions
Pennisetum glaucum	Pearl millet, bulrush millet, *mil à chandelles, petit-mil*	Throughout Africa
Eleusine coracana	Finger millet	East Africa
Eragrostis tef	Teff	Ethiopia
Digitaria exilis, D. iburua	Fonio, hungry rice	West Africa

than either sorghum or maize, as well as being more heat tolerant. In Africa it is of greatest importance in the Sahel region, where it is the principal cereal. It is the staple cereal in Namibia (Anon., 1991).

In West Africa, the principal stem borer of pearl millet is *C. ignefusalis. Busseola fusca* has been recorded attacking pearl millet in both East and West Africa. Several *Sesamia* species, as well as *C. partellus* and *E. saccharina* are also known to attack *P. glaucum.*

Like pearl millet, finger millet (*Eleusine coracana*) is now considered to have originated in Africa (Ethiopia) and later imported to Asia, where it assumed greater importance than it had in Africa (Rachie and Peters, 1977). It is still grown extensively in East Africa, especially around Lake Victoria, and also in Burundi, Rwanda and eastern Zaïre, as far south as Zimbabwe, and in Madagascar. Stem borers attacking *E. coracana* include the noctuids *B. fusca, S. botanephaga* and *S. calamistis.*

Among the less important millets, teff (*Eragrostis tef*) is almost exclusively grown in Ethiopia, and cultivation of fonio (*Digitaria* spp.) is restricted to a few West African countries, among which Guinea and Mali are the largest producers.

Major references: Acland, 1971; Rachie and Peters, 1977; Rachie and Majmudar, 1980; Spencer and Sivakumar, 1987.

RICE

The indigenous African rice, *Oryza glaberrima*, or red rice, probably originated in the Niger delta about 3500 years ago, and was widely cultivated in that area. Its cultivation is now largely restricted to northern Nigeria and Sierra Leone, having been replaced by the imported Asian *Oryza sativa* (Plate 1:3,4). *Oryza sativa* varieties from Guyana and Sri Lanka were introduced into Nigeria in about 1890, but the date and origin of the first introductions of *O. sativa* into Africa are not known. While some suggest it was brought to the west coastal regions by Portugese traders, it is equally likely that *O. sativa* came by trans-Saharan caravan routes (Grist, 1983). *Oryza sativa* is grown as many different varieties or cultivars, depending on the type of rice cultivation. Nigeria is the only African country south of the Sahara currently producing rice in any quantity, most African countries relying largely on imported rice. Within the Afrotropical region, Madagascar is another important rice producer.

There are two fundamental kinds of rice cultivation, deep-water (submerged) and upland (non-submerged), both of which are to be found in Africa. Further subdivisions of these categories to be found in Africa are as follows:

1. Deep-water (fluxial) rice.

a. Mangrove rice. Found at river estuaries and periodically invaded by salt water, mangrove rice cultivation is affected to a great extent by salinity. About 10% of all African rice is grown in mangroves along the West African coast from Gambia to Nigeria, in particular in Guinea-Bissau, Senegal and Sierra Leone.

b. Freshwater rice. Deep-water, non-irrigated rice cultivation (= lowland swamp) accounts for 23% of all African rice cultivation. Flotation rice cultivation, an extreme example of deep-freshwater rice cultivation, can be found in Mali and Niger. Deep-water irrigated rice cultivation, common in Asia, accounts for less than 5% of the total area of African rice cultivation.

2. Upland rice.

a. Pluvial (dryland) or rain-fed rice. Sixty per cent of the total area under rice cultivation in Africa is rain-fed. A few countries account for the majority of acreage: Ivory Coast, Zaïre, Sierra Leone, Liberia, Nigeria and Madagascar.

b. The remaining 2.5% of African rice is what is known as hydromorphic or phreatic, where cultivation occurs on land watered by the retreating annual high groundwater level. In West Africa, this is known as '*riziculture de nappe*'.

The most important stem borers attacking rice in Africa are the noctuid *Sesamia* species, in particular *S. calamistis. Busseola fusca* has been reported from rice very occasionally. Among the Crambidae and Pyralidae, several *Chilo* species, in particular *C. zacconius*, are rice pests. *Maliarpha separatella* can be an important pest of rice. Several *Scirpophaga* species attack rice in Africa but, although they can be locally or periodically important, they are never the major pests that *Scirpophaga* species constitute in Asian rice. *Eldana saccharina* can also be found in rice.

Major references: Angladette, 1966; Buddenhagen and Persley, 1978; Grist, 1983; Jacquot and Courtois, 1983; Anon., 1984; IITA, 1984; Alam *et al.*, 1985; Khan *et al.*, 1991.

SORGHUM

Like pearl millet and finger millet, sorghum (*Sorghum bicolor*, Plate 1:5) is a native African cereal, originating in the Sudan/Ethiopian border region, and has since spread throughout the world (De Wet, 1978; Doggett, 1988). It is drought-resistant and outyields maize in drier areas, as well as being more resistant to waterlogging than millet. Sorghum grain is used for human consumption, including brewing, stalks are used as animal feed, fuel, house and fence building and mulching.

Almost half of the world's total sorghum acreage is in Africa, the most important producers being Nigeria and Sudan, followed by Ethiopia and some of the Sahelian countries.

The most important stem borers of sorghum are the noctuids *B. fusca, S. botanephaga, S. calamistis* and *S. cretica*, and the introduced crambid *C. partellus*.

Major references: Acland (1971); De Wet (1978); ICRISAT (1984a, b, 1986a, b); Doggett (1988).

SUGAR CANE

Sugar cane (*Saccharum officinarum*) differs from the other cereals summarized here, because it is grown virtually exclusively as a cash crop, rather than as a subsistence crop for local consumption. In Africa, sugar cane is an exotic crop, originating in Asia. The first recorded cultivation of sugar cane in sub-Saharan Africa is in the Cape Verde Islands in the early 15th century.

Sugar-cane plantations producing relatively large quantities are found in the following sub-Saharan African countries: Kenya, Sudan, Swaziland and Zimbabwe, Mauritius is also an important producer of sugar cane. South Africa is by far the greatest producer of cane sugar in the continent (FAO, 1993).

The most important stem borer attacking African sugar cane is the pyralid *E. saccharina*. *Sesamia calamistis* (Noctuidae) is frequently found infesting sugar cane but is rarely of any economic importance. *Busseola fusca* can also be found occasionally in sugar cane, but never in pest numbers.

References: Artschwager and Brandes (1958); Blackburn (1984).

CITRONELLA GRASS

Citronella grass, or lemon grass (*Cymbopogon nardus*, *Cymbopogon winterianus*), has been introduced from Asia into various parts of Africa for the production of citronella oil, used in the perfume and cosmetic industries. Recently, unidentified species of stem borers have been observed attacking this crop in Ghana, causing serious damage (M. Botchey, personal communication).

WILD HOSTS

Most African cereal stem borers are generally polyphagous and have several graminaceous and other wild hosts, in addition to more than one cultivated crop. Several non-cultivated wild host plants of these stem borers have been recorded and documented by various workers. As might be expected, the spectrum of wild, non-crop hosts attacked by stem borers is vast. No doubt, the number of actual wild hosts is even greater, given that stem-borer damage to wild grasses is unlikely to be investigated. Hosts are found in the three main families – *Cyperaceae* (Plate 1:6), *Gramineae* (*Poaceae*) and *Typhaceae*. An extensive survey carried out by the International Centre of Insect Physiology and Ecology (ICIPE) scientists during 1994–1997, where every possible kind of host plant was examined at different times of the year in various agroecologies throughout Kenya, resulted in the compilation of Table 1.2. No plant was given host-plant status until any larva found in it had been reared through to the adult stage.

In the past, many host-range records did not take into account how the plant–insect association was determined. In nature, an insect locates a host plant through a sequence of behavioural responses. In the laboratory, feeding a larva in a closed environment ignores this natural feeding process and only measures the insect's acceptance of the plant as food. Also, increasing evidence indicates that the host ranges and economic importance of most insects are dynamic, and often location- and time-specific. For example, *B. fusca*, *C. orichalcociliellus*, *E. saccharina* and *S. calamistis*, the major African stem borers, originally would have attacked only wild graminaceous plants in tropical and subtropical parts of the continent. As long as there were only wild plants, these insects were of no special consequence. After the introduction and cultivation of maize and extensive planting of sorghum in vast areas of Africa where they did not occur originally, the insects followed the

Table 1.2. Wild hosts of cereal stem borers recorded from various parts of Kenya (information collected and compiled by Z.R. Khan from different agroecological zones of Kenya).

	Borer species						
Plant species	CL	BF	SC	ES	PB	BS	PS
Gramineae (Poaceae)							
Andropogon sp.	–	–	++	–	–	–	–
Cenchrus ciliaris	++	–	++	–	–	–	–
Coix lacryma-jobi	++	–	++	–	–	–	–
Dactyloctenium bogdanii	++	–	–	–	–	–	–
Echinochloa haploclada	++	–	++	–	–	–	–
Echinochloa pyramidalis	++	++	++	–	++	–	–
Hyparrhenia cymbaria	–	++	–	–	–	–	–
Hyparrhenia filipendula	++	–	++	–	–	–	–
Hyparrhenia pilgerana	++	++	–	–	–	–	–
Hyparrhenia rufa	++	++	++	–	–	–	–
Panicum deustum	++	++	–	–	–	–	–
Panicum maximum	++	++	++	++	–	–	–
Pennisetum macrourum	–	–	–	–	–	–	??
Pennisetum procerum	–	–	–	–	–	–	??
Pennisetum purpureum	++	++	++	++	–	–	++
Pennisetum trachyphyllum	++	++	–	–	–	–	??
Phragmites sp.	++	–	++	++	++	–	–
Rottboellia cochinchinensis	++	++	–	++	–	–	–
Setaria incrassata	++	++	–	–	–	–	–
Setaria sphacelata	–	++	++	–	–	–	–
Sorghum arundinaceum	++	++	++	++	–	–	–
Sorghum versicolor	++	++	++	++	–	–	–
Sorghum vulgare var. *sudanense*	++	++	++	++	–	–	–
Sporobolus marginatus	++	++	–	–	–	–	–
Sporobolus pyramidalis	–	++	–	–	–	–	–
Tripsacum laxum	–	++	++	–	–	–	–
Vossia spp.	–	–	++	–	–	–	–
Cyperaceae							
Cyperus distans	–	–	++	++	–	??	–
Cyperus immensis	–	–	++	++	–	++	–
Cyperus maculatus	–	–	–	++	–	??	–
Cyperus papyrus	–	–	++	++	–	??	–
Typhaceae							
Typha domingensis	–	++	++	–	–	–	–

??, Possible host; ++, Recorded as a host; –, not recorded as a host; CL, *Chilo* sp.; BF, *Busseola fusca*; SC, *Sesamia calamistis*; ES, *Eldana saccharina*; PB, *Phragmataecia boisduvalii*; BS, *Bactra stagnicolana*; PS, *Poeonoma serrata*.

cultivated forms of their host plants and became more widely distributed and of economic importance. In Africa, where the bulk of maize and sorghum is grown only on small plots surrounded by land occupied by wild graminaceous plants, these stem borers, which are normally associated with grasses, infest the cultivated crops in high populations, due to the presence of wild grass hosts in the immediate neighbourhood. On the other hand, *C. partellus*, a species indigenous to Asia which has become established in East and Southern Africa, is also polyphagous and attacks several wild host plants.

Some authors have argued that the presence of alternative borer hosts in cereal-growing areas is detrimental in serving as a stem-borer reservoir, whereas others have pointed out that natural enemies can persist and increase their populations during the non-growing season, thanks to these alternative borer hosts. For example, Napier grass, *Pennisetum purpureum*, which is an important pasture grass, is often encountered as a host of stem borers, and may be important in some areas as a reservoir host of various stem-borer species. Several wild sedges (*Cyperus* spp.) are known to be the natural hosts of *E. saccharina*. Table 1.2 summarizes recent surveys, by the second author, of infestation of wild graminaceous hosts by various borer species in different agroecological zones in Kenya. These data are far from complete at the time of writing, and, in particular, the presence of non-economic species of *Busseola* and *Sesamia*, provisionally identified as *B. fusca* and *S. calamistis*, needs to be investigated. In addition, several *Chilo* species, apart from *C. partellus*, are no doubt present, but have not yet been identified to species level. However, the occurrence of the borers *Phragmataecia boisduvalli* (Herrich-Shaeffer) (Cossidae) (see Plate 5:4–6), *Bactra stagnicolana* Zeller (Tortricidae) and *Poeonoma serrata* (Hampson) (Noctuidae) on some of these wild hosts, the first two never found on cereals, is of interest. Of particular interest is the fact that some of these borers can and do act as reservoir hosts for natural enemies of economically important borers; for example, the braconid wasp *Stenobracon rufus* attacks *P. boisduvalii*.

There are plenty of examples illustrating the importance of uncultivated land as a source of pests for adjacent crops. However, there is not now, nor does it seem there could be, a general theory that will predict the role of uncultivated land in insect invasions and outbreaks in cultivated crops. The overall value of uncultivated land will have to be judged on a case-by-case basis, depending on the ecological mechanisms underlying the various disadvantages and advantages. In cultivated crops, most of which are short-lived in comparison with natural communities, damage is typically caused by herbivores that migrate into fields from the outside. These wild habitats often harbour food sources for many pest-insect species, and they may encourage insect invasion and outbreaks in neighbouring agroecosystems. It is also possible that the presence of wild crop relatives in nearby habitats could result in the maintenance of genetic diversity in local, relatively non-vagile insect populations, leading to the emergence of new strains or biotypes that could overcome host-plant resistance.

On the other hand, wild hosts adjacent to cultivated crops can provide extremely important refugia for natural enemies, as well as sources of nectar, pollen and host/alternative prey. There are also several examples of adjacent wild habitats being used to suppress insect outbreaks by keeping the pest on the so-called wild trap plants. Furthermore, non-host plants intercropped with cereals have also been shown to have a significant effect on stem-borer damage. In trials at ICIPE, molasses grass (*Melinis minutiflora*) intercropped with maize or sorghum significantly decreased levels of infestation by stem borers and also increased larval parasitism of stem borers by the braconid, *Cotesia sesamiae* (Khan *et al.*, 1997).

During the off-season, when there are no cultivated crops in the field, in addition to the hibernating or diapausing populations in crop residues, stem borers remain present on wild host plants and infest the cultivated hosts after they are planted. Although some basic information on the wild habitats as hosts of cereal stem borers in Africa is available, from the applied perspective, a systematic study of the dynamic relationship between the populations of insects in wild host plants and those on cultivated crops is important for the development of a sustainable integrated pest management (IPM) approach for cereal stem borers. A more complete understanding of the role of such wild hosts in insect outbreaks will generate suitable management strategies for stem borers. Appropriate methods of detecting migrants from wild habitats to cultivated crops and the information on the precise role of wild habitats in stem-borer invasion of cereal crops are needed. We also need to update and advance our current knowledge on the phenomenon of cereal stem-borer outbreaks in Africa. By discovering what accounts for differences between natural systems and agroecosystems, we can learn much about the underlying ecological processes that create the observed patterns of distribution and abundance of stem borers in nature, and about those chemical features that protect wild plants from herbivores. Plant breeders and entomologists can use this information in developing resistant/tolerant crop cultivars.

In addition, the information on the roles of wild habitats in providing refugia for natural enemies and the possible use of wild hosts in strip cropping and as trap crops will be helpful in the management of stem borers in a cropping system. Such knowledge can be used to strengthen and sustain the impact of IPM strategies.

Further studies on wild stem-borer hosts, taking into account local conditions, are currently under way in West Africa (International Institute of Tropical Agriculture) (IITA, Benin) and East Africa (ICIPE, Kenya).

Major references to wild stem-borer hosts: Ingram (1958); Nye (1960); Milner (1967); Bowden (1976); Atkinson (1980); Seshu Reddy and Unnithan (1985); Gebre-Amlak (1988a).

2 Maize: West and Central Africa

N.A. Bosque-Pérez[1] and F. Schulthess[2]
[1]*Formerly Plant Health Management Division (PHMD), International Institute of Tropical Agriculture (IITA), PMB 5320, Ibadan, Nigeria; present address: Department of Plant, Soil, and Entomological Sciences, University of Idaho, Moscow, Idaho 83844-2339, USA;* [2]*PHMD, IITA, BP 08-0932, Cotonou, Republic of Benin*

INTRODUCTION

Lepidopterous borers are among the economically most important pests of maize in Africa. Four stem borers and one ear borer cause significant yield loss: the maize stalk borer, *Busseola fusca* Fuller (Noctuidae); the pink stalk borer, *Sesamia calamistis* Hampson (Noctuidae); the African sugar-cane borer, *Eldana saccharina* Walker (Pyralidae); the spotted stalk borer, *Chilo partellus* Swinhoe (Crambidae); and the ear borer *Mussidia nigrivenella* (Ragonot) (Pyralidae). The first three are African and are present in most countries of sub-Saharan Africa, while *C. partellus* originates in Asia and was accidentally introduced to eastern Africa approximately 60 years ago (Bowden, 1954; Harris, 1962; Appert, 1970; Brenière, 1971; Bordat *et al.*, 1977; Girling, 1978). *Mussidia nigrivenella* is present in several countries of sub-Saharan Africa but its area of origin is not clear. In addition to the above five species, other borers of minor importance which have been reported on maize in Africa include *Sesamia botanephaga* Tams and Bowden, *Chilo aleniellus* (Strand), *Chilo orichalcociliellus* (Strand), *Chilo agamemnon* Bleszynski, *Chilo diffusilineus* (de Joannis), and *Coniesta* (= *Acigona*) *ignefusalis* (Crambidae) (Harris, 1962; Endrody-Younga, 1968; Appert, 1970; Bleszynski, 1970; Bonzi, 1982a; Moyal and Tran, 1992). The ear borer *Cryptophlebia leucotreta* (Tortricidae) has also been reported as a minor pest of maize in several West African countries (Libby, 1968; Moyal, 1988; Shanower *et al.*, 1991).

Maize is an exotic crop, introduced to Africa from the Americas. The most important pests of maize are, however (with two exceptions, *C. partellus* and the larger grain borer, *Prostephanus truncatus* (Horn) (Coleoptera: Bostrichidae)), indigenous to Africa and their natural hosts are indigenous grasses and sedges. Thus, any attempts to control indigenous pests must take into consideration the close link between their ecology and that of the native grasses (Bowden, 1976; Polaszek and Khan, Chapter 1, this volume).

MAIZE IN WEST AND CENTRAL AFRICAN FARMING SYSTEMS

Maize is an increasingly important crop in West and Central Africa, where over 5 million ha of the crop are grown (Fajemisin, 1992). The crop is grown in all major ecologies of these regions, from the humid forest to the Sudan savannah and from sea level to over 2000 m altitude. Maize has diversified uses, including food, animal feed and industrial use; however, 74% of the crop in these regions is used for human consumption (Fajemisin, 1992).

Maize production in West and Central Africa is influenced by diverse agroclimatic and socioeconomic factors. A large proportion of the maize in these regions is grown in the forest zone. In this agroecology, maize production is intense, but yields are limited by numerous diseases and insects, high rainfall, low solar radiation and high temperature. Maize is traditionally intercropped and small-scale farmers predominate. Average yields are low, at about 1 t ha^{-1}. Owing to difficulties in grain drying and storage, a great proportion of the crop in the forest is consumed as green maize.

The rainy season in the West African forest zone lasts 6–9 months (March to November); in some regions the rains are interrupted by a short, unreliable 'August break'. The August break divides the rainy season into 'first' and 'second' seasons. Maize planted at the beginning of the rains is called first- or major-season maize, and maize planted after the August break is referred to as second- or minor-season maize. Maize stem borers are far more abundant in the second than in the first season, and the second crop is sometimes a complete loss (Adeyemi *et al.*, 1966; Girling, 1980). Consequently, many farmers in the forest zone of West Africa do not plant second-season maize (Tams and Bowden, 1953).

In the southern and northern Guinea savannahs of West Africa, maize production has experienced dramatic increases in the last 15 years, to a great extent due to the adoption of improved varieties of maize and the use of fertilizer, as is the case of Nigeria (Smith *et al.*, 1993a). In these areas, maize is often intercropped with other cereals, as well as with legumes, but monocropped maize is becoming more prevalent (Fajemisin, 1992). Yields are up to twice as much as those in the forest zone, due to lower pressure from diseases and insects, greater solar radiation, lower night temperatures and better rainfall distribution.

In the coastal savannahs of Benin and Togo, maize has fewer production constraints than in the forest zone of neighbouring countries, and the crop is normally grown in both the first and second seasons. Benin and Togo have the greatest per capita consumption of maize in West Africa, with 96 and 79 kg $year^{-1}$, respectively (CIMMYT, 1992).

In the Central African countries of Cameroon and Zaïre, over 50% of the maize is grown in midaltitude regions (over 800 m) and average maize yields are greater than in the forest zone of West Africa, owing to adequate rainfall, cool temperatures and richer soils (Fajemisin, 1992). Small-scale farmers predominate in these countries and they often grow maize intercropped with other cereals, legumes or root and tuber crops. However, the proportion of medium- and large-scale farmers growing solely maize is increasing (Fajemisin, 1992).

In Rwanda and Burundi, highland bananas, roots and tubers are the major crops and maize is only of moderate importance, with a combined maize area for the two countries of approximately 210,000 ha. Per capita consumption is low, at 32 and 14 kg $year^{-1}$ for Burundi and Rwanda, respectively (CIMMYT, 1992).

BIOLOGY OF MAJOR MAIZE STEM AND EAR BORERS

Sesamia calamistis (Plates 4:4,5; 5:1)

Tams and Bowden (1953) described 15 species of *Sesamia*. Of these, the two most important ones in Africa are *S. calamistis* and *S. botanephaga*. According to Bordat *et al.* (1977), the former species is present in most countries of sub-Saharan Africa, Madagascar and the Comoros, while the latter is present in West Africa, Sudan, Uganda and Kenya.

The host range of *S. calamistis* is limited to the family *Gramineae* (*Poaceae*) and includes cultivated crops, such as maize, sorghum and millet, as well as wild grasses, such as *Pennisetum purpureum*, *Panicum maximum* and *Setaria splendida* (Harris, 1962).

Adults of *S. calamistis* emerge in the late evening and mate. Females lay their eggs between the leaf sheaths of the host plant. Longevity of adults is influenced by temperature. In laboratory tests, females lived 5.7 days at 30°C and 10 days at 20°C (Shanower *et al.*, 1993a). Bosque-Pérez and Dabrowski (1989) found that, when larvae were reared on an artificial diet, females laid an average of 320 eggs each in a period of 5 days at temperatures between 24 and 26°C. Fecundity tests conducted by Shanower *et al.* (1993a) revealed that, on average, a female lays 249 at 25°C and 688 eggs at 20°C. Under field conditions, eggs hatch in 5–6 days and most larvae penetrate the stem shortly after egg hatch.

The development rates of *S. calamistis* have been studied under controlled conditions by Shanower *et al.* (1993a). On an artificial diet, *S. calamistis* required 709 degree-days (DD) to complete its development; eggs needed 122 DD above a threshold of 9.7°C, larvae 383 DD above 12.2°C and pupae 204 DD above 10.2°C. These threshold temperatures are well below the average mean daily temperatures in the maize-growing regions of West Africa; *S. calamistis* is thus theoretically capable of developing throughout the year, provided suitable host plants are available (Shanower *et al.*, 1993a). When reared on an artificial diet, insects developed faster than when reared on maize-stem cuttings (Shanower *et al.*, 1993a). Maize stems appear to be nutritionally inferior compared with an artificial diet. Development time was also influenced by temperature; immature stages took 162, 94, 50 and 38 days at 15°, 20°, 25° and 30°C, respectively.

Larval feeding might result in the destruction of the growing point, typically referred to as 'deadhearts'. At later stages, the tunnelling and girdling activities of the larvae often result in stalk breakage. In the field, development of the larvae takes 4–6 weeks and most larvae pupate within the stem or cobs.

In contrast to *B. fusca*, *S. calamistis* breeds throughout the year and has no resting stage (Harris, 1962). However, densities are low during the dry season, when its hosts are mature grasses and maize growing in small areas in hydromorphic soils. Bowden (1976) reported that *Sesamia* spp. adults that emerge at the beginning of the cropping season are smaller and less fecund than those emerging later in the year and that the combined effects of smaller numbers of less fecund adults result in lower incidence of *Sesamia* spp. in first-season maize crops. This report has been confirmed by Shanower *et al.* (1993b), who studied the effect of larval diet on the growth and development of *S. calamistis* in the laboratory. Five indigenous African grasses were compared with maize and an artificial diet for their ability to support this insect. While 95 and 30% of the larvae successfully pupated after being reared on artificial diet and maize-stem cuttings, respectively, only 7% pupated on *Sorghum arundinaceum*, 1% or less on *P. maximum*, *P. purpureum* and *Andropogon* sp. and none on *Pennisetum polystachion*. Development time was similar on maize and the grasses, but was 50% slower than on the artificial diet. Pupal weights of larvae reared on four of the grasses were much lower than for larvae reared on maize or artificial diet (Shanower *et al.*, 1993b). Given that pupal weight is an indicator of fecundity (Sétamou *et al.*, 1993), it is to be expected that, in areas where *S. calamistis* is forced to feed on wild grasses to bridge the cropping seasons, larval survivorship and adult fecundity will be greatly reduced. On the other hand, in regions where more than one crop of maize is planted, the potential for pest outbreaks is great because maize provides a suitable host for insect growth. In West Africa, the population of this borer increases until it peaks around August–September (Harris, 1962; Bosque-Pérez and Mareck, 1990). This occurs when second-season maize crops are being grown, and, as a result, *Sesamia* spp. can be a serious problem. *Sesamia* populations often crash around November, in spite of continuous oviposition by adult insects; the mortality factors responsible for this decline are not clearly understood, but natural enemies, especially egg parasitoids and pathogens, are believed to play a major role (Sétamou and Schulthess, 1995; Schulthess *et al.*, 1998).

In addition to host-plant and weather factors, soil nutrients are known to affect the bionomics of *S. calamistis*. Detailed studies in this area have been conducted by Sétamou *et al.* (1993). Increasing nitrogen levels in maize plants was found to increase larval survivorship, female fecundity and longevity; nitrogen, however, was not observed to have an effect on development time. Increasing silica applications to maize plants reduced larval survivorship but had no effect on other life parameters (Sétamou *et al.*, 1993).

Busseola fusca (Plate 2:2,4,5; 4:1–4)

Busseola fusca is considered by some authors to be the most important pest of maize in sub-Saharan Africa (Appert, 1970). However, this is mainly due to its abundance on maize in midaltitude regions of East and southern Africa, as the pest does not appear to be as important as *S. calamistis* or *E. saccharina* on maize in West Africa (Bosque-Pérez and Mareck, 1990; Shanower *et al.*, 1991; Gounou *et al.*, 1993). *Busseola fusca* is distributed from approximately 12°N to 30°S (Appert, 1970). This insect was recognized as a major pest of cereals when originally described in 1901 (Tams and Bowden, 1953).

In West Africa, *B. fusca* is abundant in the drier savannah zone, especially where sorghum is grown (Harris, 1962). Sorghum is believed by some to be the native host of *B. fusca* in West Africa (Bowden, 1976). In Ghana, Gounou *et al.* (1993) found *S. arundinaceum* to be the most common wild host of *B. fusca* in both the forest and the savannah zones. This insect is also abundant in the midaltitude regions across the continent.

The adult moths generally emerge in the evening and mate. Females lay their eggs between the leaf sheaths of the host plant; egg batches contain from 30 to 100 eggs (Harris, 1962). On average, a female lays 400 eggs. The egg incubation period is influenced by temperature and relative humidity (RH); incubation time was found to decrease with increasing temperature and RH – at 80% RH or above, egg hatching occurred in 5–6 days, while, at 70 and 60% RH, hatching took 6–7 and 7–8 days, respectively (Usua, 1968). Larval development is also influenced by temperature. Usua (1968) reported that at 22°C larvae had arrested development, while temperatures of 32°C and above were found to be lethal. He concluded that temperatures between 26°C and 30°C were the most favourable for development of *B. fusca*.

After emergence, larvae crawl over the plants, congregate in the funnel and feed on the rolled leaves. As the leaves grow away from the funnel, a characteristic pattern of holes and 'window panes' can be seen (Harris, 1962; Plate 1:1). Continuous feeding by the larvae might result in deadhearts. After killing a plant, larvae usually migrate to new plants and enter them by boring into the stem near the base. Tunnelling of the stem and ears then occurs. Larval development takes between 26 and 33 days.

During the dry season, larvae enter diapause – a period of arrested development, which usually occurs during adverse environmental conditions – and take up to 6 months to complete their development (Harris, 1962). With the initiation of the rains, the larvae pupate within the stems and 10–12 days later adult moths emerge.

Eldana saccharina (Plate 8:5,6; 9:1)

The stem borer *E. saccharina* was first described from Sierra Leone and has been known as a pest of graminaceous crops in West Africa for more than a century (Appert, 1970). It probably occurs in all suitable areas of sub-Saharan Africa, from approximately latitude 15°N to 30°S (Girling, 1978). In East and southern Africa, the wild hosts of *E. saccharina* are sedges, *Cyperus* spp., which are preferred over crops, such as sugar cane, as oviposition sites (Atkinson, 1980). On sedges, *E. saccharina* is not a stem borer but feeds in the culms and inflorescence. During numerous surveys by International Institute of Tropical Agriculture (IITA) scientists in West Africa this pest was rarely found feeding on wild sedges and grasses. The major hosts of this borer in West Africa are crop plants, such as maize, sugar cane, sorghum and millet. *Eldana saccharina* has been reported as an important pest of maize (Leyenaar and Hunter, 1977; Girling, 1978, 1980; Atkinson, 1980; Bosque-Pérez and Mareck, 1991) and is often the most abundant borer species at the end of the maize-growing season (Bosque-Pérez and Mareck, 1990). Infestations of maize plants usually start at anthesis (Carter, 1985).

The longevity of *E. saccharina* adults is influenced by temperature. In laboratory tests, females lived 6.6 days at 30°C and 15 days at 20°C (Shanower *et al.*, 1993a). Females lay eggs on debris on the soil (Atkinson, 1980) or on the hairy margins of maize leaf sheaths (Cochereau, 1985). Bosque-Pérez and Dabrowski (1989) reported that, when larvae are reared on an artificial diet, females lay an average of 380 eggs each during a 5-day period at temperatures between 24 and 26°C. Fecundity tests conducted by Shanower *et al.* (1993a) using moths from larvae reared on an artificial diet revealed that, on average, a female lays 470 eggs at 20°C and 620 eggs at 30°C. In these studies, fecundity was positively correlated with temperature, but at higher temperatures females survived fewer days (Shanower *et al.*, 1993a). Under field conditions, eggs hatch in 5 or 6 days and, after feeding on the leaf sheaths for a few days, larvae enter the stem, where they continue to feed. Larvae may move into the ears and feed on the grain.

Shanower *et al.* (1993a) studied the development rates of *E. saccharina* under controlled conditions in the laboratory. On an artificial diet, *E. saccharina* required 709 DD to complete its development; eggs required 125 DD above 9.1°C, larvae 400 DD above 10.6°C and pupae 140 DD above 8.8°C. Like *S. calamistis*, *E. saccharina* is capable of developing throughout the year, provided suitable host plants are available, as these threshold temperatures are below the average mean daily temperatures in the maize-growing regions of West Africa (Shanower *et al.*, 1993a). Larval development and growth are also known to be influenced by food source. For example, development was found to be significantly slower when larvae were fed maize stems, compared with artificial diet (Shanower *et al.*, 1993a). In further experiments by Shanower *et al.* (1993b), it was found that larval development time was similar on maize and five species of grasses but larval survivorship was < 5% on grasses, compared with 20% on maize stems and 65% on artificial diet. Temperature also influenced developmental rate, with immature stages lasting 173, 63, 44 and 33 days at 15°, 20°, 25° and 30°C, respectively (Shanower *et al.*, 1993a).

Under field conditions, pupation occurs inside the stem or cobs and the pupa is covered by a cocoon made of silk and plant debris. A good external sign of *E. saccharina* attack is the adult exit hole cut by the larvae prior to pupation, which often has a large amount of frass hanging from it. Adults emerge 7–14 days after pupation.

Although infestations by this stem borer occur relatively late in the development of the maize plants, damage as a result of their feeding can be severe, with yield losses of up to 20% (Bosque-Pérez and Mareck, 1991). Damage done by *E. saccharina* provides access to the stem and cobs for pathogens that cause rots, and infestations by this borer are associated with high levels of stalk lodging, due to tunnelling and the effect of stalk rots.

Mussidia nigrivenella

The literature on the ear borer *M. nigrivenella* is scanty. The species was first described by Ragonot in 1888, and since then it has been recorded on numerous plant species, including maize, cotton, cocoa, lima beans (*Phaseolus lunatus*), the shea butter tree (*Vitellaria paradoxa* (syn. *Butyrospermum parkii*)), *Mucuna pruriens*, *Canavalia* sp. and *Sphenostylis stenocarpa* (Moyal and Tran, 1991b). This pest infects fruiting structures of mature plants and continues feeding on the stored products (Bordat and Renand, 1987). Infestations of maize start in the field; female moths lay their eggs on the silks and the husk leaves. Eggs hatch in 5–7 days and young larvae feed within the silk channel for a few days before reaching the grain. Developing larvae feed on the distal portion of the maize ear and tunnel through the grain, causing extensive damage and often consuming the embryos in a way that it is not superficially visible. Only close inspection reveals the degree of damage. Pupation takes place within the tunnels or on the surface of the grain, and the pupae are surrounded by a silky cocoon. Damage to the grain continues during storage; hence, *M. nigrivenella* can be regarded as both a field and a storage pest, although no reproduction occurs in the store. Preliminary observations indicate that the pest can survive in stored cobs for up to 8 weeks, even at grain moisture contents of 12–15% (O. Bolaji, N.A. Bosque-Pérez and M. Ivbijaro, unpublished data).

Surveys have been conducted by IITA scientists in farmers' fields in Benin, Ghana, Ivory Coast and Nigeria to establish the geographical distribution, host-plant range and natural-enemy complex of this species in West Africa. *Mussidia nigrivenella* has been found in every country and ecological zone from the forest to the northern Guinea savannah but is rare in the Sudan savannah. Survey results from Benin indicate that each larva causes, on average, 4% ear damage; five larvae per infested ear are often found (F. Schulthess, unpublished). Surveys in south-western Nigeria demonstrated this borer to be the most abundant pest of maize at the time of harvest (O. Bolaji, N.A. Bosque-Pérez and M. Ivbijaro, unpublished data). In studies conducted at IITA, Ibadan, maize varieties with a short husk-tip extension and loose husk leaves were found to be more severely infected by *M. nigrivenella* than those with a good husk cover; additionally, the abundance of this ear borer was found to increase with delayed harvesting (O. Bolaji and N.A. Bosque-Pérez, unpublished data).

The life history of *M. nigrivenella* reared on an artificial diet in the laboratory has been studied by Bordat and Renand (1987). On average, at 25°C, the incubation period of the eggs was 4 days, larval stages lasted 18–21 days and the pupae emerged in 6–9 days. Females laid an average of 115 eggs over a period of 5 days, with a maximum of 213 eggs (Bordat and Renand, 1987).

Moyal and Tran (1991a) have also studied the life history of this borer. When reared on an artificial diet at 22°C, *M. nigrivenella* required between 46 (males) and 49 (females) days to complete its development; on average, eggs required 6.4, larvae 31.6 and pupae 8.9 days. Although dissection of females revealed over 650 eggs in the ovaries, under the conditions of their experiments the actual number of eggs laid per female (i.e. 69) was low (Moyal and Tran, 1991a).

DISTRIBUTION OF STEM AND EAR BORERS

During the past few years, surveys have been conducted by scientists from various national and international institutions in several countries of West and Central Africa to obtain information on the abundance, species composition and relative importance of lepidopterous borers and their natural enemies. The major findings of these surveys are summarized below.

Benin

In country-wide surveys conducted by Shanower *et al.* (1991), *Sesamia* spp. were found to have the widest distribution and were detected in 60% of all maize fields surveyed. *Eldana saccharina* was also common. Other borers observed were *C. ignefusalis*, *M. nigrivenella* and *Cryptophlebia* sp. Borers were found in every province, but infestations were higher in the southern part of the country compared with the north. Observed parasitization rates for *Sesamia* spp., *E. saccharina* and *C. ignefusalis* larvae and pupae were generally very low. Monthly surveys for egg parasitism in the southern provinces, however, showed mean peak parasitism of *S. calamistis* eggs by both *Telenomus busseolae* and *Telenomus isis* of over 90%.

Burundi

Regular on-station samplings, carried out by Muyango (1987) in the location of Mosso at 1200 m altitude, confirmed the presence of several stem-borer species in maize fields in Burundi. *Eldana saccharina* constituted the main stem-borer species in all seasons; 60–65% of all borer specimens collected over a 4-year period were this species, and the remaining were identified as *Sesamia* sp. and *B. fusca*. The incidence of *E. saccharina* on maize was similar in all planting seasons, with 20–25% of plants attacked. *Busseola fusca* was the only species detected in the high-altitude (2100 m) location of Gisozi. In a survey conducted by IITA and national scientists during the second season in 1991, *B. fusca* was found to be the most common stem borer on both maize and sorghum. Infestations increased with altitude, and at mid-elevations up to 80% of maize and sorghum plants were found to be infected with this borer.

Cameroon

In 1991, scientists from the National Cereals Research and Extension project studied seasonal abundance of borers on maize in several locations in humid forest, sub-humid forest and the forest/savannah transition zone. Across all elevations, the pre-dominant borer species was *B. fusca*, often accounting for more than 90% of all species, followed by *E. saccharina, S. calamistis* and *M. nigrivenella* (Cardwell *et al.*, 1997).

Ghana

Surveys conducted in 1990/91 revealed that *S. calamistis* and *E. saccharina* were the most damaging field pests of maize (Gounou *et al.*, 1993). In addition to these two species, *B. fusca, M. nigrivenella* and *C. leucotreta* were also detected. In late 1990, stem borers were more important than ear borers, with *S. calamistis* the pre-dominant species in all ecological zones. In early 1991, borer densities were lower than in late 1990. In late 1991, the predominant species was *E. saccharina*, followed by *S. calamistis*, except in the Guinea savannah where *M. nigrivenella* pre-dominated. Although the highest incidence of borers was detected in the forest zone, individual fields with high infestations were found in all areas with a bimodal rainfall pattern (Gounou *et al.*, 1993).

Ivory Coast

Surveys of *M. nigrivenella* were conducted by Moyal (1988) from 1981 to 1984. Population density of this ear borer varied with location and year and was higher in the south-eastern part of the country than in the north or west. Up to 14 larvae per maize ear were detected in a southern location in 1984 (Moyal and Tran, 1991b).

In country-wide surveys conducted by IITA and national-programme scientists, *E. saccharina* was the most important borer species, whereas *S. calamistis* and *B.*

fusca were each of equal importance. The ear borers *M. nigrivenella* and *C. leucotreta* were at this time of minor importance. The average percentage of infested plants was 72% for the rain forest, 57% for the derived savannah and 10% for the Guinea savannah.

Nigeria

In experiments conducted in six locations in southern Nigeria from August to November of 1985 and 1986, *S. calamistis* was the dominant maize borer up to 8 weeks after planting and *E. saccharina* was the most abundant from 9 weeks after planting onwards (Bosque-Pérez and Mareck, 1990). *Mussidia nigrivenella* was found at all sites, while *B. fusca* was found only in forest/savannah transition-zone locations. Other borer species encountered included *C. ignefusalis* and *Cryptophlebia* sp.

Additional surveys of borers and their natural enemies were carried out by IITA scientists in maize fields in Nigeria during 1991 and 1992 (Bosque-Pérez *et al.*, 1995, and unpublished data). *Sesamia calamistis* and *E. saccharina* were the most commonly encountered pests of maize; additionally, *B. fusca*, *C. ignefusalis*, *M. nigrivenella* and *C. leucotreta* were detected. Fifty-seven per cent and 55% of the fields sampled had evidence of borer damage in 1991 and 1992, respectively. The percentages of plants with borer damage in individual fields varied according to year and ecological zone, with a maximum of 17% for both Guinea savannahs in 1991 and up to 30% in the northern and 47% in the southern Guinea savannah in 1992. A severe infestation with *C. ignefusalis* was observed in two locations in the southern Guinea savannah in 1992.

Zaïre

The most important stem borer of maize in Zaïre appears to be *B. fusca*. Observations on the dynamics of this pest have been made by Alam (1990), who reported borer infestations of 1–2% in maize planted in early December and infestations of 25–36% in mid-January plantings during the 1988/89 cropping season in the midaltitude (1200 m) location of Kaniameshi. *Eldana saccharina* was recorded on maize in Gandajika in south-central Zaïre, but not from Kaniameshi (Alam, 1990).

SAMPLING MAIZE BORERS

Enumerative (destructive via dissection) and presence–absence sampling (non-destructive via visual examination for damage symptoms) methods with a pre-defined precision level have been developed by Schulthess *et al.* (1991) and Shanower *et al.* (1991) for various lepidopterous pests on maize. These sampling plans take into consideration the spatial distribution of a species, as defined by Taylor (1961). The species *S. calamistis.*, *E. saccharina*, *M. nigrivenella* and *C. leucotreta* all show a highly aggregated distribution in the field. Since two or more lepidopterous species are often found on the same maize plant and almost all plant parts – leaves, stems, tassels and ears – are attacked, Schulthess *et al.* (1991) recommended sampling whole plants in order to accurately estimate borer populations.

Dissection of maize plants is time-consuming and costly. For survey work, it is therefore often more practical to use a non-destructive presence–absence sampling procedure. This procedure requires the establishment of the relationship between the proportion of infested plants and the mean per-plant density for the stem and ear borers described above. For this, 40 plants are taken at random within a field and examined for borer damage (i.e. deadhearts, leaf damage, entrance/exit holes, presence of frass) (Schulthess *et al.*, 1991). Once the proportion of infested plants in a given field is established, the respective density can be determined. Tables are provided by Shanower *et al.* (1991) which can be used to obtain the borer mean density per plant, using the per cent plants infested and the species composition. To establish species composition, ten damaged plants should be dissected.

YIELD REDUCTIONS CAUSED BY MAIZE BORERS

Reported yield losses due to lepidopterous borers in Africa vary greatly (0–100%) among ecological zones, regions and seasons (i.e. first and second season in West Africa). The severity and nature of stem-borer damage depend upon the borer species, the plant growth stage, the number of larvae feeding on the plant and the plant's reaction to borer feeding. Feeding by borer larvae on maize plants usually results in crop losses as a consequence of death of the growing point (deadhearts), early leaf senescence, reduced translocation, lodging and direct damage to the ears (Appert, 1970; Brenière, 1971; Bosque-Pérez and Mareck, 1991). Infestations with stem borers have been found to increase significantly the incidence and severity of stalk rots (Bosque-Pérez and Mareck, 1991).

The degree of stem-borer attack might be assessed by determining the percentage of plants attacked, extent of plant damage – i.e. level of leaf damage, percentage of the stem tunnelled, percentage of the nodes bored, stalk breakage, extent of ear damage – and/or number of larvae or pupae per plant (Walker, 1981). It is possible to measure the relationship between infestation rate and crop loss by using natural or artificial infestations and by preventing attack, using insecticides or other methods of control (Walker, 1981).

The effect of *E. saccharina* infestations on the yield components of maize has been studied by Bosque-Pérez and Mareck (1991) through comparisons of infested and insecticide-protected plants. Yield comparisons of infested plants versus insecticide-protected ones showed that natural infestations decreased yields by 16, 15 and 28% in the dry, first and second rainy seasons, respectively. Infested plots had significantly lower 100-grain weight than insecticide-protected ones and a significant negative correlation was found between 100-grain weight and the percentage stem tunnelled ($r = -0.79$), suggesting that *E. saccharina* damage to the stems affects grain filling. The regression coefficient obtained from covariance analysis showed 100-grain weight to decrease by 0.125 ± 0.0477 g for each unit increase in the percentage of the stem tunnelled.

Using data from surveys conducted in late 1990 in farmers' fields in Ghana, Gounou *et al.* (1993) found a positive relationship between the number of *Sesamia* sp. and the extent of the maize stem tunnelled and a significant negative relationship between per cent stem tunnelling and maize yield. These authors used multiple-regression equations, which include plant parameters (stem diameter, phenological stage) and damage variables (per cent stem tunnelled) to calculate the average ear-weight losses for each ecological zone. The calculated yield-loss figures for late 1990 were 32.5, 14, 19 and 18 g plant^{-1} for the rain forest, coastal, derived and Guinea savannahs, respectively, equivalent to 27, 15, 18 and 14% yield loss.

In Zaïre, the effect of stem-borer attack (mostly *B. fusca*) on maize in farmers' fields has been studied by scientists from the IITA/US Agency for International Development (USAID)/Zaïre Applied Agricultural Research and Outreach Project (Anon., 1990). Multiple-regression analysis was used to identify the relationship between various agronomic practices, observed variables and yield of maize. In maize plots receiving fertilizer treatment, early plantings (i.e. October–November) were estimated to suffer 8–9% yield losses, while more heavily attacked, late plantings (i.e. December–January) suffered 22–25% losses (Anon., 1990).

Trials conducted by Muyango (1987) in Burundi, using insecticides and exclusion cages, revealed maize-yield reductions of 12–15% as a result of *E. saccharina* attacks. In regions between 1500 and 2100 m, attacks by *B. fusca* occasionally caused yield reductions of 30–50% (Muyango, 1987).

METHODS TO CONTROL STEM AND EAR BORERS

Control of lepidopterous borers includes biological, cultural and chemical methods, as well as host-plant resistance.

Biological Control

Natural enemies play an important role in the control of lepidopterous borers in Africa. Approximately 100 genera of parasitoids (Hymenoptera and Diptera) have been recorded attacking cereal stem borers in Africa and surrounding islands (Polaszek, 1992). In West Africa, several imported natural enemies, such as *Trichogramma* spp., *Sturmiopsis inferens*, *Cotesia flavipes*, *Amyosoma chinense* and *Tetrastichus* spp., have been released against stem borers in Ghana (Scheibelreiter, 1980), although these failed to establish. In South Africa, introduced parasitoids were first used for the control of *E. saccharina* in 1975 (Carnegie and Leslie, 1979), but this programme was not successful. Releases of exotic parasites against *S. calamistis*, *B. fusca*, *C. partellus* and *E. saccharina* have also been tried in East Africa, without much success (Ingram, 1983). It appears that the only successful biological-control campaign against stem borers in Africa was conducted in Madagascar against *S. calamistis* and *C. partellus* (Appert *et al.*, 1969). However, biological control is still considered a viable control option. It is probable that, due to lack of funds, biological-control attempts in the past have not been pursued for a sufficiently long period. In most cases, no effort has been made to understand why a natural enemy failed to establish. Before releasing a natural enemy, it is important to have a profound knowledge of the ecosystem in which it is to be released. This includes not only crops but also wild host plants, which are important overseasoning sites for pests and natural enemies. In addition, information is needed about the relative importance of a natural enemy in the original habitat, and behavioural studies (e.g. searching behaviour, functional-response studies) in the laboratory should be carried out. The results may already give some information about the strengths and weaknesses of a natural enemy and the chance of survival in the new ecosystem. After release, follow-up studies are needed to assess establishment, spread and impact on pest populations and crop yield.

Work on cataloguing natural enemies of stem borers has been initiated in several African countries by various institutions, including IITA. Emphasis is being given to parasitoids, although predators, especially ants, are known to be important natural enemies of stem borers (Girling, 1978; Bonhoff, Chapter 24, this volume). At IITA, the cataloguing is complemented with studies on abundance throughout several cropping seasons, including the dry season, to determine the relative importance of the different species in the system. The aim of this work is to compare the natural-enemy complexes in various ecological zones and regions of Africa, in order to identify promising candidates to be exchanged between regions. Results of various country-wide surveys and detailed on-station research carried out by IITA scientists in West Africa have shown that larval and pupal parasitoids of *S. calamistis* and *E. saccharina* are exceedingly rare in this part of the continent (IITA, 1993). Peaks of parasitization rates of usually less than 3% are found at the end of the second planting season, when the damage to maize is already done. Survey and on-station work by IITA scientists in Benin, Nigeria, Ghana and Ivory Coast suggest that the most common natural enemies of *S. calamistis* are the egg parasitoids *T. busseolae* and *T. isis* (Polaszek *et al.*, 1993; Sétamou and Schulthess, 1995).

Several pathogens have been detected on *S. calamistis* and *E. saccharina* in Nigeria (Bosque-Pérez and Dabrowski, 1989) and some of these appear to be important field-mortality factors (IITA, 1986; see also Poinar and Polaszek, Chapter 23, this volume). Ecological studies, selection for greater virulence and application technology of these pathogens could make them a useful pest-management tool, provided their ecology is understood better and means are developed for their adequate dispersal/perpetuation in the field.

Cultural Practices

In West and Central Africa, farmers do not usually make a conscious effort to control borer populations. Numerous authors have reported greater infestations of stem borers in late-planted maize compared with early plantings, i.e. in Benin (Shanower *et al.*, 1991), Burundi (Muyango, 1987), Cameroon (Aroga, 1988), Ghana (Girling, 1980; Gounou *et al.*, 1993), Nigeria (Harris, 1962; Adeyemi *et al.*, 1966; Carter, 1985) and Zaïre (Alam, 1990). Thus, in some areas farmers do not plant maize during the second season, because of devastating borer infestations. This may influence the population dynamics of the pest, especially in the rainforest zone, where alternative wild plant hosts, i.e. dry-season hosts, are relatively scarce. Other farmers' practices may have an effect on carry-over populations of borers, especially during the dry season. For example, crop residues left after harvest or used as building materials are important overwintering sites for diapausing species, such as *B. fusca* (Harris, 1962). Burning at the end of the dry season to clear bush fallow may have an effect on both pests and natural enemies.

As shown by Sétamou *et al.* (1995), nitrogen fertilization enhances both borer development and the plants' tolerance to borer attack. Yield losses decreased linearly from 20% with no fertilizer to 11% with 120 kg nitrogen ha^{-1}.

Survey results from Ghana and information from the literature suggest that mixed cropping does not have an important effect on borer infestations on maize (Gounou *et al.*, 1993). Occurrence and abundance of wild hosts appear to be more important in explaining fluctuations in borer populations than intercropping. In most cases, the role of wild hosts is unknown and misunderstood. They can be sources for new outbreaks of pests or reservoirs of natural enemies (and thus prevent or reduce pest outbreaks), be overwintering sites for both pests and their natural enemies, or even act as trap plants. Consequently, the effect of a sustainable intervention technique, such as biological control or resistant varieties, can only be understood in the context of the natural habitat of stem borers and their antagonists.

Chemical Control

Chemical insecticides are used in some countries of sub-Saharan Africa for the control of lepidopterous borers. It is important to realize, however, that the majority of African farmers cannot afford to buy insecticides, that these are seldom available on time and usually require application equipment and knowledge seldom available to farmers in the continent. Additionally, insecticides often have a negative impact on the environment, are a health hazard to humans and are not necessarily compatible with other means of control, such as natural enemies. Many of the insecticides originally recommended for borer control in Africa have been banned in the USA and Europe, and although their use in Africa may still continue, this approach is highly questionable.

The chemical control of stem and ear borers is more difficult than that of insects which are external plant feeders, as most of their life cycle is spent within plant tissues that cannot be reached with contact insecticides. Thus, foliar applications of insecticides for the control of stem borers have generally been found to be unsatisfactory (Jotwani, 1983). For borers that spend their larval stages in the stem, systemic insecticides might be required. For borers such as *B. fusca* and *C. partellus*, contact insecticides have been recommended for use during early plant stages, when insects are feeding in the plant funnel. Pyrethroids have been found to be effective for the control of *B. fusca* in Burundi (Muyango, 1987).

Resistant Varieties

Host-plant resistance is an ideal method of pest control, because it requires minimum input and action by the farmer (i.e. purchase of seed of the right variety only), often has a cumulative effect, is compatible with other control methods and has no negative environmental effects. Maize varieties resistant to stem borers have been suggested as one of the most promising means of control (Bowden, 1976; Girling, 1980). For the last several years, IITA has devoted efforts to developing sources of resistance to *S. calamistis* and *E. saccharina* (Bosque-Pérez *et al.*, 1989). Sources of moderate resistance to the two stem borers have been found and maize populations with moderate resistance to either *S. calamistis* or *E. saccharina* have been formed; these are undergoing selection to increase the levels of resistance (Bosque-Pérez and Mareck, 1989). Moderate levels of resistance could be combined with other methods of control to reduce the economic impact of stem borers.

In addition to the work of IITA, the International Maize and Wheat Improvement Center (CIMMYT) has been developing sources of resistance to stem borers. A maize population with tropical adaptation and resistance to borers has been formed (multiple insect resistance, tropical (MIRT)) and is undergoing further selection for resistance (Mihm *et al.*, 1988).

The opportunity for utilizing new insect-resistance genes and the ability to transfer these across plant species, through biotechnology, open new doors to the field of host-plant resistance.

3 Maize and Sorghum: East Africa

K.V. Seshu Reddy

Crop Pests Research Programme, International Centre of Insect Physiology and Ecology, PO Box 30772, Nairobi, Kenya

INTRODUCTION

In eastern African countries (Djibouti, Ethiopia, Kenya, Somalia, Sudan, Tanzania and Uganda), both maize and grain sorghum are very important traditional cereal crops in the diets of millions of people. These crops are cultivated, by and large, on smallholdings, under a wide range of environments, as intercrops with other cereals, legumes and root and tuber crops, which are subject to differences as regards date of planting, varieties, land management and amount and distribution of seasonal rainfall.

HISTORICAL BACKGROUND

Maize was first introduced into East Africa in the 17th century, but the varieties originated in the Caribbean and were only suited to the coastal strip of East Africa. Its proliferation in the East African highlands and in the medium-altitude areas, with a consequent decline, in most places, of the indigenous cereals, is a comparatively recent development. This was largely due to the introduction, by European settlers, of varieties from South Africa, which were better suited to the inland environment (Acland, 1971). Maize is grown in a wide range of rainfall regions in eastern Africa and used for food, green cobs, fodder, building and mulching.

Sorghum, in contrast, is a native African cereal and reported to have originated in the Sudan/Ethiopian border region and has since spread throughout the world (De Wet, 1978). It is the most important staple food crop grown under rain-fed conditions in semiarid areas of eastern Africa. Sorghum grain is used for food and beverages, whereas stalks are for animal feed, fuel, house and fence building and mulching in coffee and banana plantations.

CULTIVATED-AREA CONSTRAINTS

In eastern Africa, where 3.9% of the world's acreage is under maize and 11.5% under sorghum, grain yields are very low, ranging from 700 to 1800 kg ha^{-1} in maize and

455 to 1471 kg ha^{-1} in sorghum, as against 7437 kg ha^{-1} and 3951 kg ha^{-1} in maize and sorghum, respectively, in the USA (FAO, 1991). One of the major constraints responsible for low yields is insect pests. Among these, a range of stem borers has proved to be the most important insect pests of maize and sorghum.

DISTRIBUTION OF STEM BORERS

Although there are 12 stem-borer species recorded in this region, essentially *Chilo partellus, Chilo orichalcociliellus, Busseola fusca, Sesamia calamistis, Sesamia cretica* and *Eldana saccharina* are important and widely distributed.

The introduced stem borer from the Indian subcontinent, *C. partellus,* was first recorded in Africa from Malawi in 1930 (Tams, 1932), whereas all the other major stem borers are indigenous to Africa.

DAMAGE AND CROP LOSSES

Most stem-borer species produce similar symptoms on attacked maize and sorghum plants. Newly hatched larvae feed initially by scraping in leaf whorls of growing plants, producing characteristic 'window-paning' and 'pinholes'. Later, the larvae tunnel into the stems and may kill the central leaves and growing point, producing 'deadhearts'. The larvae also bore into the maize cobs and feed on the developing grains. Plants thus affected have poor growth and reduced yield and are more susceptible to wind damage and secondary infections.

In the eastern African region, reports on objective assessment of grain-yield losses in maize and sorghum attributable to the stem borers are very few.

Yield losses of 18% in maize due to *C. partellus* and *C. orichalcociliellus* were recorded in Kenya (Warui and Kuria, 1983). Seshu Reddy and Sum (1991) reported that there was a maximum grain-yield reduction and stalk damage in maize (cv. Katumani) due to *C. partellus* on 20-day-old crop, while there was insignificant larval effect on yields for 60-day-old crop. They also defined economic injury levels (EIL) of *C. partellus* as 3.2 and 3.9 larvae $plant^{-1}$ in maize 20 and 40 days after plant emergence, respectively.

Jepson (1954) reported 40–100% sorghum plants infested by *B. fusca* in Tanzania, while, in Ethiopia, movement of *B. fusca* larvae into the base of sorghum head resulted in undersized heads and a grain loss of 15% (Megenasa, 1982). In Uganda, a 56% loss of grain yield resulted when sorghum was infested with *C. partellus* 20 days after emergence (Starks, 1969), whereas, in Kenya, losses ranged from 2 to 88% (Seshu Reddy, 1988). Seshu Reddy *et al.* (1989) in their work on sorghum (cv. ICS2) recorded as high as 80% grain-yield loss with eight *C. partellus* larvae $plant^{-1}$ infested at 21 days after emergence, whereas the same density gave statistically insignificant losses when plants were infested at 60 days after emergence. They also concluded that grain-yield losses due to *C. partellus* infestation were strongly influenced by the cultivars, age of the crop at the time of infestation and also density of the borer population.

CONCLUSION

Maize and grain sorghum form a major staple food source for millions of people, and the stem borers infesting them are among the major factors limiting production in eastern Africa. Although the identification, distribution and relative importance of stem borers of maize and sorghum have been studied in some countries in the region, still there is a need for in-depth studies to evaluate the economic importance of borer infestations for devising biointensive integrated pest management strategies.

4 Maize and Grain Sorghum: Southern Africa

Rami Kfir
Plant Protection Research Institute, Private Bag X134, Pretoria 0001, South Africa

INTRODUCTION

Maize is the most important agricultural crop grown in southern Africa. Grain sorghum is also an important crop, especially in the marginal grain-producing areas, because of its drought-resistant properties (van Rensburg and van Hamburg, 1975). It is grown primarily for human consumption but surpluses are used for feeding livestock (Sibanda, 1985).

Of the various insect pests attacking maize and grain sorghum in southern Africa, the African maize stalk borer, *Busseola fusca* (Fuller) (Lepidoptera: Noctuidae), and the spotted stem borer, *Chilo partellus* (Swinhoe) (Lepidoptera: Crambidae), are by far the most important (Skoroszewski and van Hamburg, 1987). To a lesser extent, maize is also attacked by the pink stalk borer, *Sesamia calamistis* Hampson. (Lepidoptera: Noctuidae). The borer is normally kept under control by parasitoids, which prevent serious outbreaks, and additional control measures have rarely been necessary. The use of insecticides against *S. calamistis*, which also infests sugar cane, is not recommended. Another stem borer in the region is the African sugar-cane borer, *Eldana saccharina* Walker (Lepidoptera: Pyralidae). In East and West Africa, *E. saccharina* can cause severe crop losses to maize, but, in southern Africa, for unknown reasons, it attacks mainly sugar cane and very rarely infests other cereal crops. Consequently, *E. saccharina* will not be dealt with in this chapter. The emphasis is on *B. fusca* and *C. partellus; S. calamistis* is briefly discussed.

GEOGRAPHICAL DISTRIBUTION

Busseola fusca is native to Africa and occurs throughout the mainland of the continent south of the Sahara. In southern Africa, it has been recorded from Angola, Botswana, Lesotho, Malawi, Mozambique, Rwanda, South Africa, Swaziland, Zaïre, Zambia and Zimbabwe (CIE, 1988). *Busseola fusca* is the dominant stem borer at elevations above 900 m in Botswana, Lesotho, Malawi, Mozambique, South Africa and Swaziland but also occurs at lower altitudes in those countries and in Zimbabwe (Sithole, 1989b).

Chilo partellus occurs throughout the Indian subcontinent and throughout South-East Asia to Indonesia and Taiwan and southern parts of Asia (Harris, 1989a). It spread into Africa earlier this century. The first record of *C. partellus* in Africa was from Malawi in 1932 (Jepson, 1954), and in South Africa it appeared in 1958 (van Hamburg, 1979). It has since spread to most of the lowland areas (below 1500 m) of the countries in eastern and southern Africa (CIE, 1989). After its introduction, *C. partellus* became the most destructive pest of maize and grain sorghum in the warm, low-altitude regions of southern Africa (van Hamburg, 1979). Until the late 1970s *C. partellus* constituted about 10% of borer infestations in most maize-production areas of South Africa (van Rensburg *et al.*, 1988b). Recently, it was recorded that *C. partellus* is increasing its economic importance by expanding its geographical distribution to the high elevations of the western Highveld region of South Africa (Bate *et al.*, 1991), where in some areas mixed borer populations consist of 90% *C. partellus* and 10% *B. fusca* (van Rensburg and Bate, 1987). It seems that in this region *C. partellus* is competitively displacing *B. fusca* (Kfir, 1997).

Sesamia calamistis is widely distributed in Africa. In southern Africa, it has been recorded from Angola, Malawi, Mozambique, Zimbabwe, South Africa and Zambia. It is also known from the islands of Madagascar, Réunion and Mauritius, where it is mainly a pest of sugar cane (Rao and Nagaraja, 1969).

BIOECOLOGY

In southern Africa, the bioecology of *B. fusca* was studied in Zimbabwe by Smithers (1960a, b) and in South Africa by Mally (1920), du Plessis and Lea (1943), van Rensburg *et al.* (1987) and Kfir and Bell (1993). That of *C. partellus* was studied in South Africa by van Hamburg (1979, 1980) and Kfir (1992c). In general the life cycle of the two borers is similar. Adults emerge from pupae during late afternoon and early evening and are active at night. During the day, they are inactive, resting on plants and plant debris. After mating, during several nights after emergence, females lay eggs in batches: 30–100 eggs under the vertical edges of leaf sheaths by *B. fusca* females, and 10–80 overlapping eggs, parallel to the long axis of the underside of leaves, by *C. partellus* females. Several days later, the larvae hatch and disperse to adjacent plants before they move up to the leaf whorl to feed on the young leaves. Later, they penetrate into the stem tissues to feed, producing extensive tunnels in stems and in maize ears. They then pupate in the tunnels, after excavating emergence windows to facilitate the exit of moths.

In southern Africa, *B. fusca* and *C. partellus* diapause in winter, which is the cold, dry season (April–October), as mature larvae in the lower parts of the dry stalks, where they are well protected from natural enemies and adverse climatic conditions (Kfir, 1988, 1990b) (Plate 2:1,2). *Busseola fusca* diapauses throughout its distribution, whereas *C. partellus* does not enter diapause in the warm low-lying coastal regions of Natal (van Hamburg, 1976), Swaziland (Way and Kfir, 1997) and southern Mozambique (Berger, 1981).

In a 4-year study in the Transvaal, South Africa, it was recorded by Kfir (1992c) that *C. partellus* completed two and a half generations per season. The first generation occurred during October–November, infesting only early grain sorghum tillers from a ratoon crop. The second generation occurred between the beginning of December and the end of January, and the third generation was from the beginning of February to the end of March. In March–April, fifth and sixth instars of the third-generation *C. partellus* entered diapause in the dry stalks (Kfir, 1992c). Kfir and Bell (1993) recorded two generations of *B. fusca* in Natal, South Africa. The first generation, which infested the early grain sorghum plants of the ratoon, peaked in the middle of January and the second generation peaked on the crops at the beginning of April, about 10 weeks after the first peak. In April–May, the crop dried up and *B. fusca* larvae of the second generation reaching fifth and sixth instars entered diapause (Kfir and Bell, 1993).

In South Africa, *C. partellus* starts to emerge from diapause in the second part of August and it lasts until the first week of November, a period of 12 weeks. With *B. fusca*, on the other hand, pupation lasts only during October–November (Kfir, 1988, 1992c). *Chilo partellus* starts to emerge from diapause earlier and moth emergence lasts twice as long as in *B. fusca* (Kfir, 1991a). This difference in the pattern of emergence from diapause by the two borer species explains the distinct generations of *B. fusca* (Mally, 1920; du Plessis and Lea, 1943) and the overlapping generations of *C. partellus* (van Hamburg, 1987; Kfir, 1992c) found every year in the region.

PEST STATUS AND YIELD LOSSES

The profit margin of maize or grain sorghum grown by commercial farmers in the region is relatively low, due to high production costs; in South Africa, for example, pest control alone may amount to 56% of the gross margin above cost for an average yield (van Hamburg, 1987). The timing of insecticidal application is crucial, as control measures are effective against young borer larvae only. Older larvae penetrate the stalks and are difficult to control with insecticides. In addition, the overlapping generations of the borers, especially *C. partellus* (van Hamburg, 1987; Kfir, 1988; Kfir *et al.*, 1989b), result in reinfestations throughout the season, often rendering chemical control of borers unsatisfactory.

Stem tunnelling by the larvae weakens the stem and interferes with translocation of nutrients and metabolites in the plant, resulting in malformation of the grain. Stems can break and the growth of the plant may be stunted (Plate 2:3,4). Estimated yield losses from borer damage range between 10% and total loss (Mally, 1920; Matthee, 1974; van Rensburg and Bate, 1987; Plate 2:5). Infestations by *C. partellus* of up to 100% of plants, with considerable yield losses, have been recorded in the Maputo and Gaza provinces of Mozambique (Nunes *et al.*, 1985) and the Limpopo Valley (Berger, 1981). Yield losses of more than 50% are common in southern Mozambique (Sithole, 1990). *Chilo partellus* is the most important stem borer in Mozambique; larvae of the third generation were reported to infest 87% of cobs of maize planted late and to severely damage 70% of their grain (Berger, 1981). In Zimbabwe, stem borers are the most important insect pests of maize and are capable of completely destroying the crop. Borer infestations range from 30 to 70% in fields of resource-poor farmers, and are less than 30% in commercial farms, where insecticides are used for control (Sithole, 1987b). In general, losses caused by borers vary widely between farms, from almost no loss to virtually total crop loss (van Rensburg and Bate, 1987).

NATURAL ENEMIES

Early studies of *B. fusca* in South Africa, by Mally (1920) and by du Plessis and Lea (1943), identified the gregarious larval parasitoid, *Cotesia sesamiae* (Cameron) (Braconidae; Plate 2:6), as an important mortality factor. Van Rensburg *et al.* (1988a) also recorded several natural enemies of *B. fusca* in South Africa. The three studies led to the similar conclusion that natural enemies are ineffective in curtailing *B. fusca* populations. The tachinid fly, *Sturmiopsis parasitica* (Curran), was recorded as the most common parasitoid of *B. fusca* larvae in the Harare area of Zimbabwe by Smithers (1960a). In a more recent study in Natal, South Africa, Kfir and Bell (1993) recorded high parasitism levels of *B. fusca* pupae by the parasitoid *Procerochasmias nigromaculatus* (Cameron) (Ichneumonidae). Up to 100% parasitism was recorded in the middle of the growing season, during February–March, and 80% during November, when borer larvae which had terminated diapause were parasitized after pupation. Larval parasitism, mainly by *C. sesamiae*, which was reared from about 87% of parasitized larvae, peaked at 75% during January. Despite the high parasitism levels by these parasitoids, they did not prevent economically significant damage (Kfir and Bell, 1993).

In addition, the following parasitoids were recorded by Kfir (1995) from *B. fusca* in South Africa: the egg parasitoids *Telenomus busseolae* Gahan (Scelionidae) and *Trichogrammatoidea lutea* Girault (Trichogrammatidae); the larval parasitoids *Aleiodes* sp. (Braconidae), *Bracon sesamiae* Cameron (Braconidae; Plate 3:3), *Chelonus curvimaculatus* Cameron (Braconidae), *Stenobracon* (= *Euvipio*) sp. (Braconidae), *Glyptapanteles maculitarsis* (Cameron) (Braconidae), Bracon (*Habrobracon*) *brevicornis* (Wesmael) (Braconidae), *Iphiaulax* sp. (Braconidae), *Palexorista* sp. (Tachinidae), *Paradrino halli* Curran (Tachinidae), *Sarcophaga* sp. (Sarcophagidae), *Temelucha* sp. (Ichneumonidae) and *Odontepyris transvaalensis* (De Buysson) (Bethylidae); the pupal parasitoid *Pediobius furvus* Gahan (Eulophidae); and the hyperparasitoids *Aphanogmus fijiensis* (Ferrière) (Ceraphronidae), which attacks *C. sesamiae* cocoons, and *Eurytoma braconidis* Ferrière (Eurytomidae), which attacks *Stenobracon* sp. and *B. brevicornis*. The ant *Dorylus helvolus* (L.) was also frequently observed attacking and preying on *B. fusca* larvae in sorghum stems in the Transvaal (Kfir, 1988).

Watmough and Kfir (1995) recorded *D. helvolus* and the mouse *Mastomys natalensis* (Smith) as the dominant predators of stem borers in maize and grain sorghum in South Africa. Other predators, in descending order of importance, were Cardiophorinae larvae, the doryline ant *Aenictus* sp., larvae of *Astylus atromaculatus* Bl. (Coleoptera: Melyridae), tenebrionid larvae, larvae and adults of *Heteronychus arator* F., carabid larvae, staphylinids and Dermaptera (Watmough and Kfir, 1995).

In Mozambique, Gonçalves (1970) recorded 60% egg parasitism in *C. partellus*, and Berger (1981) recorded the ichneumonid *Dentichasmias busseolae* Heinrich, the scelionid *Lepidoscelio* sp. [editor's note: this latter record was almost certainly based on a misidentification of the host; see chapter 21) and the eulophid *P. furvus* (Gahan) as parasitizing 10–14% of *C. partellus* pupae. Sithole (1989b) observed ants, spiders, mites and reduviids preying on *C. partellus* in the field in Zimbabwe. In the Transvaal, South Africa, Kfir (1990a, 1992c) recorded peaks of up to 100% parasitism on *C. partellus* pupae every year in several consecutive growing seasons during the second part of January and the first part of February. The most abundant pupal parasitoids were the solitary *D. busseolae* and the gregarious *P. furvus* (Plate 3:4). Occasionally, *Conomorium* sp. (Pteromalidae) was also reared from field-collected pupae of *C. partellus* (Kfir, 1992c). Larval parasitoids were active throughout the growing season on maize and grain sorghum. Parasitism oscillated around 10%, with occasional peaks of 20–30%. The most abundant larval parasitoid of *C. partellus* was *C. sesamiae* (Plate 3:1,2). This parasitoid was active throughout the season and was reared from about 93% of all parasitized larvae (Kfir, 1990a). The hyperparasitoid, *A. fijiensis*, was frequently reared from field-collected *C. sesamiae* cocoons, sometimes reaching 50–100% parasitism in autumn. Other larval parasitoids that attacked *C. partellus* were *Iphiaulax* sp., the egg–larval parasitoids *C. curvimaculatus*, and *Chelonus* sp., *Bracon* sp., *Norbanus* sp. (Pteromalidae), *Pristomerus* sp. (Ichneumonidae) and *Palexorista* sp. The hyperparasite *Eurytoma* sp. was reared from field-collected cocoons of *Iphiaulax* sp. on several occasions (Kfir, 1990a, 1992c). During winter, parasitism of diapausing *C. partellus* larvae was very low. Two parasitoids, *C. sesamiae* and *Bracon* sp., were observed to be active during winter. Three parasitoids, *Chelonus curvimaculatus*, *Chelonus* sp. and *Pristomerus* sp., diapaused in synchrony inside the larvae, and *Iphiaulax* sp. diapaused in its own cocoon inside the dry stalks (Kfir, 1988). Phiri (1995) gave a detailed account of the natural enemies of *C. partellus* in Malawi.

All parasitoids listed above are indigenous to Africa, and their association with the exotic *C. partellus* is relatively new. These parasitoids usually develop on other indigenous Lepidoptera, and they moved from their indigenous hosts to utilize *C. partellus* as a new alternative host (Kfir, 1992c). Although larval and pupal parasitoids play an important role in reducing the population levels of *C. partellus,* they are not able to reduce it to below economic damage levels (Kfir, 1992a). The parasitoids could not prevent the dispersal and subsequent wide distribution of *C. partellus* after its introduction into Africa.

Stem-borer pathogens in southern Africa have received very little attention. High winter mortality of hibernating *B. fusca* larvae was recorded by Kfir (1988). As low activity by stem-borer parasitoids was recorded in winter (Kfir, 1988), this mortality could be partially attributed to pathogens. A microsporidium, *Nosema partelli* Walters and Kfir (Protozoa: Microsporidia: Nosematidae), was recorded in field-collected *C. partellus* in South Africa (Walters and Kfir, 1993). In addition, a recent survey of pathogens conducted in South Africa (Hoekstra and Kfir, 1997) revealed the following pathogens identified from cadavers of *B. fusca*: cytoplasmic polyhedrosis virus, granulosis virus, nuclear polyhedrosis virus, the entomopathogenic fungi *Beauveria bassiana* (Balsamo) (Fungi Imperfecti) and *Entomophthora* sp. (Phycomycetes), the bacteria *Serratia marcescens* Bizio (Gram-negative facultative anaerobic pathogen, non-spore-former) and *Bacillus thuringiensis* Berliner (endospore-forming rods and cocci, obligate pathogen), and the microsporidium *Nosema* sp. From cadavers of *C. partellus* cytoplasmic polyhedrosis virus, entomopox virus, the fungi *B. bassiana*,

Entomophthora sp. and *Aspergillus* sp. (Fungi Imperfecti), the bacteria *Streptococcus* sp. (Gram-negative cocci), *S. marcescens* and *B. thuringiensis* and the microsporidium *N. partelli* were isolated (Hoekstra and Kfir, 1997) (see also Chapter 23, this volume).

PEST MANAGEMENT AND CONTROL METHODS

Cultural Control

Crop residues are important for carrying over *B. fusca* and *C. partellus* larval populations from one growing season to the next. Ploughing in order to bury maize stubble was an effective control measure used in South Africa to control *B. fusca* (Mally, 1920). In a study done in Zimbabwe, it was shown by Jack (1918) that *B. fusca* moths emerging through 5 cm of soil were crippled and that deeper burial of the stalks under 10–15 cm of soil ensured that no adult moths emerged. It was shown in South Africa that slashing maize and sorghum stubble destroyed 70% of *C. partellus* and *B. fusca* populations and that ploughing and discing destroyed a further 24% of the pest population in sorghum and 19% in maize (Kfir *et al.*, 1989b; Kfir, 1990b) (Plate 3:6). In some rural areas where dry stems are used for fencing and building, it may not be possible to reduce carry-over populations. However, in commercially grown crops it might be possible to reduce the pest status of borers by destroying their hibernation sites. This should be done with the cooperation of all farmers in a region, because moths emerging from untreated fields can infest adjacent crops (Kfir, 1992a). Currently, this system is not practised in South Africa due to the advent of minimum tillage and the importance of winter grazing of maize to beef farmers (Kfir, 1992a).

Sowing dates could be used to avoid severe borer infestations. In Tanzania, Swaine (1957) found that maize sown early in the season was more affected by *B. fusca* than maize sown later. On the other hand, in the Highveld region of South Africa, farmers sow maize early in the growing season to avoid severe damage by second-generation *B. fusca*. Little research attention has been given to other aspects of cultural control in southern Africa (Sithole, 1990).

Attempts at Biological Control

The high cost and inefficacy of insecticides to control *B. fusca* and *C. partellus* in South Africa precipitated a biological-control programme using exotic parasitoids (Kfir, 1991c, 1992a). The parasitoid *Cotesia flavipes* (Cameron), which was responsible for the successful control of *Diatraea saccharalis* (Fabricius) in Barbados (Alam *et al.*, 1971), was introduced into South Africa from Pakistan. The parasitoid was reared in the laboratory and liberated in large numbers in infested maize fields in Transvaal and Natal. A few recoveries were made but the parasitoid did not establish (Skoroszewski and van Hamburg, 1987). Further consignments of *C. flavipes* were received from Mauritius and Indonesia, but again, after releases during 1991 and 1992, the parasitoid failed to become established. Other parasitoids that were released in South Africa were: *Lydella minense* Townsend (Tachinidae) from Trinidad, *Paratheresia claripalpis* Van der Wulp (Tachinidae) from Brazil,

Trichogramma chilonis Ishii and *Trichogramma ostriniae* Pang and Chen (Trichogrammatidae) from Taiwan, *Tetrastichus howardi* (Olliff) (Eulophidae) from the Philippines and *Xanthopimpla stemmator* Thunberg (Ichneumonidae) from Mauritius (Kfir *et al.*, 1989a, 1993; Kfir, 1991c, 1992a). Several recoveries were made of *T. chilonis* , *X. stemmator* and *T. howardi* from field-collected *C. partellus*. These recoveries were made shortly after the releases and close to the release sites, so it is not yet known if the parasitoids have become established (Kfir, 1994).

Host-plant Resistance

Resistant cultivars of maize and grain sorghum against *B. fusca* and *C. partellus* could play an important role in the implementation of an integrated pest management system in the region. The development of resistant hybrids could decrease the dependency on pesticides to control borers and enhance grain production. Host-plant resistance as a control method is environmentally safe, economically acceptable to farmers and the most compatible with other components in integrated pest management programmes (Harris and Nwanze, 1992).

There were some early attempts in South Africa to develop maize cultivars resistant to *B. fusca* (du Plessis and Lea, 1943; Walters, 1974). Several lines of maize with some resistance to first-generation (whorl-feeding) *B. fusca* larvae were identified in South Africa, but no information is available on resistance to the second-generation larvae, which cause stem and ear damage (Barrow, 1989). Barrow (1985) observed two resistance factors to *B. fusca* in maize: a short-lived but effective factor in the whorl tissue that either killed or repelled early-instar larvae, and a long-lasting factor that retarded larval development. However, both levels were too low for incorporation into the breeding material. Evidence received from maize breeding programmes indicates that inheritance of resistance appears to be additive (Barrow, 1989). Two approaches are currently used to breed maize for resistance in South Africa: population improvement and the development of inbred lines.

Screening sorghum germplasm for resistance to *B. fusca* in natural infestation (Sithole, 1987b, 1988) and for resistance to *C. partellus* in artificial infestation has been reported from Zimbabwe (Sithole, 1990). In an assessment of the level of resistance to *C. partellus* in 49 commercially available grain-sorghum hybrids in South Africa, van den Berg *et al.* (1993) concluded that the general level of resistance is low. However, pronounced differences in yield losses were observed between some hybrids, which were ascribed to tolerance to damage and the ability of the plants to compensate for damage by tiller formation. Three hybrids, CRN 7690, CRN 7692 and PAN 8501, exhibited the greatest tolerance to borer damage, and consequently also suffered relatively low yield losses (van den Berg *et al.*, 1993). Van den Berg *et al.* (1994) showed increased insecticide efficacy against stem borers in South Africa in grain-sorghum lines showing some resistance. The resistance mechanism, which is based on antibiosis and antixenosis, apparently causes stress in *C. partellus* and *B. fusca*, making them more susceptible to the insecticides (van den Berg *et al.*, 1994).

Pheromone Trapping

Hall *et al.* (1981) isolated the female sex pheromone of *B. fusca*, which contains three different tetradecenyl acetates. Initial trials in Zimbabwe and Malawi showed that a synthetic pheromone mixture caught as many males as did the natural pheromone from live females. Revington *et al.* (1984) showed that higher loading of pheromone gave higher trapping rates of *B. fusca* males in South Africa. During a 5-year study in Natal, South Africa, Kfir and Bell (1993), using delta-shaped traps (registered trade name Biotrap) and omnidirectional traps (registered trade name Omnitrap), recorded two peaks of *B. fusca* moth flights (Plate 3:5). The first flight, which resulted from moths emerging from hibernating borers, occurred from the end of October to the beginning of December, with a peak about the middle of November. The second flight, usually larger, occurred from the beginning of February to the middle of March, with a peak at the end of February. Larval peaks of *B. fusca* in the field lagged about 4–6 weeks behind the corresponding moth flight peaks (Kfir and Bell, 1993). Omnidirectional traps were found to be superior to delta traps for quantitative and qualitative estimation of *B. fusca* moth populations (van Rensburg, 1992). Revington (1987) monitored moth flights at various sites in the main maize-growing areas of South Africa and showed that trap results, correctly interpreted, could be used to forecast attacks on maize by *B. fusca.* Based on Revington's (1987) work, pheromone traps for commercial use were developed. Currently, omnidirectional pheromone traps are commercially available in South Africa and are used by farmers to determine the timing of insecticidal applications against *B. fusca*. The distributor's recommendations are:

1. For the first moth flight of the season, monitoring commences in the second week of September and continues for about 6 weeks or until the moth flight terminates.
2. For the second flight, traps are monitored from the middle of December for at least 6 weeks.
3. A group of three traps is sufficient to cover a farm of about 1000 ha.
4. The accepted action-threshold level for spraying against *B. fusca* is when the average catch from three traps per site exceeds two moths per week for 4 consecutive weeks.

The female sex pheromone of *C. partellus* was identified (Nesbitt *et al.*, 1979), but initial trials in South Africa (R. Kfir, unpublished) and in Kenya (Unnithan and Saxena, 1990) gave erratic results. More research in trap design and correlations between moth catches and infestations in the field is required before trapping *C. partellus* can be used to determine economic threshold levels.

Chemical Control

The use of pesticides for the control of *B. fusca* was first reported from South Africa and Zimbabwe, where maize crops were sprayed with Derris and carbolic dip, Pulvex (ground derris root), Kymac (a sheep dip containing derris), Cryolite, hycol solution and several other insecticides based on rotenone, an extract from the plant *Derris chinensis* (Arnold, 1928; Ripley, 1928, Ripley and Hepburn, 1928, 1929, 1934). Later studies in South Africa showed that application of the systemic insecticide, carbofuran, in granules to the planting furrows of maize (Walters and Drinkwater, 1975) and grain sorghum (van Rensburg *et al.*, 1978) results in significant reduction of infestations by *B. fusca*. Van den Berg and van Rensburg (1991a) found that a single application of endosulfan against *C. partellus* was beneficial when applied shortly before tasselling in maize and before panicle emergence in grain sorghum.

Control of *B. fusca* with endosulfan on finger millet (Leuschner, 1990) and with dipterex on grain sorghum in the communal areas (Sithole and Makombe, 1989) was also reported from Zimbabwe. It was shown in Mozambique that, when insecticides are applied against *C. partellus* in maize, yield can increase two- to fourfold, the recommendation being to apply diazinon twice into the whorl, when the plants are 3 and 5 weeks old (Berger, 1981).

Revington (1986) reported that deltamethrin alone or in a mixture with endosulfan gives effective control against *C. partellus* in maize and grain sorghum when applied 10–14 days after crop emergence. Pesticides used to control *C. partellus* are trichlorfon and synthetic pyrethroids (Sithole, 1989b). However, in the marginal rainfall areas of southern Africa, because of the low profit margin, farmers cannot afford the cost of chemical control against stem borers.

Until recently, stem-borer populations in maize and grain sorghum in southern Africa were comprised mainly of *B. fusca* (van Rensburg *et al.*, 1988b). Chemical control of borers in both crops has become more complicated in South Africa, due to the recent increase in the distribution and the relative abundance of *C. partellus* in most production areas (Bate *et al.*, 1991). Efficient chemical control against a mixed population is problematic because the two borer species differ conspicuously in characteristics such as the duration of larval development, intraseasonal infestation patterns and injuriousness (van Hamburg, 1980; van Rensburg *et al.*, 1988b; van den Berg *et al.*, 1991). As a result, control measures against mixed populations are still unresolved and inclined towards repeated applications of insecticides (Vermeulen *et al.*, 1992).

Currently, the economic threshold level for *C. partellus* and *B. fusca* in South Africa is usually accepted as 10% of plants showing visible damage. However, in a recent study on the effect of *C. partellus* on growth and yield of maize, the economic threshold was determined as 40% plants with visible damage (Bate and van Rensburg, 1992).

5 Sorghum: West Africa

O. Ajayi
ICRISAT, Sabo Bakin Zuwo Road, PMB 3491, Kano, Nigeria

INTRODUCTION

Sorghum, or Guinea corn, *Sorghum bicolor* (Plate 1:5), is the most widely cultivated cereal crop and the most important food crop in the savannah areas of West Africa. Its importance can be illustrated by the situation in Nigeria, the major producer, where it accounts for about 50% of the total cereal production and occupies about 46% of the total land area devoted to cereal production (other cereals are rice, maize, millet and wheat). The area devoted to sorghum in Nigeria is slightly more than 6 million ha and production is estimated to be about 9 million t (IAR, 1984).

SPECIES COMPLEX

Insect pests constitute an important factor limiting grain-sorghum production in West Africa. The insect pests of sorghum have been surveyed by Risbec (1950b) and Appert (1957) in Senegal and Mali; Harris (1962) and Ajayi (1978) in Nigeria; Forsyth (1966) in Ghana; Brenière (1970) for French-speaking West Africa, Bridge *et al.* (1978) in the Gambia; Gahukar (1981) in Senegal; Bonzi (1980, 1982b) and Nwanze (1988) in Burkina Faso; Gahukar (1981), Bonzi *et al.* (1986) and Ajayi (1989) for West Africa; and Seshu Reddy (1991) for Africa. Harris (1989a) reviewed available literature on the distribution, biology and ecology of *Busseola fusca*. The most important field insect pests are shoot flies, stem borers, spittle bugs, head bugs, head caterpillars and grain midges. Of these, stem borers are the most important and the most widespread.

Stem borers reported on sorghum in West Africa include *B. fusca*, *Sesamia calamistis*, *Sesamia poephaga*, *Sesamia penniseti*, *Sesamia cretica*, *Sesamia nonagrioides*, *Sesamia botanephaga*, *Coniesta ignefusalis*, *Chilo diffusilineus* and *Eldana saccharina*. *Coniesta ignefusalis* is primarily a pest of pearl millet (*Pennisetum glaucum*) but is a minor pest of sorghum in a sorghum/millet intercrop. Although details vary among countries, *B. fusca* is the most important stem borer, since it predominates in the major sorghum-growing areas, i.e. the Sudanian and northern Guinean agroecological zones.

The International Institute of Entomology (CIE, 1988) and the Inter-African Phytosanitary Council (IAPSC, 1985a) list the following countries in West Africa as being affected by *B. fusca*: Benin, Burkina Faso, Cameroon, Chad, Ghana, Guinea, Ivory Coast, Nigeria, Sierra Leone and Togo. In Burkina Faso, *B. fusca* predominates south of latitude 12°N (Nwanze, 1988). According to Libby (1968), *B. fusca* was distributed throughout Nigeria; *S. calamistis, S. poephaga* and *S. penniseti* were widely distributed throughout the major vegetation zones of Nigeria; *S. calamistis* was more abundant going south, *S. poephaga* was most numerous in the derived savannah, and *S. penniseti* occurred throughout Nigeria, while *S. nonagrioides botanephaga* was recorded from a few localities in the southern Guinea savannah and lowland rain-forest zones in Nigeria; *E. saccharina* was distributed throughout northern Nigeria and at least as far south as latitude 7°N (Ibadan), being most abundant in the riverain areas. More recently, Abu (1986a) reported that *S. calamistis* was predominant and *E. saccharina* was important in the southern Guinea savannah in Nigeria; Doumbia (1980) observed *B. fusca* on sorghum in the Masantola, Sotuba and Kogoni regions of Mali; Nwanze (1981) found *C. ignefusalis* to be more important than *B. fusca* north of latitude 11°30′N in Burkina Faso. *Sesamia cretica* and *S. calamistis* have been recorded by this author as pests of dry-season sorghum in Cameroon, Chad and Nigeria, while Ratnadass *et al.* (1992) reported *S. poephaga* (apparently misidentified as *S. penniseti*) on dry-season sorghum in Mali. In spite of the above reports, it is necessary to continue to monitor the relative importance and distribution of these insects, since changes could occur for various reasons, including, for example, deforestation, changes in crops and cropping patterns, adoption of new crop cultivars and misidentifications.

Yearly variations in borer populations have been reported (Harris, 1964; Elemo and Ajayi, 1989). There are speculations that environmental factors, especially rainfall, determine these variations (Mercer-Quarshie, 1970, Elemo and Ajayi, 1989), but detailed correlation studies are needed to confirm this.

ECONOMIC IMPORTANCE

Although stem-borer attack on sorghum is widespread in West Africa and total crop loss sometimes occurs, actual yield losses have only rarely been reported. Depending on the method used and the circumstances of the estimation, yield-loss assessment trials have had different outcomes. For example, Harris (1962) reported a significant positive correlation between the number of stems bored by *B. fusca* and yield. Mercer-Quarshie (1970) indicated that stem borers caused a yield reduction of 4% in Ghana in 1967 and 1968. Abu (1986a) showed that stem-borer control under natural infestation in the southern Guinea savannah of Nigeria, where *Sesamia* predominates, improved yields by 16–19%. Ajayi (1987) reported a yield loss of 49% in the northern Guinea savannah of Nigeria, where *B. fusca* predominates. McFarlane (1990) found that yield loss per panicle ranged from zero to 1.2% per unit of infestation by *B. fusca*.

The foregoing were the outcome of on-station trials, which relied on natural infestation. Fewer on-farm experiments on yield-loss assessment have been reported. Ajayi (1991) compared the yield of sprayed and unsprayed sorghum in 22 farmers' fields in 1990 and reported a mean yield loss of 21% averaged over three local cultivars: Kaura (32%), Farafara (31%) and Gaya Early (12%). However, the effects due to *B. fusca* and *S. calamistis* could not be separated.

Studies on the effect of stem-borer attack on grain yield in West Africa have been hampered by various factors, including the following: reliance on natural infestation limits the control over the timing and intensity of infestation; well-grown sorghum tends to compensate for stem-borer infestation by producing tillers, some of which can mature if there is sufficient rainfall; ovipositing females prefer larger, thick-stemmed plants, which are able to withstand tunnelling and still produce higher grain yield than the smaller, thinned-stemmed plants; the level of yield reduction depends on the generation of stem borer involved – the first and second generations caused more yield loss than the third generation (Harris, 1962); yield loss also varied with the time of infestation – McFarlane (1990) found that sorghum infested by *B. fusca* before the booting stage suffered greater yield losses. In addition, the parameter used to assess stem-borer damage is critical. McFarlane (1990) recorded that the percentage of internodes bored in the lower half of the stalk had a more consistent negative correlation with grain yield than did the percentage of stem tunnelled. Finally, while artificial infestation can elucidate the yield loss attributable to each species of stem borer in controlled conditions, it seldom suffices to capture the actual field situation, in which more than one species is involved. More studies in farmers' fields are required.

CONTROL MEASURES

Cultural Control

Since *B. fusca* spends the dry season in a facultative diapause within the stems, the most logical cultural-control measure would be to reduce the first generation of adult population by killing the larvae in the old stalks. This is best done through destroying old stalks by burning or composting before the onset of the rains (Harris, 1962). Where the stalks are required for building, fencing or firewood, they should be partially burned; the leaves should be burned a few days after the panicles have been removed, while the leaves are dry but the stalks are not (Adesiyun and Ajayi, 1980). Heat generated from the burning leaves is sufficient to kill the larvae within the stalks. This curing process incidentally makes the stalks stronger and more resistant to termite attack. The stalks should be kept in the open rather than in large stacks in the shade of trees, where larval survival is much higher. Where the stalks are fed to animals, the residues and all unused stalks should be destroyed before the onset of the rains.

For these cultural-control measures to be effective, they must be adopted by all farmers in the target area; otherwise, the remaining insect populations will continue to seek available hosts. One way to accomplish this is to enlighten the farmers adequately or, if possible, enforce compliance by legislation.

Sesamia can also be controlled by partial burning. In addition, however, its population should be reduced on its alternative hosts, on which it survives during the dry season. These other hosts include wheat and wild sorghum (*Sorghum halepense*) (Ajayi, 1986) and the dry-season sorghum (*muskwari* in Cameroon, *berbere* in Chad and *masakwa* in Nigeria). In the southern Guinea savannah, where sorghum ratoons support a carry-over of *Sesamia* populations through the dry season (Abu, 1986b), it is advisable to plough the ratoons into the soil.

For both *Busseola* and *Sesamia*, early planting has been found to lower stem-borer infestation (Abu, 1986a). This probably explains why peasant farmers, who normally plant as soon as the rains become established, suffer less stem-borer infestation than occurs at experimental stations, where planting is usually delayed.

Another cultural-control measure involves intercropping. Peasant farmers in West Africa almost invariably intercrop millet and sorghum. Adesiyun (1983) indicates that interplanting millet and sorghum in alternate stands within the same row reduces larval infestation of *B. fusca* on sorghum, since the adults do not effectively utilize millet for oviposition; Adesiyun (1988) has also shown that *B. fusca* larvae develop better and faster on sorghum than on millet. Elemo and Ajayi (1989) confirmed the finding by Adesiyun (1983) that intercropping sorghum and millet in alternate rows does not influence *B. fusca* attack on sorghum. Similarly, Amoako-Atta *et al.* (1983) reported that the frequency of occurrence of *B. fusca* was not influenced by the pattern of intercropping maize, cowpea and sorghum. McFarlane (1984) observed that intercropping sorghum with groundnuts decreased *B. fusca* infestation, while infestation increased with an increase in nitrogen fertilizer. Elemo and Ajayi (1989) showed that, although *B. fusca* damage increased with an increase in the level of nitrogen fertilizer applied to sorghum, neither the rate nor time of nitrogen-fertilizer application affected stem-borer damage. It therefore appears that each cropping pattern must be tested, since no general conclusion can be made.

Resistant Varieties

In West Africa, the use of pesticides is largely beyond the means of the small farmer, and recommended chemicals are usually not available at the right time. Plant resistance is, therefore, an attractive method for reducing stem-borer damage.

The search for stem-borer resistance in sorghum began in Nigeria in the early 1970s under the US Agency for International Development (USAID) JP 26 project. About 4000 sorghum lines from the world collection were screened over a 4-year period at Samaru, using natural infestation. The entries were rated as R (resistant) if they had less than 1 cm tunnelling and evidence of larval mortality or abandonment; I (intermediate) if tunnelling was 1–10 cm and there was evidence of larval mortality or abandonment; and S (susceptible) if the tunnelling was more than 10 cm and there was very little or no evidence of larval mortality (IAR, 1976). On this basis, 306 lines were tentatively selected for possible resistance to *B. fusca* (Barry, 1980). Eventually, 22 of these lines were rated as resistant to *B. fusca.* Of the 22 varieties, four were also reportedly resistant to shootfly (IS 6747, 6441, 6449 and 8910).

In 1981 and 1982, McFarlane (1984) screened 122 entries from the International Crops Research Institute for the Semi-Arid Tropics (ICRISAT) and the Institute for Agricultural Research at Samaru under natural infestation. His parameters were percentage of stem tunnelled, percentage of internodes bored and visual damage rating. Another set of 122 entries from an ICRISAT project in Burkina Faso were screened at Samaru under natural infestation.

From 1988 to 1995, entries in the International Sorghum Stem Borer Nursery and the International Sorghum Shoot Pests Nursery were conducted at Bagauda, Nigeria, under natural infestation. There are indications that a number of the lines that are resistant to *Chilo partellus* in Asia, such as IS 2205, also have low susceptibility to *B. fusca*. However, the prospects for high levels of resistance to stem borers in sorghum appear to be limited and we may have to settle for moderate levels.

A number of observations can be made from these attempts to find sources of resistance to stem borers in sorghum. One observation is that borer infestation is highly variable from year to year, making interpretation of data difficult. Another is that the amount of leaf feeding is not a reliable parameter for measuring stem-borer damage. This is probably because mixed infestations of *Busseola* and *Sesamia* usually occur and leaf feeding is more common with *Busseola*. Also, there appears to be no correlation between tunnel length, a frequently used criterion, and grain yield.

It is clear that the methodology used for evaluating stem-borer nurseries needs to be perfected, preferably using laboratory-reared larvae to artificially infest nursery plants. To help accomplish this, an artificial diet was developed at Samaru composed of wheat, soybean flour, brewers' yeast, ascorbic acid, sorbic acid, formaldehyde, water and agar. This diet yielded up to 60% pupation and adult emergence. Increasing the soybean content significantly improved the rate of adult emergence (IAR, 1975). More recently, attempts have been made to rear *B. fusca* continuously in the Centre de Cooperation Internationale en Récherche Agronomique pour le Développement (CIRAD)/ICRISAT laboratory in Mali. The rate of pupal and adult emergence was above 60% as at 1995 (A. Ratnadass, CIRAD/ICRISAT Mali, personal communication) and further improvements are anticipated, which will enhance the capacity for effective screening for stem-borer resistance in West Africa. This may need to be combined with the method used by van Rensburg and van Rensburg (1993) in South Africa, in which diapausing larvae are collected from the field and their first-generation offspring are used for artificial infestation.

Potential for Biological Control

Several parasitoids, predators and disease organisms have been reported as natural enemies of sorghum stem borers in West Africa (Harris, 1962; Maafo, 1975; Bordat *et al.*, 1977; Gahukar, 1981; Betbeder-Matibet, 1989; Polaszek and Walker, 1991). According to Harris (1962), *Tetrastichus atriclavus*, *Cotesia* (as *Apanteles*) *sesamiae* and *Pediobius furvus* are always present on *B. fusca* towards the end of the growing season and are probably its most important parasites. *Cotesia sesamiae* has been reported on *S. calamistis* in Burkina Faso, *S. cretica* in Senegal, and *Sesamia* sp. in Nigeria (Polaszek and Walker, 1991). *Bacillus thuringiensis* and *Aspergillus* spp. are also found on the larvae and pupae of *B. fusca* at Samaru. The list of natural enemies will probably increase as more efforts are made to search for them. For example, in a survey conducted in the Sudanian and northern Guinean zones of northern Nigeria by the author in 1993, the following parasitoids and hyperparasitoids were collected: *T. atriclavus*, *Exoristobia dipterae*, *Brachymeria* sp. (probably undescribed), *C. sesamiae* and some tachinid flies. Although the rate of parasitism of stem borers is low and seldom exceeds 10%, the use of natural enemies for biological control appears to be worth pursuing. Meanwhile, there have been no studies on the population, ecology and efficiency of these natural enemies. The roles of the hyperparasitoid *E. dipterae* and the facultative hyerparasitoid *T. atriclavus* also need to be determined. *Exoristobia dipterae* usually attacks Tachinidae (A. Polaszek, personal communication).

Chemical Control

The use of insecticides for sorghum stem-borer control is not attractive to most small farmers, but it may be the only effective method in a severe infestation. A number of insecticides have been tested in the field and can be recommended for use.

For the control of leaf-feeding stages of *B. fusca*, carbaryl 85 WP, granular endosulfan 5G and granular trichlorphon 5G applied into the whorl three times at weekly intervals have been recommended (Ajayi, 1978; Adesiyun, 1986). Granules of the systemic insecticide carbofuran, applied into the planting hole at planting, followed by a side-dressing 6 weeks later, controls *Sesamia* larvae (Ajayi, 1987; Uvah and Ajayi, 1991). Malathion 50 EC, basudin 10 G and basudin 60 EC are also effective (Sagnia, 1983).

REQUISITES FOR INTEGRATED CONTROL

Various measures can be used in combination to control stem borers. But there are a number of requisites before integration can be successfully employed. Some of the more obvious ones are listed below:

1. The economic threshold level for each stem borer and complex of species needs to be determined, in order to refine the use of insecticides and make them more cost-effective.
2. To develop an effective biological control programme, the proper identification, biology, ecology and efficiency of the (identified) natural enemies of the stem borers, as well as the role of hyperparasites, need to be determined.
3. There is a need to develop an accurate and standard procedure for breeding and screening sorghum lines for resistance to stem borers. Laboratory rearing and screen-house facilities are important requisites, but are generally beyond the reach of most national agricultural-research institutions in West Africa.
4. A method of predicting outbreaks of sorghum stem borers is needed. Such a method should be amenable to use by individual farmers, or groups of farmers, in view of the communication difficulties existing in West Africa. Pheromone trapping is a possibility.
5. Knowledge of the distribution of sorghum stem borers is still inadequate in some parts of West Africa, and the factors influencing changes in the intensity of infestation from year to year are poorly understood.

6. Effects of the changing cropping systems, in some parts of West Africa, on the stem-borer species complex and activity need to be elucidated. Examples are the production of wheat under irrigation in the dry season and post-rainy-season sorghum grown on residual moisture in vertisols.
7. The number of entomologists working on sorghum insects in the region is grossly inadequate, as is the level of research funding.

6 Millet: West Africa

K.M. Harris[1] and O. Youm[2]
[1]*Research Associate, CABI Bioscience UK Centre, c/o Entomology Department, The Natural History Museum, London SW7 5BD, UK;* [2]*ICRISAT Centre Sahélien, BP 12404, Niamey, Niger*

Pearl millet (*Pennisetum glaucum* L.) (Plate 1:2) is of particular importance as a staple cereal crop in West Africa, especially in the harsh environments of the Sahel. It is mainly grown as a rain-fed subsistence crop in areas of low and erratic rainfall (200–800 mm per annum) and total annual grain production in West Africa is about 8 million t, which is about a quarter of total world production. The most important areas of production lie between 10 and 17°N, and the main millet-growing countries in West Africa are Nigeria, Niger, Burkina Faso, Chad, Mali, Mauritania and Senegal. Yields on farmers' fields are low, averaging about 600 kg ha^{-1}, but yields of 2000–3000 kg ha^{-1} have been obtained on research farms, indicating great potential for future development (Kumar, 1989).

The insect pests of pearl millet in West Africa have been reviewed by a number of authors in recent years (Gahukar, 1984; Sharma and Davies, 1988; Nwanze and Harris, 1992; Matthews and Jago, 1993). They are considered to be one of the major constraints on production (together with drought, diseases, weeds and birds) and the major insect pests in the region are the millet head miner, *Heliocheilus albipunctella* De Joannis (= *Raghuva albipunctella*); the millet stem borer, *Coniesta ignefusalis* (Hampson); and grasshoppers, meloid beetles and scarabaeid beetles. The only stem borer of major importance on pearl millet in West Africa is *C. ignefusalis.* Other species, notably *Busseola fusca, Sesamia calamistis, Sesamia penniseti, Sesamia poephaga, Manga basilinea* and *Eldana saccharina,* have been recorded from this host but generally in small numbers. Harris (1962) reported that *C. ignefusalis* predominated in the main millet-growing areas of northern Nigeria above latitude 10°N and Nwanze (1989) confirmed that this is still the case. He also reported that *C. ignefusalis* was the only species recovered in surveys of farmers' millet crops in the drier northern Sahelian zone but that some *S. calamistis* and *E. saccharina* were encountered further south. Studies of *C. ignefusalis* in West Africa started in the 1950s, but as relatively isolated observations and projects. Risbec (1950b) and Appert (1957) published brief notes on this species in Senegal. In Nigeria, the Federal Department of Agricultural Research initiated general surveys of cereal stem borers and, in 1955, established a stem-borer research project based at the Regional Research Station of the Northern Nigerian Ministry of Agriculture at Samaru, Zaria (Harris, 1962).

During the late 1960s and the 1970s, some research continued in Nigeria and in the francophone countries of West Africa (especially Senegal, Mali and Niger), but it was not until the late 1970s that new impetus was given to work on this pest. Collaboration between national programmes in research on pests of Sahel food crops was then enhanced by the Food and Agriculture Organization (FAO)/Comité Permanent Inter-états de Lutte Contre la Sécheresse dans le Sahel (CILSS)/US Agency for International Development (USAID) Integrated Pest Management (IPM) Project and by the establishment of the International Crops Research Institute for the Semi-Arid Tropics (ICRISAT) Sahelian Center at Sadoré, Niamey. In Nigeria, the Institute for Agricultural Research of Ahmadu Bello University established a research programme in the 1980s to investigate possibilities of integrated control of the millet stem borer in Nigeria and local studies continued in Niger, Burkina Faso, Mali and Senegal. Recently, new interest has been generated by the Directorate General for International Cooperation (DGIS)-funded project on parasitoids of lepidopterous stem borers of cereals in Africa and by the Natural Resources Institute/ICRISAT collaborative project on millet stem-borer pheromones. Key publications, in chronological order, are those of Harris (1962), describing adult and immature stages, distribution, biology, ecology and crop-loss assessments in Nigeria; Nwanze (1989), reporting results of field surveys, biological and ecological studies and crop-loss assessments in Burkina Faso, Niger and northern Nigeria; Ajayi (1990), reporting experiments on cultural and chemical control in northern Nigeria during the 1980s and discussing the possibilities of integrated control; Youm *et al.* (1993b), reviewing biology, ecology and integrated pest management; and Youm *et al.*, (1993a), reporting the development of pheromone traps. Additional information is also available in various published and unpublished reports and papers on millet pests (see Nwanze and Harris, 1992). Although much is now known about the biology and ecology of *C. ignefusalis* (see Chapter 11), there are continuing needs for better information on population dynamics (especially on the importance of diapausing larvae and early rainfall in contributing to sporadic and unpredicted outbreaks) and on yield losses, economic injury levels and other necessary inputs to the development of effective IPM programmes. There is also a need for greater involvement of farmers in the development of such programmes.

7 Rice: West Africa

E.A. Heinrichs
Formerly West Africa Rice Development Association (WARDA), BP 2551, Bouaké, Côte d'Ivoire

INTRODUCTION

Stem borers are key pests of rice in West Africa, as in other rice-growing regions throughout the world. Rice stem borers in West Africa belong to two orders, the Diptera and the Lepidoptera. The dipterous stem borers consist of the family Diopsidae and the lepidopterous stem-borer complex consists of the Noctuidae, the Crambidae and the Pyralidae families. Although there are a number of stem-borer species attacking rice in West Africa, the dipterous borer, *Diopsis longicornis*, commonly referred to as the stalk-eyed fly, and three lepidopterous species, the white rice borer *Maliarpha separatella*, striped stem borer *Chilo zacconius* and pink stem borer *Sesamia calamistis*, are considered to be of major importance (Akinsola, 1975, 1979; Alam, 1988).

Plant damage caused by the dipterous and lepidopterous stem borers differs. *Diopsis* species occur early in the crop season and cut the interior of the stem in a slanting fashion, about 10 cm above the ground, causing the development of 'deadheart' symptoms (Brenière, 1983). The lepidopterous stem borers, however, cause two main types of damage. Attack of young plants at the tillering phase of plant development results in the development of deadheart symptoms (Plate 1:3). However, at the flowering stage, attack occurs a few centimetres below the panicle resulting in a white or dry panicle called 'whitehead' (Plate 1:4). In addition, when mature larvae lodge in the lower parts of the stems, they may reduce or interrupt panicle growth. When this occurs, during the milk stage of panicle development, it results in the drying of one or more spikelets, thus reducing the number of developed grains. This damage, although not visible as are deadhearts or white-heads, results in the reduction of grain weight. Stem-borer attack is most damaging when it occurs after tillering is completed, as plants can compensate, by producing more tillers, when attack occurs early, prior to completion of tillering.

Pioneering work on natural enemies of rice stem borers in West Africa was undertaken by Jordan (1966), and the natural enemy complex of the white rice borer, *M. separatella*, is now known in detail (Polaszek *et al.*, 1994).

The general biology of the lepidopterous stem-borer species is similar. Nocturnal adults oviposit on rice leaves or between the leaf sheath and the stem. Newly hatched larvae move on the plant surface and to neighbouring plants by means of a silk thread, which they attach to leaf tips. Larvae feed at first on the leaf but shortly thereafter penetrate through the leaf sheaths into the interior of the rice stems. Pupation occurs in the stem or in the folds of leaf sheaths, or occasionally in the soil. There are generally two generations on a given crop and five to seven generations a year in a given area, depending on the availability of suitable host plants (Brenière, 1982).

The relative abundance of rice stem-borer species at different locations in West Africa is influenced by climatic zone and ecology. *Diopsis longicornis* is equally abundant in all climatic zones of West Africa; the humid tropical, Guinea savannah and the Sudanian savannah. *Maliarpha separatella* and *Chilo diffusilineus* are both more abundant in the humid tropical and Guinea savannah than in the Sudanian savannah. *Chilo zacconius* is most abundant in the more northern zones, the Guinea savannah and the Sudanian savannah. The pink stem borer, *S. calamistis*, and *Sesamia nonagrioides botanephaga* occur only in the humid tropical and Guinea savannah climatic zones.

Chilo diffusilineus, *C. zacconius* and *Sesamia* spp. are more abundant in the upland than in the rain-fed lowland portion of the continuum. *Scirpophaga* sp. and *Diopsis* spp. have higher populations in the rain-fed lowland than in the upland fields and are most abundant in irrigated fields. *Maliarpha separatella* is also more abundant in rain-fed lowland than upland fields and is the most common stem-borer species in mangrove swamp rice.

The West African rice stem borers also differ in their preferred stage of rice plant growth (Pollet, 1977). In irrigated rice, *Chilo* spp. and *D. longicornis* generally attack plants early, during the vegetative stage of plant growth (Fig. 1). *Scirpophaga* occurs in the panicle initiation to the booting stage, while *M. separatella* and *S. n. botanephaga* are found in the rice stems from the panicle inititation stage to maturation. In general, lepidopterous stem borers are rare in nurseries and during the early vegetative stage of rice development, while dipterous stem-borer attack by *Diopsis* spp. occurs early in plant development. This is evident in the number of adult stalk-eyed flies and lepidopterous stem-borer larvae and pupae occurring in irrigated rice in Nigeria at plant growth stages from tillering to ripening (Fig. 2).

DISTRIBUTION, BIOLOGY AND ECOLOGY OF WEST AFRICAN RICE STEM BORERS

DIPTERA: DIOPSIDAE

Diopsis longicornis Macquart (stalk-eyed fly)

Distribution in West Africa: Benin, Burkina Faso, Cameroon, Ghana, Guinea, Ivory Coast, Liberia, Mali, Nigeria, Senegal, Sierra Leone, Togo.

At least three species of *Diopsis* have been reported as feeding on rice. Of the three species, *D. apicalis*, *D. collaris* and *D. longicornis*, the latter species has been reported as being the most important (Cochereau, 1978; Vercambre, 1982; Dale, 1994). *Diopsis thoracica* and *D. macrophthalma* are synonyms of *D. longicornis*. *Diopsis tenuipes* is a synonym of *D. apicalis* (Feijen, 1985). Feijen believes that, after his systematic revision of the genus *Diopsis* is completed, additional species will be found occurring on rice. Because *D. longicornis* is the most important species of the complex, the description of the biology and ecology will emphasize this species.

The stalk-eyed fly *D. longicornis* occurs in all rice ecosystems in West Africa (Alam and Efron, 1986). However, it is most abundant in lowland, irrigated and hydromorphic ecosystems. Studies conducted in the Republic of Guinea indicated that transplanted rice was much more severely damaged than direct seeded rice in the dry season, but no differences were observed in the wet season (Chiasson and Hill, 1993).

The stalk-eyed borer has not been reported to attack crop plants other than rice. However, it is likely that this pest has alternative hosts among wild rice and grasses. *Cyperus difformis,* a weed commonly found in rice fields, on which eggs, larvae, pupae and adults have been found, may be a host plant during non-rice-cropping seasons (Alghali, 1979).

Diopsis longicornis is reported as a major pest of rice in many parts of tropical Africa. Severe damage has been reported in the valley of Benue in North Cameroon and in Sierra Leone. In Senegal, it only occurs in the south. In Burkina Faso and Mali, it is a sporadic pest in areas that have sufficient humidity during the dry season. It infests irrigated rice fields in Benin and Ivory Coast and is less frequent in rain-fed fields (Brenière, 1976).

There is great variation in the yield losses reported as caused by *Diopsis.* In Ghana, Abu (1972) reported that *D. longicornis* could cause 9% yield loss. In a screen-house test conducted in Badeggi, Nigeria, Akinsola (1980a) reported yield losses of 5–19% when plants were infested at the nursery stage. In southern Nigeria, Alghali (1981) reported yield losses ranging from 2 to 97%.

Detailed studies were conducted by Alghali and Osisanya (1984) on the effect of *D. longicornis* damage on the yield components of rice. The feeding of the maggot significantly reduced the number of panicles produced (both total and mature), the percentage of tillers with panicles, grain weight and the total yield of unprotected plants, and increased the number of immature panicles and time to 50% flowering. Yield reductions were unrelated to percentage of damaged tillers. Production of new tillers did not fully compensate for damaged tillers in most cultivars tested. Photoperiod-sensitive cultivars were better able to compensate for pest damage than non-photoperiod-sensitive cultivars.

Adult *D. longicornis* are typical diopsids and are easily recognized by their characteristic eyes located at the tip of stalks, hence the name, 'stalk-eyed' flies. Flies are found in areas with water throughout the year. They occur in swarms in shady areas near streams and canals during the dry season. Gravid females lay eggs singly on the upper surface of young leaves, normally in the midrib groove. In older plants the eggs are placed on the leaf sheath (Alghali and Osisanya, 1982).

Boat-shaped, striated eggs with a characteristic anterior projection are attached to the leaf with a glue-like substance that prevents them from being washed off in heavy rains. Eggs are creamy-white when laid but later turn to tan. Each female lays about 20 eggs over a 10-day oviposition period. Peak oviposition occurs at 30 to 40 days after transplanting and practically terminates by the end of the tillering stage. Virtually no eggs are laid and no deadhearts develop on 60-day-old plants (Alghali and Osisanya, 1981).

The larva is easy to recognize, because of the two long extensions on the abdomen, which end in black hooks pointed forward. The larva is a yellowish maggot, about 18 mm long and 3 mm wide. Upon hatching, the maggot moves down inside the leaf sheath and feeds above the meristem on the central spindle of young leaves causing a deadheart symptom to develop. Every tiller that is attacked is destroyed. Larvae move readily from one tiller to another and one larva can destroy up to ten neighbouring tillers (Feijen, 1979). Later generations feed on the developing flower-head. The larval stage lasts for 25–33 days.

Pupae are flat and almost triangular because of the compression inside the stem. They are red with brown dorsal bands. During the later stages of development, the imago can be seen inside the pupal case. After a 10–12-day pupation period adults emerge. Two principal generations occur between June and October and a third, less prominent generation occurs during the off-season.

Diopsis apicalis is a polyphagous species that is often seen in rice fields. It can easily be distinguished from *D. longicornis* by its small size and the presence of an apical smoky spot at the tip of each wing. Larvae feed either on healthy plants or on decomposed plant tissue resulting from stem-borer attack. They are also predaceous, eating larvae of other species, including *D. longicornis*, but their role as predators does not make up for the damage they cause to the rice crop (Brenière, 1983).

LEPIDOPTERA: CRAMBIDAE

Chilo zacconius Blezynski, African striped rice borer (Plate 6:3–6)

Distribution in West Africa: Benin, Burkina Faso, Cameroon, Ivory Coast, Mali Niger, Nigeria, Senegal, Sierra Leone.

Host plants other than rice: *Echinochloa crus-galli, Oryza barthii, Sorghum arundinaceum, Pennisetum* spp.

Light-trap collections made in Ivory Coast by Tavakilian (1977) indicated the presence of six *Chilo* species, but he only reported three species, *C. aleniellus, C. diffusilineus* and *C. zacconius* attacking rice stems. The latter two species were the ones predominating in the humid forest and savannah zones, respectively. *Chilo zacconius* is considered the predominant striped rice borer in West Africa, while in East and North Africa *C. partellus and C. agamemnon* are the primary species, respectively (Akinsola, 1990). *Chilo zacconius, C. diffusilineus* and *C. aleniellus* are similar in appearance and in the plant damage that they cause. Bordat and Pichot (1978) report that *C. zacconius* prefers upland rice, while *C. diffusilineus* prefers lowland rice fields.

Oviposition occurs on the upper or middle leaves (Dale, 1994). Pale yellow eggs are laid in two or three overlapping longitudinal rows (Brenière, 1976). Eggs are oval, flat and imbricate, like fish scales (Plate 6:3). The number of eggs laid per female varies considerably (12 to 135) (ISRA, 1989; Akinsola, 1990). Eggs hatch in 4–6 days after oviposition. Upon hatching, the larva moves actively on the plant, feeds for a short time on the leaf and then enters the stem by penetrating the leaf sheath. Larval feeding occurs at the upper internodes and the larva can move from one stem to another when the stem decays and no longer provides nourishment.

The larvae pass through five larval instars and pupation occurs inside the stem after an exit hole for adult emergence has been cut. Larval and pupal periods are about 28 and 6 days, respectively (Akinsola, 1979).

Plant damage caused by *C. zacconius* is similar to that of other lepidopterous stem borers. Feeding inside the stem during the vegetative stage prevents the central leaf whorl from opening, and, instead, it turns brown and dries up. Although the lower leaves remain green, the apical reproductive portion of the tiller is destroyed and the tiller fails to produce a panicle. Larval attack at the panicle initiation stage, or thereafter, prevents the development of the panicle and it is thus white and is known as a 'whitehead' (Plate 1:4).

Although larvae upon hatching quickly establish themselves within the stem, the appearance of damage symptoms occurs gradually. Akinsola (1990) reported that, when plants were artificially infested with larvae, the maximum number of deadhearts occured 20 days after infestation. The particular internode in which feeding occurs is a factor in the rate at which damage symptoms appear; the higher the internode, the sooner the appearance of the deadhearts. The number of larvae per tiller also affects the percentage of tillers that show deadheart symptoms. In Akinsola's (1990) studies, four larvae per tiller were required to attain 85% deadhearts. In nature, several larvae can be found in one tiller a few days after hatching. However, they disperse to other tillers after a few days and usually only one larva remains in a tiller to maturity.

From five to seven generations of *C. zacconius* can occur within a year. The number of generations is limited by the length of the dry season and the availability of suitable host plants (Brenière, 1982). Generally, two successive generations occur in a given rice field.

The first generation causes mainly deadhearts, while damage by the second generation results in whiteheads. During the dry season, larvae may feed on weeds or in off-season rice fields. They can, however, live without food for several months (December to April in the Casamance, Senegal).

LEPIDOPTERA: CRAMBIDAE

Scirpophaga spp. (Plate 6:1,2)

Distribution in West Africa: *Scirpophaga melanoclista* Meyrick: Cameroon, Ivory Coast, Mali, Nigeria, Senegal. *Scirpophaga subumbrosa* Meyrick: Ghana, Mali.

Scirpophaga spp. are among the minor rice stem borers in West Africa. There are 33 species in this genus throughout the world. In tropical Asia, *S. incertulas,* the yellow stem borer, is a major rice pest.

Although adult *Scirpophaga* populations may be high, rice-plant damage by this genus appears to the minimal. In Mali, *Scirpophaga* spp. were 78% of the stem-borer adults collected in light traps, while in the rice field *Scirpophaga* only made up 3% of the larval population and 91% were *M. separatella* (Akinsola and Agyen-Sampong, 1982). The biology, ecology and damage caused by *Scirpophaga* spp. are described by Brenière (1982, 1983). The adult has satin-like white wings that are folded roof-like when at rest (Plate 6:2). The female has a wing-span of 30–35 mm, a long slender body, a black thorax covered with white scales and white legs. For identification, see the general section on pyralid stem borers in this volume.

Adults oviposit on young rice plants during tillering. A female lays 100–180 eggs. Eggs are laid in a mass of two or three layers and are covered with a downy mass composed of scales from the abdomen of the female (Plate 6:1). Egg masses are stuck to the upper or lower surfaces of the leaves. Eggs hatch in about 8 days.

The larva is covered by fine brown hairs, which give it its colour. The first-instar larva bores through the leaf sheath, and then into the stem, a few centimetres below the neck. As the larva develops, it passes from one internode to another and sometimes cuts the stem in two places to form a 4-cm-long section of stem, in which it lodges. Protected by this stem section, it floats on the water surface and may pass from one tiller to another tiller, which it then penetrates. It is semiaquatic, which allows it to reach stems below the water level, in floating or deep-water rice, from within the protective stem covering. One larva can destroy up to four stems. Feeding damage causes deadhearts and whiteheads. The larval period lasts 28–44 days.

The pupa is transparent pearly white. It is located at the base of the stem in the median cavity constituted by the lower internodes. The pupal period is 12–14 days.

LEPIDOPTERA: PYRALIDAE

Maliarpha separatella Ragonot (African white rice borer) (Plate 9:2–6)
Distribution: throughout West Africa.

In Madagascar and West Africa, *M. separatella* is considered to be the most common borer attacking rice (Dale, 1994). The rice plants are usually attacked during the later stages of their growth cycle (Fig. 2). Larval populations in Nigeria are highest at 90 days after transplanting (Ukwungwu, 1987). Damage caused by *M. separatella* differs from that of other rice stem-borer species. Because of the habit of remaining at the lower internodes of the tiller, the larva does not usually produce whiteheads and does not produce deadhearts. The growing apical portion of the plant is not cut off from the base, as is the case with other stem-borer species, and thus panicles can be initiated at the last node. Feeding damage mostly causes a reduction in plant vigour and tiller number and a high percentage of unfilled grains. They may cause whiteheads, but the whiteheads are not caused by a modification of the panicle spike but, rather, are due to an alteration of the base of the stem where the larva is located at the end of its growth period (Brenière, 1983). In addition to the direct damage caused by the stem borer, Pollet (1978) has reported a synergistic relationship between *M. separatella* and *Pyricularia oryzae*, the organism causing rice blast disease. The fungus preferentially attacks plants infested with the stem borer.

Studies on the biology and ecology of *M. separatella* have been conducted by Pollet (1981) in Ivory Coast, Akinsola (1975, 1979) in Nigeria and Diop (1979) in Mali. Brenière (1976, 1983) has described the eggs, larvae and pupae.

The adult moth is slender and the fore wings are yellowish or straw-coloured, with a prominent reddish brown line along the front edge. The hind wings are pale yellow with a metallic sheen and fringed with long yellow hairs. The body is covered with yellowish scales. At rest, the wings are V-shaped and cover the body.

Body length is 11–13 mm in the male and 13–15 mm in the female. For identification, see the section on pyraloid (Crambidae and Pyralidae) stem borers in this volume.

Eggs are laid on the leaves at the tillering stage. Eggs are laid in parallel rows on the upper side of the leaf (Plate 9:2) and attached by a cement-like substance, which, on drying, causes a characteristic pinching of the lamina, completely enclosing the eggs (Plate 9:3). Each oviposition site consists of 30–95 eggs. The incubation period is 7–10 days.

Newly hatched larvae are white, with dark brown heads, but later turn yellow (Plate 9:4). Larvae upon hatching are very active and move to the tip of the leaf blade, where they suspend themselves from silken threads which they produce. Suspended from the silken threads, the larvae drift in the wind and attach themselves to the base of a rice plant, where they bore into the stem. Larvae feed on the inner stem tissues once they have bored into the first or second internode. Once larvae are established, they spend the entire larval stage in one stem and do not migrate from one tiller to another. The larva digs small cavities in the stem wall, without piercing it, and passes from one internode to another by piercing the node. The larvae are not voracious feeders and, throughout the larval stage, they limit their feeding to one or two internodes. They generally cannot survive on plants without internodes. The larva has five to seven instars, and the larval stage lasts about 35–50 days.

Prior to pupation, the larva moves from the feeding site to the upper portion of the stem. There an exit hole is cut into the stem, through which the adult will later emerge. Then a conical silky channel is prepared to connect the inner wall of the stem to the adult exit hole. The larva then pupates, with its head just below the exit hole. The pupal period lasts from 32 to 65 days.

Diapause occurs in the sixth instar at the base of dried stubble. The larva begins activity with the return of humid conditions and rapidly completes its life cycle. Larval diapause may last as long as 251 days (Akinsola and Agyen-Sampong, 1984). In southern Nigeria, diapause starts in July and lasts until March of the next year. Induction of diapause usually occurs before the rice plant starts drying. Diapausing larvae are sluggish and milky-white and have a wrinkled body, while non-diapausing larvae are agile, yellowish white, with a smooth body.

Maliarpha separatella is the major species in the upland rice-growing areas of West Africa. In upland rice, the larvae are located toward the base of the plant, while, in flooded and swampy fields, the larvae tend to be located in the upper portion of the plant.

In deep-water rice, initial infestations are believed to occur before flooding. In deep-water rice at Mopti, Mali, it was observed that tillers infested with *M. separatella* were detached at their base (Akinsola, 1980b). The attack of the stem borer, plus the water pressure at 3 m depths around the base of the plant, leads to the detachment of the damaged portion of the plant. As a result, *M. separatella*-infested stems were observed floating on the water surface in the very deep zones (3 m) and 60% of the hills were missing due to *M. separatella* damage.

Natural enemies of *M. separatella* were treated in detail by Polaszek *et al.* (1994), including correction/explanation of previous records of natural enemies by Agyen-Sampong (1980), Agyen-Sampong and Fannah (1987) and Etienne (1987).

LEPIDOPTERA: NOCTUIDAE

Sesamia calamistis Hampson (Plates 4:5,6; 5:1)

Distribution in West Africa: Burkina Faso, Cameroon, Gambia, Ghana, Ivory Coast, Niger, Nigeria, Senegal.

Host plants other than rice: *Chrysopogon aciculatus, Hemarthria compressa, Hyparrhenia rufa, Paspalum conjugatum, Pennisetum* spp., *Pennisetum subangustum, Rottboellia exaltata, Saccharum officinarum, Setaria splendida, Sorghum bicolor, Sorghum halepense, Tripsacum laxum, Triticum aestivum, Vossia cuspidata, Zea mays.*

Five *Sesamia* species have been reported attacking rice in West Africa: *S. calamistis* Hampson, *S. nonagrioides botanephaga* Tams and Bowden, *S. n. penniseti* Tams and Bowden, *S. cretica* Lederer and *S. poephaga* Tams and Bowden. *Sesamia calamistis* is the most common species and occurs throughout West Africa, East Africa, Madagascar and South Africa (Brenière, 1982). *Sesamia nonagrioides botanephaga* is probably the next most common *Sesamia* species in rice and is limited to the tropical and equatorial areas of Africa.

Sesamia spp. are the most polyphagous of the rice stem borers. *Sesamia calamistis* has many hosts in addition to rice. Infestation of the crop by *S. calamistis* is highest at the flowering stage. The young larvae occupy the stalk at the base of the panicle, which causes whitehead development. As larvae become older, they feed lower in the stem. Irrigated rice is less severely attacked than rain-fed lowland crops. Upland rice is often severely attacked by *S. calamistis* (Ho *et al.*, 1983). *Sesamia calamistis* populations are especially high in upland rice fields, near maize, which is the preferred host plant of this pest.

Brenière (1976, 1983) and Dale (1994) have described the various stages and the life cycle of *S. calamistis*. The adult moth is light brown with brown stripes. The margin of the fore wings is wide, whitish and partly smoky. The rest of the fore wing is speckled with dark patches. The hind wings are pearly white with a yellowish margin. For identification, see the general section on notched stem borers in this volume. The moths are nocturnal and are capable of flying long distances. Mating of adults takes place as early as the first night after emergence and oviposition begins the same night. The female lays up to 300 eggs. Eggs are thrust within the leaf sheath surrounding the upper internodes. Eggs are subspherical, are flat at the poles and have numerous longitudinal striations. Eggs are yellow and are laid side by side, without any specific alignment. Eggs hatch in 7–10 days after oviposition.

Sesamia calamistis larvae are smooth, uniformly yellowish pink on the dorsum, with greyish lateral and dorsal stripes, and a dark brown head. After hatching, lavae first feed within the tissues of the leaf sheath, and then enter the stem through a horizontal cavity and move downward, sometimes through several internodes. The frass which fills the vertical galleries is partly pushed out through openings in the leaf sheath. The duration of the larval period is 28–35 days.

The larva pupates within the base of the stem or in folds of withered leaf sheaths. The pupation period is 10–14 days.

In wet, tropical regions, the life cycle is practically continuous throughout the year. Drought or cold temperatures may slow development. Mature larvae become inactive from the start of the dry season and remain so until the rains begin. On crops that are irrigated out of season, development continues uninterrupted. The number of generations varies from five to six in most of West Africa to three in the Sahel region.

8 Rearing Maize and Sorghum Stem Borers

M.O. Odindo and F.O. Onyango

Animal Rearing and Quarantine Unit, International Centre of Insect Physiology and Ecology, PO Box 30772, Nairobi, Kenya

INTRODUCTION

The success of any insect-pest control programme relies on a dependable supply of high-quality insects in adequate numbers, at specified times and at specific stages of development. Several areas of applied research on biological control, toxicological testing, hormone and pheromone manipulation, host-plant resistance, sterile-insect techniques, cultural practices and genetic engineering, as well as basic research in taxonomy, morphology, anatomy, physiology, biochemistry, behaviour, nutrition, ecology, genetics, etc., require quality insects.

Relying on field populations of insects is not only unpredictable and season-dependent, but also impractical where large numbers of insects are required and user specifications of quality dictate, for instance, high-volume screening of crop breeding lines and varieties, sterile-insect techniques and mass production of parasitoids and predators for field releases.

The art and science of insect rearing has advanced tremendously over the years, as the need for laboratory-reared insects for entomological research and pest control has grown in the recent past. It is in this light that the term insect-rearing management (IRM) was conceived (Singh and Moore, 1985). Insect-rearing management is defined as 'the efficient utilization of resources for the production of insects of standardized quality to meet programme goals'.

The IRM concept describes seven component elements cardinal to successful and efficient rearing, irrespective of the target insect to be reared, scale of operation or intended end-use. These include objectives, colony establishment, rearing laboratory, research and development of techniques, resources, quality control and production. A combination of these factors in optimum proportions to each other eventually leads to the supply of the target insects for their intended purpose, thereby meeting the objectives. This chapter borrows from the IRM concept, and discusses the principles and practice of colony initiation, maintenance and quality-control aspects of stem-borer rearing, with specific reference to the stem borers of maize and sorghum, *Chilo partellus, Busseola fusca, Sesamia calamistis* and *Eldana saccharina*.

SETTING OBJECTIVES

A thorough understanding of the intended use of the laboratory-reared insects and the ultimate goal of the research project are important factors to consider when setting up the objectives of rearing any insect species on which future appraisal of the production system, as well as quality control, depends. During this phase of management, pertinent decision issues arise:

- Why rear the insect?
- Which insect species should be reared?
- For what purpose is the insect species to be reared?
- How many insects and at what stage of development are they going to be required?
- What specific period are the insects to be supplied for?
- How are the insects going to be reared?
- Are the reared insects going to accomplish the specifications and standards for their intended purpose?

In light of the above, stem borers are reared at the International Centre of Insect Physiology and Ecology (ICIPE) for studies on microbial control, host-plant resistance (required in large numbers during the long- and short-rains planting seasons), semiochemicals, behavioural studies (small quantities required continuously throughout the year) and the production of parasitoids.

COLONY ESTABLISHMENT

It is advisable to initiate insect colonies with as large a founder population as possible, and also to maintain them under conditions that induce normal levels of competition and survival. It is essential to ensure genetic diversity and pathogen- and parasite-free parental stock through quarantine protocols of field-collected materials. Genetic decay has been found to be higher when small initial populations are used to initiate a colony. Mackauer (1976) suggested that biological-control projects require a founder colony of about 500 individuals collected over a wide geographical area. Inbreeding and genetic drift can be minimized by establishing a colony with the largest possible number of males and females. According to Bartlett (1985), all genotypes should be represented by collecting 100–200 individuals at each location that is separated from others by the distance over which the insects are capable of dispersing.

Colony maintenance becomes an important consideration following colony establishment, once deterioration, excessive homozygosity or divergence from the founder population results in the production of a large proportion of insects that no longer accomplish their intended purpose (Joslyn, 1984). The basic maintenance strategies are strain restoration through relaxed rearing, selection, gene infusion (or hybridization) and, if necessary, strain replacement (Leppla, 1989).

At ICIPE, mature stem-borer larvae and pupae are periodically obtained from sorghum or maize stubble collected in the farmers' field around the Mbita Point Field Station near the shores of Lake Victoria, in western Kenya. Sorghum and maize stems are dissected and borer larvae and pupae are maintained in the insectary until insects emerge. The adults (1:1 ratio) are allowed to mate in an oviposition cage, and the eggs laid are surface-sterilized, thoroughly rinsed using distilled water, and used to start the first laboratory generation. Diseased and parasitized insects are discarded.

REARING LABORATORY

The rearing laboratory is a new habitat for the insects, and should therefore have environmental conditions that are conducive to their development and eventual field performance. Moreover, they should be kept under conditions where diseases, parasites and predators, such as ants, are easily controlled and they are protected from adverse environmental conditions, e.g. wind, rain and sunshine. As far as possible, the insects should be reared under variable environmental conditions over time and space. Furthermore, the laboratory environment should concurrently cater for workers' comfort.

Insectaries currently used for rearing insects were originally designed for other purposes, but later modified to house small colonies. As a general rule, however, the structure for insect-rearing facilities should meet the building code of the area, provide sanitary and disease-control conditions, facilitate environmental needs for rearing operations, provide safety for workers, have adequate space and permit efficient arrangement of each operation sequence and work-flow pattern.

In this regard, a clean storage area and media-preparation and surface-sterilization rooms, which are critical, should be isolated from duty work areas. Materials and products should always move from duty work areas to other areas, unless the materials and products are sterilized. Unsanitary rooms (e.g. quarantine area) should never form part of an internal flow pattern. As far as possible, the rooms should be seamless and the walls made of materials that can be repeatedly washed and sterilized. Portable racks for holding insects are preferred.

Installation of appropriate equipment in the rooms in relation to each other saves time, increases efficiency and enhances safety. Depending on the availability of resources and scale of operation, several production processes can be automated. Entry into the insect-rearing facility is restricted to insectary personnel only.

The insect-rearing laboratories at ICIPE, both at Duduville and Mbita Point Field Station (MPFS), have provisions for larval rearing rooms, diet-preparation room, infestation room, adult oviposition/egg-handling room, weighing/storage of diet ingredients room, sluice/washing room and grinding room. At MPFS, insects are reared successfully under ambient laboratory conditions (25–30°C; 50–80% relative humidity (RH).; L12:D12) without any environmental-control devices in the rooms. At Duduville insectaries, however, temperature, RH and light conditions are regulated.

RESEARCH AND DEVELOPMENT OF TECHNIQUES

Insect rearing is a dynamic field where new techniques are constantly being invented and the existing techniques are improved from time to time, with changes in demand and technological evolution.

When introducing a new insect species into the laboratory, the insect is divorced from its natural habitat and is in captivity. Numerous facts about the insect otherwise unknown to the insect-rearing practitioner need to be established before an effective mass-rearing programme can take off. Some of these facts include:

- Development of a suitable artificial diet.
- Establishing the optimum threshold level for antimicrobials.
- Manipulation of the laboratory environment to suit insects' growth, development and reproduction.
- Design of suitable rearing cages.
- Suitable oviposition substrate.
- Suitable rearing containers in terms of size, shape, material and cost considerations.
- Determination of optimum insect-rearing densities at all stages of development.
- Establishment of quality criteria and protocols.
- Development of standard operating procedures (SOP).

Therefore research and development (R & D) of techniques should form an integral part of an insect-rearing system. Conlong (1989) noted that 'many people including administrators still hold the notion that insect rearing does not require much science; it is easy work; anybody can keep some insects in a jar and keep them in a rundown building where they will grow'.

The research phase is usually very extensive and may take several years to conclude. The insect rearer is usually caught in a dichotomy, having to choose between the pressure to produce the insect and provide a supply at the scheduled time, on the one hand, or to devote his/her time to R & D of new techniques of rearing or in refining the old techniques, on the other.

HISTORICAL BACKGROUND

The rearing of insects on an artificial medium is a relatively new field of entomology, and has only gathered impetus in the last two to three decades. The pioneering work was done by Bagdanow (1908), who was the first to rear an insect axenically from egg to adult when he reared the blowfly *Calliphora vomitoria* L.. Later, other insect species, e.g. *Drosophila*, cockroach, mosquito, etc., were colonized on artificial food sources.

Rearing of phytophagous insects on artificial diet was reported by Bottger (1942) in South Africa for the European corn borer *Ostrinia nubilalis*. Vanderzant and Reiser (1956) used an aseptic technique to rear the pink bollworm *Pectinophora gossypiella* on a diet devoid of plant extract. The use of wheat germ as a primary nutrient source in formulating artificial diets for lepidopterous and phytophagous insects was due to the work of Adkinson *et al.* (1960) on *P. gossypiella*.

The modification of the diet by Berger (1963) resulted in an artificial diet for rearing *Heliothis* sp. Since then, wheat germ has been used in several hundred insect diets. Shorey and Hale (1965) used bean as a primary protein source in a diet to rear nine noctuid species. Several varieties of beans have since been used to compound lepidopteran diets. A general-purpose diet (Singh, 1983, 1985) has been formulated to rear several insect species. In South Africa, Kfir (1992b) described an artificial diet for rearing *Chilo partellus*, precluding agar, vitamin E and leaf factor, and composed of (%): chick-pea (24.93); brewers' yeast (3.32); glucose (1.73); methyl-*p*-hydroxybenzoate (0.21); ascorbic acid (0.30); sorbic acid (0.14), ethanol (0.10); and distilled water (69.26). The current status of insect rearing, including the development of new artificial diets and modifications/improvements on existing techniques, is well documented (Singh and Leppla, 1984; Singh and Moore, 1985).

DIET PREPARATION

In general, to prepare the diet for stem borers, distilled water is boiled, cooled to 60°C and mixed with the weighed ingredients in the blender (see fraction A, Table 8.1) for 30 min. Agar powder is weighed and added to the distilled water separately, boiled and cooled to 60°C to make fraction B. Ingredients of fraction A and fraction B are blended together for 3 min. Lastly, formaldehyde 40% (fraction C) is measured and mixed with ingredients of fraction A and B for 3 min. The diet is dispensed while warm in appropriate rearing containers, as will be explained later in individual stem-borer species.

The ingredients are expensive, and therefore the search for cheaper substitutes, e.g. agar (Ochieng-Odero *et al.*, 1991; Gillespie, 1993a), or complete elimination of some ingredients (Kfir, 1992b), without sacrificing the quality of the insect, is a continuous process.

REARING *CHILO PARTELLUS*

The first artificial diet to rear *C. partellus* included casein, glucose, salt mixture, brewers' yeast, choline chloride, cholesterol, cellulose, leaf factor, methyl-*p*-hydroxybenzoate and water (Pant *et al.*, 1960). Chatterji *et al.* (1968) reared *C. partellus* on a wheat-germ diet, which was earlier used by Keaster and Harrendorf (1965) for rearing the southern corn borer *Diatraea grandiosella*. The major breakthrough with the rearing of this species came with the use of a Kabuli gram-based diet (Dang *et al.*, 1970). Since then, a number of diets have been used, mainly in the Indian subcontinent, by deleting, adding or modifying the quantity of one or more ingredients. Taneja and Nwanze (1990) have reviewed the evolution of artificial diets for rearing *C. partellus*. At ICIPE, the artificial diet used for the large-scale rearing of *C. partellus* (Ochieng *et al.*, 1985) is essentially a modification of the diet developed at International Crops Research Institute for the Semi-Arid Tropics (ICRISAT) (Seshu Reddy and Davis, 1978).

Dispensing the Diet

The diet is dispensed into the rearing containers while still warm. At ICIPE, the diet is dispensed into heat-sterilized (100°C for 1 h), wide-mouthed glass jars (19.0 ×

Table 8.1. Ingredients of artificial diets for rearing stem borers at ICIPE.

	Quantity per 100 g diet (ml)	
Ingredient	*Chilo/Eldana*	*Busseola/Sesamia/Eldana*
Fraction A		
Distilled water	46.47	40.41
Brewers' yeast	0.74	2.27
Sorbic acid	0.09	0.07
Methyl-*p*-hydroxybenzoate	0.15	0.10
Ascorbic acid	0.15	0.25
Vitamin E capsules	0.02	0.21
Sorghum/maize-leaf powder*	3.70	2.54
Bean (*Phaseolus vulgaris*) powder†	10.20	8.84
Sucrose	–	3.54
Fraction B		
Agar (Tech. No. 3) powder	1.12	1.26
Distilled water	37.17	40.41
Fraction C		
Formaldehyde 40%	0.19	0.10
Total	100.0	100.0

* Six-week-old leaves, dried in air oven at 60°C for 12 h till brittle and ground using mill (Posho or ultracentrifugal).
† Seeds washed with tap water, dried in the oven at 60°C for 24 h and ground to a fine powder.

7.5 cm), each jar containing approximately 200 g diet. The diet is allowed to gel and to condition overnight at room temperature while covered with a clean white cloth or paper towel.

Egg Management

Eggs oviposited on butter-paper are removed from the oviposition cage each day. Eggs in batches (50 eggs per batch) are cut, using a pair of scissors or scalpel blade, and placed in a lunch-box on moist filter-paper and covered with a well-ventilated top. Incubation takes place at ambient laboratory conditions but it is best at 27 ± 2°C and over 80% relative humidity (RH). At the blackhead stage, the eggs are surface-sterilized using a 10% formaldehyde solution, for 20 min, rinsed thoroughly, using distilled water, and dried on filter-paper. The development of the eggs can be arrested by placing blackhead eggs in the refrigerator at 10°C for up to 3 days without significant loss in hatchability.

Infestation Using Blackhead Eggs

Surface-sterilized eggs are each dropped into a jar containing the diet. The mouth of the jar is then covered with a paper towel and the metal lid is fastened.

Larval, Pupal and Adult Management

Upon hatching in the rearing jars, larvae are allowed to feed undisturbed until pupation and eventual adult emergence. Adults are removed daily from the rearing jars, using a battery-operated hand aspirator or glass vials (7.5 × 2.5 cm), and transferred to a rectangular oviposition cage (40 × 30 × 40 cm), made of wire frames covered with wire mesh. The inside of the cage is lined with butter-paper, pleated into folds running vertically to form an internal lining of the cage for the moths to deposit their eggs on and for rest. The top of the cage is fitted with a mosquito-netting sleeve to facilitate the introduction of the butter-paper, adult diet (tap water soaked into cotton wool and placed into a Petri dish) and moths into the cage. A maximum of 200 adult pairs are introduced into the cage.

REARING *CHILO ORICHALCOCILIELLUS*

At ICIPE, *Chilo orichalcociliellus* is reared using the same diet and technique as for *C. partellus*. However, using *C. partellus* diet and technique, the development has been reported to be low and mortality high, leading to poor establishment of the larvae. Delobel (1975) described an artificial diet for rearing *C. orichalcociliellus* composed of brewers' yeast 5 g; wheat germ 30 g; ascorbic acid 1.4 g; methyl-*p*-hydroxybenzoate 1.0 g; 10% formaldehyde 2.0 ml; levulose 10 g; dried milled maize stem 2.0 g; agar 7 g; and distilled water 340 g. This diet is yet to be tested for suitability at ICIPE.

REARING *BUSSEOLA FUSCA*

Research on *Busseola fusca* has been hampered by the occurrence of a 6-month-long diapause in the last larval stage by the second-generation larvae (Mally, 1920; Smithers, 1960b; Harris, 1989a; Harris and Nwanze, 1992). Factors inducing diapause in *B. fusca* under field conditions are poorly understood (van Rensberg *et al*., 1987), although its termination is considered to be associated with the onset of the rains (Swaine, 1957). Usua (1973) observed that diapause could be prevented by feeding *B. fusca* larvae on 6–8-week-old maize stems with high water content but low in carbohydrates. In contrast, older maize stems induced diapause. Unnithan (1987) reared five successive generations of *B. fusca* on young sorghum stems without any intervening larval diapause. Nevertheless, rearing on a host plant is often laborious, time-consuming and inefficient for producing the large colonies that are often required for the screening of breeding lines and varieties for pest-resistance trials, for mass production of parasitoids and predators or for the production of pathogens.

In South Africa, Barrow (1989) reported a method of collecting diapausing larvae during the late winter from maize stalks and cold-storing them in darkness until field conditions were favourable for the termination of the diapause. This method was found to be unreproducible under Mbita conditions, and 'rearing with diapause' is rather season-dependent and not all stages of the insect are available throughout the year. Onyango and Ochieng-Odero (1994) described for the first time a functional artificial diet and an associated rearing technique for rearing *B. fusca* continuously in the laboratory, thereby eliminating the problem of the 6-month-long diapause.

Dispensing the Diet

Approximately 15 g of the diet is dispensed while warm into a heat-sterilized (100°C for 1 h) glass vial, 2.5 × 7.5 cm, using a ketchup dispenser. The vials are covered with a clean white cloth or paper towel and left overnight on the bench in the laboratory to gel and then the condensate is removed.

Larval Management

Neonate larvae are introduced singly into heat-sterilized glass vials containing the diet, using a camel-hair brush no. 1 (dipped into 70% ethanol and rinsed in distilled water), such that the brush does not come into contact with the diet. The vials are closed immediately with tight-fitting cotton-wool plugs and arranged in the metal trays in batches of 100 vials per tray. The larvae are allowed to feed *ad libitum* until pupation.

Management of Pupae, Adults and Eggs

Pupae are collected weekly from the rearing vials and placed on the floor of a Perspex cage (45 × 40 × 40 cm) for the adults to emerge. Newly emerged adults are collected each morning, using a battery-operated hand aspirator or glass vial (75 cm × 2.5 cm diameter), sexed and paired in metal-frame oviposition cages (30 × 30 × 45 cm) made of five squares cm^{-2} mesh on the sides and top. The sex of each adult is determined from the antennae, female antennae being filiform and male ones pectinate. A single cage is adequate for the oviposition of up to 50 pairs of adults. Dead adults are removed from the cage. The cages are cleaned and disinfected weekly.

Adults are given water soaked in cotton wool in a Petri dish. Fresh sorghum stems with intact leaf sheaths are provided diagonally inside the cage daily. The sorghum stems are removed from the cage daily and examined for eggs. The portions of the leaf sheaths containing eggs are carefully excised, the eggs dislodged from the leaf sheath, weighed as a group after daily collection and sterilized in 1% Jik solution (commercially available household bleach containing 0.05% sodium hypochlorite) for 10 min. A few drops of liquid detergent (Teepol) are added to cause the eggs to sink to the bottom of the solution. The eggs are then rinsed in five washings of distilled water, dried between two filter-papers and placed in sterilized glass vials (7.5 cm × 2.5 cm diameter, up to 20,000 eggs estimated by weight in each), which contain a moist filter-paper at the bottom of each. The vials are tightly closed with cotton wool wrapped in tissue-paper and the eggs are allowed to hatch under prevailing laboratory conditions, although 27 ± 2°C and over 80% RH are ideal. The eggs may be stored in the refrigerator (10°C) for up to 3 days without significant deterioration or loss of hatchability.

In South Africa, Barrow (1989) reported that all egg masses are laid on waxed paper spirally wound round a broomstick. The eggs are stripped off by placing the paper on a tabletop and running a blunt knife between the paper and the eggs. At ICIPE, we get comparatively fewer eggs using this method and hence the host plant for oviposition of *B. fusca* is still preferred.

REARING *SESAMIA CALAMISTIS*

Work on the development of an artificial diet for *Sesamia calamistis* started at the International Institute of Tropical Agriculture (IITA), Ibadan, Nigeria, in 1980 (Jackai and Raulston, 1982). Several diets were evaluated and finally a soyflour–wheat-germ-based diet, similar to the one used for mass rearing of the tobacco budworm (Raulston and Lingren, 1972), was found to be the most satisfactory. However, the diet contains several expensive ingredients and is rather difficult to prepare. At ICIPE, several previous formulations of lepidopteran diets were tested for *S. calamistis* rearing, and finally the *B. fusca* diet was found to be the most suitable. The rearing-management practices for *S. calamistis* are also similar to those described for *B. fusca*.

REARING *ELDANA SACCHARINA*

The artificial diet used to rear *S. calamistis* in IITA (Jackai and Raulston, 1982) was also used to rear *Eldana saccharina*. However, Fumidil-B and Benlate (both fungicides) were subsequently eliminated from the *E. saccharina* diet, since the two compounds were found to cause deformation and adversely affected survival (Bosque-Pérez and Dabrowski, 1989).

In South Africa, *E. saccharina* was successfully reared on a diet comprising several expensive ingredients (Graham and Conlong, 1988). Since then, efforts have been directed toward reducing the quantity of more expensive ingredients, e.g. agar, or finding substitutes for them. Gillespie (1993b) reported the reduction in the use of agar, as well as the suitability of pectin (produced by processing ripe fruit) as a substitute for agar.

At ICIPE, we found *E. saccharina* to grow, develop and reproduce satisfactorily on the *C. partellus* diet. Larval, pupal and adult management follows a similar protocol to that described for *C. partellus*.

Egg Management

Eldana saccharina lays eggs on a paper towel or butter-paper folded diagonally to form spirals. Eggs are collected daily and new paper provided for oviposition. In order to collect the eggs, it is necessary to moisten the oviposition paper by spraying it with water, since the eggs are glued to the paper by the female moth during oviposition and break easily unless handled with care. After the oviposition substrate is opened and the eggs are exposed, the egg masses are cut off with a pair of scissors and surface-sterilized, as in *C. partellus*.

RESOURCE ALLOCATION

Availability of resources is a vital component in any insect-rearing system. Time, space, materials, personnel and funds are major aspects of resource management, which ultimately affect the project target. The management of resources for cost-effective insect rearing and to achieve a balance that minimizes wastage is probably the single most important element in IRM.

Funding should therefore be allocated based on priority and scale of preference. Materials for use should be purchased in bulk and well in advance (not less than 6 months before the rearing actually starts), in order to take advantage of the economies of scale and discount and in order to facilitate advance planning.

Insectary personnel should be well trained and properly motivated. As in hospitals, insect rearing is a 7-day job and only those who are willing to work during weekends and public holidays should be retained in the insectary.

QUALITY CONTROL

Inputs into the rearing system are subject to a wide range of variability. The selection of the acceptable limits of standards is useful for the production of targets for optimal biological performance. Quality assessment highlights deficiencies in the rearing system that need immediate remedies. Quality control is effectively realized when assessment (indicators or performance measures) results in appropriate adjustment or remedy (control) to meet the user specifications and standards.

Quality is a term relative to the end-user's needs. In this regard, the individual subject-matter specialist (researcher) should set the desired criteria (parameters) for quality aspects of the insects they intend to use and, in concert with the insect-rearing practitioner, they should develop procedures and protocols to test the various criteria. The insect-rearing specialist should then incorporate the set criteria and protocols for routine testing. Quantification of criteria should be limited to only those characteristics (traits) that are directly linked to the intended end-use, with the aim of obtaining the accepted quality rather than achieving insects identical to their feral counterparts. Quality refers to the degree of excellence in some trait, skill or attribute, relative to a reference. Skill refers to the performance requirement for achieving the objective. Relativity is the ranking and comparing of the degree of excellence in performing the skill. Reference refers to some predetermined standards, needs and constraints against which the skills are compared. Control is synonymous with checking, verification, comparison, regulation, periodicity and restraint, which implies an input and feedback mechanism, cross-checking, comparison, decision-making and developing protocols for the imposition of control upon quality. Over the generation time of laboratory rearing, insects in captivity lose their typical behaviour (Boller, 1972).

Subdivision of Quality Control

Leppla and Fisher (1989) identified three primary subdivisions of quality control that occur in insect mass-production systems: production control, process control and product control.

PRODUCTION CONTROL

Production control ensures that insect rearing and associated operations are performed. Performance of the rearing operations is controlled by the SOP, equipment and environment. This involves the use of operation schedules, checklists and other means of measuring the completion of each task in the rearing and maintenance operation.

PROCESS CONTROL

Process control involves the evaluation of the immediate stages of the insects and comparing them with the already established specifications and standards. Wild populations are used for relativity studies.

PRODUCT CONTROL

Product control evaluates the final insect stage produced and prepared for distribution to the end-users by comparing their performance with feral populations.

Setting Quality Criteria

The wild populations are normally used for relativity studies against the performance of the laboratory culture (product). Some traits found to be linked directly with performance of laboratory-reared insects need to be identified and simple techniques developed for quick implementation. For the first time, a preliminary list of quality of commercialized beneficial insects, namely, *Trichogramma* sp. and seven other species used commercially in green crops, was set by the International Organization for Biological Control of Noxious Weeds and Plants (IOBC) global working group on quality control of mass-reared arthropods, held at the Agricultural University of Wageningen (WAU) in the Netherlands (Biggler and Laing, 1991; van Lenteren and Steinberg, 1991).

Generally, some traits found to be linked with the quality of mass-reared stem borers and their parasitoids (where applicable) include:

- Survival rate (%):
  - survival to pupation;
  - adult emergence.
- Developmental period: larval, pupal, total duration, incubation, adult lifespan.
- Live-weight: larval, pupal and adult.
- Symptomatic observations:
  - deformities (%);
  - nature of abnormalities, e.g. wing, melanization, ecdysis failure, colour changes, etc.
- Growth index (GI): calculated as the ratio of per cent pupation over the mean larval development period, and an important indicator of assessing the suitability of larval diet for supporting survival, growth and development; thus the higher the GI the more suitable the diet.
- Reproductive capacity:
  - number of egg masses laid;
  - total number of eggs laid;
  - egg hatchability;
  - egg fertility;
  - sex ratio;
  - mating propensity (frequency, time, duration, insemination rate, etc.).

- Host finding ability: orientational response.
- Host acceptance:
  settling response;
  feeding behaviour;
  efficiency of food consumption and utilization.
- Locomotor mechanism:
  walking speed;
  flight propensity (distance, duration, velocity, frequency) (flight mill/mark–release–recapture techniques).
- Visual sensitivity: using electroretinogram.
- Frequency of morphological and biochemical markers: patterns of genetic variations (electrophoretic and chromatographic techniques).
- Chemoreception responses: response to semiochemicals (pheromones/ allelochemicals) (olfactometers and electroantenogram techniques).
- Adaptability: response to changeable environmental conditions under field situations.
- Rate of parasitization: per cent parasitization of eggs, larvae or pupae (applicable to parasitoids and predators).
- Feedback from users: plays an important role in assessing the quality of laboratory-reared insect.

PRODUCTION

The ultimate goal of rearing is to obtain insects of acceptable quality. This can be achieved by a proper combination of the factors of production, namely, objectives, colony establishment, rearing laboratory, R & D of techniques, resource allocation and quality control.

MAINTENANCE SYSTEM

This is designed primarily for the perpetuation of the laboratory population at a level that enables colony build-up as and when large numbers are required.

PRODUCTION-AND-SUPPLY SYSTEM

This comes in response to specific insect orders and therefore stems from the maintenance system. In practice, the quantity of insects produced and supplied should be relative to the quantity demanded to avoid over/underproduction, leading to wastage/scarcity, respectively. Adhering to the SOP laid down is the key to the predictability of the production-and-supply system.

Insect users are advised to place insect orders well in advance of the date required. Depending on the insect species, a 3–6 months' advance order is recommended. For a new insect species, at least 12 months' advance notice is required. Requests for insects, including specifications and standards of quality, are made on an insect order form. The information from the insect order form is transferred to a yearly planner, to facilitate planning at a glance and avoid conflict.

MICROBIAL CONTAMINATION OF INSECTARIES

Microorganisms in an artificial diet cause spoilage, alter the biological performance of insects and may harm the gut symbionts. Several species of these pathogenic organisms may be found in the diet. A good diagnostic laboratory is required for identification of microorganisms associated with insectaries (Odindo, 1994).

Sources of Contamination

Field-collected insects, the laboratory environment and the improper handling of the insects and diets during preparation, storage and use are some of the sources of contamination in the insectary.

Causes of Contamination

Contamination in insectaries is caused by pathogenic microorganisms, namely:

- Fungi, e.g. *Aspergillus* sp.
- Bacteria, e.g. *Rhizopus.*
- Viruses, e.g. nuclear polyhedrosis virus (NPV), cytoplasmic polyhedrosis virus (CPV) and entomopox virus (EPV).
- Protozoa, e.g. *Nosema* sp.

Some of these microorganisms may be transmitted from one generation to the next, either on the egg surface (transovum) or in the egg (transovarial).

Control

The main objective of control is to keep the microorganisms below the injury level. This can be achieved by the following recommended practices and enforcing the regulation procedures:

- No job is complete until all materials used are put away inside the drawers or cabinets and the work surface is sprayed with disinfectant and wiped dry.
- The work area should be cleaned thoroughly daily and mopped with germicidal disinfectant at least weekly.
- Access to the work area should be restricted to authorized personnel only and on a restricted basis.
- Individuals exposed to disease organisms – for example, when conducting field trials – should not be allowed access to the insectary.
- Laboratory coats and plastic aprons used during work in dirty areas should not be allowed to be worn elsewhere in the insectary.
- Each isolated area should have materials and equipment required to complete the jobs assigned for the area.
- Larvae and adults that have escaped from the cages should be discarded immediately.

- Undesirable insects, such as cockroaches and houseflies, should be controlled, using traps or boric acid.
- Personal hygiene of the insectary staff should be observed strictly.
- The use of antimicrobial compounds, e.g. sorbic acid, formaldehyde or commercially prepared mould inhibitors, should be encouraged.

HEALTH HAZARDS

Insect scales and toxic fumes are the major predisposing causes of health hazards in the insectary. These may result in the manifestation of various allergic reactions, pulmonary diseases and environmental pollution. Insectary workers should use protective gear, e.g. mouth masks, control air filters, fume hoods and protective clothing, e.g. laboratory coats.

Humans may be affected by microbial contaminants. Several airborne microorganisms, e.g. *Aspergillus*, *Pseudomonas*, and *Streptococcus* spp., are known to affect humans and may present some hazards to insectary personnel (Sikorowski, 1984).

II Taxonomy

9 Lepidoptera: Introduction

K.V.N. Maes
AgroBioSys International, Kl. Smetledestraat 192, B-9230 Wetteren, Belgium

WHAT IS A MOTH ?

Moths and butterflies belong to the order Lepidoptera. They are holometabolous insects: their development follows four stages and includes a complete metamorphosis. A caterpillar hatches from the egg. The caterpillar grows, passing through several moults, and then forms a chrysalis or pupa. From this pupa, the adult moth or butterfly will emerge. The caterpillar is the principal feeding and growing stage. Caterpillars are essentially herbivorous. They have mouthparts adapted for biting and chewing. In this stage, moths cause damage to agricultural crops. The adult moths are usually completely covered with scales, including the four wings. On the head they have, in general, a well-developed coiled proboscis or tongue, a structure adapted for sucking liquid. The adults form the only reproductive and dispersive stage.

SOME CHARACTERS OF ADULT LEPIDOPTERA

The wings are traversed by a number of veins. The pattern of the veins (the venation) is used in taxonomic keys to identify families. Fore and hind wing have the following basic venation. There is a free subcostal vein (Sc), followed by the stem of the radial vein (Rs), which constitutes at the same time the anterior border of the discal cell. The radial vein is five-branched (R1–R5). There are three median veins (M1–M3) present. The base of the anterior cubital vein (CuA) constitutes the posterior part of the discal cell. This vein is normally two-branched (CuA1 and CuA2). The posterior cubital vein (CuP) originates directly from the base of the wing. This vein is usually thin. Two to three anal veins are present.

The Pyraloidea and Noctuoidea have a wing-coupling apparatus of the frenate type. Near the base of the hind wing there is a strongly developed bristle, the frenulum, which locks in a retinaculum of the fore wing. The frenate type shows sexual dimorphism in these groups: thus, in the male the frenulum consists of a single stout bristle (Fig. 6) and in the female the bristles remain separate and the number may vary from two to nine (Fig. 7).

Both superfamilies have tympanal organs (Figs 8, 9). These hearing organs consist mainly of a tympanum, a thin transparent membrane, which is delimited by another opaque membrane, the conjonctivum. The tympanal organs are often hidden by scaling, different lobes or folds.

Characteristic of the male genitalia (Figs 3, 4) are the two lateral valves. The valves may have different forms or processes, such as spines, hooks or bristles. The dorsal part is called the costa and the ventral part, near the base, the sacculus. The valves articulate on a ring-shaped structure; the dorsal part is the tegumen and the ventral part the vinculum. The uncus originates on the median part of the tegumen. Underneath the uncus is the membranous tuba analis, which terminates in the anus. Sometimes part of the tuba analis is sclerotized: the dorsal part is called the scaphium and the ventral part the subscaphium. Underneath the tuba analis and attached to the tegumen is the gnathos. The vinculum is usually U- or V-shaped. The median part at the base of the vinculum, where the two arms fuse, is called the saccus. The area between the two arms of the vinculum is closed by a membrane that supports the aedeagus. This part of the membrane is the anellus, and may bear sclerotizations with specific value. A sclerotized plate, the juxta, supports the aedeagus or penis. The aedeagus consists of an invaginated membrane, the vesica, terminating in the ductus ejaculatorius. The vesica may carry sclerotizations called cornuti. These often have specific value.

The female genitalia (Fig. 5) are less diverse than those of the male. The caudal part of the abdomen ends in a pair of papillae anales. They delimit the anus and the ovipositor. The rod-like sclerotizations at the base of the papillae anales are the apophyses posteriores, the ones on the eighth segment being the apophyses anteriores. The genital aperture, or ostium bursae, is situated in the intersegmental membrane of the seventh and eighth segments. The area around the ostium bursae is the sinus vaginalis. Around the sinus vaginalis and the ostium bursae occur variously sclerotized structures of great taxonomic importance. For the whole complex of these the name sterigma is used as a general term. The first part of the copulatory duct, the ductus bursae, is usually sclerotized. This part is delimited by a ring-shaped structure, the antrum. The ductus seminalis usually originates just beneath the antrum. The ductus bursae ends in the corpus bursae, a membranous pouch often carrying sclerotizations called signa. The ductus bursae and corpus bursae together may be referred to as the bursa copulatrix.

SCHEMATIC OVERVIEW OF THE AFRICAN STEM BORERS

Superfamily	Characters
Noctuoidea	Proboscis present or reduced, no scales on the base Metathoracic tympanal organs (Fig. 8) Hind wing: Sc fused with Rs before the cell (noctuid stem borers are all in the tribe Apameini, where vein M2 is reduced or absent in the hind wing) (Fig. 10)
Pyraloidea	Proboscis present or reduced, always scales on the base. Abdominal tympanal organs in 1st sternite (Fig. 9) Hind wing: Sc fused with or approaching Rs beyond the cell (vein M2 present) (Fig. 11)

Species List

Superfamily	Family	Subfamily	Genus	Species
Noctuoidea	Noctuidae	Amphipyrinae	*Busseola*	*fusca* Fuller
			Sesamia	*calamistis* Hampson
				cretica Lederer
				nonagrioides Lefebvre
				nonagrioides botanephaga T&B
				penniseti T&B
				poephaga T&B
Pyraloidea	Crambidae	Crambinae	*Chilo*	*agamemnon* Bleszynski
				aleniellus (Strand)
				diffusilineus (de Joannis)
				orichalcociliellus (Strand)
				partellus (Swinhoe)
				sacchariphagus (Boyer)
				zacconius Bleszynski
			Coniesta	*ignefusalis* (Hampson)
		Schoenobiinae	*Adelpherupa*	*flavescens* Hampson
			Scirpophaga	*subumbrosa* Meyrick
				melanoclista Meyrick
				marginepunctella (de Joannis)
	Pyralidae	Galleriinae	*Eldana*	*saccharina* (Walker)
		Phycitinae	*Maliarpha*	*separatella* (Ragonot)

THE PREPARATION OF WINGS AND GENITALIA OF LEPIDOPTERA

Wings

The wing venation can be studied temporarily if one applies a drop of acetone or alcohol (70%) with the help of a fine paintbrush on the wing. The veins will be visible until the solvent evaporates. For permanent wing preparations, follow the procedure below.

1. Detach the wing from the thorax with forceps (pay attention to the wing-coupling apparatus).
2. Submerge the wing for a few seconds in absolute alcohol.
3. Submerge it for a few seconds in 10% hydrochloric acid.
4. Leave in household bleach for 1 min; remove from this solution when air bubbles start to form on the wing surface.
5. Wash the wing in water.
6. When the wing floats on the water surface, pass a microscope slide underneath it and pull the wing onto the slide, avoiding folds.
7. Let the water evaporate and cover the wing with a cover glass, attached only by the four corners.

Genitalia

1. Detach the abdomen from the thorax. The abdomen is easily detached from dried-out specimens when some upward pressure on the tail end of the abdomen is applied.
2. Place the abdomen in a test-tube with about 5 ml of a solution of 10% potassium hydroxide (KOH) or sodium hydroxide (NaOH). The abdomen can be left to soak in this solution for 12–24 h, according to the size of the abdomen, or heat the test-tube in a water-bath to near boiling-point for 5–10 min. In the latter case, close the test-tube with a loose plug of cotton wool to avoid spilling the KOH. Do not heat the test-tube with the KOH directly above a flame. The KOH will spurt almost immediately from the test-tube. This chemical causes permanent damage to clothes and temporarily burns the skin.
3. Place the abdomen in a cavity block with 70% ethyl alcohol (EtOH) for dissection and cleaning. Clean the specimen with an artist's brush, a pair of minute needles or snipe feathers.
4. The structures can be stained with chlorazol (1% solution chlorazol in 70% EtOH). The abdomen is placed for a few seconds in a drop of chlorazol on a microscope slide. Avoid staining too heavily; rather, stain several times and wash each time in 70% EtOH.
5. When the structures are sufficiently stained in 70% EtOH, put them in a cavity block with 95% EtOH for fixation of the staining and complete dehydration. Do some final cleaning; this stage is necessary for the permanent slides made in Euparal or Canada balsam.
6. For permanent preparation on a microscope slide, remove the last scales and hairs or any other material. The standard for mounting a slide is when it is suitable for photography. When sufficiently cleaned, place the structure in a cavity block with Euparal essence. If the abdomen is not correctly dehydrated, opaque bubbles will form. In this case, put the abdomen back into fresh absolute alcohol. If correctly dehydrated, position the genitalia on the slide and cover them with an appropriate cover glass.
7. Label the slide and the insect. Slide and insect must carry a label with the same number so that the two items will always be linked together. It is advisable to keep a slide notebook with references on the number of the slide, subfamily, identification and data on the insect.
8. The slide is kept flat and allowed to harden in a covered tray for several days at room temperature or in an oven at 40–50°C for at least 48 h. When the slide is sufficiently dried, it can be kept in a separate slide box or pinned next to the specimen in the insect drawer.

10 Noctuidae

J.D. Holloway
CABI Bioscience UK Centre, 56 Queen's Gate, London SW7 5JR, UK; present address: Department of Entomology, The Natural History Museum, Cromwell Road, London SW7 5BD, UK

INTRODUCTION

The habit of boring in the stems of grasses, reeds and rushes appears to have arisen only once in the Noctuidae, despite their being one of the largest moth families, with about 20,000 species.

Grass feeding as a specialism is found elsewhere in the family, such as throughout the diverse *Mythimna* complex of the Hadeninae, in armyworms of the genus *Spodoptera* and in some Catocalinae in the genus *Mocis*, but the larvae in these cases are always external feeders, defoliators.

Larvae boring within plant tissues, such as shoots, pods, seeds and fruit, are also found widely in the family, but these are predominantly on non-graminaceous hosts, e.g. Euteliinae of the genera *Penicillaria* and *Chlumetia* boring in the shoots of *Anarcardiaceae* (Holloway, 1985) or leguminous pod feeders in the Nycteolini. Heliothinae are seed and fruit feeders, and some *Helicoverpa* species (the 'corn-earworm' complex) are pests of maize cobs as well as of other fruits (Matthews, 1991).

The specialism of feeding within the pith of seasonally produced stems of *Gramineae* is restricted to a group of genera within the tribe Apameini. This group may be allied to the grass-feeding genus *Apamea* itself. The group is virtually restricted to the Old World, and is most diverse in Africa and temperate zones. Only a few taxa are known from either the Oriental and Australasian regions or the Americas.

The group has not yet been subjected to a modern phylogenetic analysis, or even a systematic revision based on an assessment of uniquely derived characteristics. These are character changes that occur within the process of speciation, or branching of lineages, and serve to mark and identify all descendant taxa of the common ancestor in which the change occurred. In the process of phylogenetic analysis, it is often difficult to distinguish the primitive from the changed, or derived, state in characters, or to recognize instances where parallel changes have occurred. For example, the 'streaky dead grass' wing pattern in noctuid stem borers is not unique to them, but is paralleled in many other unrelated grass-feeding taxa, such as *Mythimna* in the Noctuidae, in the *Ambadra*/*Norraca* complex of the Notodontidae (palm feeders) and in *Psalis* in the Lymantriidae. These examples are all from the superfamily Noctuoidea; the stem-boring Pyraloidea (Crambidae and Pyralidae), of course, provide another parallel example (see Chapter 11).

Similarly, the prominent, produced frons character used in the definition of the stem-boring genera *Conicofrontia*, *Sciomesa* and *Poeonoma* is not unique to them (and therefore an unsatisfactory feature for a genus group), but is common to numerous Lepidoptera taxa where the adults have to break out from a tough cocoon or from impacted soil, particularly in semiarid areas. Thus the feature in these taxa is probably more informative about their common ecological characteristics (e.g. the hardness of the stems from which they break out) than about their interrelatedness.

The main character used as indicative of the common ancestry of the stem-borer complex, the structure of the ovipositor, might also be criticized on these grounds. It is probably an adaptation for inserting eggs within a grass sheath. But other noctuids with this habit have different structures, such as the blade-like ovipositor of *Mythimna*. The character is also correlated with features of the male genitalia, such as the rather 'shouldered' tegumen, and is seen in the related *Apamea* generic complex, where the biology is somewhat different. Hence it might be considered a 'preadaptation'. The *Apamea* complex is diverse in the Palaearctic, the species being specialists on *Gramineae* and *Cyperaceae*, with the larvae feeding on roots or inflorescences (Heath and Emmet, 1979).

A phylogenetic classification is more than an academic luxury in the systematics of pest taxa. It provides an evolutionary hypothesis for the taxa concerned, a natural framework on which biological observations may be hung. Most importantly, it is predictive: groups of taxa descended from a single common ancestor are more likely to share common biological features (e.g. rapid population increase, short generation time), host plants and behavioural characteristics (e.g. tendency to disperse or migrate) and, of particular relevance to this book, common natural enemies and diseases.

Derivation of such a classification for noctuid stem borers is a task for the future. A preliminary investigation was published by Holloway (1989), but was overlooked by Harris and Nwanze (1992) in their monograph on *Busseola fusca*. This was unfortunate, as the genus *Busseola* has been diagnosed in the past (Tams and Bowden, 1953) on a suite of primitive, rather than derived, characteristics and hence may prove untenable under modern criteria (Holloway, 1989).

The species in just the tropical section of the group, including the genus *Acrapex*, number perhaps 190 (Poole, 1989), many described somewhat inadequately on the basis of a single specimen (Laporte, 1984) in *Acrapex*: these are probably misplaced, having plesiomorphic facies of the *Busseola* type. Only a few of these 190 are of economic significance, and most of these pests fall within a strict concept of the genus *Sesamia*. But a satisfactory classification will only arise from a study of the majority of these taxa, utilizing as wide a range of morphological and biochemical characteristics as possible. The tentative arrangement of Holloway (1989), with prior discussion of the characters on which it is based, is reproduced in the following paragraphs.

Tropical stem-borer taxa have the valve apex of the male genitalia more or less rounded, lacking a corona; the valve costa is angled or with an acute process dorsally (these features are also seen in the Palaearctic genus *Coenobia*). A ventral process occurs in many taxa but is more strongly associated with the sacculus than in temperate taxa, though also often linked to the costal process by a band of sclerotization. There is thus the strong possibility that it is the harpe and therefore not homologous with the ventral structure of the costa seen in temperate taxa.

Some features of the female genitalia bringing together the tropical taxa as a possible natural group are in the ovipositor lobes: these are arranged together in an acute, conical structure, perhaps more densely setose than in temperate taxa, and with the setae on the dorsal side of each (often slightly larger than the ventral ones), often directed away from the apex of the ovipositor.

These female features are shown by the type species of the following genera: *Sesamia* Guenée, *Busseola* Thurau (= *Calamistis* Hampson), *Conicofrontia* Hampson, *Poeonoma* Tams and Bowden, *Tridentifrons* Warren, *Poecopa* Bowden, *Carelis* Bowden, *Manga* Bowden, *Sciomesa* Tams and Bowden. The features are not seen in *Speia* Tams and Bowden (type species *vuteria* Stoll, South Africa). The indications are that these genera form a natural group distinct from Palaearctic stem-borer genera, defined on the female features and on differences in form of the costal process of the male valve and the position and strength of the harpe more distally on the sacculus.

The whole complex needs further review and might even be treated as *Sesamia sensu lato* until the characters within it can be more completely assessed. Some sections of it might be considered plesiomorphic and therefore possibly paraphyletic. They exhibit characters such as: more typical trifine patterning (cf. grass-like); a weak costal flange to the valve (cf. a spine-like process); band-like signa in the female bursa (loss is probably apomorphic); presence of a saccular harpe (both loss and greater development might be considered apomorphic). Consideration of these characters arranges the genera (as represented by their type species) as follows, with numbers of species worldwide currently included in each (Poole, 1989) indicated:

1. Costal process of valve acute; harpe strong, spinose; fore wing pattern grass-like; signum absent. All these characters are considered apomorphic.

Sesamia Guenée. 52 species.

2. Costal process of valve acute; harpe lost; fore wing pattern grass-like; signum weak (*Acrapex*) or lost (*Poecopa*).

Acrapex Hampson, type species *prisca* Walker (Sri Lanka). 82 species

Poecopa Bowden, type species *mediopuncta* Bowden (N. Africa). One species

3. Costal process of valve a transverse flange (plesiomorphic); harpe lost (apomorphic); fore wing pattern plesiomorphic; signum lost (except in *Sciomesa*).

Carelis Bowden, type species *albula* Bowden (West Africa). Two species.

Poeonoma Tams and Bowden, type species *serrata* Hampson (East and West Africa). Five species.

Tridentifrons Warren, type species *insularis* Warren (Java). Two species.

Sciomesa Tams and Bowden, type species *mesioscia* Hampson (South Africa). 20 species.

4. Costal process of valve a flange; harpe present on valve; fore wing pattern plesiomorphic though tending to grass-like in *Conicofrontia*; signum absent (*Conicofrontia*) or present (other genera).

Conicofrontia Hampson, type species *sesamoides* Hampson (South Africa). Three species.

Busseola Thurau, type species *sorghicida* Thurau = *fusca* Fuller. 16 species.
Manga Bowden, type species *basilinea* Bowden (West Africa). Four species.

IDENTIFICATION OF THE AFRICAN SPECIES OF MAJOR IMPORTANCE

This account will be restricted to a diagnosis of the important species, based on adult features. Larval features are discussed in Chapter 12. It is possible that the pupal cremaster may provide additional diagnostic information, but insufficient data are to hand.

As stated above, there is a tendency for grass-feeding moths of many families to have developed straw-coloured, uniform or longitudinally striate fore wings. Convergence on this pattern means that identification by wing pattern alone is rarely possible. Examination of other features, particularly of the male and female genitalia, is therefore necessary for accurate identification.

As a first step, it is necessary to sex the specimen to be identified. This is most readily done by examination of the wing-coupling mechanism, located below the wings. In the male, a single bristle, the frenulum, arises from the base of the anterior edge of the hind wing and engages in a tongue-like structure, the retinaculum, which arises from the subcostal vein of the fore wing. In the female, the frenulum consists of several bristles and engages in a tuft of scales on the fore wing directed anteriorly from the cubital vein, where it forms the posterior part of the discal cell. In dried specimens, the frenulum is often disengaged from the retinaculum, so the number of bristles (one versus more than one) is the critical feature. In addition, males sometimes, but not always, have their antennae feather-like (bipectinate). Females always have thread-like (filiform) antennae.

All noctuid genera that bore in the stems of *Gramineae* appear to be closely related within the tribe Apameini. The adults can be confused with those of the other major trifine noctuid grass-feeding group, the *Mythimna* complex, where the larvae are external feeders. *Mythimna* species have dense bristles over the surface of the eye, whereas the eye is smooth in stem-boring species.

The ovipositor in *Mythimna*, as mentioned earlier, is flattened and obliquely blade-like, and the eggs are laid in neat rows in the axils or between blades of grasses. In stem borers, the ovipositor is characteristically conical, strongly sclerotized and invested densely with very short setae. This ovipositor feature is shared with the related Palaearctic *Apamea* complex of genera, where the larvae are external feeders on grasses and sedges, often on the roots or inflorescences. Oviposition in tropical stem borers is usually within the leaf sheath of the host plant, the eggs often in a mass.

Stem borers are very diverse in Africa (well over half of the 190 or so tropical species), and no key is available that covers them all. Those of Tams and Bowden (1953) and Nye (1960) cover the most important and frequently encountered species. Stem-borer species can be divided broadly into three types: species like *B. fusca*, with dark brown fore wings, having a more typical noctuid pattern of two stigmata or spots in the discal cell and a transverse submarginal band (groups 3 and 4 above); small species of delicate build with straw-like longitudinal pattern to the fore wing, paler in a lens-like band anteriorly (mostly *Acrapex* species – the valve of the male genitalia is of characteristic shape, with a row of setae or corona along the distal margin) (group 2 above); larger, more robust species, with straw-like fore wings, mostly in the genus *Sesamia* (group 1 above).

Apart from *B. fusca*, all African species of major importance fall within the third group and belong to *Sesamia*. Five are considered here: *S. cretica* Lederer; *S. calamistis* Hampson; *S. nonagrioides* Lefebvre, with African subspecies *botanephaga* Tams and Bowden; *S. penniseti* Tams and Bowden; *S. poephaga* Tams and Bowden. They are most reliably distinguished on genitalia characters; hence, dissection is advisable in all cases. Some of the characters can be seen in fresh specimens if the genitalia are gently extruded and brushed clear of scales. Some features of male antennae and wing pattern and colour are also helpful.

Sesamia cretica Lederer (Figs 12, 13, 26, 27, 38; Plate 5:2,3)

Taxonomic notes. The taxa *Sesamia uniformis* Dudgeon and *Sesamia griselda* Warren are probably synonymous.

Diagnosis. The fore wings are a more uniform pinkish straw than in the following species, and the hind wings are pure white, as in *S. calamistis*, rather than bone-white, as in the final three species. The male antennae (Fig. 38) are simple, filiform and ciliate rather than bipectinate.

The male genitalia (Figs 12, 13) have a process in the centre of the juxta, lacking in the following species, and the anellus is diagnostically spinose. The aedeagus vesica lacks the globular cornutus of the remaining species, the costal process of the valve is a short flange rather than a long spine and the ventral spined saccular process of the other species is lacking. The female genitalia (Figs 26, 27) are simple, without sclerotized zones, resembling those of *S. calamistis* in this respect. The bursa copulatrix is relatively short and the ostium bursae is wide.

Biology. Presumably similar to that of *S. calamistis*.

Distribution and host plants. *Sesamia cretica* ranges through the savannah regions of north-east Africa, i.e. Morocco, Egypt, Sudan, Somalia (Tams and Bowden, 1953), and extends into the extreme north of Kenya (Nye, 1960). It is also recorded from northern Nigeria. Outside Africa it occurs in the south and west Mediterranean, Crete and the Near East, extending to India, Sri Lanka and Thailand. This species is a major pest of sorghum and, to a lesser extent, maize. It was also recorded from *Pennisetum typhoides* and *Sorghum halepense* by Tams and Bowden (1953).

Sesamia calamistis Hampson (Figs 14, 15, 28, 29, 42; Plate 4:5,6; 5:1)

Diagnosis. The fore wings are a streaked straw colour, as in the remaining species, but the pattern is more uniform, 'smoother' in appearance, and the distal margin is straighter. The fore wings are white rather than bone colour. The male antennae (Fig. 42) are bipectinate, the pectinations longer than the width of the antennal shaft, as in S. *nonagrioides* (shorter than the shaft in the last two species).

The male genitalia (Figs 14, 15) have the juxta diagnostically flask-shaped, almost as broad as long, with a dorsal notch; in the next three species it is more elongate, strap-like, again with a dorsal notch. In the female genitalia (Figs 28, 29), the sterigma (surrounding the opening, or ostium, of the ductus bursae) is unsclerotized.

Biology. In 3–5 days, the female lays up to 350 eggs, deposited in batches of 10–40. The eggs are arranged in two to four contiguous rows and inserted between the lower leaf sheaths and stems. Several hours after hatching, the larvae leave the oviposition site to penetrate the stems either directly or after feeding on the leaf sheath. During the larval stage, which lasts for 30–60 days, depending on climatic conditions, and usually involves five to six moults, larvae may successively attack a number of young stems. Only one immature larva is observed per young stem or tiller. Pupation generally takes place in the stem, and rarely between the sheath and stem. The pupal period lasts for 10–12 days at 25°C. Under tropical conditions five to six generations are completed a year.

Distribution and host plants. *Sesamia calamistis* is widespread in tropical and southern Africa: South Africa, Zimbabwe, Malawi, Uganda, Tanzania, Kenya, Zanzibar, Madagascar, Mauritius, Réunion, Angola, Nigeria, Ivory Coast, Cameroon, Senegal, Gambia, Ghana (Tams and Bowden, 1953).

It is recorded from rice, maize, sorghum, finger millet and sugar cane (Nye, 1960). Among wild hosts are *Pennisetum purpureum* and *Sorghum arundinaceum* (Tams and Bowden, 1953).

Sesamia nonagrioides botanephaga Lefebvre (Figs 22, 23, 34, 35, 41)

Diagnosis. This and the next two species are best distinguished on genitalic characters, although the male antennae of S. *nonagrioides* (Fig. 41) have long pectinations, as in S. *calamistis.* All three tend to have a submarginal row of dark brown flecks on the fore wing.

The male genitalia (Figs 22, 23) share the long, strap-like juxta with the next two species but differ from them in the valve processes: the costal process is more curved, tapering and acute, with a smaller subsidiary spine just subapically; the saccular process is narrower and more finely spined. The female genitalia (Figs 34, 35) have lateral sclerotization at the sterigma, but it is somewhat globular rather than squarish.

Biology. Presumably similar to that of *S. calamistis.*

Distribution and host plants. Typical S. *nonagrioides* is found around the Mediterranean, in western Turkey and south-western Iran. The African subspecies *botanephaga* is recorded from Ghana, Ivory Coast, Nigeria, Togo and Sudan (Tams and Bowden, 1953).

According to Tams and Bowden (1953), the ecological requirements of *S. n. botanephaga* and *S. calamistis* are different, S. *n. botanephaga* being the dominant species in forest and forest-fringe areas, where there is a minimum rainfall of about 50 inches (127 cm) well distributed through the year and therefore a relatively short and less severe dry season. *Sesamia calamistis* is more common, at least in

West Africa, in savannah areas that have a well-marked dry season. However, note that the two species can occur in the same field at the same time.

The species is known from maize, rice, sorghum and sugar cane. Among the wild hosts are *P. purpureum*, *S. arundinaceum*, *Rottboelia exaltata* and *Chasmopodium afzelii* (Tams and Bowden, 1953).

Sesamia poephaga Tams & Bowden (Figs 18, 19, 32, 33, 39)

Taxonomic notes. *Sesamia poephaga* may be closely related to, or conspecific with, *Sesamia epunctifera* Hampson from East and southern Africa, a smaller species with a prominent, broken marginal line on the fore wing. More material of the latter is needed to resolve this issue.

Diagnosis. The fore wings are a paler straw colour than in either *S. nonagrioides* or *S. penniseti*. The male antennal pectinations of this and the next species (Fig. 39: *S. poephaga*; Fig. 40: *S. penniseti*) are shorter than the width of the antennal shaft. The venation character used to distinguish *S. poephaga* and *S. penniseti* in the key by Tams and Bowden (1953) is unreliable and should not be used.

The species are best distinguished by genitalic features. In the male (*S. poephaga*, Figs 18, 19; *S. penniseti*, Figs 16, 17), both have the saccular process of the valve robust and coarsely spined, but the costal spine is apically bifid symmetrically in *S. poephaga* and more spatulate, irregular and finely spined in *S. penniseti*. In the female (*S. poephaga*, Figs 32, 33; *S. penniseti*, Figs 30, 31), both have the pair of sclerotizations flanking the ostium rather squarish but the distal margin is more or less straight in *S. poephaga* and irregular, somewhat bilobed in *S. penniseti*. The ostium is wide in *S. poephaga*, narrow in *S. penniseti*.

Distribution and host plants. *Sesamia poephaga* appears to be rather localized in western and central Africa. Indications from Ghana (Tams and Bowden, 1953) are that the species has a preference for savannah vegetation.

Sesamia penniseti Tams & Bowden (Figs 16, 17, 30, 31, 40)

Diagnosis. See the previous species for diagnostic features.

Distribution and host plants. *Sesamia penniseti* is recorded over a similar geographical range to *S. poephaga*, and may be conspecific with *S. poebora* Tams and Bowden (Fig. 20, 21) in East Africa. Material of the latter only reliably consists of the holotype male. In Ghana, *S. penniseti* is more frequent in forest localities than *S. poephaga* (Tams and Bowden, 1953).

Remarks. The last three species are very closely related and not always clearly distinguished from other taxa. Their interrelationships and the status of the taxon pair *S. nonagrioides/botanephaga* and the complex quartet of *S. poephaga/ epunctifera/ penniseti /poebora* require further study.

Busseola fusca Fuller (Figs 24, 25, 36, 37, 43; Plates 2:2,4,5; 4:1–4)

Diagnosis. This species may be distinguished from the many others of similar fore wing pattern in *Busseola* and related genera most reliably by characters of the male and female genitalia. The discussion here will focus on the small group of such species reviewed by Bowden (1956) and Nye (1960).

All such species have brownish grey fore wings and somewhat paler hind wings. The fore wings have an oblique submarginal shade, containing darker arrow-head (sagittate) marks, distal to which is a pale band, sometimes broken. The reniform and orbicular spots (stigmata) are present in the cell, which is otherwise sometimes a much darker brown (e.g. in *Busseola quadrata* Bowden and *Busseola phaia* (= *segeta*) Bowden). There are fine, faint, dark brown postmedial and antemedial lines (fasciae); the former is more strongly present on the pale fore wing of *Busseola quadrata*, nearer the submarginal shade. *Busseola fusca* tends to have the hind wings distinctly paler than the fore wings, whereas in *B. phaia* and *B. quadrata* they are only slightly so and, in some specimens of the latter, they are darker. In the fore wing of *Manga basilinea* Bowden, the submarginal sagittate marks are irregular (cf. increasing towards the centre) in size, with two much larger than the rest, the vein at the posterior of the cell is usually pale and there is a basal black streak.

The male genitalia (Figs 24, 25) of *B. fusca* have strong, slender, club-like processes from the valve sacculus, and these are sometimes asymmetrically developed; in *M. basilinea* these processes are broader, apically toothed, with the valve of much more even depth. In *B. phaia* these are small lobes and in *B. quadrata* a short triangular process is directed somewhat towards the base. In the female genitalia (Figs 36, 37), *B. fusca* has an asymmetric sclerotization of the minutely spined lamella vaginalis into lateral lobes, between which the ductus is inserted obliquely. In *B. quadrata* there is a transverse triangular plate and in *B. phaia* and *M. basilinea* there is no strongly sclerotized lamella vaginalis.

Biology. The female lays several hundred eggs in batches of 30–50, inserted between the sheath and stem. Incubation lasts about 1 week. After hatching, the larvae feed on the young blades of the leaf whorl and then, suspended from silk strands, spread to neighbouring plants. They penetrate into the stems by boring through the whorl base. Generally, they destroy the growing point and tunnel downward. After passing through six to eight stages in 30–45 days, they pierce an outlet for the adult and pupate in the tunnel. Pupation lasts 10–20 days. Up to four generations are produced per year. At the end of the rainy season, the larvae of the last generation enter into diapause in maize and sorghum stubble or in wild grasses. They pupate a few months later, just before the start of the following rainy season.

Distribution and host plants. *Busseola fusca* is widespread in Africa south of the Sahara but has not been recorded from Madagascar. Of the close congeners mentioned above, *B. phaia* is widely sympatric and extends to Madagascar and Mauritius, whereas *B. quadrata* is only known from Ghana.

This species has been recorded from maize, sorghum, rice, sugar cane and millet (Harris, 1962).

11 Pyraloidea: Crambidae, Pyralidae

K.V.N. Maes
AgroBioSys International, Kleine Smetledestraat 192, B-9230 Wetteren, Belgium

INTRODUCTION

The Pyraloidea contain about 13 species that are important stem borers of cereal crops in Africa, and many species that are of importance as pests elsewhere in the world.

Adults are small to medium-sized moths, with a wing length of about 10–20 mm. The main diagnostic characters of the superfamily are the presence of paired abdominal tympanal organs directed anteroventrally of the basal sternum and fusion (or at least close approximation) of the subcostal vein (Sc) and the stem of the radial vein (Rs) in the hind wing (see Fig. 11). Larvae have circles or ellipses of crochets on the abdominal prolegs, distinguishing them from noctuid larvae, which have arcs of crochets (see Chapter 12, this volume, for a detailed account of both noctuid and pyralid stem-borer larvae). The Pyraloidea contain two families, the Crambidae and Pyralidae, and they consist of several subfamilies.

The African pyralid stem borers are distributed over both families: in the Pyralidae with the subfamilies Galleriinae (*Eldana*) and Phycitinae (*Maliarpha*), and in the Crambidae with the subfamilies Crambinae (*Chilo, Coniesta*) and Schoenobiinae (*Scirpophaga, Adelpherupa*).

Characters observed in the tympanal organs make an easy distinction between the families Pyralidae and Crambidae possible.

Tympanum and conjonctivum lying along the same plane, not making a clear angle (Fig. 80) .... Pyralidae: Phycitinae and Galleriinae
Tympanum and conjonctivum making a clear angle, not lying along the same plane (Figs 81, 82) .... Crambidae: Crambinae and Schoenobiinae.

The main diagnostic characters of these subfamilies are given by Holloway *et al.* (1987). The descriptions of these subfamilies are adapted here to the species listed as stem borers in Africa.

Crambinae: Fore wings relatively long and narrow and somewhat truncated distally. Hind wings broad and much folded. Wings always smooth-scaled and the hind wing with a pecten of hairs on the upper side. Labial palps typically very long and porrected straight forward, often exceeding three times the width of the eye. Maxillary palps triangularly scaled. Chaetosemata absent. (Genera: *Chilo* and *Coniesta*.)

Schoenobiinae: Adults generally white or yellowish white, sometimes with dark spots and other markings. Some species are brown and resemble Crambinae. Sexual dimorphism present. Females larger than males and often with a bulky anal tuft of long deciduous hairs used to cover egg masses. Tongue obsolescent. Ocelli and chaetosemata present. (Genera: *Scirpophaga* and *Adelpherupa*.)

Galleriinae: Fore wings elongate, broadly rounded distally and with sombre, subdued markings. Tongue rudimentary or obsolete. Labial palps sexually dimorphic. Maxillary palps obsolete in males, distinct but small in females. Chaetosemata and ocelli absent. Male genitalia without a gnathos and with broad uncus, rounded terminally. (Genus: *Eldana*.)

Phycitinae: Fore wings generally elongate and with grey-black–brown-white coloration, sometimes tinged with red, yellow or green. Vein R5 always absent. Hind wing with cubital pecten. Labial palps usually curved upwards, often sexually dimorphic and well developed but occasionally reduced or absent. Chaetosemata present. (Genus: *Maliarpha*.)

IDENTIFICATION

For a correct identification of most pyraloid stem borers of Africa it is necessary to make genitalia dissections. The male and female genitalia of all species mentioned in the following text were dissected and compared with type material. If specimens should be found that are not within the range of species mentioned here, it is strongly advised to submit the material to specialists who have access to type material and large series of species. I strongly advise against creating a new name without proper comparison with large series of known material, since this will create further taxonomic confusion and, as such, deception among the field workers. The aim of this work is to avoid just this kind of problem.

CRAMBINAE

The genus *Chilo* contains many species that are stem borers of cereal crops and grasses in Africa and Asia. The taxonomy of these species is difficult but was considerably clarified by the publication of a comprehensive revision by Bleszynski (1970). He provided identification keys and descriptions, with particular emphasis on characters of the male and female genitalia and head profiles. It has to be emphasized that the adults of this family are superficially similar and can only be identified with certainty by examination of the genitalia. Descriptions of external characters of the adults are deliberately omitted here, since they will lead to misidentifications. A special effort was made to present plates composed of photographs of male and female genitalia from specimens that were compared with the types of each species mentioned.

Chilo partellus (Swinhoe): spotted stem borer (Figs 44, 45, 68; Plate 2:1,3; Plate 7:1–3)

Chilo partellus was accidentally introduced and has become widely established in eastern and southern Africa. It is one of the most important pyralid stem borers of cereal crops in these parts of Africa.

Taxonomic notes. Bleszynski (1970) clarified the synonymy of this species, which has often been incorrectly referred to as *Chilo zonellus* or, in the earlier literature, as *Chilo simplex*. It is also known by different common names, including 'durra stalk borer', 'maize stem borer' and 'pink stem borer', but 'spotted stem borer' is now generally accepted as the common name of this species.

Diagnosis. Adults and larvae of other species of *Chilo* and of *Coniesta ignefusalis* are very similar in general appearance and colour and may occur on the same host plants. The male genitalia (Figs 44, 45) have a submedian, strong, tapering projection on the costa of the valve; the juxta plate has a large triangular central plate, two basal notches and lateral arms not extending beyond the costa of the valve. The aedeagus has a bulbous basal projection and a ventral arm. The female genitalia (Fig. 68) have a heavily sclerotized and swollen ostial pouch, which is longitudinally wrinkled and deeply notched caudally between two semicircular lobes; the corpus bursae has a lamellate signum with a median ridge.

Biology. Published information on the biology and ecology of this species has been reviewed by Harris (1989a, 1990). Adults emerge from pupae in late afternoon and early evening and are active at night. During the day they rest on plants or plant debris and are seldom noticed. Females mate soon after emergence and lay eggs on two to three subsequent nights, laying batches of 10–80 overlapping eggs on the undersides of leaves, mainly near the midribs. Adults live for about 2–5 days and do not normally disperse far from emergence sites. Eggs hatch in the early morning (06.00–08.00 h), 4–8 days after laying, and young larvae ascend plants to enter the leaf whorls, where they start to feed. Older larvae tunnel into stem tissues and, after feeding for 2–3 weeks, pupate in the stems for 5–12 days. The life cycle is completed in 25–50 days when conditions are favourable, and five or more successive generations may develop during a growing season. In cold and/or dry conditions, larvae enter a resting stage (diapause) in stems, stubbles and other crop residues, where they may spend up to 6 months before pupating when favourable conditions recur during the next growing season.

Distribution and host plants. *Chilo partellus* is indigenous to Asia and has been recorded from Afghanistan, Bangladesh, Cambodia, India, Indonesia, Laos, Nepal, Pakistan, Sikkim, Sri Lanka, Thailand and Vietnam. According to Nye (1960), the first published record of this species from Africa was from Malawi (Nyasaland) in 1932 (as *Argyria lutulentalis* Tams). Since the early 1950s, it has become well established as a pest in Ethiopia, Kenya, Malawi, Mozambique, Somalia, South Africa, Sudan, Tanzania and Uganda. There are unconfirmed reports of its occurrence in Cameroon and Togo. Further spread in Africa is certainly possible.

Chilo partellus is an important pest of maize, sorghum and pearl millet in Asia and in Africa. It also attacks wheat, sugar cane, rice, foxtail and finger millets, and grass hosts include *Sorghum halepense*, *Sorghum verticilliflorum*, *Panicum maximum*, *Pennisetum purpureum*, *Hyparrhenia rufa* and *Rottboelia compressa*.

In Africa it is of greatest importance as a pest of maize in East and southern Africa but could become established as a pest of this and other cereal crops in West Africa.

Chilo zacconius Bleszynski (Figs 46, 47, 75; Plate 6:3–6)

Chilo zacconius is a pest of rice throughout West Africa. Its pest status has not been accurately assessed. In the field it may be confused with *Adelpherupa flavescens* Hampson.

Taxonomic notes. Bleszynski (1970) described this species. It resembles *Chilo diffusilineus* in external appearance but has distinctive male and female genitalia. To a lesser extent, it can also be confused with *A. flavescens* Hampson. Earlier records of *Proceras africana* Aurivillius may be based on misidentifications of this species.

Diagnosis. Detailed examination of male and female genitalia is essential for accurate identification of this species. The male genitalia (Figs 46, 47) have distinctive long, thin arms arising from the juxta plate. The juxta itself has a median circular incision. The aedeagus has a distinct bulbous basal projection and a subapical thorn or cornutus on a long base, and lacks a ventral arm. The female genitalia (Fig. 75) have a broad ostial pouch, partly heavily sclerotized, with a well-demarcated, twisted ductus bursae. The signum is lacking on the corpus bursae.

Biology. Females lay batches of up to 80 overlapping, scale-like eggs on leaves of rice plants on up to five nights following emergence. Eggs hatch after 4–5 days and young larvae first feed on epidermal tissues and then bore into the stems. Larval development is completed in 3–5 weeks and larvae then pupate in a cocoon within the stem. Two or more generations may develop on one rice crop and development continues on suitable hosts throughout the dry season. No larval diapause has been recorded.

Distribution and host plants. *Chilo zacconius* occurs throughout West Africa and has been recorded from Benin, Burkina Faso, Cameroon, Ivory Coast, Mali, Niger, Nigeria, Senegal and Sierra Leone.

Cultivated and wild rices are the main host plants but this species has also been recorded from *Echinochloa stagnina* and *Sorghum arundinaceum*.

Chilo diffusilineus (de Joannis) (Figs 48, 49, 72)

Chilo diffusilineus seems to be mainly a stem borer of rice but, although it occurs widely in tropical Africa, there is little published information on its pest status. Brenière (1983) mentions it but does not treat it as an important pest of rice in West Africa.

Taxonomic notes. This species was first described by de Joannis in 1922. Bleszynski (1963) redescribed it and synonymized it later with *Chilo phaeosema* Martin (Bleszynski, 1970).

Diagnosis. Adult and immature stages are very similar to those of *C. zacconius* and other species of *Chilo*, and positive identification is mainly based on characters of the male (Figs 48, 49) and female (Fig. 72) genitalia. The male genitalia have two arms of different lengths projecting from the juxta. In general, these are shorter than those of *C. zacconius*. The juxta plate has no circular incision as in *C. zacconius*. The aedeagus is similar to that of *C. zacconius* but has a small projection near the middle and lacks the cornutus of *C. zacconius*. Female genitalia are very similar to those of *C. zacconius* but the ostium is relatively smaller compared with the overall length of the ductus bursae and the corpus bursae. It is well demarcated from the ductus bursae and has a heavily sclerotized rod into the ductus bursae. The signum on the corpus bursae is absent.

Biology. Presumably similar to that of *C. zacconius*.

Distribution. *Chilo diffusilineus* occurs throughout tropical Africa and has been recorded from Ethiopia, Guinea, Malawi, Mozambique, Nigeria, Senegal, Sierra Leone, Sudan and Zimbabwe.

Chilo aleniellus (Strand) (Figs 50, 51, 70)

Chilo aleniellus has been observed on maize and rice.

Taxonomic notes. The species was first described by Strand in 1913 from Spanish Guinea. Bleszynski (1970) noted morphological differences in the female genitalia from specimens originating from Zaïre. This may indicate the existence of a distinct subspecies or species.

Diagnosis. The male genitalia (Figs 50, 51) have narrow, distinctly tapered valves. The juxta plate bears two long arms, which are variable in length but always longer than those of *Chilo orichalcociliellus*. The aedeagus is very similar to that of *C. orichalcociliellus*. The female genitalia (Fig. 70) have a large, heavily sclerotized plate on the seventh sternum and the ostial pouch is lightly sclerotized, at least in specimens from West Africa.

Biology. Moyal and Tran (1992) reported that in Ivory Coast *C. aleniellus* larvae start to feed at the end of the maize-growing season, during the milky stage at about 60 days after emergence of the plants. They mainly attack stems (80%) but also damage cobs (20%). Larval development takes about a month and the pupal stage lasts 11–12 days at 22°C.

Distribution and host plants. Recorded by Bleszynski from Cameroon, Equatorial Guinea (Rio Muni), Ghana, Nigeria, Zaïre and Uganda. Moyal and Tran (1992) recorded that, in the Ivory Coast during 1988/89, it was regularly found only in the western part of the forest area, between 6° and 8°N. Recorded from maize and rice.

Chilo orichalcociliellus (Strand): coastal stalk borer (Figs 52, 53, 74; Plate 7:4)

This species is mainly known as a stem borer of maize and sorghum in East Africa, especially in coastal areas.

Taxonomic notes. *Chilo orichalcociliellus* was originally described by Strand in 1911 in the genus *Diatraea*. Bleszynski (1962) put it into synonymy with *Diatraea argyrolepida* Hampson and later published (Bleszynski, 1970) a detailed taxonomic description and stated that it was easily distinguishable from other species of *Chilo* by characters of the male and female genitalia.

Diagnosis. In the male genitalia (Figs 52, 53), it can be easily distinguished from *C. aleniellus* and other related species by the proportionately short, scarcely tapering valve with a broadly rounded apex. The female genitalia (Fig. 74) have two additional triangles lateral to the median sclerotized plate on the seventh sternum.

Biology. Mathez (1972) and Warui and Kuria (1983) have studied the biology of this and other species of stem borer in Kenya. It is generally similar to that described above for *C. partellus*.

Distribution and host plants. Bleszynski (1970) recorded this species from Kenya, Madagascar, Malawi, South Africa, Tanzania and Zimbabwe and noted that it does not occur in West Africa. Nye (1960) noted that specimens reared during his surveys in East Africa were invariably obtained from localities below 2000 ft (305 m). It is a pest on maize and sorghum.

Chilo sacchariphagus (Bojer) (Figs 56, 57, 69; Plate 7:5,6)

Chilo sacchariphagus is a pest of sugar cane, possibly introduced into East Africa.

Taxonomic notes. The identity of this species has for a long time been uncertain and several names have been used. It was first published by Bojer in 1856. A complete discussion is given by Bleszynski (1970). A certain variability of the genitalia is pointed out and he distinguishes three subspecies: the nominate subspecies described from Mauritius, the subspecies *C. s. stramineellus* (Caradja) from China and the subspecies *C. s. indicus* (Kapur) from India. Externally the different subspecies cannot be distinguished. They show differences in the male and female genitalia.

Diagnosis. The genitalia of *C. s. indicus* are illustrated here. The male (Figs 56, 57) has an aedeagus with a series of heavily sclerotized cornuti. The juxta has no projecting lateral arms. The valves are blunt near the apex. The female genitalia (Fig. 69) have the part of the ductus bursae just posterior to the ostium bursae with a longitudinal fold.

It has to be noted that considerable variation in the genitalia exists; therefore it is advisable to send specimens to specialists for identification and comparison with type material.

Biology. There are no biological data available for this species in Africa.

Distribution and host plants. This species has a large distribution in Asia and South-East Asia. The species was introduced, probably by man, into Madagascar, Réunion and Mauritius (Bleszynski, 1970). There are unconfirmed reports of this species from Mozambique and Zanzibar on sugar cane.

Betbeder-Matibet (1971) described the biological control of this species in Madagascar.

Chilo agamemnon Bleszynski (Figs 54, 55, 73)

Taxonomic notes. The identity of *Chilo agamemnon* was clearly established by Blezynski (1970). It was easily confused with *C. diffusilineus* and *C. zacconius*, which are well-established species in Africa, and *C. agamemnon* has been easily overlooked or confused with the former two. *Chilo agamemnon* is restricted to the Palaearctic region and only a few data exist from Africa.

Diagnosis. Externally this species cannot be distinguished from *C. diffusilineus* and *C. zacconius*. It is easily identified using the male and female genitalia. The male genitalia (Figs 54, 55) have a broad plate-like sclerotization on the base of the costa near the tegumen. The juxta has two lateral arms, which are rather short compared with those of *C. diffusilineus* and *C. zacconius*. The aedeagus usually bears a series of short cornuti. In the female genitalia (Fig. 73), the ostium bursae is membranous and shaped like a cup. There is no signum.

Biology. Details on the biology of this species are given by Rivnay (1963, 1967). Adults were captured in Sudan during the month of February and in Cameroon during the month of November.

Distribution and host plants. This species is a pest of maize and other cereal crops, mainly in the Middle East. It has been reported from Israel, the north of Egypt, Sudan and Uganda. Nonveiller (1984) reported it from the north of Cameroon.

It is recorded as a pest on maize but also occurs on sugar cane and rice.

Coniesta ignefusalis (Hampson): millet stem borer (Figs 58, 59, 71; Plate 8:1–4)

This crambid is indigenous to Africa and has probably been associated with pearl millet since the crop was first cultivated, about 5000 years ago. It is a major pest of pearl millet in West Africa but seems to be of less importance in eastern and southern Africa.

Taxonomic notes. Hampson described this species and placed it in the American genus *Diatraea*. Since then, taxonomists have transferred it to other genera, including *Acigona*, *Chilo*, *Donacoscaptes*, *Eoreuma*, *Haimbachia* and *Coniesta*. Its correct placing has still not been definitely determined and current taxonomic opinion is that it should be retained in the genus *Coniesta* pending further taxonomic research. Despite the confusion caused by these frequent name changes, taxonomists have always treated the millet stem borer as a distinct species. Related species probably occur on various grasses in West Africa but have not been adequately studied.

Diagnosis. *Coniesta ignefusalis* can be identified with certainty by examination of the male and female genitalia. Harris (1962) published illustrated descriptions of all stages, based on specimens collected in northern Nigeria; Betbeder-Matibet (1987) published a data sheet with colour photographs of all stages and of bored stems; and Matthews and Jago (1993) published brief descriptions, with colour illustrations of adults, larvae and attacked stems.

The genitalia are clearly different from those of *Chilo* species and no confusion is possible. The male genitalia (Figs 58, 59) have a beak-like uncus and gnathos, a complex trilobed valve with the lobes progressively larger ventrally, a stout aedeagus and a vesica with many spicular cornuti. The female genitalia (Fig. 71) consist of a laterally compressed ovipositor with a sinuate, sclerotized caudal margin, and the corpus bursae has a pair of pincushion-like signa.

Biology. Adult moths emerge from pupae after dark, usually before midnight. Females release a pheromone to attract males and mate either on the night of emergence or during the following nights. During daytime, adults rest in concealed positions on plants and on crop debris and are seldom seen unless disturbed, when they fly briefly before resettling. Females lay eggs at night, inserting batches of 20–50 under leaf sheaths (Plate 8:1). These eggs hatch within about 1–2 weeks and larvae immediately tunnel into the leaf sheaths and stems (Plate 8:2). They then feed for a month or more before pupating in the tunnels that they have excavated in damaged stems (Plate 8:3). Adults emerge about 1–2 weeks after pupation, giving a total development period of about 2 months. Three generations may develop during a 6-month growing season, with adult emergence peaks in about May, July and September in northern Nigeria, but, as the rains cease and the dry season begins, an increasing number of larvae stop feeding and enter a dormant period (diapause). These larvae remain in dry stems and stubble throughout the dry season and do not pupate until well into the following rainy season, by which time the new crop is well established and available for oviposition by the first generation of adults in the new season (Plate 8:4). The best-known natural enemies of *C. ignefusalis* are parasitic Hymenoptera. The only published record of a pathogen is *Metarhizium anisopliae* from larvae in Nigeria (Harris, 1962).

Distribution and host plants. *Coniesta ignefusalis* is restricted to mainland Africa south of the Sahara and occurs in suitable habitats throughout West Africa, where it has been formally recorded from Benin, Burkina Faso, Chad, Gambia, Ghana, Guinea Bissau, Mali, Mauritania, Niger, Nigeria and Senegal. It has also been recorded from Sudan, Ethiopia and Angola and is probably more widely distributed in tropical Africa than published records indicate. In Nigeria it occurs up to altitudes of about 1000 m.

This stem borer is a pest on pearl millet but has also been recorded from sorghum, maize and sugar cane and from a number of grasses, including *P. purpureum, P. maximum* and *Andropogon gayanus*. Youm (1990) reported 14 different host plants, including some non-graminaceous plants, that harbour diapause larvae during the dry season, but these will probably not sustain full development from egg to adult.

SCHOENOBIINAE

Scirpophaga species (Figs 66, 67, 76, 82; Plate 6:1,2)

Although species of this genus are important pests in other parts of the world, no species is of any real importance in Africa at present. Brenière (1983) recorded *Scirpophaga melanoclista* Meyrick and *Scirpophaga subumbrosa* Meyrick as pests of rice in Africa and noted that these are probably not the only species involved.

Taxonomic notes. Very little taxonomic research has been done on African species of *Scirpophaga*. Lewvanich (1981) revised the Old World species and recorded *Scirpophaga occidentella* (Walker) from rice in Nigeria and *S. subumbrosa* Meyrick from rice in Nigeria and Sierra Leone. She noted that these two species gave rise to many problems in rice pest control but did not give details. Brenière (1983) listed *S. melanoclista* from Ivory Coast, Mali, Nigeria, Angola, Cameroon, Senegal, Zaïre and Madagascar and *S. subumbrosa* from Mali and Ghana, so it would seem that there are at least three different species that attack rice in Africa. Although scientific collections retain a large number of *Scirpophaga* species and even some undescribed ones, it is a pity that few agricultural field workers rear and collect material on crops with the necessary data.

Diagnosis. Adult *Scirpophaga* have distinctive pure white silky wings, but there are many similar species and expert identification should be based on examination of male and female genitalia by specialists, who can compare the slides with type material. Good taxonomic descriptions are not available for African species and positive identification is therefore usually impossible in our present inadequate state of knowledge. In this book I illustrate the male (Figs 66, 67) and female (Fig. 76) genitalia of *Scirpophaga marginepunctella* (de Joannis), a species collected from rice in Malawi.

Biology. Brenière (1983) has described the biology of species attacking rice. The eggs are laid in batches of 100–180 on upper or lower leaf surfaces of rice plants and are covered by a light brown, downy mass of hairs and scales. Young caterpillars bore through the leaf sheaths into stems and, when older, tunnel from one internode to another. They are semiaquatic and may cut stems into separate 4 cm lengths, in which they float from stem to stem. Larval development is completed in 28–44 days and larvae usually pupate in the lower internodes at the base of the stems. Adults emerge after 12–14 days and live for about a week.

Distribution and host plants. Species of *Scirpophaga* occur throughout tropical Africa. Brenière *et al.* (1962) recorded *S. melanoclista* from Angola, Cameroon, Ivory Coast, Madagascar, Mali, Nigeria, Senegal and Zaïre and *S. subumbrosa* from Ghana and Mali but noted that these are probably not the only two species concerned. *Scirpophaga marginepunctella* was collected in Malawi.

The species are reported from rice.

Adelpherupa species (Figs 60, 61, 77)

This genus was mentioned by Descamps (1956) from cultivated rice. The record does not mention the species. Material collected from rice fields in Benin was identified as *A. flavescens* Hampson.

Taxonomic notes. *Adelpherupa flavescens* was described by Hampson in 1919 from Sudan. At this moment, no recent taxonomic study of this genus exists and a revision should be undertaken.

Diagnosis. Externally this species may be confused with *C. zacconius* and both species occur together in West Africa.

The male genitalia have a membranous gnathos, covered apically with spines. The uncus bears some simple setae on the apex. *Adelpherupa flavescens* is illustrated here (Figs 60, 61). The female genitalia lack the setae on the edge of the eighth segment. Figure 77 illustrates the female genitalia of *Adelpherupa albescens*.

Biology. Descamps (1956) described the biology of an unidentified *Adelpherupa* species from the north of Cameroon. The eggs are laid on the leaves of rice and the young caterpillar bores in the stem near the internodes, feeding on the soft tissues. A cylindrical swelling up to 5 cm will occur near the base of the stem. The plant shows no other signs of the attack. The larval stage takes between 20 and 25 days. The adult caterpillar pupates in the stem near the base in an oval chamber. The moth emerges after 14–15 days and can be found on the surrounding vegetation. Descamps observed two generations a year. The caterpillar goes into a diapause.

Distribution and host plants. I have examined material from Sudan, Tanzania, Malawi, Nigeria, Ghana and Senegal. The species was also collected in Benin, from rice fields.

GALLERIINAE

Eldana saccharina (Walker): African sugar-cane borer (Figs 64, 65, 79, 80; Plates 8:5,6; 9:1)

Eldana saccharina is mainly a pest of sugar cane but it also attacks maize, sorghum and other cereal crops.

Taxonomic notes. *Eldana saccharina* was described by Walker in 1865. It is well characterized, easily recognizable and unlikely to be confused with any of the other stem borers of African cereals.

Diagnosis. This is the only galleriine pyralid species boring stems of African cereal crops and it is therefore easily distinguished from other stem-borer species in both adult and larval stages. Betbeder-Matibet (1985) published a brief illustrated data sheet. The fore wings are elongate, almost elliptical, with a rounded distal edge, which is typical of many Galleriinae. The longitudinal veins are brown against a lighter background and there are two distinct dark spots in the anterior half of the fore wings (Plate 9:1). At rest, the wings are folded over the abdomen and cover the lighter-coloured hind wings. There seems to be considerable variation in the size and colour of adults from different localities in Africa.

Descriptions and illustrations of the male and female genitalia have not been published before. The male genitalia (Figs 64, 65) have the valva rounded near the apex, protrusions are lacking. The juxta is oval and the triangular uncus bears a row of spines. The aedeagus is slightly bent near the middle. The female genitalia (Fig. 79) are strongly elongated. The apophyses posteriores are very long. The ostium is V-shaped and the ductus bursae is narrow, with the ductus seminalis just beneath the antrum. The corpus bursae has a large appendix bursae.

Although there is a large phenotypical variation, there were no major differences observed in the genitalia in specimens from different locations all over the African continent.

Biology. Atkinson (1980) published a detailed account of the biology, distribution and natural host plants of this species in Natal, South Africa. Girling (1978) reported similar studies in Uganda, and Sampson and Kumar (1985) have studied this species in Ghana. Females lay batches of 50–100 eggs on dry leaves at the bases of plants and may lay up to 500 eggs over a period of about a week. Females prefer to oviposit on dry dead leaves and this partly accounts for its tendency to infest mature crops. Eggs hatch after about 6 days and the young larvae feed externally on epidermal tissues before penetrating the stems and eating out galleries. The length of larval development is very variable and may take up to 2 months, and larvae pupate within the stems. Atkinson (1980) published tabulations of durations of egg, larval and pupal stages at different temperature regimes and calculated thermal constants in day-degrees. Egg development ranged from 4 days at 32°C to 12 days at 17°C; larval development averaged 20 days at 30°C to about 50 days at 19°C and pupal development was completed in about 7 days at 30°C and about 16 days at 19°C. Up to six generations may develop in a year and there is no larval diapause. Betbeder-Matibet (1985) noted that more than 95% of each generation is destroyed by predators (ants and spiders), parasites and adverse weather (particularly wind and rain). For South Africa, both indigenous and imported natural enemies were listed by Carnegie (1991).

Distribution and host plants. This species occurs throughout tropical Africa from 15°N to 30°S and has been recorded from Burundi, Chad, Ghana, Kenya, Mozambique, Nigeria, Rwanda, Sierra Leone, Somalia, South Africa, Tanzania, Uganda and Zaïre. Atkinson (1980) stated that it is almost certainly present in the swamp regions of north Botswana, south Angola and Zambia. Detailed studies by Atkinson (1980) in Natal, South Africa, showed that *E. saccharina* is restricted to a narrow coastal belt coinciding with the 16°C isotherm for the winter month of July.

It is found on sugar cane and maize, but also on rice and other cultivated cereals, as well as on sedges (*Cyperus* spp.), its presumed original hosts.

PHYCITINAE

Maliarpha separatella Ragonot: African white rice borer (Figs 62, 63, 78; Plate 9:2–6)

Taxonomic notes. *Maliarpha separatella* is the only important stem borer of cereals in Africa of this subfamily and is easily recognized in adult and larval stages. The taxonomic status of the species was recently discussed in a revision of the genus by Cook (1997). It has a number of synonyms, notably *Enosima vectiferella* Ragonot (= *Rhinaphe vectiferella* Ragonot) and *Anerastia pallidicosta* Hampson (= *Ampycodes pallidicosta* Hampson).

Diagnosis. Both the adults and the larvae are quite unlike any other stem borers of rice in Africa and identification of this pest is therefore relatively easy. Photographs of eggs, larvae, pupae and adults have been published by Brenière (1983) and Betbeder-Matibet (1986). Usua (1987) has published detailed technical descriptions of the final-instar larva and Tran (1977b) illustrated the male and female genitalia and wing venation. Pollet (1981) also described eggs, larvae and adults. The male genitalia (Figs 62, 63) have two large setaceous subtriangular arms as lateral extensions of the uncus. The gnathos is comparatively greatly reduced. The valves are elongate and spatulate, and the saccus is indistinct. The juxta is a round shield-like plate. The female genitalia (Fig. 78) have the papillae anales dorsally fused in a lobe. The ostium bursae is membranous with a sclerotized part in the antrum. The corpus bursae is elongate and has a small rounded signum.

Biology. Adults become sexually mature 1–2 days after emergence and, after mating, females lay seven to eight batches of about 40 overlapping eggs on the upper sides of the leaves of mature plants (Plate 9:2). The eggs are attached to the leaf surfaces by a strong cement, which dries and contracts to draw the leaf lamina over the eggs (Plate 9:3). Caterpillars hatch after about 8 days and are active in the mornings, when they disperse from plant to plant swinging on silken threads. By the end of the day, they penetrate between the leaf sheath and stem and then bore down through the sheath and enter the stem above a node (Plate 9:4). They then feed within the stem cavity and eat through the nodes to pass from one internode to the next. Larval development takes 20–70 days and larvae pupate within the stems, having first prepared an exit hole for adult emergence. It is unusual to find more than one larva in a stem and larvae generally occupy maturing stems and not tillers. Pupal development takes from 10 to 60 days and several successive generations develop during the growing season. During the dry season, final-instar larvae diapause in stems at the bases of dried stubbles and resume activity during the following rains. A complex of natural enemies, mainly parasitic Hymenoptera, regulate populations towards the end of the growing season. These species have been described in detail by Polaszek *et al.* (1994). Pollet (1981) made detailed studies of the biology and population dynamics of this species over 5 years in Ivory Coast. Using *k* factor analysis over 14 crop cycles, he found that a cytoplasmic polyhedrosis virus was an important factor regulating populations.

Brenière *et al.* (1962) published a detailed study of this species in Madagascar.

Distribution and host plants. According to Cook (1997), the confirmed distribution of this species is mainland Africa (primarily West Africa) between 20°N and 20°S. The species also occurs on the Comoro Islands, Madagascar, Mauritius and Réunion. A specimen has also been found that is labelled as having been collected in Burma and the species may also occur in China.

The species is probably restricted to cultivated and wild rice.

12 Larvae (Morphology)

L. Meijerman and S.A. Ulenberg
Institute of Systematics and Population Biology (Zoological Museum Amsterdam), University of Amsterdam, Plantage Middenlaan 64, 1018DH, the Netherlands

INTRODUCTION

The larvae of moths belonging to the families Crambidae, Pyralidae and Noctuidae of the order Lepidoptera are pests of graminaceous crops in many parts of Africa. The damage done to crops by these stalk or stem borers, as they are commonly known, can be considerable. They bore into the stems of maize, millets, sorghum, rice and sugar cane, damaging or even killing the plant. Either way, they lead to a reduction in yield. These crops are the staple food for the population in most parts of Africa, grown by smallholders to feed themselves and their families, and therefore reduction in yield affects the population directly.

In the past, control of stem borers has been based mainly on the use of insecticides. Today, control has shifted towards a more integrated approach, relying on cultural methods, development of resistant varieties and encouragement of both classical and natural biological control. The first step towards any pest-management strategy is the correct identification of the pest. Many of these stem borers are closely related and are difficult to distinguish from each other. Misidentifications of these pests have occurred frequently, resulting in the publication of false data, which are perpetuated, often for decades (Polaszek, 1992).

Species belonging to the genus *Chilo* Zincken and *Scirpophaga* Treitschke present most of the taxonomic problems at species level. Bleszynski (1970) solved most of the problems concerning the adults of *Chilo* species, as did Lewvanich (1981) for the taxonomic problems concerning the adults of *Scirpophaga* species. Keys to the adults of the African species belonging to the Noctuidae are provided by Tams and Bowden (1953) and Bowden (1956), although not all taxonomic problems within this family have been solved.

The taxonomy of the larvae was studied by Usua (1969, 1987), Mathez (1972), Tran (1977a) and Moyal and Tran (1989, 1992). None of the above-mentioned studies incorporated all African pest species. The present chapter is an account of a study of the external morphology of the larvae of African species of stem borers, with a discussion on earlier publications about this subject.

Chaetotaxy studies have been the usual way to distinguish between the larvae, and the key presented in this chapter is based mainly on differences in chaetotaxy of the various segments of the body, complemented by some notes on the arrangement of the crochets on the abdominal prolegs. Chaetotaxy of the head capsule was also studied, but, since it is not essential for the identification of the genera and did not provide information for the identification of different species within a genus, it is not incorporated in the descriptions of the larvae of the different species.

Chaetotaxy studies did not provide characters for the identification of all species of *Sesamia* and *Chilo* and therefore mouthparts, i.e. spinnerets and mandibles, were studied in *Chilo* and *Sesamia*, using a scanning electron microscope.

The following species are included in the key: *Chilo aleniellus* (Strand), *Chilo diffusilineus* (de Joannis), *Chilo orichalcociliellus* Strand, *Chilo partellus* (Swinhoe), *Chilo zacconius* Bleszynski, *Coniesta ignefusalis* (Hampson), *Scirpophaga* sp. (Crambidae), *Eldana saccharina* Walker, *Maliarpha separatella* Ragonot (Pyralidae), *Busseola fusca* (Fuller), *Sesamia calamistis* Hampson, *Sesamia cretica* Lederer, *Sesamia nonagrioides* (Lefebvre), *Sesamia penniseti* Tams & Bowden and *Sesamia poephaga* Tams & Bowden (Noctuidae).

Some remarks are made on the morphology of *Chilo agamemnon*, but this species is not incorporated in the key, since no material of this species was available and no publications of previous chaetotaxy studies have been found.

MATERIAL AND METHODS

In this chapter we focus on African species of stem borers but, when available, we also studied material of these species from localities outside Africa.

Larvae preserved in 70% alcohol were studied using a stereobinocular microscope. The drawings were made using a camera lucida. For this purpose, larvae were opened ventrally and kept in 10% potassium hydroxide (KOH) for 24 h. The cuticle was cleaned with water and alcohol. Setal nomenclature of Hinton (1946), which divides the setae into dorsal (D), subdorsal (SD), lateral (L), subventral (SV) and ventral (V) groups, was used.

To find characters to distinguish between the different species of one genus, the length of all macroscopical setae, the distance between all pairs of neighbouring setae and the angles between all three neighbouring setae were measured on all segments.

To check the identity of samples of live larvae, a number of the larvae were reared. The genitalia of the reared adults served for identification. *Chilo* species were identified using Bleszynski (1970), and *Sesamia* species were identified using Tams and Bowden (1953).

Mouthparts (i.e. spinnerets, mandibles) were studied, using a scanning electron microscope. For this purpose, they were dissected and dehydrated, using absolute alcohol.

Spinnerets and mandibles were studied in five species of *Chilo* and in *S. calamistis* and *S. nonagrioides botanephaga.*

In *Chilo* spp, the tip of the spinneret is extended into two narrow extremities. The length of these extremities varies greatly between, but also within, the species. In general, one can state that the length of the extremities was much longer in the examined larvae of *C. orichalcociliellus*, *C. aleniellus* and *C. diffusilineus* than in

C. partellus, while these extremities were extremely short in *C. zacconius*. This character can probably not be used for identification, since the length of these extremities varies greatly between, but also within, species. The mandibles of all examined species of *Chilo* have six teeth, plus a ventral inner tooth. They provided no information for the separation of the species.

In *S. calamistis* and *S. n. botanephaga*, the tip of the spinneret is rounded, without extremities. The mandibles have five teeth. No characters were found to distinguish between the two species.

The diagnostic characters mentioned in this chapter separate species (or groups of species) that have been studied. It is not known to us if these characters are diagnostic on other taxonomic levels, such as genera or subfamilies. Incorporation of other species of the studied genera into the key may mean that some of the characters will no longer be reliable for the separation of the species.

The figures show the chaetotaxy of 'normal' larvae. 'Normal' in that, within almost all species, we found larvae that had either an additional microscopical, or even macroscopical seta, or setae missing in various groups on various segments. In general, the number of microscopical setae seemed to vary more than the number of macroscopical setae. Especially larvae up to the third instar can lack some microscopical setae.

Key to Common African Stem-borer Larvae Belonging to the Genera *Chilo* Zincken, *Coniesta* Hampson, *Scirpophaga* Treitschke (Crambidae), *Eldana* Walker, *Maliarpha* Ragonot (Pyralidae), *Sesamia* Guenée and *Busseola* Thurau (Noctuidae)

Identifications based on the colour of pinacula or on the presence of asetose tubercles can only be applied when dealing with non-diapause larvae.

1. Crochets on ventral prolegs in a circular row or mesal penellipse, triordinal or biordinal (Fig. 92a–i); six macroscopic setae on prothoracic shield (Figs 83a–88a); two lateral setae on prothorax (Figs 83a–88a); one dorsal microscopic seta on metathorax (Figs 83b–88b) 2

– Crochets on ventral prolegs in mesoseries, uniordinal (Fig. 92j); five macroscopic setae on prothoracic shield (Figs 89a, 90a); one lateral seta on prothorax (Figs 89a, 90a); two dorsal microscopic setae on metathorax (Figs 89b, 90b) 9

2. Lateral pinaculum on prothorax ventral to the spiracle (Figs 83a–85a); large, dark-coloured dorsal asetose tubercle on meso- and metathorax (Figs 83b–85b) (may be divided in very young larvae); D2 smaller than seta D1 on the first to eighth abdominal segment (Figs 83c–f, 84c–f, 85c–f); SD1 smaller than seta D1 on the ninth abdominal segment (Figs 83g–85g) 3

– Lateral pinaculum on prothorax anterior or anteroventral to the spiracle (Figs 86a–88a); no large dorsal asetose tubercle on meso- and metathorax (Figs 86b–88b); D1 smaller than seta D2 on the first to eighth segment (Figs 86c–f, 87c–f, 88c–f); D1 smaller than seta SD1 on the ninth abdominal segment (Figs 86g–88g) 7

3. Either one or no lateral asetose tubercle on the first to seventh abdominal segment (Figs 83c–e, 84c–e); two subventral setae on meso- and metathorax (Figs 83b, 84b) .... *Chilo* spp. 4

– Three lateral asetose tubercles on the first to seventh abdominal segment, occasionally two on some segments (Fig 85c–e); one subventral seta on meso- and metathorax (Fig. 85b) .... *C. ignefusalis*

4. On mesothorax and usually metathorax an additional small asetose tubercle anterior to the large dorsal asetose tubercle (Fig. 84b) ...... *C. orichalcociliellus* .... *C. aleniellus*

– On meso- and metathorax no additional small dorsal asetose tubercle (Fig. 83b) 5

5. Lateral asetose tubercles on the first to seventh abdominal segment, and subventral asetose tubercles on meso- and metathorax (Figs 83b–e); four longitudinal stripes along the body, on the ninth abdominal segment. D1, D2 and SD setae of both sides of the body, usually together on one large pinaculum (Fig. 83g); dark-coloured pinacula .... *C. partellus*

– No lateral asetose tubercles on abdominal segments and subventral asetose tubercles on meso- and metathorax; five longitudinal stripes along the body; on the ninth abdominal segment both D2 seta on one pinaculum, and D1 and SD1 together on a smaller pinaculum (Fig. 84g); no dark-coloured pinacula, the larvae have a striped appearance 6

6. Crochets on ventral prolegs arranged in a complete circle (Fig. 92d) .... *C. diffusilineus*

– Crochets on ventral prolegs arranged in an incomplete circle or mesal penellipse (Fig. 92c) .... *C. zacconius*

7. Both subdorsal setae are macroscopic on the first to eighth abdominal segment, SD2 dorsal to SD1 (Fig. 87c–f); three lateral setae on the ninth abdominal segment (Fig. 87g) .... *E. saccharina*

– One macroscopic subdorsal seta (SD1) and one microscopic subdorsal seta on the first to eighth abdominal segments, SD2 ventral to SD1 (Fig. 88c–f); one or two lateral setae on the ninth abdominal segment (Figs 86g, 88g) 8

8. Larvae white; two lateral setae on the ninth abdominal segment (Fig. 88g); two subventral setae on the seventh abdominal segment (Fig. 88e); three macroscopical lateral setae on meso- and metathorax and first to eighth abdominal segment (Fig. 88b–f) .... *M. separatella*

– Larvae light brown to dark grey; one lateral seta on the ninth abdominal segment (Fig. 86g); three subventral setae on the seventh abdominal segment (Fig. 86e); two macroscopical lateral setae on meso- and metathorax and first to eighth abdominal segment (Fig. 86b–f) .... *Scirpophaga* spp.

9. The angle between setae D2, D1 and SD1 usually less than 120°, MD1 anteroventral to D1 on the ninth abdominal segment (Figs 89h, 91b); the angle between setae MD1, D1 and D2 usually less than 150° on the eighth abdominal segment (Figs 89g, 91a) 10

– The angle between setae D2, D1 and SD1 usually more than 140° on the ninth abdominal segment (Fig. 90h); the angle between setae MD1, D1 and D2 usually more than 150° on the eighth abdominal segment (Fig. 90g) ....*S. calamistis*; *S. nonagrioides*; *S. penniseti*; *S. poephaga*

10. Larvae cream to brown; the distance between the setae D2 and D1, divided by the distance between D1 and MD1, smaller on the third abdominal segment than on the seventh abdominal segment (Fig. 89e, f); usually on meso- and metathorax seta SD1 smaller than SD2 (Fig. 89b) .... *B. fusca*

– Larvae cream to pale pink; the distance between the setae D2 and D1, divided by the distance between D1 and MD1, smaller on the seventh abdominal segment than on the third abdominal segment (Fig. 90e, f); on meso- and metathorax seta SD2 smaller than SD1 (Fig. 90b) .... *S. cretica*

Crambidae

Chilo Zincken (Figs 83, 84, 92a–d)

The larvae of *Chilo* spp. are characterized by two macroscopical subventral setae on the meso- and metathorax (Fig. 83b). *Chilo* species and *C. ignefusalis* are different from other pyraloid species included in this study in the length of D1 and D2 on the first to eighth abdominal segment. In *Chilo* and *Coniesta*, D1 is longer than D2 (Figs 83c–f; 84c–f; 85c–f), while D2 is the longer seta in all other examined pyraloid species.

Chilo partellus (Swinhoe) (Figs 83, 92a)

Literature on morphology of larvae. Isaac and Rao, 1941 (as *Chilo zonellus*); Mathez, 1972; Delobel, 1975; Usua, 1987.

Diagnosis. Non-diapause larvae cream-coloured with a spotted appearance caused by large brown pinacula, four longitudinal stripes along the body. Diapause larvae either completely pale or striped. Head capsule, prothoracic shield and suranal plate brown. Spiracles oval-shaped, black. Internal tracheal system visible. Dorsal surface of the body with four reddish brown or purple longitudinal stripes. Larger number of asetose tubercles compared with other *Chilo* larvae. In addition to the pinacula-bearing setae, one large dorsal and a smaller subventral asetose tubercle on the meso- and metathorax, and lateral asetose tubercles on the first to seventh abdominal segment. Crochets on abdominal prolegs at least partly triordinal (Fig. 92a), in a complete circle, sometimes smaller towards the lateral side than towards the meson. Very young or diapause larvae also biordinal.

Host plants. Maize, sorghum, rice, sugar cane, *Eleusine coracana, Hyparrhenia rufa, Panicum maximum, Pennisetum purpureum, Rottboellia compressa, Sorghum verticilliflorum, Vossia cuspidata* (Bleszynski, 1970).

Remarks. On the ninth abdominal segment, setae D1, D2 and SD1 of both sides of the body are usually situated together on one large pinaculum, while in the other examined *Chilo* species only the two D2 setae shared one pinaculum and D1 and SD1 shared a smaller pinaculum ventral to the larger D2 pinaculum. Note that this is not a completely reliable diagnostic character, because exceptions were found, i.e. one out of 43 larvae examined from Kenya had D1 and SD1 separated from D2.

Usua (1987) mentioned, when comparing *C. partellus* and *C. orichalcociliellus*, that *C. partellus* can be recognized by having SD2 situated outside the SD1 pinaculum on the second abdominal segment. This was not the case in six out of 43 examined larvae from Kenya; thus we would not consider this character to be reliable. Usua also described a third species of *Chilo*, *Chilo* sp., which could be recognized by two macroscopical SV setae on the eighth abdominal segment, compared with *C. orichalcociliellus* and *C. partellus*, which bear only one. Occasionally, we also found two macroscopical SV setae in *C. partellus*, so this character should not be used for identification.

Chilo orichalcociliellus Strand (Figs 84, 92b)

Literature on morphology of larvae. Mathez, 1972; Delobel, 1975; Usua, 1987 (misidentification?).

Diagnosis. Non-diapause larvae cream-coloured with a spotted appearance caused by large brown pinacula, four longitudinal stripes along their body. Diapause larvae either completely pale or striped. Head capsule, prothoracic shield and suranal plate brown. Oval-shaped black spiracles, internal tracheal system visible. Dorsal surface of the body with four reddish brown or purple longitudinal stripes. Meso- and metathorax with a small asetose tubercle anterior to the large dorsal asetose tubercle (Fig. 84b). Crochets on abdominal prolegs at least partly triordinal (Fig. 92b), in a complete circle, sometimes smaller towards the lateral side than towards the meson.

Host plants. Maize, sorghum, sugar cane, *P. maximum, P. purpureum* (Mathez, 1972).

Remarks. The larvae of *C. orichalcociliellus* have already been studied by Mathez (1972), Delobel (1975) and Usua (1987). All mention that, compared with *C. partellus*, the non-diapause larvae can be recognized by the occurrence of a smaller dorsal asetose tubercle anterior to the larger asetose tubercle, which latter is present in all *Chilo* species on the meso- and metathorax.

Chilo orichalcociliellus from Nigeria was drawn by Usua with a subventral asetose tubercle on the mesothorax, which we, like Mathez, did not find on *C. orichalcociliellus*. We suspect him to have drawn *C. aleniellus*, which is a West African species that looks extremely like *C. orichalcociliellus*. *Chilo aleniellus* also has dorsal asetose tubercles on the meso- and metathorax and occasionally also subventral asetose tubercles on the same segments. However, since *C. aleniellus* does not always have this subventral asetose tubercle, this character cannot be used for separating *C. orichalcociliellus* from *C. aleniellus*.

Chilo aleniellus Strand

Literature on morphology of larvae. Tran, 1981; Moyal and Tran, 1992.

Diagnosis. Non-diapause larvae cream-coloured with a spotted appearance caused by large brown pinacula, four longitudinal stripes along their body. Diapause larvae either completely pale or striped. Head capsule, prothoracic shield and suranal plate brown. Oval-shaped black spiracles, internal tracheal system visible. Dorsal surface of the body with four reddish brown or purple longitudinal stripes. Meso- and metathorax with a small asetose tubercle anterior to the large dorsal asetose tubercle (as in Fig. 84b). Crochets on abdominal prolegs at least partly triordinal (as in Fig. 92a, b), in a complete circle, sometimes smaller towards the lateral side than towards the meson.

Host plants. Maize, rice, *Pennisetum subangustum* (Tran, 1981; Moyal and Tran, 1992).

Remarks. No stable characters have been found to separate the larvae of *C. aleniellus* from those of *C. orichalcociliellus*. Some of the larvae studied have an additional subventral asetose tubercle on the meso- and metathorax, but this cannot be used for separating the two species.

Chaetotaxy studies of *C. aleniellus* were carried out by Moyal and Tran (1992). They assumed, based on the drawings of *C. orichalcociliellus* made by Mathez (1972) and Delobel (1975), that the two species could be separated by counting the microscopical setae on the mesothorax, the third and seventh abdominal segment and the head. However, these drawings are incomplete; not all microscopical setae were drawn on these segments. We found no difference in the number of both microscopical and macroscopical setae on the larvae of the two species. In chaetotaxy drawings of *C. aleniellus* made by Moyal and Tran (1992), MV3 is drawn anterodorsal to SV1 on the ninth abdominal segment. Since we always found this seta anteroventral to SV1, we believe this to be an exception to the usual position of this seta.

Chilo zacconius Bleszynski (Fig. 92c)

Literature on morphology of larvae. Brenière, 1976; Tavakilian, 1977; Tran, 1977a, 1981.

Diagnosis. Non-diapause larvae cream-coloured with large cream-coloured or, especially on the thorax segments, light brown pinacula. Head capsule brown. Prothoracic shield and suranal plate slightly darker than the cuticle. Dorsal surface of the body with five reddish brown longitudinal stripes. Crochets on abdominal prolegs biordinal, in an incomplete circle or mesal penellipse (Fig. 92c).

Host plant. Rice (Bleszynski, 1970).

Remarks. The setal arrangement in *C. zacconius* is the same as in *C. aleniellus*, but the larvae have no additional asetose tubercles besides the large dorsal tubercle on meso- and metathorax, characteristic for *Chilo* spp. and *C. ignefusalis*. Their appearance is more striped, but one should keep in mind that usually species like *C. partellus* and *C. orichalcociliellus* and probably also *C. aleniellus* have longitudinal stripes too, and diapause larvae of these species may therefore also have a striped appearance. Also larvae that have just entered a new larval instar do not have dark pinacula. However, *C. partellus* and *C. orichalcociliellus* have four stripes, while *C. zacconius* has five stripes.

Tran (1977a, 1981) mentioned that *C. zacconius* is characterized by having the crochets on the ventral prolegs arranged in an incomplete circle, with an open space on the lateral side. This idea was adopted by Vercambre *et al.* (1990). We examined *Chilo* larvae from rice in Senegal which we have not been able to rear to adults for identification. However, electrophoretic studies showed a genetic difference between the larvae with an open space between the crochets and the larvae that had the crochets in a closed circle. Only one out of 36 larvae of *C. zacconius*, identified by a study of allozyme variation, has the crochets arranged in a closed circle. Therefore we think that this character is fairly accurate for the distinction between *C. zacconius* and *C. diffusilineus*, whose larvae are difficult to separate otherwise. Note that in other *Chilo* species occasionally young larvae and diapause larvae were observed with an open space between the crochets.

Tran (1977a) also found *C. zacconius* to bear four L setae on one joined pinaculum on the prothorax. None of the examined larvae of *C. zacconius* have the setae, referred to in this chapter as two L setae and two SV setae, on one joined pinaculum.

Chilo diffusilineus (de Joannis) (Fig. 92d)

Literature on morphology of larvae. Tran, 1977a, 1981.

Diagnosis. Non-diapause larvae cream-coloured with large cream-coloured or, especially on the thorax segments, light brown pinacula. Head capsule brown. Prothoracic shield and suranal plate slightly darker than the cuticle. Dorsal surface of the body with five reddish brown longitudinal stripes. Crochets on abdominal prolegs biordinal, in a complete circle (Fig. 92d). Can be very small towards the lateral side.

Host plants. Rice, maize, sorghum, *Panicum* sp., *Paspalum scrobiculatum, Pennisetum typhoides, Oryza longistaminata* (Bonzi, 1982a).

Remarks. The larvae of *C. diffusilineus* have, like those of *C. zacconius*, a striped appearance and no asetose tubercles besides the large dorsal asetose tubercle on meso- and metathorax. Tran (1977a) mentions that the larvae of *C. diffusilineus* bear fewer microscopic setae on the abdominal segments. However, according to Tavakilian (1977), there is no difference between the chaetotaxy of *C. diffusilineus* and *C. zacconius*. We could only study chaetotaxy in one larva; four larvae were deep-frozen to be studied using electrophoresis. There was no difference in the number of microscopic setae between this larva and the larvae of *C. zacconius*.

Chilo agamemnon Bleszynski

Literature on morphology of larvae. Mostafa (1992).

Diagnosis. No material of this species was available to us, nor did we find publications of thorough studies of chaetotaxy.

Host plants. Maize, sorghum, sugar cane, *V. cuspidata* (Bleszynski, 1970; Mostafa, 1992).

Remarks. Mostafa (1992) studied the morphology of *C. agamemnon* larvae, but did not make a complete study of the chaetotaxy. Based on his drawings we conclude that this species has the two dorsal setae on the ninth abdominal segment on one pinaculum. This would allow for separation of this species from *C. partellus*, the larvae of which usually have the D1, D2 and SD1 setae of both sides on one large pinaculum. According to Mostafa, the mandibles of this species have four teeth. We found the mandibles of all examined species of *Chilo* to have six teeth plus an inner tooth.

Coniesta ignefusalis (Hampson) (Figs 85, 92e)

Diagnosis. Non-diapause larvae cream-coloured with large dark brown pinacula. Head capsule dark brown. Prothoracic shield and suranal plate light brown. Spiracles dark brown, oval-shaped, broader than in *Chilo*. Meso- and metathorax with a large dorsal and a smaller subventral asetose tubercle and one macroscopical subventral seta (Fig. 85b). Abdomen with a large number of dark brown asetose tubercles. First to seventh abdominal segments each with three lateral dark brown asetose tubercles (Fig. 85c–e). First to eighth abdominal segments with each seta D2 smaller than seta D1 (Fig. 85c–f). Crochets on abdominal prolegs biordinal or (partly) triordinal, occasionally uniordinal, in a complete circle. Usually smaller towards the lateral side than towards the meson (Fig. 92e).

Host plants. Finger millet, maize, sorghum, sugar cane, *Andropogon gayanus, P. maximum, P. purpureum*.

Remarks. The asetose tubercles and the dark brown pinacula-bearing setae give *C. ignefusalis* a very spotted appearance. In several larvae, one of the three dark brown lateral asetose tubercles on the first to seventh abdominal segment has been observed to be absent or to be joined into one larger asetose tubercle, on one or more segments. The chaetotaxy is similar to that of *Chilo*, but *C. ignefusalis* only bears one macroscopical SV seta on the meso- and metathorax, a character that can be used to distinguish diapause larvae of *C. ignefusalis* from the larvae of *Chilo*. Furthermore, on the ninth abdominal segment, the D2 pinaculum is less broad in this species than it is in *Chilo* species. An important character to distinguish *C. ignefusalis* and *Chilo* species from other pyraloid species included in this study is the length of D1 and D2 on the first to eighth abdominal segments. In *Coniesta* and *Chilo*, D1 is longer than D2, while D2 is the longer seta in all other examined pyraloid species.

***Scirpophaga* spp.** (Figs 86, 92f)

Literature on morphology of larva. None.

Diagnosis. Creamy yellow larvae. Head capsule, prothoracic shield and spiracles slightly darker than the cuticle. Thoracic segments without ventral setae. Meso- and metathorax and first to eighth abdominal segments each with two macroscopical lateral setae (Figs 86b–f). Seventh abdominal segment with three subventral setae (Fig. 86e). Crochets arranged in a biordinal (Fig. 92f) or partly triordinal oval. Setae are small and indistinct.

Host plant. Rice.

Remarks. The *Scirpophaga* sp. examined is characterized by three subventral setae on the seventh abdominal segment. This is also the case in the Asian rice borers *Scirpophaga incertulas* (Usua, 1987) and *Scirpophaga nivella* (Isaac and Rao, 1941), so this may be characteristic for the whole genus. Also characteristic for *Scirpophaga* sp., compared with the other examined species, is that it bears only two macroscopical lateral setae on the meso- and metathorax and first to eighth abdominal segment. According to the drawings by Isaac and Rao (1941), this is also the case in *S. nivella*, at least on the abdominal segments; they did not describe the thorax segments. According to the description of *S. incertulas* by Usua (1987), this species bears three lateral setae on the meso- and metathorax and the first to eighth abdominal segments. Furthermore, on *Scirpophaga* sp. we have not been able to find the V1 setae on the thorax segments, which apparently are present in *S. incertulas*. The microscopic setae can hardly be discerned and, since they are not essential for the identification of the genus, they were not included in the description of this species.

Pyralidae

Eldana saccharina Walker (Pyralidae) (Figs 87, 92g)

Literature on morphology of larvae. Brenière, 1976; Tran, 1981; Usua, 1987; Moyal and Tran, 1989.

Diagnosis. Light brown- to dark grey-coloured larvae with brown pinacula. Covered with very small dark-coloured spots. Head capsule reddish brown to dark brown. Prothoracic shield and suranal plate brown. Spiracles black and oval-shaped. First to eighth abdominal segments each with two macroscopical subdorsal setae (Fig. 87d–g). Ninth abdominal segment with three lateral setae (Fig. 87h). Crochets on ventral prolegs arranged in ovals, triordinal (Fig. 92g).

Host plants. Rice, maize, sugar cane, *Cyperus* spp., *P. typhoides, Rottboellia exaltata, Sorghum vulgare* (Harris, 1962; Brenière, 1976).

Remarks. *Eldana saccharina* can easily be recognized by the two macroscopical subdorsal setae on the first to eighth abdominal segments, which were not observed in any other species. Furthermore, it bears three lateral setae on the ninth abdominal segment.

In all larvae examined, the crochets were arranged triordinally. Moyal and Tran (1989) described them as biordinal. They also found only one microscopical subventral seta on the prothorax, while we found two microscopical setae. Usua (1987) found only one mandibular seta, whereas we found two setae per mandible. Furthermore, he mentions the position of MXD1 behind the prothoracic shield as a character to distinguish this species from *M. separatella*. We found MXD1 also on the prothoracic shield in *E. saccharina*.

Maliarpha separatella Ragonot (Fig. 88, 92h, i)

Literature on morphology of larvae. Brenière, 1976; Tran, 1981; Usua, 1987.

Diagnosis. Slender white larvae. Head capsule light brown, prothoracic shield light yellow-brown. Spiracles light brown, oval-shaped to almost round. Ninth abdominal segment with two lateral setae, seta L2 very small (Fig. 88g). Crochets on ventral prolegs arranged in an oval, triordinal or biordinal (Fig. 92h, i).

Host plants. Cultivated and wild rices.

Remarks. On the ninth abdominal segment, this species bears two lateral setae. However, the smaller L2 setae can easily break off. In the latter case, a small puncture can still be observed. Usua (1987) did not draw two lateral setae on the ninth abdominal segment. We consider this feature to be characteristic of *M. separatella*. In material from Ivory Coast and Nigeria, the crochets were arranged triordinally. In material from Senegal and Ghana, they were arranged biordinally. In the latter case, fewer crochets were present.

Noctuidae

Busseola fusca (Fuller) (Fig. 89)

Literature on morphology of larvae. Tran, 1981; Usua, 1987; Moyal and Tran, 1989.

Diagnosis. Cream- to brown-coloured larvae with brown pinacula. Spiracles black, oval-shaped, internal tracheal system clearly visible. Head capsule, suranal plate and prothoracic shield brown. Distance between the setae D2 and D1, compared with the distance between D1 and MD1, is smaller on the third abdominal segment than on the seventh abdominal segment (Fig. 89e, f). Angle between D2, D1 and SD1 on the ninth abdominal segment always smaller than 120° (Fig. 89h). Crochets arranged in uniordinal mesoseries (Fig. 92j), can be curved to resemble a penellipse.

Host plants. Maize, sorghum, pearl millet, finger millet, sugar cane, *H. rufa, P. maximum, P. purpureum, Phragmites* sp., *R. exaltata, Sorghum arundinaceum, S. verticilliflorum* (Harris and Nwanze, 1992).

Remarks. *Busseola fusca* has a similar chaetotaxy to *Sesamia* spp. Distinguishing between the genera *Busseola* and *Sesamia* is therefore much more difficult than separating the different genera of Pyraloidea (Crambidae and Pyralidae) involved. It is not very difficult to separate *B. fusca* from the widely spread *S. calamistis*, but separating *B. fusca* from *S. cretica* can be a problem, since the chaetotaxy of the larvae of *S. cretica* is more similar to that of the larvae of *B. fusca* than it is to that of the other *Sesamia* species. The larvae of *S. cretica* are usually much paler and the internal tracheal system is not as conspicuous as it is in *B. fusca*. On the ninth abdominal segments of *B. fusca* and *S. cretica*, the angle between the setae D2, D1 and SD1 is less than 120°, while in all other examined *Sesamia* species these setae are almost in line. On the eighth abdominal segment the angle between MD1, D1 and D2 is less than 150°, while this is more in one line in all other examined *Sesamia* species.

To separate *B. fusca* from *S. cretica*, one can examine the length of the subdorsal setae on the meso- and metathorax. In *B. fusca*, SD2 is usually the longer seta on the mesothorax and very often also on the metathorax. In only one examined larva was SD1 found to be longer than SD2. The longer seta is SD1 on both segments of all examined larvae of *S. cretica.* This character showed more variation in *S. calamistis.*

Only one character was found to be characteristic for *B. fusca*: the distance between the setae D2 and D1, divided by the distance between D1 and MD1, is smaller on the third abdominal segment than on the seventh abdominal segment (Fig. 89e, f), whereas in *Sesamia* this ratio is smaller on the seventh abdominal segment. Occasionally, there was no difference between the two segments. The distance between the setae was examined at a magnification of ×54. The ratio was rounded off to one decimal place; only one exception in *Sesamia* was found after taking the second decimal place into account.

***Sesamia* Guenée (Figs 90, 91, 92j)**

All examined *Sesamia* species are characterized by having the distance between the setae D2 and D1, compared with the distance between D1 and MD1, smaller on the seventh abdominal segment than on the third abdominal segment (Fig. 90d, e). Occasionally there was no difference between the two segments. Although we consider the position of seta L2 on the seventh abdominal segment as drawn in Fig. 90f to be the normal position of this seta, we frequently examined larvae that had this seta situated posterior to the spiracle, as it is on the eighth abdominal segment, usually on just one side of the body.

No reliable characters have been found to distinguish between *S. calamistis*, *S. penniseti*, *S. poephaga* and *S. nonagrioides.*

Note: In this chapter, *Sesamia botanephaga* Tams & Bowden will be referred to as a subspecies of *S. nonagrioides*, after Nye (1960). Usua (1969) and Tran (1981) treated *S. botanephaga* Tams & Bowden from West Africa as a species. Therefore, when referring to their articles, *S. botanephaga* will be substituted for *S. n. botanephaga.*

Sesamia calamistis Hampson (Figs 90, 92j)

Literature on morphology of larvae. Williams, 1953; Usua, 1969, 1987; Mathez, 1972; Brenière, 1976; Tran, 1981; Moyal and Tran, 1989.

Diagnosis. Pink-coloured larvae with small brown pinacula. Spiracles black and oval-shaped, internal tracheal system visible, but not as clearly as in *B. fusca.* Head capsule and prothoracic shield brown. Suranal plate yellow-brown. Angle between the setae D2, D1 and SD1 on the ninth abdominal segment more than 140°. Crochets arranged in uniordinal mesoseries (Fig. 92j), can be curved to resemble a penellipse.

Host plants. Maize, sorghum, rice, millet, sugar cane, *P. maximum, P. purpureum, R. exaltata, Setaria splendida, S. arundinaceum, Tripsacum laxum* (Tams and Bowden, 1953; Harris, 1962).

Remarks. *Sesamia calamistis* is easily separated from *S. cretica* and *B. fusca* by the difference in the angle between the setae D2, D1 and SD1 on the ninth abdominal segment, which is more than 140° in *S. calamistis*. No characters have been found to distinguish between the larvae of *S. calamistis* and *S. n. botanephaga*, *S. poephaga* and *S. penniseti*.

Usua (1987) and Moyal and Tran (1989) compared *S. calamistis* with *B. fusca*. Moyal and Tran (1989) mention the occurrence of a small puncture on the L pinaculum on the prothorax of *B. fusca*, which they did not find on *S. calamistis*. Williams (1953) did find this puncture on *S. calamistis* when he studied this species. Moyal and Tran (1989) assumed that this was a result of geographical variation within the species, since they studied larvae of *S. calamistis* from Ivory Coast and Williams (1953) studied larvae from Madagascar. The material that we studied from Madagascar and West Africa supported their idea. Usua (1987) mentions that the microscopical seta MD1 on the ninth abdominal segment is situated anteroventral to seta D1 in *B. fusca* and anterodorsal to seta D1 in *S. calamistis*. We occasionally found them anteroventral to D1 in *S. calamistis* too.

Usua (1969) found *S. calamistis* to have the two dorsal setae on a joined pinaculum on the mesothorax. This enabled him to separate this species from *S. n. botanephaga*, which has these setae arranged on separated pinacula. Furthermore, he found that in *S. calamistis* SD1 is longer than SD2 on the meso- and metathorax, which enabled him to separate this species from both *S. penniseti* and *S. n. botanephaga*. We found both characters to vary within samples of *S. calamistis* and, although the greater part of the larvae matched the description by Usua (1969), we do not consider these characters to be reliable.

Sesamia cretica Lederer (Fig. 91)

Diagnosis. Crochets and colour of cuticle and pinacula similar to those in *S. calamistis*, can be somewhat paler. Mesothorax and usually also metathorax with SD1 longer than SD2. Distance between the setae D2 and D1, divided by the distance between D1 and MD1, smaller on the seventh abdominal segment than on the third abdominal segment. Angle between the setae MD1, D1 and D2 on the eighth abdominal segment less than 150°. Angle between the setae D2, D1 and SD1 on the ninth abdominal segment less than 120° (Fig. 91b).

Host plants. Sorghum, maize, *Pennisetum typhoideum, Sorghum halepense* (Tams and Bowden, 1953).

Remarks. *Sesamia cretica* can be distinguished from all other examined species of *Sesamia* by the angle between the setae D2, D1 and SD1 on the ninth abdominal segment, which is less than 120° in *S. cretica*, while in all other examined *Sesamia* species these setae are almost in line (Fig. 90h). Furthermore, the angle between MD1, D1 and D2 on the eighth abdominal segment is less than 150° in *S. cretica*, while this is more in all other examined *Sesamia* species.

These characters, however, do not provide information for the separation of *S. cretica* from *B. fusca*. The most obvious difference between these two species is probably the colour of the larvae, *B. fusca* generally being much darker and having a more clearly visible internal tracheal system. However, there are small differences in the chaetotaxy of the third and seventh abdominal segment. The distance between the setae D2 and D1, divided by the distance between D1 and MD1, is smaller on the seventh abdominal segment than on the third abdominal segment in *S. cretica*. Furthermore, SD1 is usually longer than SD2 on the meso- and metathorax. On the eighth abdominal segment, SD1 is usually situated anterodorsal to the top of the spiracle in *S. cretica*, while this seta is situated slightly ventral to the top of the spiracle in most individuals of *B. fusca*. Note that this is not a reliable character.

Sesamia penniseti Tams & Bowden

Literature on morphology of larvae. Usua (1969).

Host plants. Maize, millet, sugar cane, *P. maximum, P. purpureum, S. splendida, S. vulgare* (Harris, 1962).

Remarks. No material of *S. penniseti* was available for our study. On the basis of drawings of Usua (1969), we conclude this species to have the same chaetotaxy as *S. calamistis, S. nonagrioides* and *S. poephaga*. Usua found a small difference in the chaetotaxy of the head capsule, which we did not find when comparing *S. poephaga*, a species closely related to *S. penniseti* (Tams and Bowden, 1953) with *S. calamistis* and *S. nonagrioides botanephaga*. Unfortunately, this character could not be examined in *S. penniseti*.

Sesamia poephaga Tams & Bowden

Diagnosis. Crochets, colour of cuticle and pinacula, and chaetotaxy similar to those of *S. calamistis*.

Host plants. Maize, sorghum, sugar cane, *P. maximum, P. purpureum* (Harris, 1962).

Remarks. No reliable characters have been found to separate this species from *S. calamistis, S. nonagrioides* and *S. penniseti*. On the ninth abdominal segment, we found the majority of the larvae to have the microscopic MD1 seta anteroventral to D1, while, in the three above-mentioned *Sesamia* species, this seta is situated anterodorsal to D1 in most individuals. Since both variations occur in all *Sesamia* species, with the exception of *S. cretica*, in which it was always found antero-ventral, this character cannot be used for identification.

Sesamia nonagrioides (Lefebvre)
Sesamia nonagrioides botanephaga Tams & Bowden

Literature on morphology of larvae. Usua, 1969; Tran, 1981. (Both authors treat *S. n. botanephaga* Tams and Bowden from West Africa as a species.)

Diagnosis. Colour and chaetotaxy similar to those of *S. calamistis*.

Host plants. Maize, rice, sorghum, sugar cane, *Chasmopodium afzelii, P. purpureum, R. exaltata, S. arundicaceum* (Tams and Bowden, 1953).

Remarks. No reliable characters have been found to distinguish *S. nonagrioides* from *S. calamistis*, *S. poephaga* and *S. penniseti*.

Of the African subspecies *S. n. botanephaga*, no reliable alcohol-preserved material was available. Only reliable deep-frozen material – collected for electrophoretic studies – was available, which is inappropriate for thorough chaetotaxy studies.

Tran (1981) separated this species from *S. calamistis* by the colour of the cuticle. We doubt the reliability of this character. According to the differences in coloration, as suggested by Tran, we tried to separate the larvae that we had collected in Cape Coast, Ghana, before rearing. After hatching of the moths, it appeared that we had not separated the two species correctly. Separation on minor colour differences is subjective and diet can influence the colour of the larvae.

Usua (1969) found SD1 longer than SD2 on the metathorax and shorter on the mesothorax in *S. n. botanephaga*, while in *S. penniseti* SD1 is the shorter seta on the metathorax and the longer on the mesothorax. We examined one larva which we consider to be *S. n. botanephaga* that had SD1 shorter on both segments.

Unfortunately, the SD setae were frequently missing in the examined alcohol material, and always missing in the reliable deep-frozen material, because they had broken off. However, in all other examined noctuid species, the length of the SD setae was, although reasonably stable on the mesothorax, less stable on the metathorax. Therefore we think one should not rely on this character for identification.

13 Larvae (Electrophoresis)

L. Meijerman, W.E. van Ginkel and
S.A. Ulenberg
Institute of Systematics and Population Biology (Zoological Museum Amsterdam), University of Amsterdam, Plantage Middenlaan 64, 1018DH, the Netherlands

INTRODUCTION

The larvae of some species of *Chilo* and especially of *Sesamia* are morphologically difficult to separate or even inseparable (see Chapter 12, this volume). In order to find characters for the separation of the species, allozyme variation was studied. Electrophoretic variation has proved to be useful for identification of species. Menken and Ulenberg (1987) give a list of relevant literature up to 1986 on the use of biochemical characters in agricultural entomology.

Starch-gel electrophoresis was carried out in order to search for diagnostic allozyme characters in *Chilo* and *Sesamia*. Five species of *Chilo* were studied: *C. diffusilineus* (J. de Joannis), *C. partellus* (Swinhoe), *C. orichalcociliellus* (Strand), *C. zacconius* Bleszynski, and *C. aleniellus* (Strand). Since only one larva of *C. aleniellus* was available, this species is not included in the biochemical key, but is included for comparison with the closely related species *C. orichalcociliellus*. Two species of *Sesamia* were studied: *S. calamistis* Hampson and *S. nonagrioides botanephaga* Tams & Bowden.

MATERIAL AND METHODS

Individual larvae analysed electrophoretically originated from the following locations (the number of larvae examined is given in parentheses):

Chilo diffusilineus	Senegal: Zinguinchor (4)
Chilo partellus	South Africa: Pretoria (5), Mount Edgecombe (1) Malawi: Thondwe (1)
Chilo orichalcociliellus	Kenya: Nairobi (12)
Chilo zacconius	Senegal: Zinguinchor (36)
Chilo aleniellus	Nigeria: Owo (1)
Sesamia nonagrioides	Ghana: Abure Dunkwe (1), Kumasi (1), Cape Coast (1)
Sesamia calamistis	Ghana: Cape Coast (1) South Africa: Mount Edgecombe (2); Nigeria: Akure (2)

Table 13.1. Enzyme essays of *Chilo* and *Sesamia* species. See text for composition of buffers I–IV.

Enzyme system	EC number	Locus abbreviation	Buffer system
Oxidoreductases			
Glyceraldehyde-3-phosphate dehydrogenase	1.2.1.12	*Gapdh*	II
Glycerol-3-phosphate dehydrogenase	1.1.1.8	*Gpd*	III
Hydroxybutyrate dehydrogenase 1	1.1.1.30	*Hbdh* 1	I + IV
Hydroxybutyrate dehydrogenase 2	1.1.1.30	*Hbdh* 2	I + IV
Isocitrate dehydrogenase	1.1.1.42	*Idh*	I
Lactate dehydrogenase	1.1.1.28	*Ldh*	II
Malic enzyme	1.1.1.40	*Me*	I
Transferases			
Hexokinase	2.7.1.1	*Hk*	III
Glutamic–oxaloacetic transaminase 1	2.6.1.1	*Got* 1	II
Glutamic–oxaloacetic transaminase 2	2.6.1.1	*Got* 2	II
Hydrolases			
Esterase	3.1.1.2	*Est*	I + II
Lyases			
Fructose-biphosphate aldolase	4.1.2.13	*Ald*	II
Isomerases			
Glucose phosphate isomerase	5.3.1.9	*Gpi*	II
Mannose-6-phosphate isomerase	5.3.1.8	*Mpi*	II
Phosphoglucomutase	5.4.2.2	*Pgm*	III

In *Chilo* 13 loci and in *Sesamia* 11 loci coding for soluble enzymes were examined. The four buffer systems used for the enzyme assays (listed in Table 13.1) are as follows:

I Electrode buffer: 0.233 M tris and 0.086 M citric acid
  Gel buffer: 0.008 M tris and 0.003 M citric acid, pH 6.7
II Electrode buffer: 0.687 M tris and 0.157 M citric acid, pH 8.0
  Gel buffer: 0.023 M tris and 0.005 M citric acid
III Electrode buffer: 0.216 M tris and 0.1 M maleic acid, pH 7.4
  Gel buffer: 0.011 M tris and 0.005 M maleic acid
IV Electrode buffer: 0.04 M citric acid, adjusted to pH 6.1 with 3-morpholinopropylamine
  Gel buffer: 0.002 M citric acid, adjusted to pH 6.1 with 3-morphelinopropylamine

The larvae were ground in a homogenization buffer (0.01 M tris, 0.01 M maleic acid, 0.001 M ethylenediamine tetra-acetic acid (EDTA) and 0.001 M magnesium chloride, 0.05 mM NADP and pH adjusted to 7.4 with NaOH). The homogenate was run on a starch gel. Practical details of electrophoretic techniques can be found in Hillis and Moritz (1990). Standard staining methods were applied; information can be obtained from the second author.

RESULTS AND DISCUSSION

Chilo

Of most species only a few larvae were available. Furthermore, the larvae of *C. orichalcociliellus* all came from one culture and for most other species all larvae came from one location. Therefore potential (geographical) variation could not be studied, which might affect the reliability of the biochemical key presented. However, we believe it can serve as a basis for further research.

According to this preliminary study, six loci of four enzyme systems (namely, *Ldh, Got* 1, *Got* 2, *Gpi, Hbdh* 1, *Hbdh* 2) are the most useful for the separation of the species. Kioko *et al.* (1995a) also found *Ldh, Got* and *Gpi* to be diagnostic loci for the separation of *C. partellus* and *C. orichalcociliellus*. Locus *Ldh* separated *C. partellus* from *C. orichalcociliellus, C. diffusilineus, C. aleniellus* and *C. zacconius*, between which we found no variation.

Locus *Got* 1 separated *C. zacconius* from *C. orichalcociliellus, C. partellus* and *C. diffusilineus*, between which there were only slight differences.

Locus *Got* 2 showed very little variation. It allows for separation of *C. orichalcociliellus* from all other species treated, because it showed only very low activity in *C. orichalcociliellus*. Kioko *et al.* (1995a) did not find expression of this locus in *C. orichalcociliellus*.

Locus *Gpi* separated *C. partellus* and *C. zacconius* from *C. orichalcociliellus* and *C. diffusilineus*. Variation in this enzyme can probably also be used to separate *C. orichalcociliellus* from *C. aleniellus*.

Locus *Hbdh* allowed for separation of *C. zacconius* from all other species treated here, because of strong expression of what we consider to be a second locus: *Hbdh* 2. Expression of this locus was stronger than that of *Hbdh* 1 in this species, which led to clear bands close to the origin (just above the origin on buffer system I and just below the origin on buffer system IV).

No variation or a broad overlap was found for the following enzymes: *Pgm, Est, Mpi, Me, Hk, Gpd* and *Ald*. Nine more enzymes gave no activity at all or demonstrated uninterpretable patterns, namely, malate dehydrogenase, carbonate dehydratase, β-glucosidase, 6-phosphogluconate dehydrogenase, fucose dehydrogenase, glutamate pyruvate transaminase, alcohol dehydrogenase, arginine kinase and nicotinamide adenine dinucleotide (NADH) dehydrogenase.

In Table 13.2, the differences in migration distance for the loci coding for *Ldh, Got* 1, *Got* 2, *Gpi, Hbdh* 1, *Hbdh* 2 and *Idh* between the species are given. At each locus, the most common electromorph of *C. partellus* is designated as 100. The electromorphs of all other examined species are numbered according to their differences in migration distance from this standard in millimetres. For *Hbdh* 2 in *C. zacconius*, the difference in migration distance is given between *Hbdh* 1 and *Hbdh* 2, since we found no expression of this locus in the standard.

Table 13.2. Electromorphs in *Chilo; Chilo partellus* is used as standard.

Enzyme system	*C. partellus*	*C. orichalcociliellus*	*C. zacconius*	*C. diffusilineus*	*C. aleniellus*
Ldh	100	110	110	110	110
Got 1	100	99	90	98, 121	*
Got 2	100, 95	95†	95	95	*
Gpi	100, 95	108, 111, 117	99	107, 113	91, 105
Idh	100, 105	105	100	100	100
Hbdh 1	100, 91, 85	91	100	91, 84	85
Hbdh 2	‡	‡	90	‡	‡

* No band formed, characterized by smearing.
† Activity was very low.
‡ No expression.
For *Hbdh* 2 the difference in migration distance between *Hbdh* 1 and *Hbdh* 2 on buffer system I is given.

Biochemical Key to Four Species of *Chilo; C. partellus* is Used as Standard

1. Hydroxybutyrate dehydrogenase: no expression of a second locus
Glucose phosphate isomerase: notably faster than standard
Glutamic–oxaloacetic transaminase 1: mobility same, faster or somewhat slower than standard 2

– Hydroxybutyrate dehydrogenase: strong expression of a second locus close to the origin
Glucose phosphate isomerase: mobility somewhat slower or somewhat faster than standard
Glutamic–oxaloacetic transaminase 1: mobility notably slower than standard .... *C. zacconius*

2. Glutamic–oxaloacetic transaminase 2: only very low activity, mobility same or slower than standard
Isocitrate dehydrogenase: mobility same or faster than standard .... *C. orichalcociliellus*

Glutamic–oxaloacetic transaminase 2: strong activity, mobility same or slower than standard
Isocitrate dehydrogenase: mobility same or slower than standard .... *C. diffusilineus*

Sesamia

Loci *Est, Hbdh* and *Idh* seem to be diagnostic. However, since so little material was available, potential variation could not be studied. Loci coding for *Gapdh, Got* (two loci), *Pgm, Hk, Gpd* and *Gpi* were the same in the two species or a broad overlap occurred. Ten more enzymes gave no activity at all or demonstrated uninterpretable patterns, namely, lactate dehydrogenase, mannose-6-phosphate isomerase, carbonate dehydratase, 6-phosphogluconate dehydrogenase, aldolase, NADH dehydrogenase, malate dehydrogenase, glucose-6-phosphate dehydrogenase, fucose dehydrogenase and alcohol dehydrogenase.

Table 13.3. Electromorphs in *Sesamia; S. calamistis* is used as standard.

	S. calamistis	*S. nonagrioides*
Est	100	108
Hbdh	100	106, 112
Idh	100	98, 104

In Table 13.3 the differences in migration distance for the loci coding for *Est, Hbdh* and *Idh* between the two species are given. At each locus, the electromorph of *S. calamistis* is designated as 100. The electromorphs of S. *nonagrioides* are numbered according to their differences in migration distance from this standard in millimetres.

CONCLUSION

Although interspecific variation could not be studied, due to a lack of material, we conclude that a study of allozyme variation can provide diagnostic characters for the identification of *Sesamia* spp. and *Chilo* spp. Since many larvae can be studied at the same time, this technique may prove to be less time-consuming than rearing and identifying adults. Especially when monitoring populations of morphologically inseparable stem-borer larvae, a study of allozyme variation is recommended.

III Natural Enemies

14 Hymenoptera: Introduction

G. Delvare[1] and A. Polaszek[2]
[1]*CIRAD-CA, Laboratoire de Faunistique, BP 5035, 34032 Montpellier, France;* [2]*CABI BIOSCIENCE UK Centre, c/o Entomology Department, The Natural History Museum, London SW7 5BD, UK*

INTRODUCTION

The Hymenoptera are among the most important orders of insects, coming just behind the Coleoptera in numbers of described species. However, recent research in tropical humid areas has revealed enormous numbers of as yet undescribed species, especially among the parasitoid Hymenoptera.

Hymenoptera are essentially terrestrial animals, the nutritional requirements of which differ from group to group. The most primitive Hymenoptera, the Symphyta, are phytophagous (phyllophages, stem borers or xylophages). Only the Orussidae may be the exception to this rule. They therefore include several pest species, especially in temperate regions. The Apocrita contain some phytophagous taxa, but the majority are parasitoids or predators of other insects. Finally, most Apoidea are nectar feeders.

Many Hymenoptera consequently play a considerable role in the maintenance of equilibrium in nature, and some of them have been used successfully in biological control against pests of cultivated plants. Some species, such as trichogrammatids, are produced for this end under industrial conditions.

Hymenoptera are holometabolous insects, having complete metamorphosis and a larval stage that differs greatly from the adult. In fact, the hymenopteran larva has a completely different life from that of the adult. The adult, or imago, is principally concerned with the preparation of the existence of the larva: mating and reproduction, searching for the host or collecting pollen, in some cases, nest-building. The role of the male in these activities is limited solely to coupling and insemination of the female.

On the subject of reproduction, it is worth noting a peculiarity of this order: sex determination is unlike that of any other insects. In Hymenoptera, males are haploid while females are diploid. In those species displaying arrhenotokous parthenogenesis, which are the most common, unfertilized females produce males only. Species having thelytokous parthenogenesis produce only females. Finally, there are a few species, called deuterotokous, in which unfertilized females produce females almost exclusively, but also produce the occasional male.

Hymenoptera larvae differ morphologically according to the group to which they belong. Symphyta larvae are of the eruciform type. They can be relatively easily distinguished from caterpillars (Lepidoptera larvae) by the presence of six to eight pairs of prolegs (five pairs or fewer among Lepidoptera). Apocritan larvae are generally vermiform and apodous, although certain first-instar larvae display specialized morphologies.

ADULT MORPHOLOGY

The hymenopteran head is usually orthognathous (hypognathous), very occasionally prognathous. The mouthparts of most species are of the simple mandibulate type. In most bees, the mouthparts have become somewhat specialized and elongated for taking nectar. The number of antennal segments is very variable. In those species having a reduced number of antennal segments (fewer than 16), the antenna is often elbowed between the scape (or scapus) and the pedicel (or pedicellus) and the scape is elongated (Fig. 93).

Regarding the thorax, in the Apocrita a unique adaptation has taken place concerning the position of the first abdominal segment. This has become fused with the thorax, of which it has become an integral part, and is known as the propodeum. The second abdominal segment is often very narrow and transformed into a petiole, which enables the abdomen to be orientated in virtually all directions. In the Symphyta the first abdominal segment is in its normal position, and there is no constriction between the propodeum and the remaining abdominal segments. In order to maintain the distinction between the true thorax and the apparent thorax in Apocrita (i.e. including the propodeum), the term mesosoma is used for thorax + propodeum. The remainder of the abdomen is therefore called the metasoma, or often the gaster.

The pronotum appears transverse in dorsal view. It is well developed laterally (Figs 95–98). The well-developed mesonotum is divided into an anterior mesoscutum and a posterior scutellum (or mesoscutellum) (Figs 94, 95, 97). The mesoscutum often has a pair of posteriorly convergent impressed lines, the notauli, dividing the mesoscutum into two side lobes and a midlobe. Lateral to the notauli may be found a second pair of impressed marks, the parapsidal lines. Lateral to the scutellum are the axillae. The mesopleuron is well developed and fused with the mesosternum, the latter being barely visible or completely invisible. A triangular or subrectangular sclerite, the prepectus, is sometimes visible, especially in the Chalcidoidea, below the first thoracic spiracle (Fig. 98). In some families it has become secondarily fused with the pronotum. In these cases, the lateral part of the pronotum reaches the tegula. The metanotum is transverse and reduced.

The Hymenoptera normally possess four membranous wings. Occasionally these may be reduced or secondarily absent (e.g. female Mutillidae, workers of Formicidae). The fore wings are coupled together with the hind wings by hamuli on the leading edge of the hind wing (Figs 99, 100). The hind wing is always smaller than the fore wing. At the base of the hind wing, the presence of a jugal lobe (Fig. 101) can often be useful for distinguishing Aculeata (in which it is present) from Parasitica (in which it is almost always absent). The wings of Hymenoptera generally present a system of longitudinal and transverse veins, enclosing a very variable number of discrete areas, the cells. About two-thirds along the leading edge of the fore wing, a large, often pigmented, non-cellular area may be present, the pterostigma (Figs 99, 100). In Symphyta the venation is at its most complete among the Hymenoptera, and the complex system for naming the veins and cells is shown in Figs 99 and 100. When identifying stem-borer parasitoids using the keys in the body of the text, most of the morphological terms, especially those relating to wing venation, should be self-explanatory, using the figures provided with the keys. In many Parasitica the wing venation is extremely reduced, and a slightly different, simpler, nomenclature is adopted (Fig. 102).

The ovipositor is of a relatively primitive type, but the valves are well developed. In Symphyta and most Parasitica they are visible in ventral view. Some Parasitica (Proctotrupoidea, Scelionoidea and Ceraphronoidea) resemble Aculeata in having the valves retracted into the interior of the metasoma, able to extend out of its extremity. In aculeates, the valves of the female are not used during the act of oviposition itself, but have been transformed into a sting, either for paralysing prey or for defending the colony.

CLASSIFICATION OF HYMENOPTERA

The Hymenoptera have traditionally been divided into Symphyta and Apocrita, the latter category being further divided into Parasitica and Aculeata. Recent authors who have studied the evolution of the order (Königsmann, 1977, 1978a, b; Rasnitsyn, 1980, 1988) have arrived at different conclusions concerning which divisions should be adopted. Gibson (1985), in a study of the morphology of the hymenopteran thorax, has provided some evidence to support the existing classification. We have adopted here an original classification that largely follows currently accepted systems (Table 14.1) (Delvare and Aberlenc, 1989).

SPECIMEN PREPARATION: MALE GENITALIA

In several cases, parasitoids of African stem borers can be reliably identified to species level solely by examination of the male genitalia. This is the case in the following genera or families: *Cotesia* (Braconidae), *Telenomus* (Scelionidae), *Aphanogmus* (Ceraphronidae) and most Trichogrammatidae. The following procedure, from Polaszek and Kimani (1990), describes the process of making a slide preparation of male genitalia. This is always preceded by card-mounting, a technique appropriate for the correct mounting of both male and female specimens of small parasitoids.

We recommend that (where possible) a reasonably long series (10–20 individuals) of specimens, containing both sexes and reared from a reliably identified host, is used. In a number of cases, closely related hosts are found sympatrically, often with two or more closely related parasitoid species, and for this reason it is extremely important that cocoon clusters (*Aphanogmus*, *Cotesia*) or egg masses (*Telenomus*, Trichogrammatidae) collected should be kept individually, separate from other egg masses. Surveys of stem-borer species present in the area studied should also be carried out by rearing through larvae.

Table 14.1. Classification of Hymenoptera.

Suborders	Infraorders	Superfamilies	Families	Subfamilies
Symphyta		Xyeloidea	Xyelidae	
		Megalodontoidea	Pamphilidae Megalodontidae	
		Tenthredinoidea	Pergidae Argidae Cimbicidae Diprionidae Tenthredinidae Blasticotomidae	
		Cephoidea	Cephidae	
		Siricoidea	Syntexidae Siricidae Xiphydriidae	
		Orussoidea	Orussidae	
		Stephanoidea	Stephanidae	
Apocrita	Parasitica (= Terebrantia)	Gasteruptioidea	Gasteruptiidae Aulacidae	
		Ceraphronoidea	Ceraphronidae Megaspilidae	
		Chalcidoidea	Leucospidae Chalcididae Agaonidae Torymidae Ormyridae Perilampidae Eucharitidae Eurytomidae Pteromalidae Eupelmidae Tanaostigmatidae Encyrtidae Tetracampidae Eulophidae Elasmidae Aphelinidae Signiphoridae (= Thysanidae) Trichogrammatidae Mymaridae Rotoitidae	
		Mymarommatoidea	Mymarommatidae	
		Ichneumonoidea	Ichneumonidae Braconidae	
		Trigonalyoidea	Trigonalyidae	
		Cynipoidea	Ibaliidae Liopteridae Eucoilidae Figitidae Cynipidae Charipidae	
		Proctotrupoidea	Proctotrupidae Pelecinidae Heloridae	
		Evanioidea	Evaniidae	
		Scelionoidea	Scelionidae Platygastridae	

Table 14.1. Classification of Hymenoptera (*continued*).

Suborders	Infraorders	Superfamilies	Families	Subfamilies
Apocrita (*continued*)	*Parasitica* (= Terebrantia) (*continued*)	(*incertae sedis*)	Monomachidae Austroniidae Roproniidae Peradeniidae Diapriidae Megalyridae	
	Aculeata	Chrysidoidea	Plumariidae Scolebythidae Bethylidae Sclerogibbidae Embolemidae Dryinidae Chrysididae	
			Tiphiidae	Anthoboscinae Thynninae Myzininae Methochinae Tiphiinae Brachycistidinae
			Sapygidae Mutillidae	
			Scoliidae	
		Vespoidea	Formicidae	Formicinae Dolichoderinae Pseudomyrmecinae Myrmicinae Cerapachyinae Ponerinae Dorylinae
			Pompilidae Rhopalosomatidae	
			Vespidae	Euparagiinae Masarinae Eumeninae Stenogastrinae Vespinae Polistinae
		Apoidea	Sphecidae	Sphecinae Ampulicinae Larrinae Mellininae Pemphredoninae Astatinae Philanthinae Nyssoninae
			Colletidae Andrenidae Oxaeidae Halictidae Melittidae Megachilidae Anthophoridae Apidae	
		(*incertae sedis*)	Sierolomorphidae	
			Bradynobaeniidae	Typhoctinae Chyphotinae Apterogyninae Bradynobaeninae

All emergent specimens should be carefully examined for the possible occurrence of two or more species emerging from a single cocoon cluster or egg mass. We have seen a number of samples containing more than one species, *apparently* reared from the same host egg mass. However, it is not always clear from collectors' data whether this was actually the case. In some instances, a few individuals, sometimes of only one sex, of a second species are present in samples consisting otherwise of a single species.

Card-mounting

Select at least five individuals of each sex. The sexes can be easily distinguished as follows:

Cotesia, Trichogrammatidae	Females with ovipositor clearly visible; males usually with genitalia visible. Antennae of *Cotesia* males are much longer than those of the females.
Aphanogmus, *Telenomus*	In these genera the ovipositor is concealed (see above). Sexes can be differentiated as follows: *Telenomus*: males with 12-segmented moniliform antennae; females with 11-segmented clavate antennae. *Aphanogmus*: males with 11-segmented moniliform antennae; females with 10-segmented clavate antennae.

Specimens will normally have been collected into alcohol and should be first carefully dried on porous white card, before being mounted individually on small card rectangles. Each specimen should be mounted longitudinally, tilted slightly on its side at about a 45° angle, as recommended by Noyes (1982) for chalcid wasps. Each specimen should, at this stage, be labelled with all the relevant data.

Slide-mounting of Male Genitalia

We strongly recommend using Canada balsam mounting medium in preference to any of the water-soluble mountants.

1. Remove the metasoma from some or all of the card-mounted male specimens. Ensure at this stage that the metasomata are kept separate from each other, giving each a number associating it with the original specimens. This is especially necessary if the series turns out to be a mixture of two or more species.
2. Macerate the metasoma in hot potassium hydroxide (KOH) for 5 min and then transfer to a glass cavity block. Remove the KOH and add a few drops of glacial acetic acid.
3. After a few minutes, remove glacial acetic acid, add distilled water, leave for 10 min.
4. Add an equal volume of 70% alcohol, leave for 4 min.
5. Remove liquid, add 70% alcohol, leave for 4 min.
6. Add an equal volume of absolute alcohol, leave for 4 min.
7. Remove liquid, add absolute alcohol, leave for 4 min.

8. Add 5–10 drops clove oil. Leave until most of the alcohol has evaporated.
9. Transfer the metasoma from the cavity block to a small drop of fresh Canada balsam on a slide.
10. Carefully dissect out male genitalia, including the basal ring (see Fig. 103). This can be done in a number of ways, most of which usually result in the metasoma itself being destroyed. With practice, however, it is possible to remove all the genitalia intact, simply by using the viscosity of the balsam to draw the genitalia carefully out of the metasoma without damaging them. The undamaged metasoma can then be washed in alcohol and remounted with the original specimen.
11. Arrange the genitalia in the balsam, trying to keep the basal ring attached, and add a small, clean coverslip.
12. Label slide with all relevant data.

15 Hymenoptera: Key to the Families

A. Polaszek
CABI Bioscience UK Centre, c/o Entomology Department, The Natural History Museum, London SW7 5BD, UK

INTRODUCTION

The following key is designed to be used for specimens examined with a stereo-binocular ('dissecting') microscope at magnifications of up to at least ×60. For the correct identification of some families, it may be necessary to make slide preparations of certain parts of the body for subsequent examination with a compound microscope at higher magnifications. The range of morphological diversity of stem-borer parasitoids extends from relatively large species (up to 20 mm), such as some Ichneumonidae, down to Trichogrammatidae (less than 1 mm), and includes groups from the section Aculeata (Bethylidae) and Parasitica (remaining families). For specimen preparation and morphological terminology the user is therefore referred to Gauld and Bolton (1996), Borror *et al.* (1989) or Goulet and Huber (1993), which are also recommended for a detailed taxonomic treatment of the entire Hymenoptera.

Detailed treatments of the genera and, in some cases, species within each family are given in the chapters following this key, listed alphabetically by family. The families belonging to the superfamily Chalcidoidea (Chalcididae, Elasmidae, Encyrtidae, Eulophidae, Eupelmidae, Eurytomidae, Pteromalidae, Trichogrammatidae) are treated in a single chapter under Chalcidoidea (Chapter 19).

This key is designed *exclusively* for the identification of specimens which are known to have been reared as primary or secondary (hyper)parasitoids of cereal stem borers in Africa.

Key

1. Fore wing without closed cells, or brachypterous (Figs 104–108) 2
– Fore wing with at least one closed cell (Figs 109–118) 12
2. Tarsi three-segmented; antennae always with fewer than eight segments (excluding anelli, Figs 121, 122, 123). Fore wing resembling Fig. 104 ....(Chalcidoidea) Trichogrammatidae
– Tarsi four- or five-segmented; antennae usually with more than eight segments 3

3. Pronotum reaching tegulae (Fig. 124); prepectus ('pr' in Figs 119, 120) not apparent 4
– Pronotum not reaching tegulae, prepectus clearly present between pronotum and tegula (Figs 119, 120, 125, 126) 5
4. Stigmal vein distinctly curved, fore wing venation as in Fig. 107 (hyperparasitoids, emerging from cocoons) .... Ceraphronidae
– Fore wing venation different, stigmal vein straight (Fig. 108) (egg parasitoids) .... Scelionidae
5. Tarsi four-segmented 6
– Tarsi five-segmented 7
6. Hind coxa greatly flattened and expanded ('c' Fig. 126) .... (Chalcidoidea) Elasmidae
– Hind coxa normal .... (Chalcidoidea) Eulophidae
7. Hind femur greatly expanded and toothed (Fig. 127) .... (Chalcidoidea) Chalcididae
– Hind femur not greatly expanded or toothed 8
8. Mesopleuron forming a strongly convex shield (Figs 119, 120, 128) 9
– Mesopleuron different 10
9. In lateral view, midcoxae close to fore coxae; at centre of mesopleuron (Fig. 119). Marginal and postmarginal veins very short (Fig. 105) .... (Chalcidoidea) Encyrtidae
– In lateral view, midcoxae close to hind coxae, behind centre of mesopleuron (Fig. 120). Marginal and postmarginal veins well developed, longer than stigmal vein (see Fig. 102) .... (Chalcidoidea) Eupelmidae
10. Pronotum rectangular in dorsal view, with coarse punctures (Fig. 129) .... (Chalcidoidea) Eurytomidae
– Pronotum different 11
11. Notauli incomplete, not extending backwards to reach the hind margin of the mesoscutum (Fig. 131) .... (Chalcidoidea) Pteromalidae
– Notauli complete, extending backwards to reach the hind margin of the mesoscutum .... (Chalcidoidea) Eupelmidae males
12. Hind wing without veins, fore wing venation as in Figs 116–118 .... Bethylidae
– Hind wing with veins, fore wing venation different 13
13. Fore wing with a single recurrent vein (that is, 2m-cu absent; Figs 109–111); first cubital cell almost always separated from the discoidal cell by a vein (Braconidae, except *Chelonus*, Fig. 110) .... Braconidae
– Fore wing with two recurrent veins (that is, 2m-cu present; Figs. 112–115); first cubital cell always fused with the discoidal cell .... Ichneumonidae

16 Bethylidae

A. Polaszek
CABI Bioscience UK Centre, c/o Entomology Department, The Natural History Museum, London SW7 5BD, UK

INTRODUCTION

Among the various groups of parasitoids of lepidopterous cereal stem borers in Africa, the Bethylidae are the only representatives of the section Aculeata of the Hymenoptera, the section containing the stinging wasps, as well as the ants and bees. Bethylidae associated with cereal stem borers are represented by two genera, *Goniozus* and *Odontepyris,* in a single subfamily, the Bethylinae. Most Bethylidae are parasitoids of coleopterous larvae but bethylines are almost exclusively parasitoids of lepidopterous larvae, including many species of economic importance as crop pests. In particular, many *Goniozus* species have been used in the attempted biological control of Lepidoptera pests, including cereal stem borers. Note that the genus *Trissomalus* Kieffer, described originally from South Africa, was recently synonymized with *Odontepyris* Kieffer (Polaszek and Krombein, 1994).

Key to the Species of Bethylidae Attacking Cereal Stem Borers in Africa

1. Propodeum with well-developed lateral carinae; scutellum with large foveae (Fig. 133). Wing venation as in Fig. 118. Petiole ventrally with a complete keel (Fig. 134) .... *Odontepyris transvaalensis* (De Buysson)

– Propodeum without well-developed lateral carinae; scutellum without large foveae, with small grooves (Fig. 135). Wing venation different. Petiole ventrally with a reduced, forked keel (Fig. 136) 2

2. Closed discoidal cell present (areola, areolet) (dc in Fig 117) .... *Goniozus garouae* (Risbec)

– Closed discoidal cell absent (Fig. 116) .... *Goniozus indicus* Ashmead

Odontepyris transvaalensis Kieffer (Figs 118, 133, 134)

Diagnosis. Scutellar foveae well developed, median, discal and posterior transverse carinae present on propodeum (Fig. 133). Petiole with a complete ventral keel (Fig. 134). Prostigma absent, radial vein smoothly curved (Fig. 118). Fore wing largely infuscated brown.

Hosts. Known so far only from *Busseola fusca*, but also found in association with a lepidopterous larva on sugar cane in South Africa (Natal), possibly *Eldana saccharina*. A long series was reared recently from a maize stem in Kenya, but the actual host on this occasion was unknown.

Distribution. Ivory Coast, Kenya, Lesotho, Namibia, South Africa, Sudan, Zaïre.

Biology. Not known in detail, but presumably a gregarious ectoparasitoid like the closely related *Goniozus*. Fewer individuals appear to be capable of developing from a single host larva than is the case with *Goniozus* species.

Goniozus garouae (Risbec) (Fig. 117)

Diagnosis. Length 2.5–4.0 mm. Entirely chestnut-brown to black, with the following usually paler: antennae, fore tibiae and tarsi. Mandibles dark brown in female, yellow in male. Head prognathous, as in all Bethylidae, body appreciably dorsoventrally flattened. Fore wing (Fig. 117), with both pterostigma and prostigma present, with a closed discoidal cell (areolet). Propodeum without a transverse carina separating the dorsal (horizontal) surface from the posterior (vertical) surface; with fine, reticulate sculpture apart from a smooth area centrally.

Hosts. *Chilo* sp.; *E. saccharina, Etiella zinckenella, Sesamia calamistis.*

Distribution. Benin, Cameroon, Ivory Coast, Nigeria, Senegal, Uganda.

Biology. Probably similar to that of *G. indicus.*

Goniozus indicus Ashmead (Figs 135–138; Plate 12:3–5))

Diagnosis. Length 3.0–6.0 mm. Entirely chestnut brown to black, with the following usually paler: antennae, all tibiae and tarsi, fore femora. Head prognathous, as in all Bethylidae, body appreciably dorsoventrally flattened. Head in dorsal view elongate, distance from the hind margin of the eyes to the posterior end of the head slightly greater than the maximum length of the eyes. Ocelli, forming an approximately equilateral triangle, separated from the posterior end of the head by about the maximum length of the ocellar triangle. Clypeus evenly rounded, not pointed or truncate. Sculpture of the head and body smooth, shining, with a few shallow punctures. Fore femora enlarged. Fore wing (Fig. 116), with both pterostigma and prostigma present, without a closed areolet. Propodeum without any trace of a transverse carina separating the dorsal (horizontal) surface from the posterior (vertical) surface; with fine, reticulate sculpture apart from a smooth, triangular area tapering anteroposteriorly on the dorsal surface. Male genitalia and subgenital plate as in Figs 137 and 138, respectively.

Hosts. Pyralidae: *Chilo infuscatellus* Snellen, *Chilo partellus* (Swinhoe), *Chilo sacchariphagus* (Bojer), *Chilo suppressalis* (Walker) (laboratory host), *Chilo zacconius* Bleszynski, *Chilo* sp., *Coniesta ignefusalis* (Hampson), *Corcyra cephalonica* (Stainton) (laboratory host), *Diatraea centrella* Moeschler, *Diatraea lineolata* (Walker), *E. saccharina* (Walker), *Emmalocera depressella* (Swinhoe), *Maliarpha separatella* Ragonot, *Scirpophaga excerptalis* (Walker), *Scirpophaga* *incertulas* (Walker), *Scirpophaga nivella* (F.), *Scirpophaga occidentella* (Walker), *Scirpophaga* sp. (see Biology, below).

Distribution. All of sub-Saharan Africa, Mauritius and Madagascar. Asia: Bangladesh, India, Pakistan.

Biology. Like most *Goniozus*, a gregarious larval ectoparasitoid, usually attacking concealed larvae of Lepidoptera. Oviposition follows temporary anaesthesia and is into the dorsum of the larva. The biology of *G. indicus* (as *Goniozus procerae*) under laboratory conditions is given in detail by Ly (1976), Ndoye (1980) and Coquard (1987). Takasu and Overholt (1996) described oviposition behaviour in this species.

Gordh and Moczar (1990) list *Cryptophlebia carpophaga* (Walsingham) and *Cnaphalocrocis medinalis* (Guenée) among the hosts of *G. indicus.* We have examined *Goniozus* specimens reared from *C. medinalis* and from *Cryptophlebia* sp., from India. They are not *G. indicus*, and these species seem unlikely to be (natural) hosts of *G. indicus.*

Goniozus indicus has been the subject of several studies in relation to biological control of the stem borer *E. saccharina* in South Africa (Conlong *et al.*, 1988; Graham and Conlong, 1988; as *Goniozus natalensis*). It is also a well-known parasitoid of the millet stem borer *C. ignefusalis* in the West African Sahel (Ndoye, 1980; as *G. procerae*). *Goniozus indicus* was introduced into Madagascar (where it may well have already been present) from Senegal in 1973, for the attempted biological control of *M. separatella* (Appert, 1975) and into Réunion in 1972 (Betbeder-Matibet, 1977). *Goniozus indicus* was imported into Trinidad for the biological control of sugar-cane borers in 1960. It successfully attacked two *Diatraea* species (see hosts, above), but it is not known whether it became established (Bennett, 1965).

Note. *G.natalensis* Gordh and *G. procerae* Risbec were synonymised with *G. indicus* by Polaszek *et al.* (1994).

17 Braconidae

C. van Achterberg[1] and A.K. Walker[2]
[1]*Nationaal Natuurhistorisch Museum, Postbus 9517, 2300 RA Leiden, the Netherlands;* [2]*International Institute of Entomology, 56 Queen's Gate, London SW7 5JR, UK*

INTRODUCTION

The Braconidae are one of the largest families of Hymenoptera, and all except a few rare species are primary parasitoids. Together with their sister group, the Ichneumonidae, they account for over half the total number of parasitoid species associated with stem borers in Africa. Braconidae associated with African cereal stem borers have been treated in detail by Walker (1994, Microgastrinae) and van Achterberg and Polaszek (1996, all species). This account is based largely on the latter treatment. With very few exceptions, species are either solitary or gregarious larval parasitoids.

While the single records from *Amicrocentrum*, *Mesobraconoides* and *Meteorus* have never been repeated, and are therefore possibly doubtful, the remaining genera have confirmed associations with stem borers, and in some cases (e.g. *Tropobracon*) appear to be obligatorily associated with borers.

There are a great many accounts of the attempted use of braconid wasps for stem-borer biocontrol, though relatively few for Africa. In recent years, however, a major biological control project to introduce *Cotesia flavipes* against *Chilo partellus* has been undertaken in East Africa by the International Centre of Insect Physiology and Ecology (ICIPE) in collaboration with Wageningen Agricultural University (Overholt *et al.*, 1994a, b; Ngi-Song *et al.*, 1995, 1996; Omwega *et al.*, 1995, 1997). In South Africa, intensive field surveys, laboratory studies and some releases were undertaken, although no parasitoid was established (Kfir, 1990a, 1992c, 1994).

For the morphological terminology used in this chapter, see van Achterberg (1988, 1993a). However, terms should be self-explanatory using the figures provided. Species are treated by genus and by subfamily, subfamilies being listed alphabetically. Genera belonging to each subfamily are as follows:

Amicrocentrinae	*Amicrocentrum*
Agathidinae	*Bassus*
Braconinae	*Amyosoma, Bracon, Digonogastra, Iphiaulax, Mesobraconoides, Stenobracon, Tropobracon*
Cheloninae	*Chelonus, Phanerotoma*
Doryctinae	*Parallorhogas, Allorhogas, Rhaconotus*
Euphorinae	*Meteorus*
Macrocentrinae	*Macrocentrus*
Microgastrinae	*Cotesia, Dolichogenidea, Glyptapanteles*
Orgilinae	*Orgilus*

Key to the subfamilies and genera of Braconidae parasitizing African cereal stem borers

1. Marginal cell of fore wing extremely narrow and rather long (fig. 1a); vein m-cu of fore wing more or less diverging posteriorly, from direction of anterior half of vein 1-M (fig. 1b); trace of vein 2-CU of hind wing usually present (fig. 1c). Subfamily **Agathidinae/ Bassinae** .... ***Bassus*** Fabricius

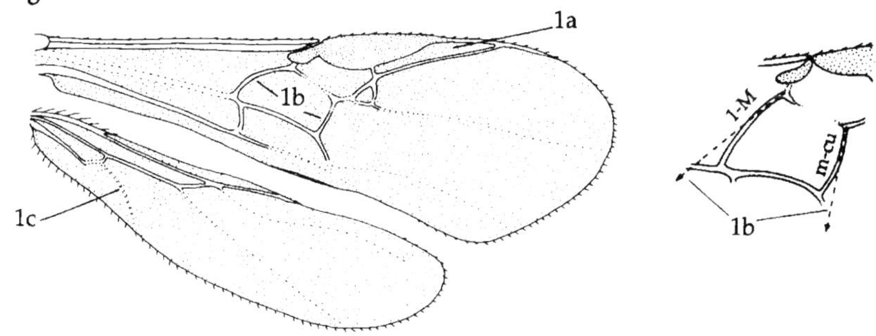

- Marginal cell of fore wing of different shape, wider (fig. 1aa); vein 2-CU of hind wing absent (fig. 1bb) 2

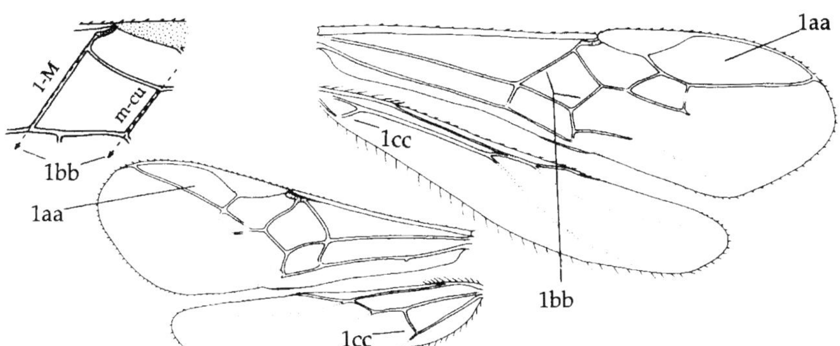

2. Hypoclypeal depression deep and wide, and middle of apparent ventral margin of clypeus distinctly above upper level of mandibular bases (fig. 2a); bottom of hypoclypeal depression consists of the concave labrum and a depressed part of clypeus (fig. 2b: "cyclostome braconids") 3

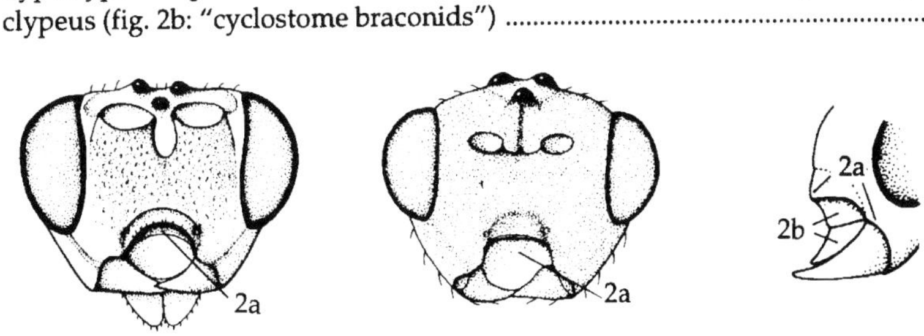

- Hypoclypeal depression absent, medio-ventral margin of clypeus close to upper level of mandibular bases (fig. 2aa); labrum flat and ventral part of clypeus not part of a hypoclypeal depression (fig. 2bb: "non-cyclostome braconids") 12

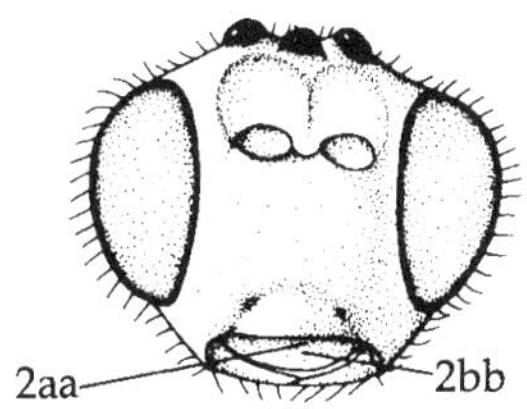

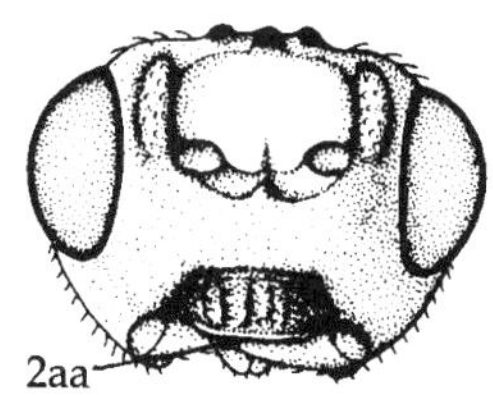

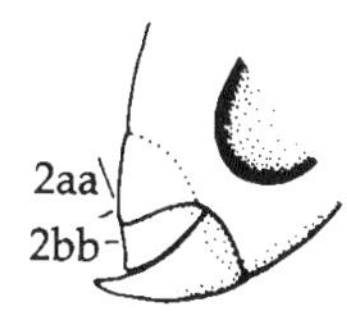

3. Occipital carina absent (fig. 3a); posterior flange of propleuron (fig. 3b) and prepectal carina of mesopleuron (fig. 3c) absent; maxillary palp with 5 segments (fig. 3d); first metasomal tergite with lateral parts flattened (fig. 3e). Subfamily **Braconinae** 4

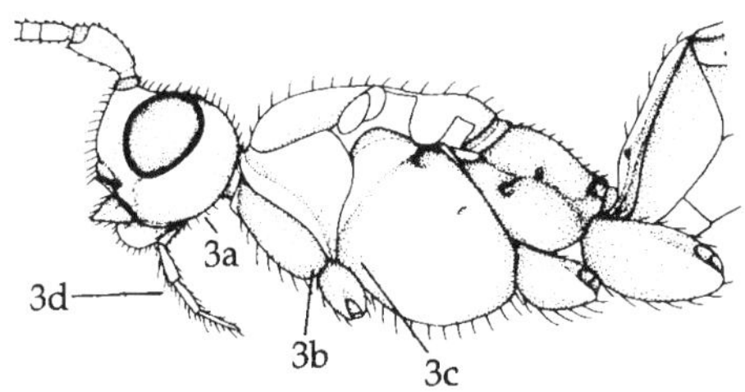

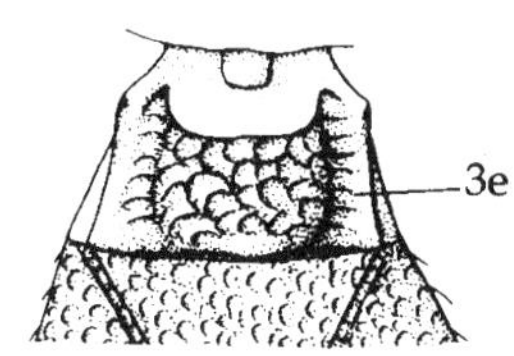

- Occipital carina present (fig. 3aa); posterior flange of propleuron (fig. 3bb) and prepectal carina (fig. 3cc) present; maxillary palp with 6 segments (fig. 3dd); first tergite with convex lateral parts (fig. 3ee). Subfamily **Doryctinae** 11

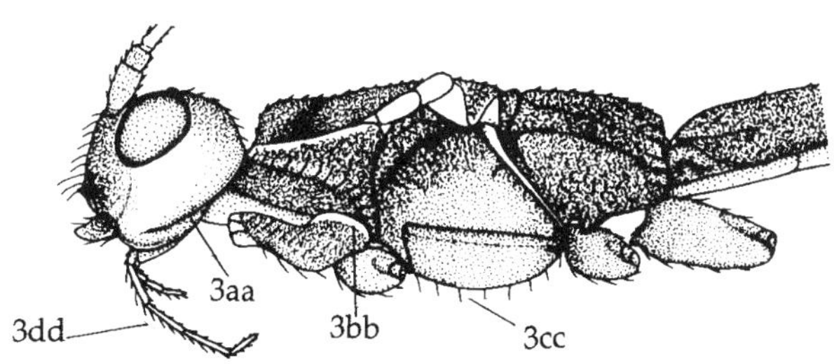

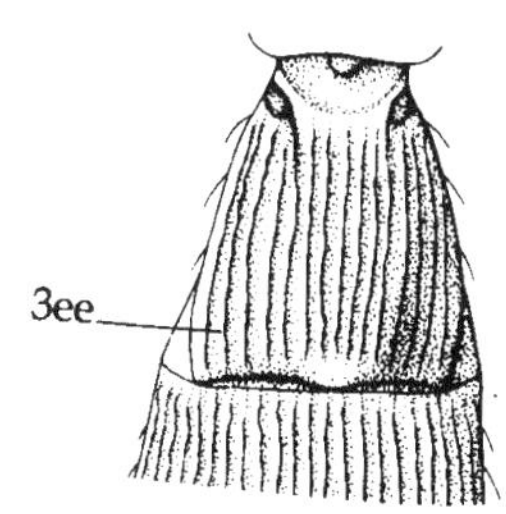

4. Second metasomal tergite with pair of converging grooves, and tergite completely sculptured (fig. 4a); tarsal claws with subbasal lobe (fig. 4b) 5

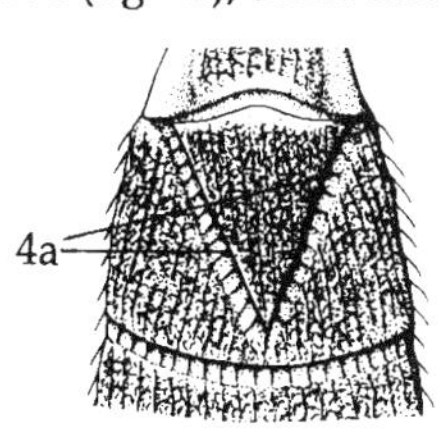

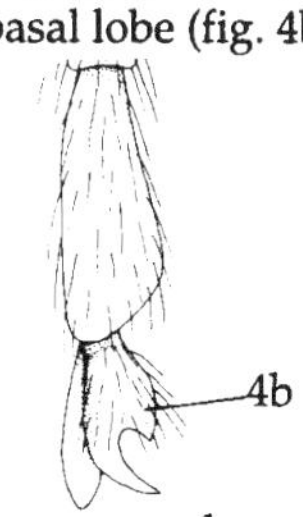

- Second tergite without pair of converging grooves, and no extensive sculpture of tergite (fig. 4aa); tarsal claws simple, without lobe (fig. 4bb) 6

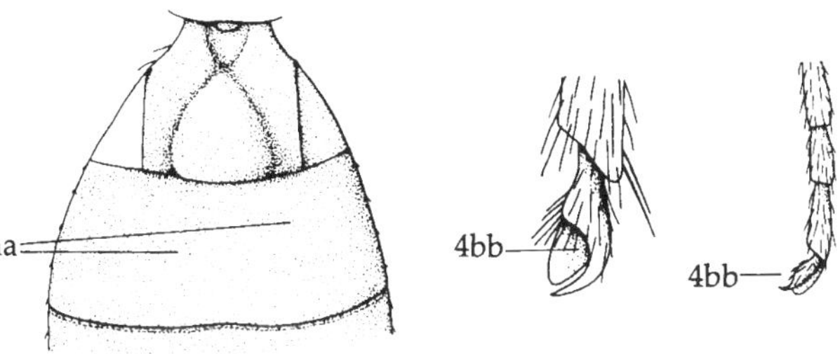

5. Second submarginal cell of fore wing slightly longer than high (fig. 5a); grooves of second metasomal tergite narrow (fig. 5b); at least posterior half of propodeum distinctly sculptured (fig. 5c); vein 1r-m of hind wing short (fig. 5d) .... ***Tropobracon*** Cameron

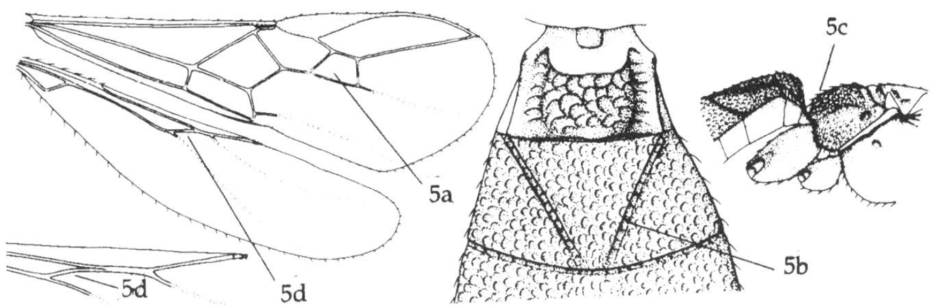

\- Second submarginal cell of fore wing much longer than high (fig. 5aa); grooves of second metasomal tergite wide (fig. 5bb); propodeum completely smooth (fig. 5cc); vein 1r-m of hind wing comparatively long (fig. 5dd) .... ***Mesobraconoides*** Sarhan & Quicke

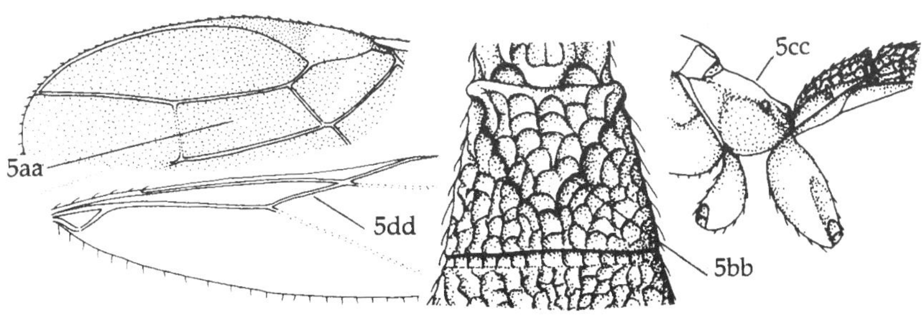

6. Scapus distinctly concave apically, ventrally longer than dorsally in lateral view (fig. 6a); vein 1r-m of hind wing comparatively long (fig. 6b); third metasomal tergite with antero-lateral grooves (fig. 6c) 7

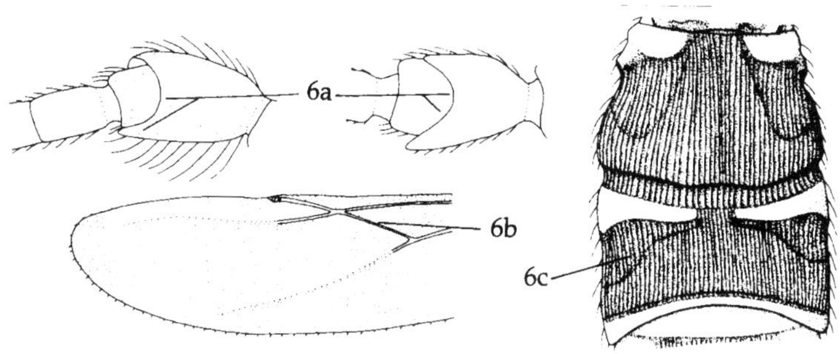

- Scapus truncate apically, in lateral view about as long ventrally as dorsally or shorter (fig. 6aa); vein 1r-m of hind wing short (fig. 6bb); third metasomal tergite without antero-lateral grooves (fig. 6cc) 9

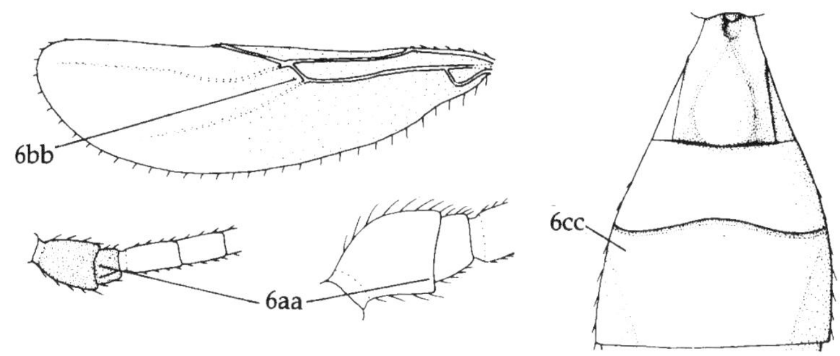

7. Vein 1-SR of fore wing distinctly angled with parastigma, because angle between veins 1-SR and C+SC+R is 60°-80° (fig. 7a); vein 3-M of fore wing terminating at about same level as vein SR1 or more basally (fig. 7b) 8

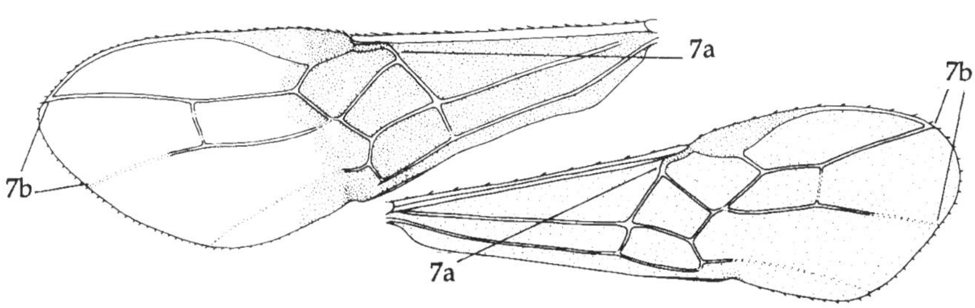

- Vein 1-SR of fore wing gradually merging into parastigma, because angle between veins 1-SR and C+SC+R is about 30° (fig. 7aa); vein 3-M of fore wing usually terminating distinctly more distad than vein SR1 (fig. 7bb) .... ***Stenobracon*** Szépligeti

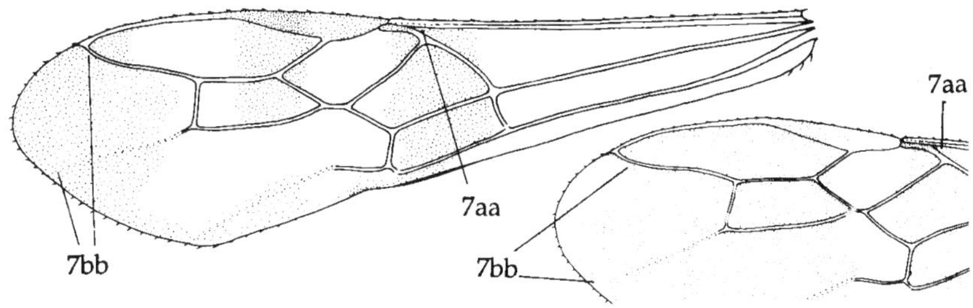

8. Second metasomal tergite with distinctly differentiated medio-basal area (fig. 8a); ovipositor normal, its upper valve not enlarged, and ventrally with some minute teeth (fig. 8b); second tergite (largely) smooth .... ***Digonogastra*** Viereck

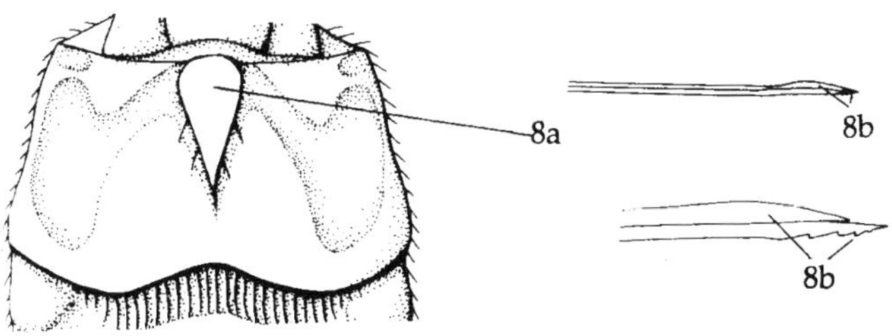

\- Second metasomal tergite without medio-basal area (fig. 8aa); ovipositor aberrant, its upper valve enlarged, and ventrally without teeth (fig. 8bb); second tergite at least partly sculptured .... ***Iphiaulax*** Foerster

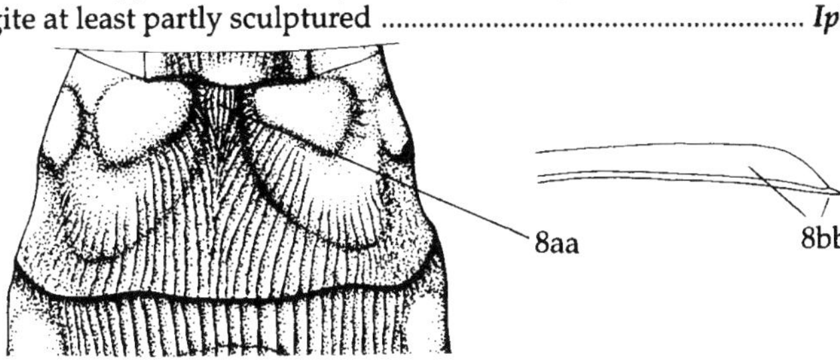

9. First tergite parallel-sided behind spiracles or diverging posteriorly, with distinct flat lateral areas (fig. 9a); antero-lateral corners of second tergite equally sclerotized as middle of tergite (not depressed in dried specimens, fig. 9b); vein 1r-m of hind wing straight (fig. 9c) 10

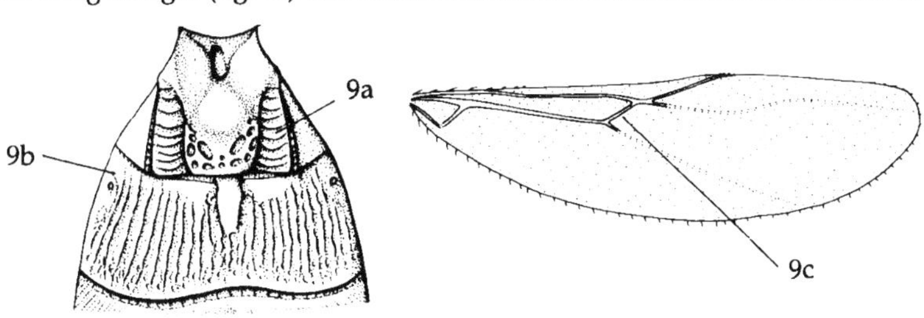

- First metasomal tergite more or less narrowed behind spiracles, without distinct flat lateral areas (fig. 9aa); antero-lateral corners of second tergite membranous, almost unsclerotized (depressed in dried specimens, fig. 9bb); vein 1r-m of hind wing more or less curved (fig. 9cc) .... ***Amyosoma*** Viereck

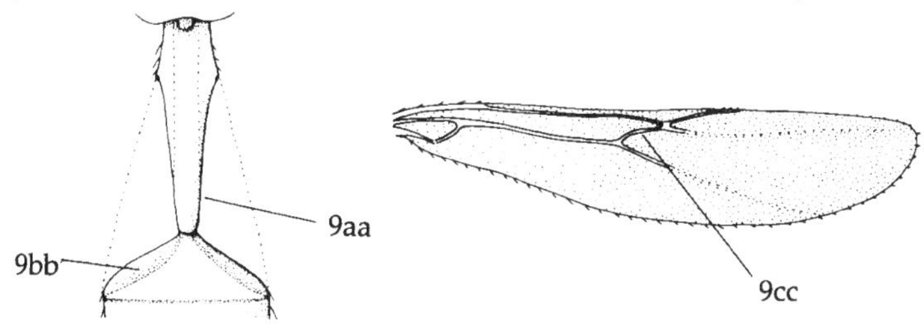

10. Vertex coriaceous (fig. 10a), dull **and** dorso-lateral carinae of first metasomal tergite absent (fig. 10b); vein 1-SR of fore wing very short or absent (fig. 10c) .... ***Bracon*** (subgenus ***Habrobracon*** Ashmead)

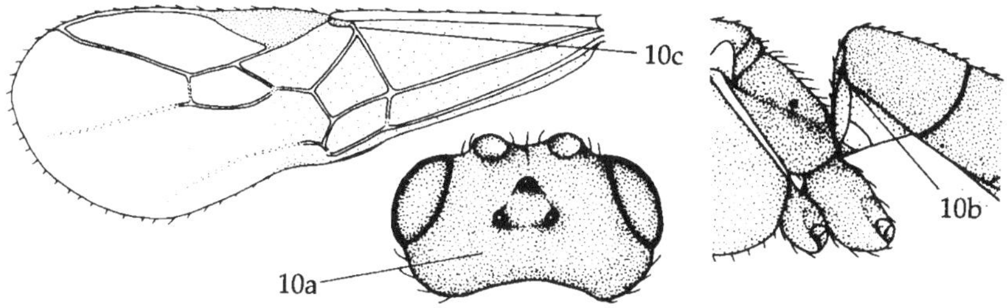

- Vertex largely smooth (fig. 10aa), shiny; if coriaceous then dorso-lateral carinae of first metasomal tergite present (fig. 10bb); vein 1-SR of fore wing medium-sized (fig. 10cc) .... ***Bracon*** (subgenus ***Bracon*** Fabricius)

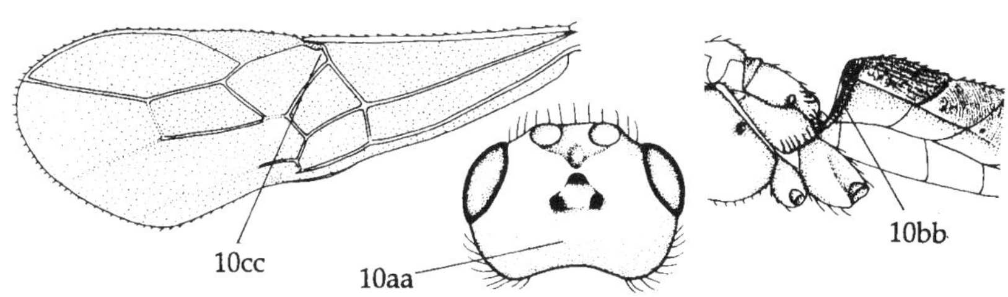

11. Third and fourth metasomal tergites smooth; first subdiscal cell of fore wing open distally and less slender (fig. 11a); propodeum distinctly areolate (fig. 11b); mesoscutal lobes largely smooth .... ***Parallorhogas*** Marsh

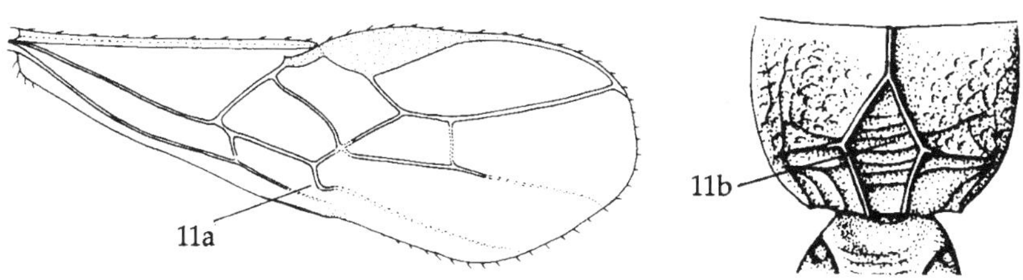

- Third and fourth tergites sculptured; first subdiscal cell of fore wing closed distally and slender (fig. 11aa); propodeum without distinct areolation (fig. 11bb); mesoscutal lobes granulate .... ***Rhaconotus*** Ruthe

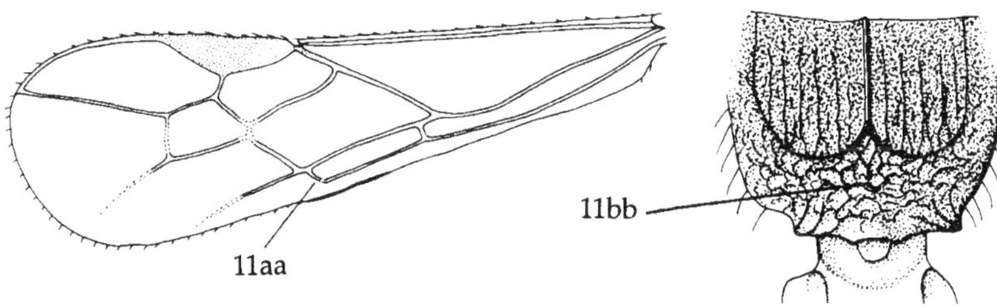

12. Metasoma with a carapace (fig. 12a); postpectal carina complete in front of middle coxae ventrally (fig. 12b). Subfamily **Cheloninae** 13

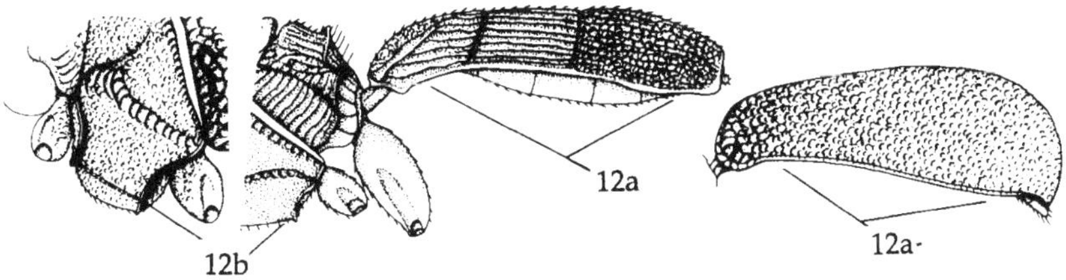

- Metasoma without a carapace (fig. 12aa); postpectal carina absent (fig. 12bb) ... 14

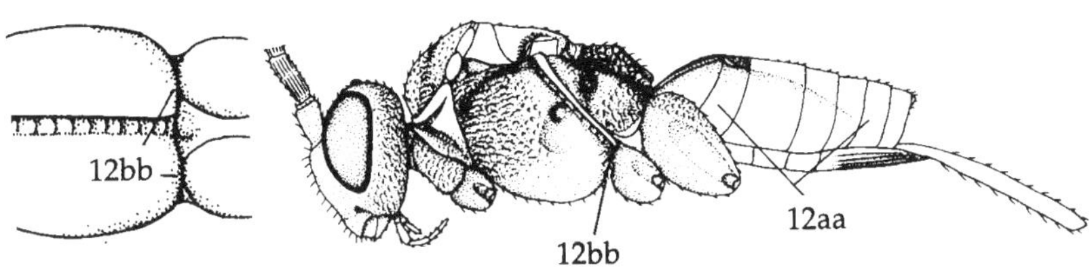

13. Metasomal carapace without transverse grooves (fig. 13a); vein 1-SR+M of fore wing absent (fig. 13b); occipital carina remains separated from hypostomal carina (fig. 13c) .... ***Chelonus*** Panzer

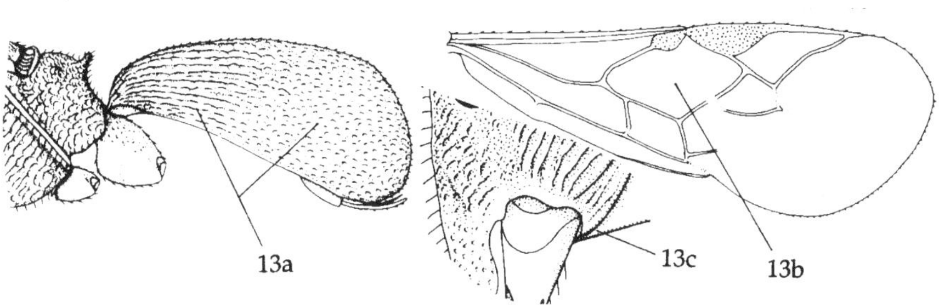

- Metasomal carapace with two transverse grooves (fig. 13aa); vein 1-SR+M of fore wing present (fig. 13bb); occipital carina just meeting hypostomal carina ventrally (fig. 13cc) .... ***Phanerotoma*** Wesmael

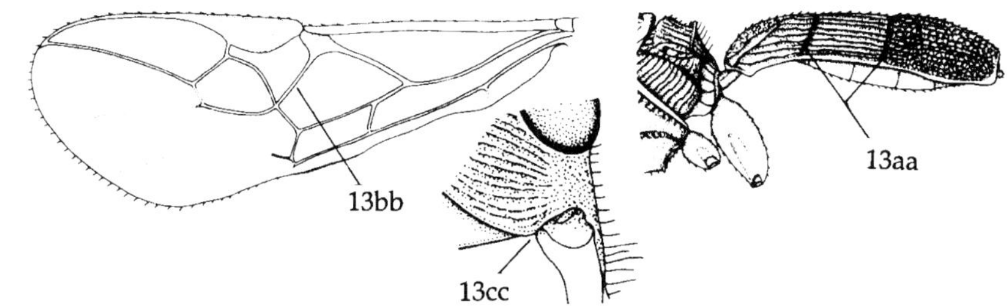

14. Vein SR1 of fore wing partly or completely unsclerotized, resulting in a distally open marginal cell distally (fig. 14a); metasoma often short (fig. 14b); plical lobe of hind wing may be large (fig. 14c). Subfamily **Microgastrinae** 15

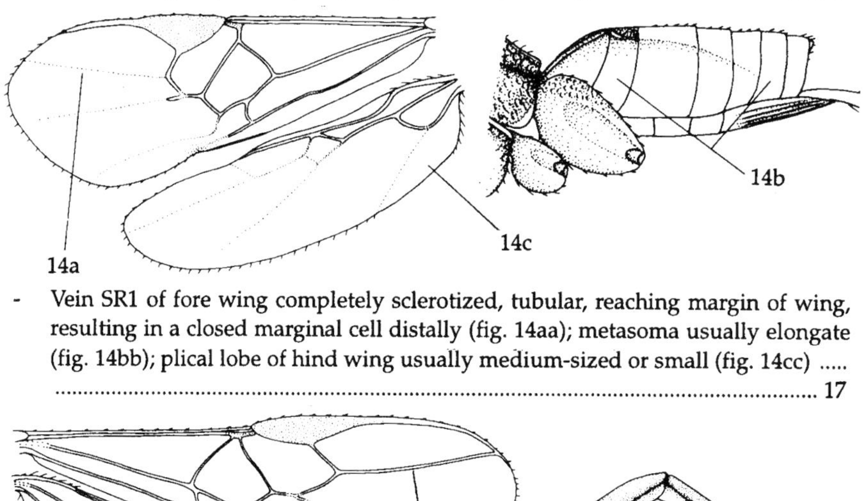

- Vein SR1 of fore wing completely sclerotized, tubular, reaching margin of wing, resulting in a closed marginal cell distally (fig. 14aa); metasoma usually elongate (fig. 14bb); plical lobe of hind wing usually medium-sized or small (fig. 14cc) 17

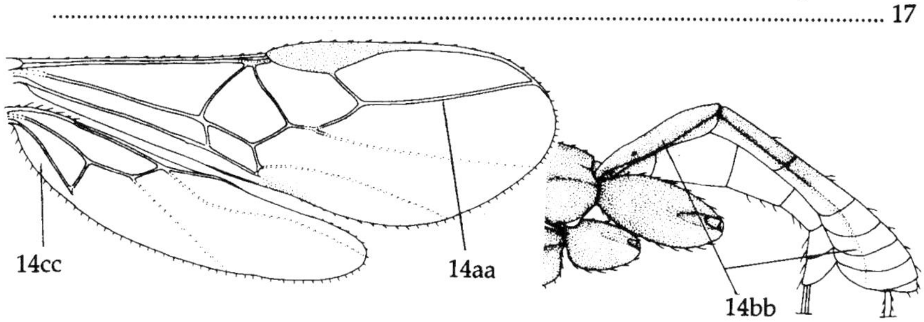

15. Ovipositor usually short, hardly protruding beyond apex of metasoma (shorter than basitarsus of hind leg; fig. 15a) and its sheath only apically setose; basal half of hypopygium of ♀ evenly sclerotized, and at most regularly curved (fig. 15b; *Cotesia* group) 16

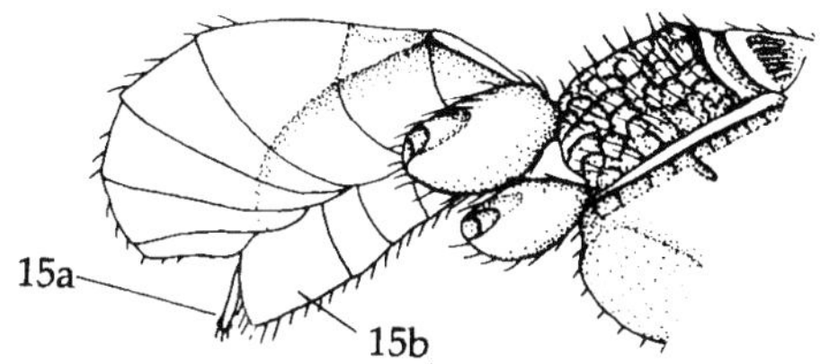

- Ovipositor usually medium-sized to long, distinctly protruding beyond apex of metasoma, at least as long as basitarsus, and more setose (fig. 15aa); basal half of hypopygium of ♀ usually folded, V-shaped at least, and often membranous medially (and creased in dead specimens; fig. 15bb) .... ***Dolichogenidea*** Viereck

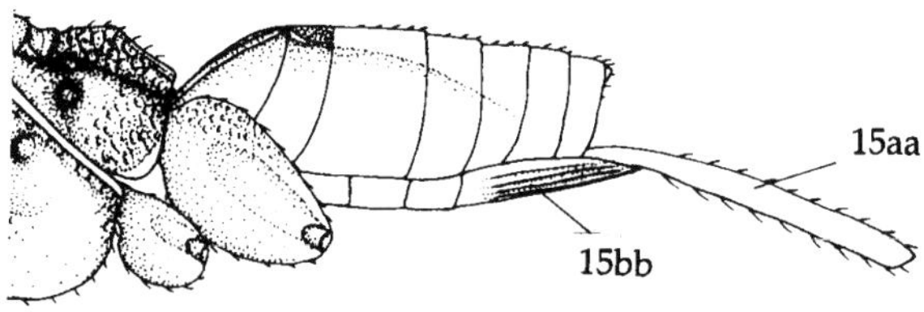

16. First metasomal tergite comparatively slender and gradually narrowed apically (fig. 16a); width of second tergite more than 1.5 times apical width of first tergite (fig. 16b); propodeum largely smooth, but may be reticulate posteriorly or largely rugulose-punctate (fig. 16c) .... ***Glyptapanteles*** Ashmead

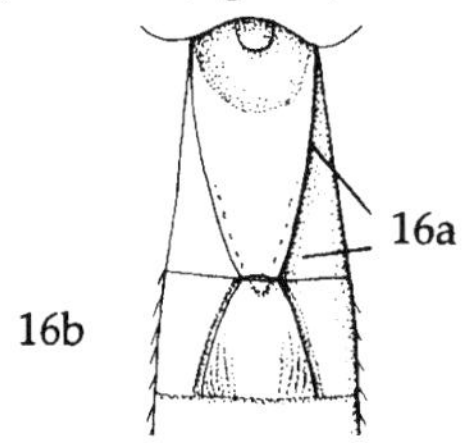

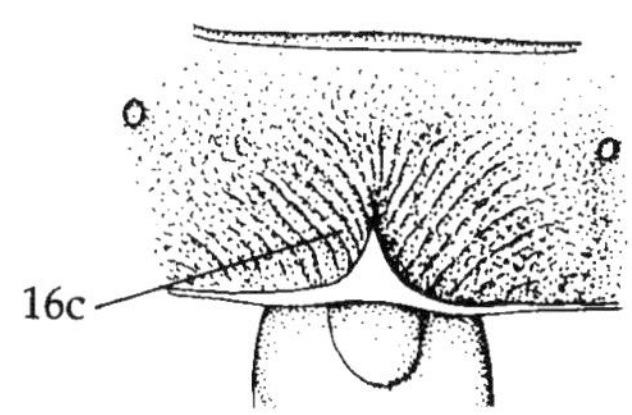

- First tergite robust, parallel-sided or widened apically, at most subapically narrowed (fig. 16aa); width of second tergite 1.5 times apical width of first tergite or less (fig. 16bb); propodeum (except anteriorly) largely coarsely rugose (fig. 16cc) . ... ***Cotesia*** Cameron

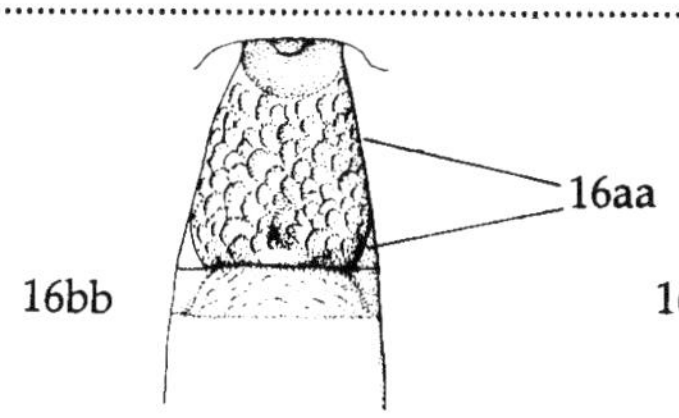

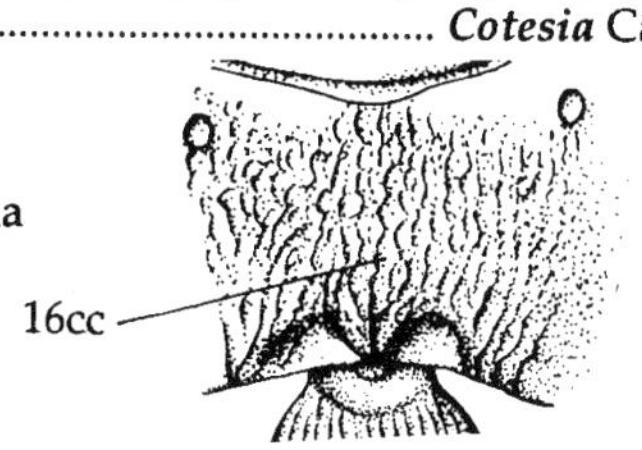

17. First metasomal tergite with a deep medio-basal hole behind its medio-basal depression (fig. 17a); maxillary palp consists of one segment (fig. 17b); plical lobe of hind wing large and reticulate setose (fig. 17c). Subfamily **Amicrocentrinae** .... ***Amicrocentrum*** Schulz

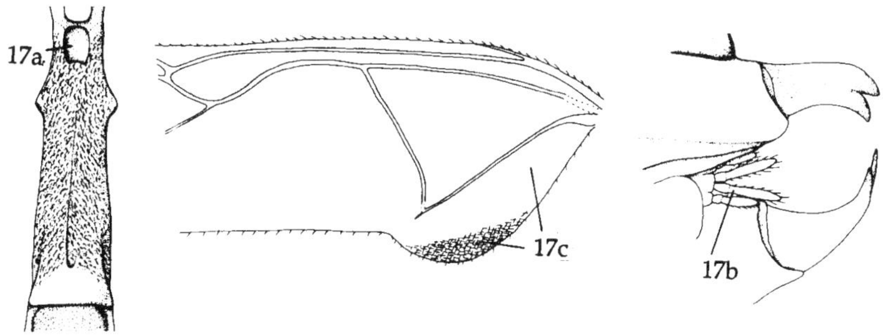

- First metasomal tergite without medio-basal hole behind its medio-basal depression (fig. 17aa); maxillary palp consists of 6 segments (fig. 17bb); plical lobe of hind wing at most medium-sized and unidirectional setose (fig. 17cc) 18

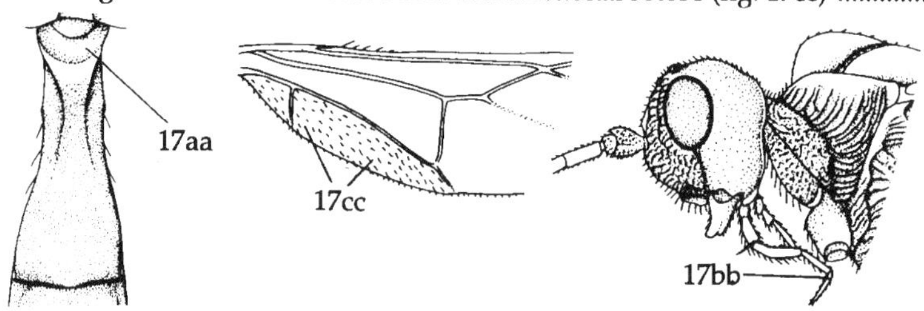

18. Anterior side of trochantelli with (sub)apical comb of short pegs (fig. 18a); metasoma connected to propodeum somewhat above hind coxae (fig. 18b); occipital carina absent (fig. 18c). Subfamily **Macrocentrinae** .... ***Macrocentrus*** Curtis

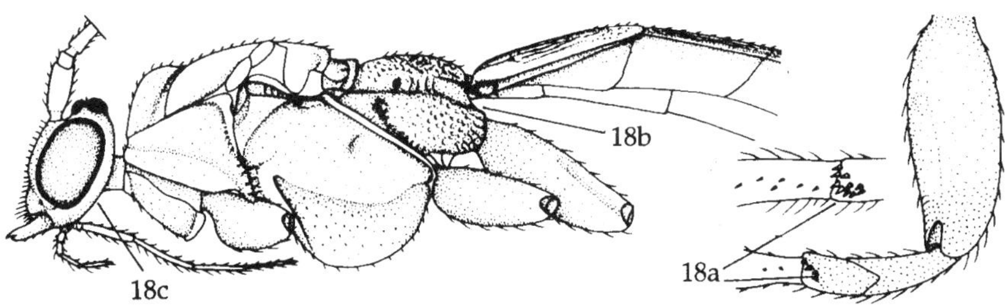

- Trochantelli without pegs (fig. 18aa); metasoma inserted near hind coxae (fig. 18bb); occipital carina present laterally (fig. 18cc) 19

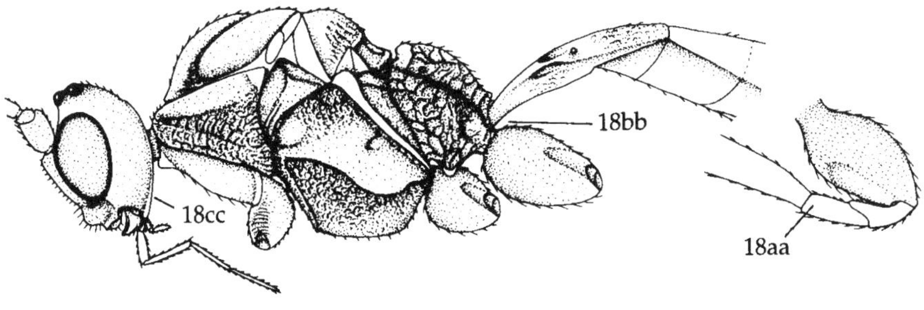

19. First metasomal tergite sessile (fig. 19a); hind tibia with pegs near base of spurs (fig. 19b); scutellum without crenulate depression medio-posteriorly (fig. 19c); vein r-m of fore wing absent (fig. 19d). Subfamily **Orgilinae** .... ***Orgilus*** Haliday

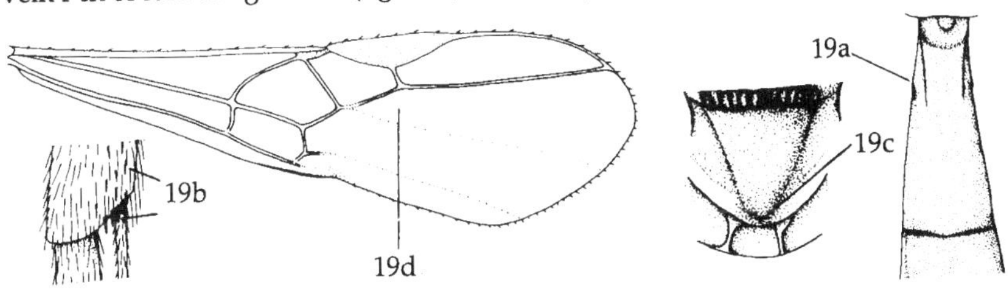

- First tergite petiolate (fig. 19aa); hind tibia without pegs near base of spurs (fig. 19bb); scutellum with transverse crenulate depression medio-posteriorly (fig. 19cc); vein r-m of fore wing present (fig. 19dd). Subfamily **Euphorinae** .... ***Meteorus*** Haliday

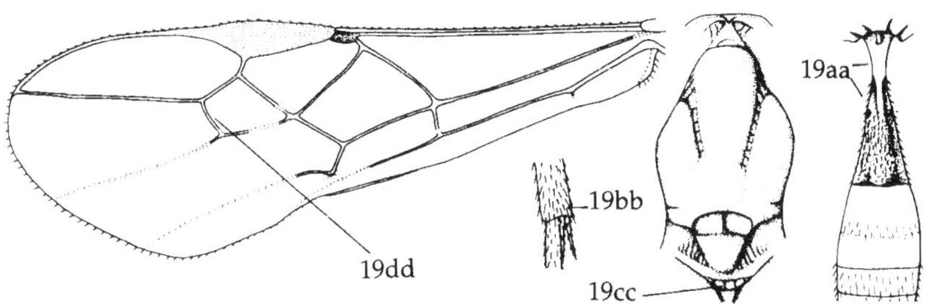

AMICROCENTRINAE

Genus *Amicrocentrum* Schultz (Figs 139–150)

For full synonymy of this genus, and all genera treated here, see van Achterberg and Polaszek (1996).

Diagnosis. Length of body 9–27 mm, of fore wing 7–21 mm; notauli present; plical lobe of hind wing reticulate–setose (Fig. 139); vein 1-M of hind wing strongly sinuate (Fig. 139); first metasomal tergite with deep depression subbasally (Fig. 150); third and following tergites normally setose.

Biology. Endoparasitoids of concealed lepidopterous larvae. Rarely collected, usually at light.

Distribution. Four species are known, one restricted to Madagascar and three to continental Africa. For a taxonomic revision, see van Achterberg (1979).

Amicrocentrum curvinervis (Cameron) (Figs 139–150)

Diagnosis. Length of body 16–24 mm, of fore wing 10–18 mm; antennal segments 51–59; clypeus distinctly convex and densely punctate; pterostigma less slender than in other species (Fig. 139); posterodorsal corner of pronotum more or less punctate; length of first tergite 3.1–4.4 times its apical width (exceptionally in males 2.8 times); sides of second metasomal tergite parallel; length of second tergite 2.0–2.8 times its width (Fig. 150), exceptionally 1.8 times; fourth tergite yellow, not contrasting with other segments; metasoma of female strongly compressed apicad; length of ovipositor sheath 1.6–2.0 times fore wing, usually 1.8–1.9 times. For the redescription, see van Achterberg (1979).

Host(s). Noctuidae: *Busseola fusca* (van Achterberg, 1979).

Distribution. Benin, Uganda, Zaïre.

Remarks. *Amicrocentrum curvinervis* is included here on the basis of a single host record from Uganda (van Achterberg, 1979: 15). Further observations are necessary to confirm the accuracy of this observation.

BASSINAE/AGATHIDINAE

Genus *Bassus* Fabricius (Figs 151–171)

Diagnosis. Head slightly or not elongated ventrally, only gradually narrowed ventrally, in frontal view trapezoid (Fig. 157); area between antennal sockets triangular to nearly flat, or with a shallow or deep median groove; antenna shorter than body and with 23–39 segments, its apex without apical spine (Fig. 154); clypeus at least partly flattened (Fig. 157), transverse or quadrate (Figs 157, 169), truncate or concave ventrally; labiomaxillary complex not or hardly protruding (Fig. 162) and its galea not longer than wide; area behind antennal sockets slightly to moderately depressed (Fig. 156); frons without lateral carinae (Fig. 156); frons usually without strong median ridge; depression in front of anterior ocellus usually absent, but may be shallowly impressed; malar space distinctly longer than basal width of mandible (Fig. 162); lateral pronope (= subpronope) rather shallow or absent; precoxal sulcus at least partly present, narrow and crenulate (Fig. 162); middle lobe of mesoscutum without pair of longitudinal depressions anteriorly (Fig. 158); notauli present (but may be reduced posteriorly) and usually narrow (Fig. 158); propodeum granulate, reticulate, rugose or punctate, and without extensive areolation but an elongate medial areola may be present; vein r-m of fore wing present (in typical *Bassus*), obsolescent (Fig. 171) or absent (Figs 151, 164, '*Camptothlipsis*/*Baeognatha* auct.') – its presence may be infraspecifically variable; second submarginal cell without ramellus; vein 1-SR+M of fore wing at least medially absent (Fig. 164); vein r of fore wing issues near middle of pterostigma (Figs 164, 171); vein 1r-m of hind wing shorter than vein SC+R1; tarsal claws not bifurcate, medium-sized and usually with a lobe (Fig. 155); fore tibial spur normal (Fig. 159), 0.4–0.5 times as long as fore basitarsus; length of inner spur of middle tibia 0.4–0.6 times middle basitarsus; apical half of middle tibia with peg(s) present above subapical cluster of pegs (Fig. 153); hind trochantellus with its lower edges rounded; hind coxa shiny, punctulate or punctate, or rather mat and granulate; length of inner spur of hind tibia 0.4–0.6 times hind basitarsus; propodeal foramen usually distinctly removed from metasternal hind coxal cavities and separated by a sclerotized bridge; first metasomal tergite sessile and robust, its length 0.9–1.4 times its apical width and laterope present (Fig. 162); first to third tergites smooth, partly or completely sculptured; second suture distinctly impressed, but sometimes shallow; second and third tergites frequently with a transverse depression; ovipositor medium-sized to long and straight; ovipositor sheath subparallel-sided and subtruncate apically, its length 0.6–2.0 times fore wing.

Biology. Endoparasitoids of concealed larvae of Tortricidae, Gelechiidae, Pyralidae, Crambidae, Scytrididae, Tineidae, Coleophoridae and Blastobasidae.

Distribution. Cosmopolitan. The only key to African species is given by Granger (1949) for Madagascan species, under *Camptothlipsis* and *Microdus*.

Key to Species of the Genus *Bassus* Parasitizing African Cereal Stem Borers

1. Second metasomal tergite elongate and granulate (Fig. 168); second submarginal cell of fore wing absent (Fig. 164); vein SR1 of fore wing straight or nearly so (Fig. 164); propodeum without areolation .. *Bassus sublevis* (Granger)

– Second tergite short, transverse and completely smooth (Fig. 170); minute second submarginal cell of fore wing present (Fig. 171); vein SR1 of fore wing distinctly sinuate (Fig. 171); propodeum more or less areolate .... *Bassus* sp. A

Bassus sublevis (Granger) (Figs 164–169)

Diagnosis. Length of fore wing 4–5 mm; face 1.8–1.9 times wider than high (Fig. 169); length of malar space 1.8–2.0 times basal width of mandible; frons with median carina more or less developed; notauli complete; precoxal sulcus finely crenulate anteriorly; median carina of propodeum weak; sculpture of remaining part of mesopleuron varies from absent (lectotype) to distinctly microsculptured;

vein 1-R1 of fore wing somewhat longer than half length of pterostigma (Fig. 164); first and second metasomal tergites finely granulate, second tergite superficially so; first tergite distinctly widened posteriorly, about 1.5 times as long as wide apically; length of second tergite about equal to its basal width (Fig. 168); length of ovipositor sheath about 1.5 times fore wing; body (except metanotum) completely yellowish brown; antenna and pterostigma dark brown, pterostigma darker anteriorly than posteriorly; wing membrane evenly subhyaline; hind tarsus and sometimes metanotum and scutellum narrowly posteriorly, infuscate. Males frequently have metanotum, propodeum, hind tibia and first tergite dark brown and wing membrane evenly infuscate.

Host(s). Pyralidae: *Eldana saccharina*; Crambidae: *C. partellus.*

Distribution. South Africa, Madagascar, Kenya.

Bassus sp. A (Figs 170, 171)

Diagnosis. Because of lack of material, we have refrained from diagnosing this species.

Host(s). Solitary endoparasitoid of crambid stem borer: *C. partellus.*

Distribution. Mozambique.

BRACONINAE

Genus *Amyosoma* Viereck (Figs 172–193)

Diagnosis. Length of fore wing 2.5–5.5 mm; scapus truncate apically, ovoid; vein 1r-m of hind wing short (Fig. 182) to medium-sized (Fig. 186); angle between veins 1-SR and C+SC+R of fore wing about 80° (Figs 182, 186); first discal cell of fore wing comparatively transverse (Figs 182, 186); hind femur and tibia compressed; hind femur, tibia and tarsus rather sparsely and whitish or pale yellowish setose; first metasomal tergite narrowed behind spiracles medially, 1.8–2.7 times longer than apically wide, without or with narrow flattened areas laterally (Figs 179, 187, 190); unsclerotized parts of second tergite triangular or nearly so (Figs 192, 193); second metasomal sulcus rather deep, straight or nearly so (Figs 187, 192); metasoma smooth.

Biology. Ectoparasitoids of lepidopterous stem borers.

Distribution. Palaeotropical, south-east Palaearctic. The only known African species is described by Quicke and Wharton (1989), a second species has been introduced from the Oriental region.

Remarks. This genus is closely related to *Bracon*, and species have often been misplaced in that genus or in the closely related New World genus *Myosoma* Brullé, 1846. The genus *Amyosoma* shares with *Myosoma* the narrowed first metasomal tergite and, to some degree, the compressed hind femur and tibia, but it lacks the peculiar dense setosity of the legs, especially of the hind femur and tibia, and the colour of the setae is pale yellowish or whitish instead of black. It has the notauli complete (but shallow posteriorly) and the vein 1r-m of hind wing straight (but in *Amyosoma nyanzaense* it is frequently weakly curved). As pointed out by Mason (1978), the setosity of some Nearctic species of *Myosoma* (as well as of some Neotropical species) may be less developed, but it still remains much denser than in *Amyosoma*, and the setae are black (van Achterberg and Polaszek, 1996).

Key to Species of *Amyosoma*

1. Apical half of fore wing largely subhyaline, distinctly paler than its dark brown basal half (Fig. 186); antennal segments of female (♀) 28–31, of male (♂) about 24; lateral areas of notum of first metasomal tergite distinctly developed along whole length of tergite; length of tergite 1.8–2.1 times its apical width (Fig. 187); hindleg (usually except infuscate apex of hind tibia and tarsus) and metasomal tergites of ♀ orange-brown (but frequently fourth and fifth tergites dark brown); Afrotropical .... *A. nyanzaense* (Quicke & Wharton)

– Fore wing unicoloured or nearly so, or apical half of fore wing darker than basal half of wing; antennal segments of ♀ 33–37, of ♂ 30–35 (unknown of *Amyosoma rufescens*); lateral areas of notum of posterior half of first metasomal tergite largely absent or narrow (except posteriorly), length of tergite 1.9–2.5 times its apical width (Figs 190, 192, 193); colour of hindleg and of metasoma of ♀ variable; Indo-Australian, south-east Palaearctic, one species introduced in Afrotropical Pacific regions 2

2. Length of ovipositor sheath about 0.55 times fore wing and about twice length of hind tibia; pterostigma yellow (except small basal spot); sixth and seventh tergites yellowish brown; basal half of fore wing mainly yellowish, and its apical half largely infuscate; Australian .... *Amyosoma flavistigma* van Achterberg

– Length of ovipositor sheath 0.25–0.45 times fore wing and 1.7 times length of hind tibia or less; pterostigma dark brown; sixth and seventh tergites mainly black or dark brown; wing membrane dark brown or subhyaline 3

3. First to fifth tergites bright orange; length of ovipositor sheath 0.25 times fore wing and shorter than hind tibia; face black; Australian ....*Amyosoma rufescens* (Quicke & Ingram)

– First to fifth tergites mainly dark brown or black; length of ovipositor sheath 0.35–0.45 times fore wing and 1.0–1.7 times as long as hind tibia; colour of face variable; Oriental, south-east Palaearctic, one species introduced in Afrotropical region 4

4. Wing membrane subhyaline; length of ovipositor sheath about three times length of first metasomal tergite and 1.7–2.0 times as long as hind tibia; face dark brown; third to fifth antennal segments of ♀ robust (Fig. 189); parasitoid of coffee stem borer ....*Amyosoma leuzerae* Rohwer

– Wing membrane evenly dark brown; length of ovipositor sheath 2.2–2.7 times length of first tergite and 1.0–1.3 times as long as hind tibia; face yellowish or black; third antennal segment of ♀ less robust (Fig. 191) 5

5. Face yellowish; length of ovipositor sheath about as long as hind tibia, 2.2–2.5 times length of first tergite; third to sixth tergites pale yellowish posterolaterally; parasitoid of stem borers in *Gramineae* .... *Amyosoma chinense* (Szépligeti)

– Face black; length of ovipositor sheath about as long as hind tibia and hind basitarsus combined, about 2.7 times length of first tergite; third to sixth tergites black posterolaterally; parasitoid of stem borers in trees and shrubs .... *Amyosoma yanoi* (Watanabe)

Amyosoma chinense (Szépligeti) (Figs 191–193)

Diagnosis. Length of fore wing 2.8–4.0 mm and of body 2.9–3.9 mm; antennal segments of ♀ 33–36, of ♂ 30–35; third antennal segment of ♀ moderately robust (Fig. 191); face largely granulate and yellowish; lateral areas of notum of posterior half of first metasomal tergite largely absent, length of tergite 2.1–2.5 times its apical width (Figs 192, 193); length of ovipositor sheath 0.32–0.40 times fore wing, 2.2–2.5 times length of first tergite, and about as long as hind tibia; wing membrane evenly dark brown; head (including face) and mesosoma largely yellowish brown; metasoma mainly dark brown or black; third to sixth tergites pale yellowish posterolaterally; foreleg yellowish (but femur may be partly dark brown); middle and hindlegs blackish or dark brown; head of female usually largely yellowish brown, but may be largely dark brown dorsally; palpi and tegulae yellowish. Head of male dorsally, propodeum and metasoma (but unsclerotized parts of first and second tergites and posterior margins of following tergites whitish) of male dark brown or blackish.

Host(s). Gregarious ectoparasitoid of a variety of pyralid, crambid and noctuid stem borers in *Gramineae.*

Distribution. Oriental, south Palaearctic, introduced into Afrotropical and Pacific regions. *Amyosoma chinense* was used extensively in biological control attempts against sugar-cane stem borers earlier this century, and was introduced into Madagascar and Mauritius (Moutia and Mamet, 1947; Appert, 1973); though probably without establishing. During the present study, it was found to occur in Oman, and its distribution may eventually be found to overlap that of *A. nyanzaense.*

Amyosoma nyanzaense (Quicke & Wharton) (Figs 186–188)

Diagnosis. Length of fore wing 2.4–4.1 mm and of body 2.3–4.0 mm; antennal segments of ♀ 28–31, of ♂ about 24; third to fifth antennal segments comparatively slender (Fig. 188); face finely superficially granulate, but mediodorsally smooth; lateral areas of notum of first metasomal tergite distinctly developed along whole length of tergite, its length 1.8–2.1 times its apical width (Fig. 187); length of ovipositor sheath 0.33–0.40 times fore wing, twice as long as first tergite and 1.0–1.3 times as long as hind tibia; apical 0.4–0.5 of fore wing subhyaline, distinctly paler than dark brown remaining part of wing; head mainly black, but orbits, clypeus and malar space yellowish; hindleg (except usually infuscate parts of hind tibia and of tarsus), mesosoma mainly and metasomal tergites (but sometimes apical tergites dark brown) of ♀ orange-brown; pterostigma and veins dark brown; colour of fourth and fifth tergites and mesosoma varies from mainly dark brown or black to completely yellowish brown; hindleg may be completely yellowish, except its dark brown telotarsus. Males are usually darker than females, having only second and third tergite laterally and entire first tergite of metasoma yellowish brown.

Host(s). Crambidae: *C. partellus*; Noctuidae: *B. fusca*.

Distribution. Comoro Islands, Kenya, Uganda.

Genus *Bracon* Fabricius (Figs 194–231, 286, 288)

Diagnosis. Small to medium-sized species; scapus not modified, distinctly longer than pedicellus, its inner apex simple, its outer apex truncate (Figs 194, 210, 221); face and clypeus differentiated, without reticulate sculpture or carinae and more or less convex; mesoscutum usually partly glabrous; vein r of fore wing shorter than vein 2-SR (Figs 194, 213, 219); vein 1r-m of hind wing short (Figs 201, 219); first discal cell of fore wing comparatively high; vein CU1a of fore wing far below level of vein 2-CU1; veins 1- and 2-CU1 of fore wing straight; vein 1-R1 of fore wing (metacarp) longer than pterostigma (Fig. 219), exceptionally about as long or shorter; vein CU1b of fore wing slender; angle between veins 1-SR and C+SC+R of fore wing 75°–90° (Figs 194, 213, 219); tarsal claws simple (Fig. 209) or with lobe (Fig. 226); propodeum with or without median carina; first metasomal tergite movably joined to second tergite, its length less than 1.7 times its apical width, and without strong constriction behind spiracles (Figs 195, 206, 227); second tergite without distinct V-shaped area or grooves medioanteriorly (Figs 195, 227); basal width of third tergite usually less than 2.7 times its median length; third tergite without anterolateral grooves (Figs 205, 222); upper valve of ovipositor normal and its lower valve with minute teeth.

Biology. The genus *Bracon* contains (partly gregarious) idiobiont ectoparasitoids of larvae with a more or less concealed way of life, belonging to Lepidoptera (Crambidae, Pyralidae, Tortricidae, Gelechiidae, Sesiidae, Coleophoridae, Gracillariidae and Cochylidae), Coleoptera (Curculionidae, Anobiidae and Bruchidae), Diptera (Cecidomyiidae, Lonchaeidae, Chloropidae, Tephritidae, Anthomyiidae, Agromyzidae and Scatophagidae) and, less commonly, also Hymenoptera (Cephidae, Tenthredinidae, Eurytomidae and Cynipidae). Four species of *Bracon* are known to be parasitoids of cereal stem borers in Africa; two belong to the subgenus *Habrobracon*, having a wide range of hosts (including several pyralid hosts of the genera *Ephestia, Plodia* and *Galleria* in stored products), and the status of one of the two as a distinct species is questioned by Tobias (1986), but it is here considered to be a valid species. Both the other species belong to *Bracon sensu stricto*, and are found to be definitely parasitoids of cereal stem borers: *Bracon sesamiae* Cameron and *Bracon testaceorufatus* Granger.

Distribution. Cosmopolitan and common. *Bracon* is one of the largest known genera of Braconidae, currently containing about 1700 described species. No revision or compilation of the Afrotropical species of the genus *Bracon* is available, but Granger (1949) treated the Madagascan species.

Remarks. The genus *Habrobracon* is sometimes regarded as distinct from the related genus *Bracon*, but even a combination of characters proved to be insufficient to come to a satisfactory separation. Therefore, in this chapter we prefer to include it as a subgenus of the genus *Bracon sensu lato*, as has been done before by Tobias (1986).

Key to Species of the Genus *Bracon* Parasitizing African Cereal Stem Borers

1. Antenna of ♀ with 14–18 segments, short, not (or slightly) longer than length of head and mesosoma combined (Figs 197, 217); first metasomal tergite completely smooth (Figs 195, 206); second submarginal cell of fore wing comparatively short (Figs 194, 213); vertex coriaceous, dull (Fig. 203), and dorsolateral carinae of first metasomal tergite absent (Fig. 205); vein 1-SR of fore wing usually short or absent (Figs 194, 213); subgenus *Habrobracon* Ashmead 2

– Antenna of ♀ with 26–36 segments, distinctly longer than length of head and mesosoma combined (Figs 221, 222, 231); first tergite distinctly sculptured medioposteriorly (Fig. 227); second submarginal cell of fore wing comparatively long (Fig. 219); vertex largely smooth, shiny; if coriaceous (Fig. 224), then dorsolateral carinae of first metasomal tergite present (Figs 222, 227); vein 1-SR of fore wing medium-sized (Fig. 219); subgenus *Bracon* 3

2. Antenna of ♀ with 13–14 segments, antenna shorter than length of head and mesosoma combined or of equal length (Fig. 197); setae of vertex erect, reaching above upper level of posterior ocellus (Fig. 198); vein 3-SR of fore wing 0.9–1.2 times vein r, rarely up to 1.4 times (Fig. 194); antennal segments of ♂ 20–23 (Fig. 212) .... *Bracon hebetor* Say

– Antenna of ♀ with 15–18 segments, antenna slightly longer than length of head and mesosoma combined (Fig. 217); setae of vertex more adpressed, mostly not reaching above upper level of posterior ocellus (Fig. 218); vein 3-SR of fore wing 1.2–1.8 times vein r (Fig. 213); antennal segments of ♂ 22–27 (Fig. 215) .... *Bracon brevicornis* (Wesmael)

3. Fourth and fifth metasomal tergites (except posterior margin) distinctly granulate; pterostigma of ♀ dark brown or brown medially (Fig. 219); hind tibia dorsally and laterally similarly coloured; rugae and striae of propodeum and of second tergite mediodorsally fine and regular (Fig. 227) or absent; mesoscutum anteromedially usually similarly coloured to lateral lobes anteriorly; Afrotropical (excluding Madagascar) 4

– Fourth and fifth tergites usually smooth or superficially granulate; pterostigma of ♀ pale yellowish medially (Fig. 219); hind tibia dorsally (except subbasally) usually darker than laterally, sometimes only basally dark brown; rugae and striae of propodeum and of second tergite mediodorsally coarse and more or less vermiculate (Figs 286, 288); mesoscutum anteromedially usually darker than lateral lobes anteriorly; Afrotropical (including Madagascar) .... *B. testaceorufatus* Granger

4. Lateral lobes of mesoscutum largely setose (Fig. 222); posterior third of notauli area more or less coriaceous, rather mat; base of hind tibia pale yellowish, distinctly contrasting with darker middle part of tibia; propodeum and second tergite medially with some fine striae or rugae (Fig. 227); apical half of hind tibia and hind tarsus mainly dark brown; notaulic area of mesoscutum broadly yellowish posteriorly (but in ♂ this area frequently largely black) .... *B. sesamiae* Cameron

– Lateral lobes of mesoscutum largely glabrous; posterior third of notauli area smooth, shiny; base of hind tibia brownish yellow, similar to colour of middle part of tibia; propodeum and second tergite only medially granulate; apical half of hind tibia and hind tarsus (except telotarsus) yellowish brown; notaulic area of mesoscutum narrowly yellowish or black posteriorly .... *Bracon* spec. A

Bracon (Habrobracon) brevicornis (Wesmael) (Figs 213–218)

Diagnosis. Length of fore wing 1.9–2.6 mm, of body 1.8–2.6 mm; antennal segments of ♀ 15–18, antenna slightly longer than length of head and mesosoma combined (Fig. 217), its segments submoniliform (Fig. 217); antennal segments of ♂ 22–27 (Fig. 215); setae of vertex comparatively adpressed, mostly not reaching above upper level of posterior ocellus (Fig. 218); clypeus nearly flat to convex (Fig. 216); clypeus, face and frons finely granulate; setae of face rather short (Fig. 216); vertex mainly smooth; mesoscutum largely setose, shiny, smooth; scutellar sulcus narrow and distinctly crenulate; vein 3-SR of fore wing 1.2–1.8 times vein r (Fig. 213); veins 1-SR and 3-SR short, but 1-SR may be rather long (Fig. 213); marginal cell of fore wing medium-sized, ending distinctly removed from apex of wing (Fig. 213); tarsal claws with small acute lobe (cf. Fig. 199); propodeum superficially granulate; length of first tergite about 0.8 times its apical width, its surface smooth; dorsolateral carinae of first tergite absent; second tergite without mediobasal area; second metasomal suture rather deep, nearly straight and finely crenulate; second to seventh segments superficially granulate, shiny; length of ovipositor sheath 0.20–0.29 times fore wing and about twice length of first tergite. Colour is very variable: pterostigma rather dark brown, with sometimes a (faint) pale basal spot; body nearly completely yellowish brown to largely dark brown or black, with frons, orbits, vertex anteriorly, face laterally, pronotum dorsally, notaulic area, scutellum (except usually medioanteriorly), first tergite (except medioposteriorly) and following tergites laterally pale yellowish; basal two-thirds of wings more or less infuscate, remainder of wings subhyaline, but sometimes infuscate apically.

Host(s). An extremely polyphagous ectoparasitoid, attacking Crambidae and Pyralidae in stored products and in the field; in the field other lepidopterous families may also be attacked.

Distribution. Cosmopolitan.

Bracon (Habrobracon) hebetor Say (Figs 194–201, 212)

Diagnosis. Length of fore wing 1.7–2.6 mm, of body 1.3–2.7 mm; antennal segments of ♀ 13–14, submoniliform (Figs 196, 197), antenna shorter than length of head and mesosoma combined or of equal length (Fig. 197); antennal segments of ♂ 20–23 (Fig. 212); setae of vertex erect, partly reaching above upper level of posterior ocellus (Fig. 200); clypeus nearly flat to convex (Fig. 200); clypeus, face and frons finely granulate; face with some long setae (Figs 197, 200); vertex mainly smooth; mesoscutum largely setose, shiny, smooth; scutellar sulcus narrow and distinctly crenulate; vein 3-SR of fore wing 0.9–1.2 times vein r, rarely up to 1.4

times (Fig. 194); veins 1-SR and 3-SR short (Fig. 194); marginal cell of fore wing medium-sized, ending distinctly removed from apex of wing (Fig. 194); tarsal claws with small acute lobe (Fig. 199); propodeum superficially granulate or smooth; length of first tergite about 0.8 times its apical width, its surface smooth; dorso-lateral carinae of first tergite absent (Fig. 195); second tergite without mediobasal area; second metasomal suture shallow to rather deep, nearly straight and usually smooth (Fig. 195); second to seventh segments superficially granulate to (mainly) smooth, shiny; length of ovipositor sheath 0.21–0.30 times fore wing and about twice length of first tergite. Colour: very variable and similar to that of *B. brevicornis*: pterostigma rather dark brown, with sometimes a large pale basal spot; body nearly completey yellowish brown to largely dark brown or black.

Host(s). An extremely polyphagous ectoparasitoid, like *B. brevicornis*; attacking Crambidae and Pyralidae in stored products, and in the field other lepidopterous families (Noctuidae, Tortricidae, Gelechiidae) may also be attacked.

Distribution. Cosmopolitan.

Bracon sesamiae Cameron (Figs 219–228; Plate 3:3)

Diagnosis. Length of fore wing 2.8–3.7 mm, of body 3.2–4.1 mm; antennal segments of ♀ and ♂ 26–36, antenna distinctly longer than length of head and mesosoma combined, moderately slender (Fig. 221); frons granulate, rather mat and with longitudinal groove (Fig. 220); vertex smooth, shiny; clypeus narrow (Fig. 224); face granulate; mesoscutum largely setose, especially near notauli densely so; posterior third of notaulic area more or less coriaceous, and rather mat; mesoscutum and scutellum punctulate; notauli narrow, shallow, indistinct posteriorly; scutellar sulcus deep and distinctly crenulate (Fig. 228); veins 1-SR and 3-SR medium-sized (Fig. 219); marginal cell of fore wing elongate (Fig. 219); tarsal claws with large acute lobe (Fig. 226); propodeum with short median carina posteriorly and with fine curved rugae in front of it (Fig. 228), flattened mediobasally; dorso-lateral carinae of first tergite strong (Figs 222, 227); length of first tergite about 0.9 times its apical width, its surface distinctly rugose posteriorly and without distinct lateral carinae (Fig. 227); second to sixth tergites (except posterior margin) mainly coriaceous, second tergite also with some comparatively fine longitudinal rugae and no mediobasal area (Fig. 227); second metasomal suture deep, sinuate and finely crenulate (Fig. 227); length of ovipositor sheath 0.35–0.39 times fore wing and about 2.5 times length of first tergite. Colour of lectotype of *B. sesamiae* Cameron: head and mesosoma mainly dark brown; palpi, pedicellus, orbits (except for interruption medioposteriorly), clypeus, malar space, notaulic area, scutellum, pronotum dorsally, tegulae, fore- and middle legs (except middle tarsus), hind trochanter and trochantellus, hind femur, base of hind tibia, most of veins, metanotum, propodeum anteromedially and laterally narrowly, and metasoma brownish-yellow; mesopleuron and metapleuron largely rather pale brown; pterostigma brown; hind femur often partly or largely dark brown. Males have metasoma dark brown apically. The holotype of *B. sesamiae* Risbec is much paler: head with only pair of patches on frons and on occiput, and stemmaticum dark brown, base of antenna mainly brown, and basal half of hind tibia pale yellowish;

hind coxa brownish yellow and pterostigma dark brown. Third and following tergites of male paratype more or less dark brown. One female from Tanzania (Lushoto) has vein r-m of fore wing absent.

Host(s). Noctuidae: *B. fusca*; *Sesamia* spp. Crambidae: *C. partellus*.

Distribution. Cameroon, Ethiopia, Senegal, South Africa, Tanzania, Uganda, Zimbabwe.

Bracon testaceorufatus Granger (Figs 229–231, 286, 288)

Diagnosis. Length of fore wing 1.5–3.5 mm, of body 2.0–4.5 mm; antenna with 26–38 segments, distinctly longer than length of head and mesosoma combined, rather slender (Fig. 231); clypeus smooth, flat, shiny, narrow, dorsal carina present or absent; frons and face finely granulate; vertex smooth; mesoscutum largely glabrous, shiny, smooth, including posterior half of notaulic area; scutellar sulcus comparatively wide and distinctly crenulate; veins 1-SR and 3-SR medium-sized (Fig. 229); marginal cell of fore wing elongate (Fig. 229); tarsal claws with small truncate lobe (Fig. 230); propodeum coarsely vermiculate–rugose (Fig. 286) or at least with some rugae; length of first tergite about 0.9 times its apical width, no distinct dorsal carinae and its surface coarsely rugose medioposteriorly (Fig. 288); dorsolateral carinae of first tergite coarse, not lamelliform; second tergite coarsely (vermiculate–)rugose mediodorsally (Fig. 286) and remainder distinctly granulate; second metasomal suture rather deep, nearly straight and finely crenulate; third and fourth segments superfically granulate, shiny, following tergites smooth or nearly so; length of ovipositor sheath 0.35–0.42 times fore wing and 2.5–2.9 times length of first tergite; pterostigma pale yellowish (brown) medially, contrasting with darker lateral parts; hind tibia with dorsal dark brown streak (interrupted by sub-basal whitish band), rarely indistinct; body mainly orange-brown; antenna, stemmaticum, most of veins, ovipositor sheath, telotarsi, mesoscutum medially (laterally variable, may be orange-brown to dark brown), propodeum, first tergite, second tergite medially, and sometimes third and fourth tergites largely, and ovipositor sheath dark brown or black. Males have the pterostigma largely dark brown, and some have apical half of metasoma largely dark brown.

Host(s). A gregarious, idiobiont ectoparasitoid of larvae. Crambidae: *Chilo diffusilineus*, *Chilo zacconius*, *Chilo* sp.; *Scirpophaga* sp. Pyralidae: *Maliarpha separatella*.

All hosts attacked in rice. With six to 14 specimens living on one host caterpillar, an average fecundity of 25 eggs and a developmental time of 12 days (Polaszek *et al.*, 1994). The larva pupate within the rice stem, a few millimetres below the host's remains. Emergent adults leave the stem via the aperture made by the host larva. In Madagascar, at least part of the population undergoes larval diapause, whereas the remainder is active throughout the year (Bianchi *et al.*, 1991).

Distribution. Cameroon, Ghana, Ivory Coast, Kenya, Madagascar, Mali, Mauritania, Mozambique, Nigeria, Senegal, Tanzania, Uganda.

Bracon sp. A

This species was reared from Tanzania and Uganda from *B. fusca* by J. Milner. Unfortunately, because of a lack of adequate material and because this species is very similar to *B. sesamiae*, it has not been described. In addition to the differences given in the key, it has the wings slightly darker than in *B. sesamiae*.

Host(s). Noctuidae: *B. fusca*.

Distribution. Tanzania, Uganda.

Genus *Digonogastra* Viereck (Figs 232–238)

Diagnosis. Length of fore wing 3–16 mm; scapus distinctly concave apically, ventrally longer than dorsally in lateral view, and medium-sized (Fig. 234); second (= pedicellus), third and fourth antennal segments normal, cylindrical (Fig. 234); face without protuberances and without reticulate sculpture, usually largely smooth or coriaceous; frons usually densely setose anterolaterally; vein 1-SR distinctly angled with parastigma, angle between veins 1-SR and C+SC+R 60°–80° (Fig. 235); vein 1-SR+M of fore wing nearly straight to distinctly angled basally; vein 3-M of fore wing terminating at about same level as vein SR1 or more basally (Fig. 237); vein 1r-m of hind wing comparatively long and angle with vein 1-SC+R very sharp (Fig. 237); fore basitarsus normal, hardly compressed, subcylindrical; tarsal claws simple, without acute lobe (Fig. 238); second metasomal tergite (largely) smooth, without converging grooves, its mediobasal area distinctly differentiated and usually posteriorly connected to a median carina (Fig. 236); third tergite with anterolateral grooves (Fig. 236); third to fifth tergites without transverse groove posteriorly (but present in *Digonogastra ornatus* (Provancher)); upper valve of ovipositor normal, not enlarged, straight, and its lower valve with some minute teeth (Fig. 233); rarely upper valve of ovipositor widened and without preapical nodus and ventral tooth reduced; ovipositor sheath usually about as long as metasoma or longer, but as short as 0.25 times length of metasoma occurs.

Distribution. Large genus, with a New World distribution; one species has been introduced in Africa to control crambid and pyralid stem borers. The genus has never been revised.

Biology. Solitary or gregarious idiobiont ectoparasitoids of concealed larvae of Coleoptera (Cerambycidae) and of Lepidoptera (Crambidae, Pyralidae, Psychidae).

Remarks. May be confused with the Old World genus *Iphiaulax*, but differs by the very sharp angle between vein 1-SC+R and 1r-m of hind wing (Fig. 237 versus Figs 241, 242), the more or less developed mediobasal area of second tergite (usually connected to a median carina posteriorly; Fig. 236 versus Fig. 250) and the normal upper valve of the ovipositor (Fig. 233 versus Fig. 251). If, rarely, the upper valve is enlarged, then the frons is densely setose anterolaterally and the fourth and fifth tergites lack the transverse posterior grooves. In *Iphiaulax*, the upper frons is sparsely setose anterolaterally and the fourth and fifth tergites possess transverse grooves posteriorly (Fig. 245). The males of *Digonogastra* have the digitus with three to four tooth-like processes, but males of *Iphiaulax* have only one such process (Quicke, 1988).

Digonogastra kimballi Kirkland (Figs 232–238)

Diagnosis. Length of fore wing 3.3–7.5 mm, of body 3.7–9.0 mm; antenna with 39–61 segments (♀ and ♂); face nearly smooth to rugulose–punctate; clypeus with distinct curved ventral carina, and its dorsal carina absent; head strongly narrowed posteriorly; frons silvery setose and largely smooth; propodeum densely (partly long) silvery setose; metasoma largely smooth; median carina of second tergite absent or nearly so (Fig. 236); second metasomal suture wide, sinuate, smooth or more or less crenulate (Fig. 236); length of ovipositor 1.0–1.2 times length of metasoma, and about 0.7 times fore wing; head completely black (rarely with red patch near eye ventrally); mesoscutum reddish, remainder of mesosoma more or less dark brown, propodeum usually with a pale median streak; metasoma of ♀ completely reddish, of ♂ marked with black apically; pterostigma, parastigma and wing membrane dark brown (Fig. 237).

Host(s). Gregarious ectoparasitoid. Crambidae: *Diatraea grandiosella* Dyar, *Diatraea lineolata* (Walker), *Diatraea magnifactella* Dyar, *Diatraea saccharalis* (Fabricius), and *Diatraea considerata* Heinrich. Reared from the pyralid *E. saccharina* only in the laboratory (South Africa; see Smith, J.W., *et al.*, 1993b). The essentially identical *Digonogastra solitaria* Wharton & Quicke, differs mainly from *Digonogastra kimballi* by its biology. It is a solitary parasitoid of the crambid *Eoreuma loftini* (Dyar) and two *Diatraea* species in southern Texas and north-eastern Mexico (Wharton *et al.*, 1989). Its males have the apex of the metasoma red rather than black.

Distribution. Nearctic: Mexico and USA, and introduced into the Afrotropical region: South Africa.

Remarks. *Digonogastra kimballi* was introduced into South Africa against the pyralid *E. saccharina* in 1985. Apparently, the species did not establish, but it is included here should it ever do so.

Genus *Iphiaulax* Foerster (Figs 239–253, 360, 361)

Diagnosis. Length of fore wing 4–14 mm; scapus distinctly concave apically, ventrally longer than dorsally in lateral view, short ovoid (Fig. 243), inner apex simple or partly double-margined; second (= pedicellus), third and fourth antennal segments normal, cylindrical; apex of antenna setose and acute; inner side of eyes notched (Fig. 247); face without protuberances, and without reticulate sculpture, usually largely smooth, punctulate or coriaceous (Fig. 247); frons mainly glabrous; clypeus bordered dorsally by a carina composed of three more or less straight segments (Fig. 239) or carina evenly curved (Fig. 247); vein 1-SR distinctly angled with parastigma, angle between veins 1-SR and C+SC+R about 80° (Figs 241, 242); first discal cell of fore wing subrectangular (Fig. 241); vein 3-M of fore wing terminating at about same level as vein SR1 or more basally (Figs 241, 242); vein 1r-m of hind wing comparatively long and angle with vein 1-SC+R moderately sharp (Figs 241, 242); tarsal claws simple, without lobe (Fig. 248); second metasomal tergite (largely) smooth, without converging grooves, and no mediobasal area differentiated and no strong median carina (Fig. 250); third to fifth tergites with distinct anterolateral grooves (Fig. 245); sixth tergite largely retracted and smooth (Fig. 245); upper valve of ovipositor enlarged, much wider than lower valve, and its lower valve without teeth (Fig. 251); ovipositor sheath about as long as metasoma or shorter.

Biology. Solitary idiobiont ectoparasitoids of concealed larvae of Coleoptera (Cerambycidae, Buprestidae) and of Lepidoptera (Crambidae, Pyralidae).

Distribution. Rather large genus with an Old World distribution (mainly restricted to the subtropical and tropical parts), with several species in the Afrotropical region. The report from the Nearctic region by Quicke (1987) is due to confusion with *Digonogastra*. The genus has never been revised; Fahringer (1935) gives a compiled key to the Afrotropical species, mixed up with many non-*Iphiaulax* species.

Remarks. Both species reared from stem borers in Africa belong to the group with the vein C+SC+R of fore wing and the pterostigma conspicuously red, the occiput and the mesosoma completely reddish yellow, the third tergite distinctly striate, the fourth and fifth tergites without the median part strongly bulging and the fifth metasomal tergite smooth.

Key to Species of the Genus *Iphiaulax* Parasitizing African Cereal Stem Borers

1. Wing membrane nearly completely dark brown (Fig. 242); mesoscutum and temples with conspicuous long blackish setae (Fig. 245); vein r of fore wing dark brown; dorsal carina of clypeus evenly curved (Fig. 247); mesoscutum comparatively slender (Fig. 253); fifth metasomal tergite nearly flat (Fig. 360) .... *Iphiaulax pilisoma* van Achterberg.

– Wing membrane largely pale yellowish, only broad apical band and band below parastigma (and including a darker stigmal spot) dark brown (Fig. 241); mesoscutum with medium-sized brownish setae, setae of temples somewhat shorter; vein r of fore wing yellow; dorsal carina of clypeus angled (Fig. 239); mesoscutum robust to moderately slender (Fig. 240); fifth tergite distinctly convex near transverse groove (Fig. 361) .... *Iphiaulax dubiotusus* Shenefelt

Iphiaulax dubiotusus Shenefelt (Figs 239–241, 361)

Diagnosis. Length of body 5–8.5 mm and of fore wing 4–7.5 mm; inner apex of scapus slightly double-margined; antennal segments 56–68, antenna 1.2–1.3 times as long as fore wing; face sparsely punctate laterally, remainder superficially granulate (but sometimes nearly whole face granulate), rather dull; ♂ has face narrower than ♀ and completely coriaceous–granulate; dorsal carina of clypeus angled (Fig. 239) or nearly so; pronotum medioposteriorly sparsely punctate laterally; mesoscutum with medium-sized brownish setae, and setae of temples somewhat shorter than of *I. pilisoma*; mesoscutum robust to moderately slender (Fig. 240); third tergite distinctly striate; fourth and fifth tergites distinctly convex medially, weakly bulging near subposterior groove (Fig. 361), of ♀ finely aciculate anteriorly, remainder usually smooth, of ♂ largely finely and densely sculptured;

length of ovipositor sheath 0.35–0.40 times fore wing; stemmaticum and sometimes immediate surroundings of it, antenna (except radix) and ovipositor sheath black; head (including occiput and frons), mesosoma and metasoma completely reddish yellow; veins C+SC+R of fore- and of hind wings, and more or less M+CU1 of fore wing and M+CU of hind wing, and pterostigma conspicuously red; wing membrane largely pale yellowish, only broad apical band and band below parastigma (latter including a darker stigmal spot) dark brown (Fig. 241); vein r of fore wing yellow, more or less reddish basally.

Host(s). Crambidae: *C. zacconius*. Pyralidae: *E. saccharina*.

Distribution. Niger, Senegal, South Africa, Tanzania.

Iphiaulax pilisoma van Achterberg (Figs 242–253, 360)

Diagnosis. Length of body about 8 mm and of fore wing about 7.5 mm; inner apex of scapus slightly double-margined; face mainly finely punctate, shiny, only near antennal sockets somewhat granulate (Fig. 247); mesoscutum and temples with conspicuous long blackish setae (Fig. 249); dorsal carina of clypeus evenly curved (Fig. 247); pronotum medioposteriorly sparsely punctulate laterally; mesoscutum slender (Fig. 253); third tergite distinctly striate (Fig. 245); medial part in front of subposterior transverse groove of fourth and fifth tergites slightly convex (Fig. 360); fifth metasomal tergite smooth; length of ovipositor sheath about 0.3 times fore wing; head (including occiput, but stemmaticum and area around it black), mesosoma and metasoma completely reddish yellow; radix of antenna mainly yellowish; vein C+SC+R of fore wing and pterostigma conspicuously red; vein r of fore wing dark brown; wing membrane nearly completely dark brown (Fig. 242).

Host(s). Noctuidae: *Sesamia* sp.

Distribution. Senegal.

Remarks. In Africa, larger species (fore wing length about 13 mm) occur which are very similarly coloured. Besides the difference in size they differ from *I. pilisoma* in the shorter setae of the mesoscutum, the frons is more concave, the inner apex of the scapus has a distinctly double margin, the face is finely granulate between the punctures, the dorsal carina of the clypeus is angled (consisting of three straight parts), the radix of the antenna is completely black, the fifth tergite is distinctly sculptured medially and the medial part of fourth and fifth tergites is distinctly bulging in front of transverse subposterior groove.

Genus *Mesobraconoides* Sarhan & Quicke (Figs 254–267)

Diagnosis. Scapus short, robust, dorsally slightly longer than ventrally (Fig. 264); second, third and fourth antennal segments normal, cylindrical; face without protuberances and without reticulate–rugose sculpture, usually largely coriaceous (Fig. 255); second submarginal cell of fore wing much longer than high (Fig. 254); angle between veins 1-SR and C+SC+R of fore wing about 40° (Fig. 261); vein cu-a of fore wing distinctly postfurcal (Fig. 254); vein 1r-m of hind wing rather long and slightly curved (Fig. 254); propodeum completely smooth (Fig. 256); tarsal claws with simple apical tooth and large submedial lobe (Fig. 263); second metasomal tergite with wide converging grooves (Fig. 266); first to sixth tergite coarsely reticulate–rugose, not serrate laterally and posteriorly, and with acute lateral crease (Fig. 258); sixth tergite with truncate lamella posteriorly (Fig. 258); upper valve of ovipositor normal, its lower valve with small teeth, and its sheath shorter than metasoma, about 0.3 times fore wing.

Distribution. Afrotropical.

Biology. Parasitoids of concealed lepidopterous and coleopterous larvae.

Mesobraconoides psolopterus (Wilkinson) (Figs 254–267)

Diagnosis. Length of body 6–7.5 mm, of fore wing 7–8 mm; antennae with 57–62 segments, median segments considerably longer than wide; frons flattened, finely granulate and with shallow median groove (Fig. 259); body yellowish brown; antenna (but scapus and pedicellus brown), frons (except narrow rim near eyes) and vertex black; hind tarsus infuscate; wings largely rather dark brown; pterostigma yellow to orange basally, dark brown apically, but sometimes largely yellowish.

Host(s). Pyralidae: *M. separatella* (Sarhan and Quicke, 1990). Scolytidae (Coleoptera): *Xyleborus* sp.

Distribution. Nigeria, Sierra Leone.

Remarks. The inclusion of this species is based on two published records of its attacking *M. separatella* in Sierra Leone (Sarhan and Quicke, 1990). No material reared from *M. separatella* was examined. This record is considered questionable by Polaszek *et al.* (1994).

Genus *Stenobracon* Szépligeti (Figs 268–279, 292–304; Plate 10:4)

Diagnosis. Scapus ovoid and protruding ventrally (Fig. 273); second, third and fourth antennal segments normal, cylindrical; face without protuberances and without reticulate sculpture, usually largely smooth or coriaceous (Fig. 269); inner side of fore tibia with band of bristles or slender spines; angle between vein 1-M and vein C+SC+R of fore wing about 30° (Fig. 292); fourth metasomal tergite with distinct anterolateral grooves (its width is variable and intermediates occur; narrow especially in the subgenus *Stenobracon* (Fig. 304), and usually much wider in the subgenus *Euvipio* (Fig. 279)); ovipositor sheath much longer than metasoma.

Distribution. Afrotropical, Oriental, south Palaearctic: eight species recognized.

Key to African Species of the Genus *Stenobracon* Szépligeti

1. Fourth tarsal segments truncate or nearly so apically (Fig. 303); telotarsi in lateral view straight (Figs 274, 299, 316); notauli complete or nearly so (Figs 275,

300, 312); vein SR1 (slightly) sinuate (Figs 268, 292, 305); face usually largely smooth (Figs 293, 306) 2

– Inner corner of fourth tarsal segments protruding apically (Figs 329–331); telotarsi in lateral view curved (Fig. 333); notauli absent posteriorly (Fig. 324); vein SR1 (except apically) straight (Fig. 318); sculpture of face variable, sometimes densely sculptured .... [*Bathyaulax* Szépligeti]

2. Vein cu-a of fore wing vertical, or posterior half curved to base of wing (Figs 268, 292); anterolateral grooves of second and third tergites distinct, usually deep (Figs 279, 304); propodeum largely smooth; clypeus comparatively narrow (Figs 269, 293); vein 1r-m of hind wing long (Figs 268, 292) 3

– Vein cu-a of fore wing straight and distinctly reclivous (Fig. 305); anterolateral grooves of second and third tergites absent or shallow (Figs 309, 317); propodeum usually extensively sculptured; clypeus wide (Fig. 306); vein 1r-m of hind wing comparatively short (Fig. 305) .... [*Odesia* Cameron]

3. Second metasomal tergite without mediobasal area, or area indistinct (Fig. 279); vein 3-SR of fore wing 1.8–2.7 times vein r (Fig. 268), rarely up to 1.5 times; anterolateral grooves of third tergite distinctly crenulate and rather wide (Fig. 279); sixth tergite yellowish brown; Afrotropical. Subgenus *Euvipio* Szépligeti ... 4

– Second tergite with (minute) smooth mediobasal area, often weakly indicated and transverse (Fig. 304); vein 3-SR of fore wing 1.2–1.8 times vein r, rarely up to 2.2 times (Fig. 292); anterolateral grooves of third tergite comparatively narrow or obsolescent, often weakly crenulate or smooth (Fig. 304); sixth tergite frequently black; Oriental, south Palaearctic; introduced in Afrotropical region (Madagascar, Mascarenes, Mauritius, Réunion, Sudan). Subgenus *Stenobracon* Szépligeti 5

4. Fourth hind tarsal segment dark brown or black dorsally, similar to colour of telotarsus; setae on outer side of hind tibia dark brown or black; frons of ♀ with (large) black patch or wide black band posteriorly, enclosing stemmaticum, exceptionally completely yellowish or reddish; basal half of fore wing usually dark brown, except small area near base of wing ....
.... *Stenobracon rufus* (Szépligeti)

– Fourth hind tarsal segment yellowish brown dorsally and usually distinctly paler than telotarsus; if similarly coloured, then setae on outer side of hind tibia pale yellowish; frons of ♀ yellowish or with a narrow black band, rarely broadly black posteriorly; basal half of fore wing usually partly yellowish ....
.... *Stenobracon unifasciatus* (Brullé)

5. Vertex of ♀ (largely) yellowish; sixth tergite of ♀ yellowish; vein cu-a of fore wing may be far postfurcal; face of ♂ somewhat widened and head dorsally partly yellow; third tergite distinctly sculptured medially ....
.... *Stenobracon deesae* (Cameron)

– Vertex of ♀ largely blackish or with black band; sixth (and fifth) tergite of ♀ blackish medially; vein cu-a of fore wing interstitial or just postfurcal by about width of vein (Fig. 292); face of ♂ somewhat narrower and dorsally head completely blackish, except near antennal sockets; third tergite variable medially 6

6. Apical width of second metasomal tergite of ♀ 0.7–1.0 times its median length, of ♂ 0.9–1.2 times; third tergite of ♀ frequently smooth anteromedially, and usually anterolateral grooves rather weak or absent (Figs 294, 304) ....
.... *Stenobracon oculatus* Szépligeti

– Apical width of second tergite of ♀ 1.1–1.5 times its median length, of ♂ 1.2–1.6 times; third tergite of ♀ usually rugose–striate anteromedially, and with distinct anterolateral grooves .... *Stenobracon nicevillei* (Bingham)

Subgenus *Euvipio* Szépligeti (Figs 268–279)

Diagnosis. Second submarginal cell of fore wing usually slightly widened distad (Fig. 268); vein 3-SR of fore wing 1.8–2.7 times vein r (Fig. 268), rarely 1.5 times; first tergite robust, distinctly widened posteriorly; second metasomal tergite without mediobasal area, or area indistinct (Fig. 279); sixth tergite yellowish brown.

Biology. Solitary ectoparasitoids of Noctuidae, Crambidae, Pyralidae and Cossidae.

Distribution. Africa and south-west Palaearctic.

Stenobracon (*Euvipio*) *rufus* (Szépligeti) (Figs 268–279)

Diagnosis. Length of body of ♀ 10–13 mm (of ♂ 7–12 mm), and length of fore wing of ♀ 7.6–10.5 mm (of ♂ 5.7–9.1 mm), antennal segments of ♀ 83–100 (of ♂ 71–74); body reddish or dark yellowish brown; frons of ♀ with (large) black patch, or wide black band posteriorly, enclosing stemmaticum, exceptionally frons completely yellowish or reddish; antenna, ovipositor sheath and hind tarsus black(ish), but hind tarsus may be largely infuscate and only telotarsus dark brown; face of male pale yellow; fore tarsus and middle tarsus (dark) brown; wings with yellowish band below base of pterostigma, basal half of wings yellowish or dark brown (but frequently completely dark brown), apical 0.3–0.4 of pterostigma and remainder of wings largely dark brown except pale patch around veins 2-SR+M and r-m of fore wing; fourth hind tarsal segment dark brown or black dorsally, similar to colour of telotarsus; setae on outer side of hind tibia dark brown or black; mesosoma (including face of propodeum largely, at most some rugulosity posterolaterally) smooth; length of first tergite 1.0–1.3 times its apical width; first and second (largely, except anterolaterally), third and fourth tergites medially and largely posteriorly, longitudinally striate or rugulose; anterolateral depressions of second and third tergites wide, deep, and of fourth tergite comparatively narrow; third and fourth tergites with deep, finely crenulate transverse grooves, and apices of tergites slightly concave; fifth to seventh tergites smooth, without grooves; length of ovipositor sheath 1.20–1.38 times fore wing.

Hosts. Cossidae: *Phragmataecia boisduvalii*. Noctuidae: *B. fusca*, *Sesamia calamistis*, *Sesamia* sp. Crambidae: *C. partellus*; *Chilo* sp. Pyralidae: *E. saccharina*.

Distribution. Afrotropical (including Madagascar).

Stenobracon (*Euvipio*) *unifasciatus* (Brullé) (Plate 10:4)

Diagnosis. Length of body of ♀ 9–14.5 mm (of ♂ 6.5–11.5 mm), and length of fore wing of ♀ 7.0–11.6 mm (of ♂ 5.2–8.6 mm), antennal segments of ♀ 72–96 (of ♂ 67–87); body reddish or dark yellowish brown; frons of ♀ yellowish or with a narrow black band, rarely broadly black posteriorly; basal half of fore wing usually partly yellowish and fore wing frequently with yellowish patch above vein r-m; pterostigma may be completely yellow; face of male pale yellow; fourth hind tarsal segment yellowish brown dorsally and usually distinctly paler than telotarsus, if similarly coloured, then setae on outer side of hind tibia pale yellowish; length of first metasomal tergite 1.0–1.2 times its apical width; length of ovipositor sheath 1.20–1.37 times fore wing. Further very similar to *S. rufus*.

Host(s). Crambidae: *C. partellus*, *Chilo* sp., *Coniesta ignefusalis*. Noctuidae: *B. fusca*, *S. calamistis*, *Sesamia cretica*, *Sesamia nonagrioides*, *Sesamia* sp. Pyralidae: *E. saccharina*.

Distribution. Afrotropical (including Madagascar) and Arabian Peninsula.

Subgenus *Stenobracon* Szépligeti

Stenobracon (*Stenobracon*) *deesae* (Cameron)

Host(s). Reported to be a parasitoid of Crambidae: *C. partellus*, *Chilo sacchariphagus*, *Corcyra cephalonica*. Noctuidae: *Sesamia* sp.

Distribution. Oriental (India, China, Pakistan), south Palaearctic (Oman) and Afrotropical (introduced Madagascar, Mauritius, Sudan).

Stenobracon (*Stenobracon*) *nicevillei* (Bingham)

Host(s). Crambidae: *Chilo* sp., *Scirpophaga* sp.; Noctuidae: *Sesamia* sp.

Distribution. India, Nepal, Sri Lanka. Introduced into Madagascar and Réunion, but apparently without establishing. An aberrant female from Mozambique has the four basal metasomal tergites coarsely sculptured, the second to fourth tergites with a weak transverse impression, the seventh tergite black and the propodeum finely rugulose. It may belong to a new species near *S. nicevillei*.

Stenobracon (*Stenobracon*) *oculatus* Szépligeti (Figs 292–304)

Host(s). Crambidae: *C. sacchariphagus*, *Chilo* sp., *Scirpophaga* sp., ?*Scirpophaga excerptalis* (Walker) (= *Scirpophaga nivella* auct.). Noctuidae: *Sesamia* sp.

Distribution. Oriental (Indonesia, Malaysia, India, Philippines, Taiwan), south-east Palaearctic (Japan) and Afrotropical (introduced into Madagascar and Mauritius).

Genus *Tropobracon* Cameron (Figs 287, 289, 334–359)

Diagnosis. Head granulate, but sculpture may be superficial or absent dorsally; scapus suboval, gradually narrowed basally, its apex slightly protruding dorsally (Figs 336, 344), and its inner apical margin simple; apex of antenna normally setose, without spine (Fig. 335); mesoscutum largely glabrous, only near notauli with some setae (Fig. 287); notauli complete; antescutal depression present (Fig.

337); side of scutellum with round depression (Fig. 345); metapleural flange absent, except for a thickened rim; metanotum protruding medioposteriorly and its median carina absent (Fig. 345); propodeum granulate, reticulate or anteriorly largely smooth and shiny; propodeal spiracle round, just behind middle of propodeum and without lateral tubercles above it (Fig. 336); angle between veins 1-SR and C+SC+R of fore wing 30–40° (Figs 334, 346, 357); vein 1-SR short (Figs 346, 354); vein 1-SR+M of fore wing straight; vein 3-SR of fore wing straight, about as long as vein 2-SR or shorter (Figs 334, 346, 354); vein 1-R1 of fore wing somewhat longer than pterostigma, ending slightly basally of level of apex of vein 3-M (Fig. 354); vein 3-CU1 of fore wing slender; vein 1r-m of hind wing short, straight or slightly curved (Figs 334, 348, 356); tarsal claws with lobe (Fig. 341), setose; tarsal segments normal, with medium-sized setae; first metasomal tergite movably joined to second tergite, flat basally, with curved area enclosed by dorsal carinae (Fig. 343), in lateral view depressed basally (Fig. 336), without median carina or differentiated medioposterior area; second tergite with pair of converging narrow grooves (Fig. 343; which usually meet posteriorly in males, Figs 355, 359) and no mediobasal area; second and following tergites without anterolateral grooves; second to sixth tergites with sharp lateral crease; ovipositor normal, with ventral teeth subapically, no nodus; hypopygium of female acute apically, not surpassing apex of metasoma (Fig. 336); length of ovipositor sheath 0.3–0.4 times fore wing.

Biology. Gregarious ectoparasitoids of larvae of stem borers in *Gramineae*, belonging to the Crambidae, Pyralidae and Noctuidae (Lepidoptera). The genus appears to be exclusively parasitic on cereal stem borers in the Old World tropics (van Achterberg, 1993a). Five described species are known, of which two (*Tropobracon antennatus* (Granger), and *Tropobracon comorensis* van Achterberg) are known to be parasitoids of cereal stem borers in Africa (van Achterberg, 1993a; Polaszek *et al.*, 1994).

Distribution. Afrotropical and Oriental: five described species.

Key to Afrotropical Species of the Genus *Tropobracon* Cameron

1. Pterostigma yellowish; hind tibia pale yellowish; third to sixth tergites equally brownish yellow or dark brown, exceptionally with pale median stripe; vein cu-a of fore wing subinterstitial (Fig. 354), rarely antefurcal; free part of vein 1r-m of hind wing 0.4–0.6 times as long as vein 1-SC+R and combined part of vein 1r-m together (Figs 350, 354); continental Africa and Madagascar 2

– Pterostigma brown, medially paler than laterally; subbasal ring of hind tibia yellowish brown; third to sixth tergites with pale yellowish median stripe or triangular area medially; vein cu-a of fore wing slightly postfurcal (Fig. 357); free part of vein 1r-m of hind wing 0.6–0.7 times as long as vein 1-SC+R and combined part of vein 1r-m together (Fig. 356); Comoro Islands .... *T. comorensis* van Achterberg

2. Mesoscutum distinctly granulate and rather dull (Fig. 352); mesopleuron superficially granulate dorsally; scutellum with small pit medioanteriorly and granulate (Fig. 352); middle lobe of mesoscutum with shallow median groove medially (Fig. 352); vein r of fore wing somewhat shorter compared with vein 3-SR (Fig. 354); hind femur more robust (Fig. 351) .... *T. antennatus* (Granger)

– Mesoscutum smooth and strongly shiny; mesopleuron smooth; scutellum without pit medioanteriorly (or obsolescent) and only sparsely punctate; median groove of middle lobe of mesoscutum absent or nearly so; vein r of fore wing somewhat longer compared with vein 3-SR (Fig. 346); hind femur usually more slender (Fig. 349, but in holotype comparatively robust (Fig. 347)) .... *Tropobracon persimilis* (Szépligeti)

Tropobracon antennatus (Granger) (Figs 287, 289, 348, 351–353)

Diagnosis. Length of body 2.5–6.5 mm, of fore wing 2–6 mm; antenna with 50–65 segments; body yellowish (yellowish orange), but antenna and part of telotarsi dark brown; stemmaticum and second metasomal tergite brownish yellow; mesoscutum distinctly granulate and rather dull middle lobe of mesoscutum with shallow median groove medially (Fig. 352); scutellum with small pit medioanteriorly and granulate (Fig. 352); mesopleuron superficially granulate dorsally; vein r of fore wing somewhat shorter compared with vein 3-SR (Fig. 354); free part of vein 1r-m of hind wing about half as long as vein 1-SC+R and combined part of vein 1r-m (Figs 353, 354); vein cu-a of fore wing subinterstitial (Fig. 354), exceptionally antefurcal; pterostigma yellowish; hind femur more robust (Fig. 351); hind tibia pale yellowish, without whitish subbasal ring; third to sixth tergites equally brownish yellow or dark brown, exceptionally with pale median stripe. The male has second metasomal tergite much more slender than female, and consequently the triangular area is more slender and the triangle may be complete posteriorly (Fig. 355). Also the grooves on the second tergite are often much wider.

Host(s). A gregarious (idiobiont) ectoparasitoid. Noctuidae: *S. cretica*, *Sesamia* sp.; Crambidae: *C. zacconius* (= *Proceras africana* auct.), *C. ignefusalis*, ?*Scirpophaga* sp; Pyralidae: *M. separatella*. Risbec (1956a) listed also *Diopsis curva* Bertolini (Diopsidae), but this record requires confirmation (Polaszek *et al.*, 1994). In Madagascar, this species is uncommon on rice borers (P. Bousses, cited in Polaszek *et al.*, 1994).

Distribution. Cameroon, Ivory Coast, Kenya, Madagascar, Malawi, Mali, Mozambique, Niger, Nigeria, Senegal, Sierra Leone, Somalia, South Africa, Sudan, Togo, Uganda.

Remarks. *Tropobracon antennatus* is largely similar to the Oriental *Tropobracon luteus* Cameron. It differs in the sculpture of the mesoscutum and propodeum, and in its colour, of which the yellowish pterostigma is the easiest character to use.

Tropobracon comorensis van Achterberg (Figs 356–359)

Diagnosis. Length of body about 4.5 mm, free part of vein 1r-m of hind wing about 0.7 times as long as vein 1-SC+R and combined part of vein 1r-m together (Figs 356, 357); vein cu-a of fore wing just postfurcal (Fig. 357); third to sixth tergites with pale yellowish median stripe or triangular area medially; colour yellowish brown (including stemmaticum and telotarsi); palpi and tegulae pale yellowish; whole hind tibia yellowish brown, without (sub)basal ring; eighth and following antennal segments rather dark brown; wing membrane subhyaline; pterostigma and veins brown, but middle of pterostigma somewhat paler than its lateral parts.

Host(s). *Chilo partellus* (on rice).

Distribution. Afrotropical: Comoro Islands.

Remarks. Very similar to *T. luteus*, but differs in the sculpture of the propodeum and the sculpture of the second metasomal tergite. The longer vein 1-SC+R of hind wing (Figs 356, 357), the pit of scutellum, the yellowish brown stemmaticum and the slightly postfurcal vein cu-a of fore wing (Fig. 226) indicate that it belongs to the Afrotropical group of species, despite having a slightly shorter vein 2-SC+R, a comparatively slender hind femur and dark pterostigma. Therefore, it is likely that the parental stock came from Madagascar or the African continent.

Tropobracon persimilis (Szépligeti) (Figs 346, 347, 349, 350)

Diagnosis. Length of fore wing 4–4.5 mm; antennal segments of ♀ 49–53; stemmaticum and second metasomal tergite brownish yellow; mesoscutum smooth and strongly shiny; median groove of middle lobe of mesoscutum absent or nearly so; mesopleuron smooth; scutellum without pit medioanteriorly (or obsolescent) and only sparsely punctate; vein r of fore wing somewhat longer compared with vein 3-SR (Fig. 346); free part of vein 1r-m of hind wing about half as long as vein 1-SC+R and combined part of vein 1r-m together (Figs 348, 350); vein cu-a of fore wing subinterstitial (cf. Fig. 354), exceptionally antefurcal; pterostigma yellowish; hind tibia pale yellowish, without whitish subbasal ring; third to sixth tergites equally brownish yellow or dark brown, exceptionally with pale median stripe; hind femur usually more slender (Fig. 348, but in holotype comparatively robust (Fig. 347)).

Host(s). Unknown, but included here as it is almost certainly a parasitoid of stem borers, as are the other members of the genus.

Distribution. Afrotropical: Malawi, Tanzania, Zambia.

Remarks. Very similar to *T. antennatus*, but differs by the smooth and strongly shiny mesoscutum, the largely smooth scutellum without a medioanterior pit and the median groove of mesoscutum absent or nearly so.

CHELONINAE

Genus *Chelonus* Panzer (Figs 362–380, 394, 395)

Diagnosis. Number of antennal segments (of both sexes) 17–44, but in females of one species group (‘*Microchelonus/Chelonella*’) frequently fixed at 16; eyes setose; occipital carina remains separated from hypostomal carina, complete medio-dorsally; clypeus without medioventral teeth; propodeum with protruding carinae and more or less rugose; vein 1-SR+M of fore wing absent (Fig. 367); vein r of fore wing usually rising near middle of pterostigma or somewhat behind middle; vein 2-R1 of fore wing absent, or as a short stub (Fig. 363); vein CU1b present, resulting in a closed first subdiscal cell apicoposteriorly (Fig. 367); vein r of hind wing absent (Fig. 367); vein M+CU of hind wing about equal to vein 1-M or longer (Fig. 367); metasomal carapace without transverse grooves (Fig. 368), at most with one partial groove; carapace of male may have an apical opening (foramen: Figs 394, 395); third metasomal tergite obtuse or elongate; fourth and following metasomal tergites largely retracted.

Biology. Solitary ovolarval endoparasitoids of Lepidoptera. Milner (1967) recorded *Chelonus curvimaculatus* on two occasions from pupae of *B. fusca*, presumably erroneously.

Distribution. A large, cosmopolitan genus (about 450 spp.). Only two parts of the Afrotropical fauna of this genus have been monographed: for the species from Zaïre, see de Saeger (1946) and, for the species from Madagascar, see Granger (1949).

Key to Species of the Genus *Chelonus* Parasitizing African Cereal Stem Borers

1. Antennal segments 16 (♀) or 25–27 (♂); metasoma in lateral view distinctly widened apically (Fig. 366); carapace of ♂ with wide slit-like foramen posteriorly (Figs 394, 395); length of body 3–4.5 mm .... *C. curvimaculatus* Cameron

– Antennal segments 28–29 (♀) or 31 (♂); metasoma in lateral view less widened apically (Fig. 362); carapace of ♂ without slit-like foramen posteriorly; length of body 6–8 mm .... *Chelonus maudae* Huddleston

Chelonus curvimaculatus Cameron (Figs 366, 394, 395)

Diagnosis. Length of body 3.0–4.5 mm, of fore wing 2.1–3.1 mm; antenna with 16 (♀) or 25–27 (♂) segments, antenna parallel-sided; eyes rather setose; face densely and finely rugose, with its setae directed downwards; clypeus densely and finely punctate; mesoscutum with coarse median carina, and its notaulic area coarsely scrobiculate sculptured, remainder of mesosocutum comparatively finely and densely punctate; scutellum mainly finely and densely punctate; ovipositor sheath about as long as apical height of metasoma (Fig. 366, but largely retracted); posteroventrally metasoma of ♀ not impressed medially; metasoma truncate and rather high posteriorly, distinctly widened apicad (Fig. 366), of ♂ less widened in lateral view, and with wide slit-like aperture or foramen (Figs 394, 395); body black, but curved subbasal wide band of metasoma, parastigma (of ♀), palpi, middle and hind tarsi (except dark brown telotarsi, and hind tarsus may be largely dark brown except its basitarsus), spurs, basal half of hind tibia (except its dark brown base), middle tibia (except minute subbasal patch), and fore tibia ivory; coxae, middle femur largely and tegulae dark brown or black; hind trochanter and trochantellus, remainder of foreleg (but tarsus largely infuscate) and basal third of antenna mainly yellowish brown; remainder of antenna blackish; remainder of hindleg blackish; pterostigma, veins 1-R1, r, 2-SR, 3-SR, 2-M, 1-CU1, cu-a, and C+SC+R (except basally) of fore wing dark brown; wing membrane subhyaline; remainder of veins (including vein 1-M of fore wing) (pale) yellowish. Male has parastigma and most veins of fore wing dark brown.

Host(s). Parasitoid of Noctuidae (*B. fusca* and *S. calamistis*), and of Crambidae (*C. partellus*, *C. zacconius*).

Distribution. Congo, Kenya, Madagascar, Mauritius, Senegal, Somalia, South Africa, Sudan, Tanzania, Uganda, Zambia, Zimbabwe.

Chelonus maudae Huddleston (Figs 362–365)

Diagnosis. Length of body 6–8 mm, of fore wing 4–5 mm; antennal segments 28–29 (♀) or 31 (♂); metasoma in lateral view hardly widened and truncate apically (Fig. 231); carapace of ♂ without slit-like foramen posteriorly (Fig. 365); carapace of ♀ with rather narrow depresssion posteriorly (Fig. 364). This species was treated in detail by T. Huddleston in Polaszek *et al.* (1994).

Host(s). So far only reared from the pyralid *M. separatella.*

Distribution. Afrotropical: Senegal.

Genus *Phanerotoma* Wesmael (Figs 290, 291, 381–393, 396–401)

Diagnosis. Number of antennal segments (of both sexes) fixed to 23, rarely up to 27; eyes glabrous; clypeus with more or less developed medioventral teeth (but often minute; Fig. 387); occipital carina just meeting hypostomal carina ventrally, complete mediodorsally; propodeum with protruding carinae (Fig. 382) and more or less rugose; vein r of fore wing arising far behind middle of pterostigma (Figs 381, 397, 401); vein 2-R1 of fore wing absent, or as a short stub (Fig. 397); vein CU1b usually present, resulting in a closed first subdiscal cell apicoposteriorly (Fig. 401); vein 1-SR+M of fore wing present (Fig. 381); vein 1-SR of fore wing absent or nearly so (thus first discal cell sessile; Fig. 381); vein cu-a of fore wing distinctly postfurcal (Fig. 401); vein r of hind wing usually present (Fig. 381); vein M+CU of hind wing equal to vein 1-M or longer (Figs 381, 401); metasomal carapace with two transverse grooves (Fig. 382); third metasomal tergite about as long as second tergite or longer; fourth and following metasomal tergites largely retracted; carapace of male without an apical opening (foramen).

Biology. Solitary ovolarval endoparasitoids of larvae of Lepidoptera (Crambidae, Pyralidae, Noctuidae, Gelechiidae, Coleophoridae, Tortricidae). Frequently collected in light traps.

Distribution. Large, nearly cosmopolitan genus (except northern boreal areas; with nearly 200 species known). The Afrotropical fauna of this genus has never been monographed: de Saeger (1948) gives an incomplete compiled key, and Granger (1949) gives an incomplete key to the Madagascan species. Zettel (1990) described two species from South Africa.

Key to Species of the Genus *Phanerotoma* Parasitizing African Cereal Stem Borers

1. Pterostigma entirely pale yellowish; vein 2-SR of fore wing straight (Fig. 410); middle tibia without blister and completely yellowish brown (Fig. 411); hind tibia yellowish brown, without dark brown parts; antenna (except scapus and pedicellus) dark brown; propodeum with transverse carina; length of fore wing 4–7 mm, of body 5.5–8 mm .... *Phanerotoma saussurei* Kohl

– Pterostigma (except basally) dark brown; vein 2-SR of fore wing moderately to weakly curved (Fig. 401); middle tibia with distinct blister and pale yellowish or whitish subbasally (Fig. 398); apical third of hind tibia infuscate, more or less dark brown, remainder pale yellowish or whitish, except for weak infuscate ring near its basal 0.3; antenna largely yellowish brown; propodeum without distinct transverse carina, at most weakly developed; length of fore wing 2–4 mm, of body 3.7–5.5 mm .... *Phanerotoma leucobasis* Kriechbaumer

Phanerotoma leucobasis Kriechbaumer (Figs 398–404)

Diagnosis. Length of body 3.7–5.5 mm, of fore wing 2–4 mm; face coarsely obliquely rugose; vein 2-SR of fore wing moderately to slightly curved (Fig. 401); vein r of fore wing distinctly shorter than half length of vein 3-SR (Fig. 401); middle tibia with distinct subbasal pale yellowish blister (Fig. 398); metasoma of ♀ somewhat widened posteriorly in lateral view and widely emarginate apically (Fig. 400); metasoma slightly longer than mesosoma, and distinctly shorter than length of head and mesosoma combined; propodeum without distinct transverse carina, at most weakly developed, rugose and intermingled with finer sculpture; third tergite distinctly convex, but may be apically rather flat; body yellowish brown, with first and second tergites (second tergite not laterally) pale yellow; stemmaticum black; antenna largely yellowish brown; middle tibia pale yellowish or whitish subbasally; apical third of hind tibia more or less dark brown, remainder pale yellowish or whitish, except for weak infuscate ring near its basal 0.3; vein 1-M of fore wing, parastigma and pterostigma (except basally) dark brown.

Host(s). Noctuidae: *B. fusca*; Pyralidae: *Maruca testulalis* Geyer. According to de Saeger (1948; under *Phanerotoma ornatulopsis*), also parasitoid of *E[a]rias biplaga* Walker and of Pyralidae (*Edulicodes hylobatis* Ghesquière, and *Desmis horaria* Meyrick). Reared in Europe from Pyralidae: *Paramyelois transitella* (Walker), *Ectomyelois ceratoniae* (Zeller), *Cadra calidella* (Guenée) and in the laboratory from *Ephestia kuehniella* (Zeller). In Egypt from Gelechiidae: *Platyedra gossypiella* (Saunders) (van Achterberg, 1990).

Distribution. Afrotropical (Kenya, Cape Verde, Zaïre), Palaearctic (see van Achterberg, 1990) and Nearctic (introduced).

Phanerotoma saussurei Kohl (Figs 290, 291, 405–411)

Diagnosis. Length of body 5.5–8 mm, of fore wing 4–7 mm; face punctate–rugose; propodeum coarsely and densely reticulate, without distinct carinae dorsally; vein 2-SR of fore wing straight (Fig. 410); vein r of fore wing distinctly shorter than half length of vein 3-SR (Fig. 410); middle tibia without subbasal blister (Fig. 411); metasoma depressed (Fig. 406) and not emarginate posteriorly (Fig. 407); metasoma distinctly longer than mesosoma, and nearly as long as head and mesosoma combined; body orange/yellowish brown; legs entirely yellowish brown; stemmaticum black; antenna (except scapus and pedicellus) and occasionally metasoma posteriorly, dark brown; vein 1-M of fore wing, parastigma and pterostigma pale yellowish brown.

Host(s). Pyralidae: *M. separatella*; Crambidae: *C. zacconius*, ?*Scirpophaga* sp. in rice stems. See also Polaszek *et al.* (1994).

Distribution. Benin, Kenya, Madagascar, Mali, Senegal, Sierra Leone, Tanzania.

DORYCTINAE

Genus *Parallorhogas* Marsh (Figs 421–431; Plate 10:3)

Diagnosis. Length of body 3–5 mm; antennal segments 22–32; frons flat; mesopleuron largely smooth (Fig. 423); mesoscutal lobes largely smooth and glabrous, and mesoscutum with long mediolongitudinal carina (Fig. 426); propodeum distinctly areolate (Fig. 426); vein r arising between middle and base of pterostigma (Fig. 421); first subdiscal cell of fore wing open distally and rather robust (Fig. 421); vein m-cu of hind wing straight (Fig. 421); outer hind tibial spur discernible at ×80 (Fig. 428); outer aspect of hind tibia setose and striate (Fig. 430); second suture of metasoma present, but may be obsolescent (Fig. 429); second metasomal tergite without diverging grooves, distinctly striate; third and fourth metasomal tergites smooth; length of ovipositor sheath about half times fore wing.

Biology. Ectoparasitoids of stem-boring lepidopterous (Crambidae) or coleopterous (Bruchidae, Bostrichidae) larvae (Marsh, 1993).

Distribution. A small genus, with all five species known from the Afrotropical region, but *Parallorhogas pyralophagus* has been introduced into Africa (South Africa, Mauritius) from the Nearctic region (Mexico).

Parallorhogas pyralophagus (Marsh) (Figs 421–431; Plate 10:3)

Diagnosis. Length of body 2.8–4.5 mm, of fore wing 2.1–3.5 mm; antennal segments 22–28; frons partly and vertex largely transversely aciculate (Fig. 422); length of first metasomal tergite 0.7–0.8 times its apical width; first and second tergites mainly sublongitudinally striate, striae of second tergite distinctly converging medially; second suture straight and shallow; third tergite with weak transverse groove near basal 0.3 (Fig. 429), third and following tergites smooth; length of ovipositor sheath 0.42–0.48 times fore wing; head, legs and metasoma largely brownish yellow; mesosoma more or less yellowish brown; outer side of scapus, apex of antenna, sometimes mesoscutum partly, precoxal sulcus and its surroundings, telotarsi, propodeum medially and metasoma medially dark brown or infuscate; pterostigma pale yellowish; veins brown; wing membrane subhyaline.

Host(s). *Chilo auricilius* Dudgeon, *C. sacchariphagus* Bojer, *D. grandiosella* Dyar, *E. saccharina*, *E. loftini* (Dyar), *Maruca vitrata* (F.) and *Argyroploce schistace* Snellen.

Distribution. Nearctic: Mexico. Introduced (probably unsuccessfully) into Afrotropical (Mauritius, South Africa) and Oriental (India and Indonesia (Sumatra)) regions.

Remarks. Bennett *et al.* (1983) discussed this species (as *Allorhogas* sp.) as a potential biocontrol agent of stem borers. It was subsequently introduced into both South Africa and Mauritius, though apparently without success.

Genus *Rhaconotus* Ruthe (Figs 280–285, 412–420, 485–488)

Diagnosis. Length of fore wing 2–4 mm; mesoscutal lobes granulate, and mesoscutum without long mediolongitudinal groove (Fig. 416); propodeum without distinct areolation (Fig. 413); vein r arising near middle of pterostigma (Fig. 412); first subdiscal cell of fore wing closed distally and slender (Fig. 412); outer hind tibial spur distinctly discernible at ×80 (Fig. 412); outer aspect of hind tibia moderately to long setose; second suture of metasoma complete (Fig. 420); second metasomal tergite without diverging grooves (Fig. 413), with or without distinct subposterior transverse groove; third and fourth metasomal tergites sculptured.

Biology. Gregarious ectoparasitoids of stem-boring lepidopterous, dipterous or coleopterous larvae.

Distribution. Medium-sized Old World genus (with more than 80 species known): mainly from the Palaeotropics, with some species occurring in the Palaearctic region. Both species keyed below lack a distinct more or less straight transverse subposterior groove of the second tergite in front of the curved second metasomal suture (Figs 284, 285, 420).

Key to Species of the Genus *Rhaconotus* Parasitizing African Cereal Stem Borers

1. At least medial third of pterostigma largely dark brown (Fig. 486); antennal segments of ♀ 26–33; head and mesosoma less densely setose (Figs 282, 283, 488); second and third tergites more robust, widened posteriorly (Fig. 369) .... *Rhaconotus carinatus* Polaszek

– Pterostigma entirely pale yellowish (Fig. 485); antennal segments of ♀ 39–45; head and mesosoma densely setose (Figs 280, 281, 487); second and third tergites slender, parallel-sided posteriorly (Fig. 284) ....*Rhaconotus scirpophagae* Wilkinson

Rhaconotus carinatus Polaszek (Figs 282, 283, 285, 486, 488)

Diagnosis. Length of body 3–5 mm and of fore wing 2.0–3.5 mm; length of antenna 1.3–1.4 times fore wing; antennal segments of female 26–33; head and mesosoma comparatively sparsely setose (Figs 282, 283, 488); mesosoma rather robust (Fig. 282) and lobes largely glabrous; second metasomal suture distinctly curved (Fig. 285); body dark reddish brown; at least medial third of pterostigma dark brown (Fig. 486); second and third tergites more robust, widened posteriorly (Fig. 285); length of ovipositor sheath 0.5–0.6 times fore wing.

Host(s). Crambidae: *C. zacconius*. Pyralidae: *M. separatella*. For additional details, see Polaszek *et al.* (1994).

Distribution. Cameroon, Ghana, Madagascar, Nigeria, Senegal, Sierra Leone, Tanzania, Togo.

Rhaconotus scirpophagae Wilkinson (Figs 280, 281, 284, 485, 487)

Diagnosis. Length of body 4–6 mm and of fore wing 3–4 mm; length of antenna 1.7–1.9 times fore wing; antennal segments of female 39–45; head and mesosoma densely setose (Figs 280, 281, 487); mesosoma slender (Fig. 280) and lobes largely setose; second metasomal suture simple and slightly curved; body dark brown; pterostigma uniformly yellowish; second and third tergites slender, parallel-sided posteriorly (Fig. 284); length of ovipositor sheath 0.40–0.55 times fore wing.

Host(s). Crambidae: *C. partellus*, *Chilo* sp., *S. ?excerptalis*, *S. nivella*, *Scirpophaga* sp.; Pyralidae: *M. separatella;* Noctuidae: *B. fusca*.

Distribution. Ghana, Ivory Coast, Kenya, Nigeria, Senegal, Sierra Leone, Tanzania. Also Oriental (Pakistan, India and Indonesia (Java)) region.

EUPHORINAE

Genus *Meteorus* Haliday *Meteorus* sp. A

Diagnosis. The small, dark brown and damaged male voucher specimen does not allow an identification and its biology is doubtful; therefore we refrain from giving a diagnosis.

Host(s). Reported as parasitoid of *B. fusca* (Noctuidae), but the record is doubtful (Milner, 1967).

Distribution. Kenya.

MACROCENTRINAE

Genus *Macrocentrus* Curtis (Figs 432–447)

Diagnosis. Antenna often about as long as or somewhat longer than body (but sometimes distinctly shorter), with 24–61 segments; occipital carina absent; middle lobe of mesoscutum in lateral view usually distinctly protruding over lateral lobes (Fig. 435), but in several species hardly or not protruding; vein 2-CU1 of fore wing straight or nearly so; subbasal cell of fore wing usually not or slightly widened apically (but some Australian species have this cell distinctly widened distally), often with elongate yellowish or brownish patch (Fig. 433); vein cu-a of fore wing vertical, usually slender (but may be widened basally or apically), and only exceptionally distinctly bent; vein r-m of fore wing sometimes absent (*Macrocentrus* is the only genus of the Macrocentrinae in which this condition occurs); vein CU1a of fore wing without a faint brownish spot; first subdiscal cell of fore wing elongate to robust and partly glabrous or setose; angle between veins 1-SR+M and 1-M about

90° (Fig. 432); vein 3-M of fore wing normal, usually longer than twice vein 3-SR; marginal cell of hind wing narrow, subparallel-sided or somewhat widened apically (Fig. 432); vein SR of hind wing at most weakly curved basally and unsclerotized; vein 1r-m of hind wing straight and short to medium-sized; vein 2-SC+R of hind wing horizontal (= longitudinal) and marginal cell slightly or not widened basally; vein SC+R1 of hind wing straight or evenly bent, exceptionally abruptly bent towards anterior wing margin; vein r of hind wing absent; vein R1 of hind wing slender; inner spur of hind tibia 0.3–0.5 times hind basitarsus; spur of fore tibia 0.2–0.6 times fore basitarsus; tarsal claws with (Fig. 444) or without ventral lobe; inner hind claw similar to outer hind claw; hind coxa with at most a few transverse striae; all trochantelli with teeth; first metasomal tergite largely smooth or longitudinally striate or rugulose, exceptionally (some Afrotropical species) with transverse striation, its length 1.5–3.4 times its apical width, usually widened posteriorly; laterope of first tergite deep, distinctly differentiated from glymma; first tergite more or less shallowly concave mediobasally; length of ovipositor sheath 0.2–2.7 times fore wing; apex of ovipositor variable, with notch more or less developed subapically.

Biology. Solitary or gregarious endoparasitoids of Tortricidae, Gelechiidae, Oecophoridae, Pyralidae, Crambidae, Sesiidae, Noctuidae and Lycaenidae. Polyembryony (usually resulting in gregarious broods) frequently occurs in this genus.

Distribution. Cosmopolitan. Within the Macrocentrinae the dominant genus in the Holarctic and Palaeotropical regions, with nearly 100 species known.

Macrocentrus sesamivorus van Achterberg (Figs 432–447)

Diagnosis. Body and pterostigma completely yellowish; area between posterior ocellus and eye (= OOL) flat or largely so; clypeus nearly straight ventrally (Fig. 440); temple compared with eye in dorsal view comparatively short (Fig. 441); middle lobe of mesoscutum distinctly protruding above level of lateral lobes; metapleuron densely punctate–rugose ventrally, partly mingled with microsculpture and dorsally mainly punctate (Fig. 435); precoxal sulcus densely punctate, mingled with granulate microsculpture; vein 3-SR of fore wing 2.0–2.3 times vein 2-SR; vein cu-a of fore wing distinctly postfurcal, vertical; vein SC+R1 of hind wing slightly curved (Fig. 438); subbasal cell of fore wing partly glabrous distally (Fig. 433); vein 1-CU1 of fore wing nearly always distinctly widened; marginal cell of hind wing more or less widened apically (Fig. 432); hind coxa granulate dorsally; length of fore wing 5–6.5 mm; length of ovipositor sheath 1.5–1.9 times fore wing.

Host(s). Noctuidae: *S. calamistis;* Crambidae: *C. partellus.*

Distribution. Kenya, Somalia, Zaïre.

MICROGASTRINAE

Walker (1994) lists three genera of Microgastrinae so far known from African cereal stem borers: *Cotesia, Dolichogenidea* and *Glyptapanteles.* These three genera belong to a group traditionally regarded as '*Apanteles*' *sensu lato* (i.e. in the broadest sense as used before the publication of Mason's (1981) paper). Actually, *Dolichogenidea* is not always well separable from the genus *Apanteles* Foerster, because intermediates occur, and it may be at most a subgenus of the genus *Apanteles*, and also the recognition of both other genera may be problematical, especially of males.

Genus *Cotesia* Cameron (Figs 448–465; Plates 2:6, 3:1,2; 10:2)

One of the largest genera of the Microgastrinae, with more than 2500 species; occurring worldwide and often common. The Afrotropical species are not monographed, except for those from Zaïre (de Saeger, 1944) and Madagascar (Granger, 1949), both under *Apanteles*.

The taxonomy of the species comprising the '*flavipes* complex' (*C. flavipes*, *C. sesamiae* and *C. chilonis*) has been studied in some detail, initially by Sigwalt and Pointel (1980) and Polaszek and Walker (1991), and more recently by Kimani and Overholt (1995), Kimani-Njogu and Overholt (1997) and Kimani-Njogu *et al.* (1997).

Biology. More than half the species in the genus are gregarious; mostly parasitizing a wide variety of larvae of Macrolepidoptera.

Key to Species of the Genus *Cotesia* Parasitizing African Cereal Stem Borers

1. Mesosoma distinctly flattened in lateral view, 1.7–1.8 times as long as high (Fig. 450); antennal segments of ♀ robust (Figs 450, 459); mesoscutum comparatively sparsely punctate, interspaces wider than diameter of punctures (Fig. 457); sculpture of second tergite covering only part of notum of tergite (Fig. 457); mesosternum and hind coxa largely smooth (Fig. 450), but hind coxa may be coarsely punctate dorsally 2

– Mesosoma normal, in lateral view 1.3–1.4 times as long as high; antennal segments of ♀ slender (Fig. 458); mesoscutum densely punctate, interspaces less than diameter of punctures; sculpture of second tergite covering entire notum of tergite; mesosternum and hind coxa densely punctate .... *Cotesia ruficrus* (Haliday)

2. Aedeagus of male rounded at apex (Fig. 463); hind coxa usually largely light brown, dark brown or black; face less convex, and antennal sockets at level of middle of eye in lateral view (Fig. 459); antenna usually completely dark brown, with pedicellus yellow, and considerably paler than scapus; outer hind spur distinctly shorter than inner spur (Fig. 460) .... *C. sesamiae* (Cameron)

– Aedeagus of male truncate at apex (Fig. 464); hind coxa usually largely yellowish; face more convex, and antennal sockets below level of middle of eye in lateral view (Fig. 450); basal half of antenna usually completely (yellowish) brown, with pedicellus (partly) darker than scapus; outer hind spur slightly shorter than inner spur (Fig. 462) .... *C. flavipes* Cameron

Cotesia flavipes Cameron (Figs 448–457, 462, 464, 465)

Diagnosis. Length of fore wing about 1.5 mm; antenna robust (♀) or slender (♂), about 0.7 times (♀) or 1.3 times (♂) as long as body, both penultimate segments

1.3–1.5 times (♀) or about three times (♂) longer than wide; face more convex than in *C. sesamiae*, and antennal sockets below level of middle of eye in lateral view (Fig. 450); mesosoma depressed, 1.7–1.8 times as long as high; mesoscutum and scutellum sparsely punctate; mesosternum largely smooth; sculpture of second tergite covering only part of notum of tergite (Fig. 457); hind coxa shiny and nearly smooth; outer hind spur slightly shorter than inner spur (Fig. 462); aedeagus of male truncate at apex (Fig. 464); body black or dark brown; second and third metasomal tergites only laterally with yellow markings; basal half of antenna usually completely yellowish brown, except for more or less darkened pedicellus, which is (partly) darker than scapus; legs brownish yellow, hind tarsus more or less infuscate; pterostigma brown; antenna of ♂ yellowish brown. For additional description, see Walker (1994).

Host(s). Polyphagous on noctuid, crambid and pyralid stem borers. Used as biological control agent; for a review, see Walker (1994) and Potting (1996). Potting *et al.* (1997a, b) have described foraging and host-selection behaviour in this species.

Distribution. Oriental (South and South-East Asia); introduced and apparently established in Kenya, Madagascar, Mauritius and parts of the neotropics.

Cotesia ruficrus (Haliday) (Fig. 458)

Diagnosis. Length of fore wing 2.0–2.5 mm; antenna slender, about as long as body, the two penultimate segments each about 1.7 times longer than wide (Fig. 458); mesosoma normal, 1.3–1.4 times as long as high; mesoscutum and scutellum coarsely punctate; hind coxa dull and rugose–punctate; body black; combined second and third metasomal tergites distally to basal area with yellow markings; legs bright reddish or yellowish, hind femur darkened apically; sculpture of second tergite covering entire notum of tergite; mesosternum and hind coxa densely punctate.

Host(s). A gregarious and extremely polyphagous endoparasitoid; e.g. of Pyralidae: *M. separatella*; of Crambidae: *C. zacconius*; of Gelechiidae: *Pectinophora gossypiella* (Saunders); and of Noctuidae: *S. nonagrioides*, *S. calamistis*, *Mythimna loreyi* (Duponchel), *Chrysodeixis chalcites* (Esper) and *Agrotis ipsilon* (Hufnagel) (Polaszek *et al.*, 1994; Walker, 1994). *Cotesia ruficrus* accepts hosts feeding both internally and externally on plants.

Distribution. Cosmopolitan: in the Afrotropical region known from Cameroon, Ivory Coast, Madagascar, Mauritius, Nigeria, Senegal, Somalia, South Africa, Sudan, Uganda and Zimbabwe (Walker, 1994).

Cotesia sesamiae (Cameron) (Figs 459–461, 463; Plates 2:6; 3:1,2; 10:2)

Diagnosis. Length of fore wing about 1.8 mm; antenna robust (♀) or slender (♂), about 0.6 times (♀) or 1.4 times (♂) as long as body, both penultimate segments about 1.2 times (♀) or about 3.2 times (♂) longer than wide; face less convex than in *C. flavipes*, and antennal sockets below level of middle of eye in lateral view (Fig. 459); mesosoma depressed, 1.7–1.8 times as long as high; mesoscutum and scutellum sparsely punctate; mesosternum largely smooth; hind coxa shiny and coarsely punctate dorsally; outer hind spur slightly shorter than inner spur (Fig. 460); sculpture of second tergite covering only part of notum of tergite; aedeagus of male rounded at apex (Fig. 463); body black or dark brown; second and third metasomal tergites largely brown; basal half of antenna usually completely yellowish brown, except for more or less darkened pedicellus, which is (partly) darker than scapus; legs yellowish, only hind coxa blackish basally and usually largely yellowish; pterostigma dark brown. For additional description, see Walker (1994).

Host(s). Polyphagous on crambid, noctuid and pyralid stemborers. Used as biological control agent; for a review, see Walker (1994).

Distribution. Afrotropical region, including Madagascar, Mauritius and Réunion. The east Palaearctic and Oriental *C. chilonis* (Matsumura) from Japan (and introduced in Java) is very similar to *C. sesamiae* (Polaszek and Walker, 1991; Walker, 1994).

Genus *Dolichogenidea* Viereck (Figs 466–475)

A very large genus, with more than 1000 species; differs mainly from the genus *Cotesia* in their rather long ovipositors, used to reach more or less concealed living hosts, and the more or less modified hypopygium of the female. In this respect, very similar to *Apanteles* Foerster, but the latter differs in the shape of the plical (or anal) lobe of the hind wing, partly lacking a fringe of setae. The Afrotropical species are not revised, except for those from Zaïre (de Saeger, 1944) and Madagascar (Granger, 1949).

Biology. Mostly solitary parasitoids of larvae of small moths, partly leaf-miners, but occasionally gregarious and sometimes parasitizing Macrolepidoptera.

Distribution. Cosmopolitan.

Key to Species of the Genus *Dolichogenidea* Parasitizing African Cereal Stem Borers

1. First metasomal tergite coarsely sculptured and very robust, with a distinct tubercle anteromedially (Fig. 466); second tergite completely rugose (Fig. 466); third tergite yellowish .... *Dolichogenidea oryzae* Walker

– First tergite largely smooth or finely sculptured and less robust, without or with weak tubercle anteromedially (Figs 468, 469); second tergite (largely) smooth (Figs 468, 470); third tergite black 2

2. Pterostigma pale yellow or white, sometimes almost hyaline; ovipositor sheath longer than metasoma, 1.5–1.8 times length of hind tibia (Fig. 467) .... *Dolichogenidea polaszeki* Walker

– Pterostigma (light) brown, exceptionally partly yellowish; ovipositor sheath distinctly shorter than metasoma, about as long as hind tibia or shorter (Figs 473, 474) 3

3. Ovipositor sheath about as long as hind tibia; first tergite comparatively slender (Fig. 469) .... *Dolichogenidea fuscivora* Walker
– Ovipositor sheath less than half the length of hind tibia; first tergite comparatively robust (Figs 470, 475) 4
4. Propodeum indistinctly sculptured, mainly smooth and without costulae; length of fore wing 1.6–1.8 mm .... *Dolichogenidea cameroonensis* Walker
– Propodeum partly sculptured, and usually costulae present, but frequently only weakly developed; length of fore wing 2.0–2.3 mm ....*Dolichogenidea aethiopica* (Wilkinson)

Dolichogenidea aethiopica (Wilkinson) (Figs 471, 472, 474, 475)

Diagnosis. Length of fore wing 2.0–2.3 mm; mesoscutum densely and coarsely punctate; scutellum flat and smooth medially, punctate laterally; vein r of fore wing distinctly angled and vein 2-SR rather slender (Fig. 471); propodeum with costulae more or less developed, largely smooth posteriorly, partly rugose dorsally; first and second metasomal tergites smooth, only some superficial sculpture on first tergite; length of first tergite about equal to its apical width, and largely parallel-sided and without distinct tubercle anteromedially (Fig. 475); second tergite with diverging grooves shallow or virtually absent (Fig. 475); second metasomal suture distinct, shallow (Fig. 475); ovipositor sheath moderately wide and acute apically (Fig. 474), sparsely setose; length of ovipositor sheath about 0.15 times fore wing and about 0.5 times hind tibia; hypopygium of ♀ acute apically (Fig. 474); body blackish or dark brown; foreleg (except basal half of femur and coxa), basal half of middle tibia, middle tarsus (except telotarsus) and basal third of hind tibia pale brownish; remainder of legs dark brown; pterostigma light brown. For additional details, see Walker (1994).

Host(s). Externally feeding hosts belonging to the Arctiidae, Noctuidae, Saturniidae and Zygaenidae, and one internal feeding crambid host: *C. zacconius*.

Distribution. Afrotropical, widespread: Ivory Coast, Kenya, Sierra Leone, Somalia, South Africa, Sudan, Tanzania, Uganda (Walker, 1994).

Dolichogenidea cameroonensis Walker (Figs 470, 473)

Diagnosis. Length of fore wing 1.6–1.8 mm; mesoscutum densely and rather coarsely punctate; scutellum flat and smooth; vein r of fore wing distinctly angled and vein 2-SR rather robust; propodeum without costulae, smooth or nearly so and at most partly microsculptured, first and second metasomal tergites smooth, at most with some punctures on first tergite (Fig. 470); length of first tergite 1.0–1.1 times its apical width, largely parallel-sided and without distinct tubercle anteromedially (Fig. 473); second tergite with shallow diverging grooves; second metasomal suture absent or nearly so; ovipositor sheath moderately wide and acute apically (Fig. 473), sparsely setose; length of ovipositor sheath about 0.15 times fore wing and about 0.5 times hind tibia; hypopygium of ♀ acute apically (Fig. 473) or nearly so; body (dark) brown; foreleg (except basal half of femur), basal half of middle tibia and basal third of hind tibia, tarsus (except telotarsus) yellowish; remainder of legs dark brown; pterostigma light brown. For additional details, see Walker (1994).

Host(s). Unknown, reported to be collected from a caterpillar on rice (Walker, 1994).

Distribution. Afrotropical: only known from Cameroon (Walker, 1994).

Remarks. Van Achterberg and Polaszek (1996) mentioned that paratypes of *D. cameroonensis* could not be separated from those of *D. aethiopica*, and the two species may therefore be synonymous. Further investigations on the limits of variation in these species need to be undertaken before synonymizing them formally.

Dolichogenidea fuscivora Walker (Fig. 469)

Diagnosis. Length of fore wing about 2 mm; mesoscutum rather coarsely punctate, but sparsely so posteriorly; scutellum flat and smooth; vein r of fore wing distinctly angled, and vein 2-SR rather wide; propodeum without costulae, largely smooth, sparsely punctate and no complete areola; first tergite finely longitudinally rugose; length of first tergite about 1.5 times its apical width, tergite largely parallel-sided and without weak tubercle anteromedially (Fig. 469); second tergite smooth, with only shallow diverging grooves anteriorly; second metasomal suture absent; ovipositor sheath moderately wide and acute apically; ovipositor sheath about as long as hind tibia; body, hind coxa and antenna black; middle coxa, dorsal face of all femora, hind tiba (except basally), all tarsi and mandible light brown; fore femur ventrally, fore tibia, middle tibia ventrally, tegulae and palpi yellow; pterostigma brown. For additional details, see Walker (1994).

Host(s). Noctuidae: *B. fusca.*

Distribution. Ethiopia.

Dolichogenidea oryzae Walker (Fig. 466)

Diagnosis. Length of fore wing 2.5–3.1 mm; mesoscutum rather coarsely and densely punctate; scutellum flat and sparsely punctate; vein r of fore wing distinctly angled; propodeum with distinct costulae, and more or less complete areola and coarsely rugose–punctate; first and second metasomal tergites coarsely and densely reticulate–rugose; length of first tergite 0.7–0.9 times its apical width, widened apically and with tubercle medioanteriorly (Fig. 466); second tergite completely sculptured, without diverging grooves; second metasomal suture distinct, crenulate (Fig. 466); ovipositor sheath wide and truncate apically, densely setose; length of ovipositor sheath 0.5 times fore wing and 1.2 times hind tibia; head, mesosoma, first and second (and sometimes third) tergites black; fore and middle legs, apex of hind coxa, hind femur and basal two-thirds of hind tibia yellow; third usually and sometimes fourth tergites (largely) yellow; antenna, remaining tergites, apical third of hind tibia and tarsus and base of hind coxa (dark) brown; pterostigma brown. For additional details, see Walker (1994).

Host(s). A solitary parasitoid; Crambidae: *C. diffusilineus*, *C. zacconius* and *Chilo* spp.

Distribution. Afrotropical: West Africa (Ivory Coast, Niger, Senegal; Walker, 1994).

Dolichogenidea polaszeki Walker (Figs 467, 468)

Diagnosis. Length of fore wing 2.9–3.3 mm; mesoscutum rather coarsely and densely punctate; scutellum flat and sparsely punctate; propodeum without distinct costulae, coarsely punctate and partly rugose posteriorly; vein r of fore wing hardly angled, forming an even curve with vein 2-SR, and vein 2-SR slender; first tergite punctate–rugulose, parallel-sided (Fig. 468); length of first tergite about 1.3 times its apical width, parallel-sided and with indistinct tubercle medioanteriorly (Fig. 468); second tergite smooth, transverse, with short curved grooves anteriorly (Fig. 468); second metasomal suture shallow, smooth; ovipositor sheath moderately wide and acute apically, densely setose (Fig. 467); length of ovipositor sheath 0.5 times fore wing and 1.5–1.8 times hind tibia; body black, except basal sternites laterally; sternites laterally (often with brown markings), legs (except brown fore and middle coxae and black hind coxa) and tegulae yellow; antenna brown, sometimes apical segments paler; pterostigma pale yellow or whitish, sometimes almost hyaline. For additional details, see Walker (1994).

Host(s). Noctuidae: *B. fusca, Poeonoma* sp., *S. calamistis, Sesamia botanephaga, Sesamia* sp.; Crambidae: *C. partellus*; Pyralidae: *E. saccharina.*

Distribution. Benin, Ghana, Kenya, Malawi, Nigeria, Uganda, Zambia.

Genus *Glyptapanteles* Ashmead (Figs 476–482)

A very large genus with about 1000 species; differs mainly from the genus *Cotesia* in the shape of the first and second tergites. The Afrotropical species are not revised, except for those from Zaïre (de Saeger, 1944) and from Madagascar (Granger, 1949).

Biology. Mostly gregarious parasitoids of larvae of, especially, Macrolepidoptera.

Distribution. Cosmopolitan.

Key to Afrotropical Species of *Glyptapanteles* Parasitic on Cereal Stem Borers

1. Vein r of fore wing distinctly angled with vein 2-SR of fore wing and vein 2-SR more robust (Fig. 480); ovipositor sheath thick and rounded apically (Fig. 476); hind femur completely yellow; first tergite gradually narrowed apically (Fig. 476) .... *Glyptapanteles maculitarsis* (Cameron)

– Vein r of fore wing hardly angled, more or less forming with vein 2-SR of fore wing a weak, even curve and vein 2-SR slender (Fig. 482); ovipositor sheath thin and acute apically (Fig. 479); hind femur at least partly (dark) brown; first tergite only apically narrowed (Fig. 481) .... *Glyptapanteles africanus* (Cameron)

Glyptapanteles africanus (Cameron) (Figs 479, 480, 482)

Diagnosis. Length of fore wing about 2 mm; mesoscutum finely punctate; scutellum flat and smooth; vein r of fore wing hardly angled, more or less forming with vein 2-SR of fore wing a weak, even curve and vein 2-SR slender (Fig. 482); propodeum, first and second metasomal tergites smooth; length of first tergite about 2.2 times its apical width, and only apically narrowed (Fig. 481); second tergite with shallow diverging grooves (Fig. 481); second metasomal suture indistinct; ovipositor sheath thin and acute apically (Fig. 481); length of ovipositor sheath 0.1 times fore wing and 0.2 times hind tibia; body, coxae and antenna black; hind femur, tibia, tarsi and basal half of middle femur usually dark brown; hind femur and apical half of hind tibia sometimes yellow; pterostigma brown. For additional details, see Walker (1994).

Host(s). Crambidae: C. *partellus* and *C. ignefusalis*. Also several externally feeding hosts belonging to the Lymantriidae and Notodontidae.

Distribution. Widespread species, occurring in the Afrotropical (Ghana, Kenya, Mali, Mozambique, Nigeria, South Africa, Uganda, Zimbabwe) and Oriental regions (India, Pakistan).

Glyptapanteles maculitarsis (Cameron) (Figs 476–478, 480)

Diagnosis. Length of fore wing 2.7–2.9 mm; mesoscutum finely punctate; scutellum convex and smooth; vein r of fore wing distinctly angled and vein 2-SR rather robust (Fig. 480); propodeum mainly rugulose; apical half of first metasomal tergite mainly punctate–rugulose, remainder smooth; length of first tergite about 2.5 times its apical width and gradually narrowed (Fig. 477); second tergite with distinct diverging grooves and punctate–rugulose near grooves (Fig. 477); second metasomal suture distinct; ovipositor sheath thick and rounded apically (Fig. 477); length of ovipositor sheath 0.1 times fore wing and 0.2 times hind tibia; body, hind coxa and antenna black; fore and middle legs, hind femur, hind tibia (except apex), tegulae, lateral compressed areas of three basal sternites yellow; apex of hind tibia, apical half of hind basitarsus and remainder of tarsus light brown; sometimes membranous lateral areas of first and second tergites yellow; pterostigma light brown. For additional details, see Walker (1994).

Host(s). Noctuidae: *B. fusca*. Several externally feeding hosts belonging to the Lasiocampidae, Noctuidae and several species of Saturniidae (Peigler, 1996).

Distribution. Kenya, Nigeria, Sierra Leone, South Africa, Tanzania, Uganda.

ORGILINAE

Genus *Orgilus* Haliday (Figs 483, 484)

Diagnosis. Length of fore wing less than 10 mm; scapus robust, apically subtruncate; clypeus without tubercles, apically straight; malar suture absent or only as a shallow depression; propleuron convex ventrally; scutellum without medio-posterior depression; vein 1-SR of fore wing absent; vein 1-M of fore wing more or less curved (Fig. 483); vein r-m of fore wing absent, at most with a minute remnant; posterior margin of hind wing convex or nearly straight subbasally (Fig. 483); vein cu-a of hind wing vertical; vein M+CU of hind wing subequal to vein 1-M or longer (Fig. 483); basal cell of hind wing medium-sized to large (Fig. 483); first metasomal tergite sessile or subsessile (Fig. 484); second tergite usually sculptured and without depressions.

Biology. Endoparasitoids of lepidopterous larvae: Pyralidae, Coleophoridae, Gelechiidae, Oecophoridae, Psychidae, Gracillariidae and Tortricidae.

Distribution. Cosmopolitan.

Orgilus bifasciatus Turner (Figs 483, 484)

Diagnosis. Length of body 4–5 mm, and of fore wing about 3 mm; antennal segments about 34, length of antenna about 1.4 times fore wing; mesoscutum rather coarsely and densely punctate; scutellum strongly convex, rather sparsely punctate; propodeum densely and finely rugulose; length of first tergite 1.1–1.2 times its apical width; first and second metasomal tergites densely rugulose, second tergite more coarsely sculptured than first one; second tergite 0.7 times longer than basally wide; second and third tergites with shallow transverse depression submedially, especially laterally (Fig. 484); third and fourth tergites finely and densely aciculate, except posteriorly (Fig. 484), fourth tergite more finely sculptured than third one; second to fourth tergites with lateral crease; fifth and following tergites largely smooth; length of ovipositor 1.15–1.20 times fore wing, somewhat shorter than body; black, with apical 0.4 of first tergite, base of middle and hind tibiae, and hind spurs ivory; hind wing subhyaline; fore wing of ♀ fasciate, membrane dark brown interrupted by a pale band below parastigma and base of pterostigma (Fig. 483), fore wing of ♂ mainly subhyaline; basal half of antenna dark brown or brown, and its apical third blackish or dark brown, and antenna of ♀ with narrow or wide yellowish submedial band. Males have second tergite partly ivory, antenna without yellowish band and wings mainly subhyaline

Host(s). Endoparasitoid; Pyralidae: *E. saccharina.*

Distribution. Afrotropical: South Africa.

18 Ceraphronidae

A. Polaszek
CABI Bioscience UK Centre, c/o Entomology Department, The Natural History Museum, London SW7 5BD, UK

INTRODUCTION

The family Ceraphronidae is a small family of parasitic Hymenoptera. A single genus, *Aphanogmus*, containing three species, is known in association with African cereal stem borers. Of these, one species is both widespread and polyphagous (*Aphanogmus fijiensis*), the second is known so far only from Africa (*Aphanogmus reticulatus*) and the third (*Aphanogmus trasides*) is known from Benin and Ivory Coast only. A detailed treatment of the *Aphanogmus* species, which are secondary parasitoids of African stem borers, has been published by Polaszek and LaSalle (1995). Since publication of that paper, the host of *A. trasides* has been discovered.

Genus *Aphanogmus* Thomson

Aphanogmus species are known as both primary and secondary parasitoids (hyperparasitoids), attacking a wide range of hosts (Dessart, 1989). Those associated with Lepidoptera are generally hyperparasitoids, attacking bethylid, braconid or ichneumonid primary parasitoids. Three species of *Aphanogmus* are now known to occur in Africa as hyperparasitoids of cereal stem borers.

The most reliable way of identifying these three species is by reference to the male genitalia, which should be mounted on microscope slides using the method described by Polaszek and Kimani (1990; see also Chapter 14, this volume).

Key to *Aphanogmus* Species Reared from African Cereal Stem Borers

1. Mesopleuron (mp in Fig. 489) smooth anteriorly, lateral striations present only posteriorly (Figs 489, 490). Male genitalia with parameres (A) divided (Figs 499, 500) 2

– Mesopleuron with evident lateral striations anteriorly (Fig. 491). Male genitalia (Fig. 501) with parameres undivided, elongate .... *A. trasides* Polaszek

2. Male genitalia as in Fig. 501, with parameres longitudinally divided and digiti (B) not extending close to the ends of the parameres. Middle club segment of female antenna (A9) evidently slightly shorter than first club segment (A8) (Fig. 503).

Head broader in facial view (Fig. 495) than *A. fijiensis* (Fig. 493), clypeus broadly truncate medially, with lateral excisions (Fig. 496) ....*A. reticulatus* (Fouts)

– Male genitalia as in Fig. 499, with parameres lobed and digiti extending close to the end of the parameres. Middle club segment of female antenna (A9) equal to first club segment (A8, Fig. 499). Head narrower in facial view (Fig. 493) than *A. reticulatus* (Fig. 495), clypeus pointed medially, without lateral excisions (Fig. 494) .... *A. fijiensis* (Ferrière)

Aphanogmus fijiensis (Ferrière) (Figs 489, 493, 494, 499, 502)

Diagnosis.

Male. Genitalia (Fig. 499) with the parameres slightly bilobed, short. Mesopleuron (Fig. 489) largely smooth, with some lateral striations posteriorly.

Female. As for male, except genitalia and the following: antenna (Fig. 502) with A4 more than half the length of A3; A7 quadrate; A8 approximately equal to A9. Head relatively narrow in frontal aspect (Fig. 493), clypeus setose, pointed medially (Fig. 494).

Hosts. Braconidae: *Apanteles tirathabae*, *Apanteles* sp., *Cotesia flavipes* (Cameron), *Cotesia kazak* (Telenga), *Cotesia sesamiae* (Cameron), *Dolichogenidea fuscivora* Walker, *Stenobracon* sp. Ichneumonidae: *Syzeuctus* sp. Reared from the following African stem-borer species: Noctuidae: *Busseola fusca* (Fuller), *Sesamia calamistis* Hampson, *Sesamia* sp. Crambidae: *Chilo partellus* (Swinhoe), *Chilo saccariphagus* (Bojer), *Coniesta ignefusalis* (Hampson). Pyralidae: *Maliarpha separatella* Ragonot.

Distribution. Africa: Burkina Faso, Cameroon, Ethiopia, Ghana, Ivory Coast, Kenya, Madagascar, Mali, Mauritius, Niger, Nigeria, Réunion, Senegal, South Africa, Uganda, Zaïre, Zambia. Asia: Fiji, India, Indonesia, Malaysia, Pakistan, Solomon Islands, Sri Lanka. Americas: Barbados.

Remarks. *Aphanogmus fijiensis* is an extremely widespread and rather polyphagous species.

Aphanogmus reticulatus (Fouts) (Figs. 490, 495, 496, 500, 503)

Diagnosis.

Male. Genitalia (Fig. 500) with the parameres clearly divided longitudinally, elongate. Mesopleuron (Fig. 490) largely smooth, with some lateral striation posteriorly.

Female. As for male, except genitalia and the following: antenna (Fig. 503) with A4 less than half the length of A3; A7 transverse; A8 clearly slightly longer than A9. Head in frontal aspect relatively broad (Fig. 495), clypeus truncate and laterally indented (Fig. 496), sparsely setose.

Distribution. Africa: Chad, Ivory Coast, Malawi, São Tomé, Somalia, Zaïre.

Hosts. Braconidae: *Apanteles* sp., *C. flavipes* (Cameron), *C. sesamiae* (Cameron), *Stenobracon* (*Euvipio*) sp. Ichneumonidae: *Syzeuctus* sp. Reared from the following African stem-borer species: Noctuidae: *B. fusca* (Fuller), *Sesamia* sp. Pyralidae: *C. partellus* (Swinhoe), *C. ignefusalis* (Hampson).

Remarks. *Aphanogmus reticulatus*, unlike *A. fijiensis*, has so far been recorded only from Africa.

Aphanogmus trasides Polaszek (Figs. 491, 497, 498, 501, 504)

Diagnosis.

Male. Genitalia (Fig. 501) with parameres simple, elongate, each with a single long seta laterally. Digiti elongate, sharply pointed. Anterior mesopleuron with evident lateral striations (Fig. 491). Scutellum without lateral carinae (arrowed in Fig. 492).

Female. As for male except genitalia characters and the following: antenna as in Fig. 504, dimensions of A7 much closer to those of A8 (first club segment) than in either *A. fijiensis* or *A. reticulatus*. Head narrow in frontal aspect (Fig. 497), similar to that of *A. fijiensis* (Fig. 493). Clypeus simple, sparsely setose (Fig. 498).

Hosts. *Goniozus garouae* (Risbec) *ex Eldana saccharina* (Walker) (Pyralidae) (Benin). The original series (from Ivory Coast) was labelled '*ex Eldana* pupa'. There are no host remains with the series, and emergence from the pupa seems unlikely.

Distribution. Africa: Benin, Ivory Coast.

Remarks. *Aphanogmus trasides* is very close to *Aphanogmus manilae* (Ashmead). Like *A. fijiensis*, *A. manilae* is a common and widespread hyperparasitoid of Braconidae in the Oriental region, which one might expect to turn up eventually in the Afrotropical region. *Aphanogmus trasides* differs from *A. manilae* in male genitalia characters, fully illustrated by Polaszek and Dessart (1996). In the female sex, *A. manilae* can be separated from *A. trasides* by having a carina on each side of the scutellum (Fig. 492).

19 Chalcidoidea

A. Polaszek,[1] J. LaSalle[1] and Y. Jongema[2]
[1]*CABI Bioscience UK Centre, c/o Entomology Department, The Natural History Museum, London SW7 5BD, UK;* [2]*Department of Entomology, Wageningen Agricultural University, PO Box 8031, 6700EH, Wageningen, the Netherlands*

INTRODUCTION

Among the parasitic Hymenoptera, the Chalcidoidea are an extremely large superfamily, found in abundance in all parts of the world, with currently about 18,000 described species (J.S. Noyes, personal communication). It is also one of the least studied groups of wasps, due mainly to the very small size of most of the species, and current estimates put the *actual* number of species at around 100,000! Most species are parasitoids, and their hosts include many pests. The Chalcidoidea have been the source of most of the parasitoids in successful biological control programmes. Because of the abundance, small size and morphological diversity of chalcids, there are still a great many unsolved taxonomic problems, at every level, from the number and systematic arrangement of included families, to the identification of closely related, morphologically identical, or very similar, species. There are currently about 20 families included in the superfamily, of which only eight are known in association with African cereal stem borers, as both primary and hyperparasitoids. The families have already been keyed in the general key to the families of hymenopterous parasitoids of cereal stem borers. The following account keys the genera and, in some cases, species associated with African stem borers. The families are treated alphabetically.

Chalcididae

A moderately large, cosmopolitan family (about 1500 species), with the majority of species in the tropics. Chalcidids are generally easily recognizable from the greatly enlarged, toothed or lobed, hind femora, present in both sexes (Figs 505, 506), the strongly punctate thorax and the pronounced carina bordering the gena posteriorly. Although some of these characters occur in certain genera belonging to other families, none of those are associated with cereal stem borers. Chalcidids tend to be non-metallic in colour, often mainly black, sometimes with red or yellow markings. Most species are primary larval/pupal endoparasitoids of Lepidoptera and Diptera, although Hymenoptera, Coleoptera and Neuroptera are also attacked. The family also contains hyperparasitoids and ectoparasitoids, and, while most are solitary, a few can be moderately gregarious.

Five genera of chalcidid primary or hyperparasitoids are known so far in association with African cereal stem borers. A species in the genus *Hockeria* was recorded as a hyperparasitoid of a stem borer, attacking the tachinid primary parasitoid *Diatraeophaga striatalis*, in Java, from where the tachinid was imported into Mauritius (Ghani and Williams, 1962). So far, no *Hockeria* species has been recorded in association with stem borers on the African mainland.

Key to Genera of Chalcididae Associated with African Cereal Stem Borers

1. Hind tibia truncated at apex and with two spurs (Fig. 505) 2
– Hind tibia tapering (or obliquely truncated) and extending sometimes as a strong spine extended beyond insertion of tarsus (Fig. 506) 4
2. Fore wing with marginal vein at the wing margin, postmarginal and stigmal vein distinctly present, sometimes short but often long (as in Fig. 507). Frons with a strong carina at inner sides of each eye and uniting above behind front ocellus (Figs 509, 510). Anterior pronotal carinae mostly with two tubercles in middle .... *Antrocephalus*
– Fore wing with marginal vein slightly removed from anterior margin, postmarginal vein absent (Fig. 508) 3
3. First metasomal (gastral) tergite without lateral carinae delimiting the basal fovea (Fig. 515). Hind femora often with longitudinal rugae on the basal tooth (Fig. 514) .... *Psilochalcis*
– First metasomal tergite with two distinct carinae delimiting the basal fovae on sides (Fig. 512). Often weaker sublateral carinae present .... *Proconura*
4. Head with strong projecting horns between eyes and antennal scrobes (Fig. 511). Ventral margin of hind femur with comb of teeth (Fig. 513) .... *Dirhinus*
– Head without such horns. Hind femur without such a comb; often with yellow and/or red spots (Fig. 506) .... *Brachymeria*

Antrocephalus Kirby

A single species of *Antrocephalus*, *A. mitys* (Walker), has been recorded once as a parasitoid of *Eldana saccharina* in Senegal, and once from *C. partellus* in Kenya (Skovgaard and Päts, 1996, under Eulophidae). Given the fact that these host associations have never been repeated, they should be treated with some doubt.

In Africa, the limits of this genus, with respect to *Hockeria* and *Kriechbaumerella*, are extremely vague, and further research is necessary to prove whether or not *Antrocephalus* is synonymous with *Hockeria*.

Brachymeria Westwood (Figs 506, 507, 513)

Brachymeria is the genus of Chalcididae most commonly associated with African cereal stem borers. It is a large genus, with many species both described and undescribed (315 valid species worldwide). During the current study, several species have been recorded in association with stem borers, including *Brachymeria albicrus* Klug, *Brachymeria feae* Masi (also recorded by Harris, 1962), *Brachymeria inornata* Masi, *Brachymeria kassalensis* Kirby, *Brachymeria olethria* Waterston (La Croix, 1967) and *Brachymeria sesamiae* Gahan (Gahan, 1928), in

Table 19.1. *Brachymeria* spp. associated with African cereal stem borers.

Brachymeria sp.	Host(s)	Distribution	Remarks
B. albicrus Klug	? Stem borer	Ghana	IITA survey
B. feae Masi	*Eldana saccharina; Sesamia ?calamistis; Syllepte derogata*	Chad; Nigeria	Referred to by the following: Harris, 1961, 1962; Ajayi, 1985a, b; Silvie, 1991; also new records during this study, identified by Dr Z. Bouček
B. inornata Masi	*Chilo partellus*	Malawi	Reared during this study by Dr G. Phiri; identified by Dr Z. Bouček
B. kassalensis Kirby	*Chilo partellus; S. derogata*	Chad, Kenya	Recorded from Chad (*S. derogata*) by Silvie (1991); from Kenya (*C. partellus*) by La Croix (1967) Latter record confirmed as *B. kassalensis* by Dr Z. Bouček
B. olethria Waterston	*C. orichalcociliellus; E. saccharina*	Kenya	La Croix, 1967 (confirmed during the current study by Dr Z. Bouček)
B. sesamiae Gahan	*Sesamia cretica*	Sudan	Gahan, 1928
Brachymeria spp.	*C. partellus*	Kenya	Chacko and Ogedah, 1991
	E. saccharina	Senegal	Overholt, 1991
	Busseola fusca	Nigeria	Diémé, 1982
			This study

IITA, International Institute of Tropical Agriculture.

addition to several undescribed species. It has not been possible to construct a species-level key to those *Brachymeria* associated with African cereal stem borers, but available information concerning *Brachymeria* spp. associated with African cereal stem borers is summarized in Table 19.1.

Dirhinus Dalman (Fig. 511)

A single species, *Dirhinus excavatus* Dalman, has been recorded from African stem borers as hyperparasitoid, attacking the primary tachinid (Diptera) parasitoids *Sturmiopsis parasitica* and *Siphona* sp. (Mohyuddin and Greathead, 1970). Three females in the collections of the Natural History Museum (NHM), London, represent at least part of the material on which this record was based. These are *Dirhinus anthracia* (Walker), previously misidentified as *D. excavatus* by G.J. Kerrich in 1967. Although referred to in the introduction to their review of African stem-borer hyperparasitoids (Polaszek and LaSalle, 1995), *Dirhinus* was not included in the key. It would key out to the genus *Eurytoma*, from which it can be easily distinguished by the greatly expanded hind femora. *Dirhinus anthracia* is a cosmopolitan, polyphagous parasitoid of various Diptera.

Proconura Dodd (Fig. 512)

A single female *Proconura*, probably an undescribed species, has been recorded from *E. saccharina* from Nigeria. It seems wise to wait until more material becomes available before describing it.

Psilochalcis Kieffer

This genus, of which *Hyperchalcidia* and *Invreia* are synonyms, contains *Psilochalcis soudanensis* (Steffan), the most important and widespread chalcidid parasitoid of cereal stem borers. A single male of a very closely related species, almost certainly reared from *Maliarpha separatella* at Tanandava, Madagascar, has been examined. It seems wise to wait until more material, reliably reared, becomes available before describing it. The description of a new species reared from *E. saccharina* is currently in press. The species is referred to below as *Psilochalcis* sp. A.

Psilochalcis soudanensis (Steffan) (Figs 508, 514, 515; Plate 10:1)

Diagnosis. Length 3.0–7.0 mm. Wing venation and hind femur as in Figs 508 and 514.

Host(s). A solitary larval–pupal endoparasitoid of Noctuidae: *Busseola fusca* (Fuller); Crambidae: *Chilo infuscatellus*, *Chilo partellus*, *Chilo zacconius*, *Coniesta ignefusalis*; Pyralidae: *E. saccharina*, *M. separatella*.

Distribution. Africa: Cameroon, Ghana, Kenya, Mali, Mozambique, Niger, Nigeria, Senegal, Sudan, Uganda. Asia: India, Pakistan.

Remarks. Despite the widespread occurrence of *P. soudanensis*, we have not seen any specimens from Madagascar. Appert (1973) referred to the proposed introduction of *P. soudanensis* into Madagascar and the Comoros against *Chilo* sp. and *M. separatella*. Presumably this introduction never took place.

Psilochalcis sp. A

Diagnosis. Length 4.5–5.4 mm.

Coloration. Black; coxae and outer face of hind femora except top and base black. Scape, pedicellus and ring segment reddish brown. Scape pedicellus and front femora variable in colour, from reddish brown to black in some parts. Tegulae with lower corners reddish brown, but somewhat variable.

Morphology. Head transverse, viewed from in front, measured from lower margin of clypeus (63 : 57). Cheeks shorter than height of eye, seen from side

(22 : 30). Frontogenal carina present, but absent in the upper part. Clypeus three times as wide as long, with roof-like projection and rounded corners, its side-margins slightly elevated. Upper surface of clypeal projection smooth. Inter-antennal callus excavated. Scrobe shallow, not reaching anterior ocellus, its three-quarters with transverse wrinkles. Scape not reaching the front ocellus. Right mandible with three teeth, left with two. Flagellum (not including ring segment) slightly shorter than height of head (55 : 57). Scape : flagellum 33 : 55. Antennal joints 86 : 22 : 13 : 18 : 18 : 18 : 16 : 15 : 15 : 26. OOL : POL = 11 : 37. Thorax length 102; mesoscutum 32; scutellum 35; width of pronotum 56; mesoscutum 56; scutellum 37.

Pronotum seen from above with intervals between the pits about equal in length to the pit diameter. Intervals smooth in posterior part, weakly shagreened in anterior half. Sides of pronotum densely punctate. Mesoscutum with the same punctation as pronotum, more dense anteriorly where the intervals are weakly shagreened. Scutellum with punctations and intervals smooth. Propodeum with accessory carinae only in anterior half; areolae more or less shagreened between the transverse carinae. Median areola ellipsoidal. Legs robust, disc of hind femur smooth to very weakly reticulate between the small punctations. Hind femora less than twice as long as broad (60 : 37). Basal teeth not strong, without carinae. Length of marginal vein 30; stigmal vein 8. Gaster ovate, less than twice as long as broad (81 : 49). Length of first tergite exceeding half the length of the gaster, its disc finely and densely punctate. Tergite II and remaining tergites with finely engraved reticulation. Epipygium short, not visible from above.

Male. Length 4.5 mm. Head, viewed from in front, shorter than in female. Width : height 60 : 49. Cheeks : height of eye (side-view) 19 : 25. Tegulae black, fore femora black (in female red to almost completely black). Clypeal projection much shorter than in female. Flagellum (excluding ring segment) much longer than head height.

Host. A solitary larval–pupal endoparasitoid of *E. saccharina.*

Distribution. Burkina Faso, Ivory Coast.

Remarks. This species was recorded previously as *Invreia* aff. *propinqua* by Betbeder-Matibet (1981).

Elasmidae

A small family (200 species in a single genus *Elasmus*). Most species are black with pale markings, and are easily recognizable from the massively enlarged hind coxa (Fig. 533).

Most species for which the biology is known are gregarious primary parasitoids of Lepidoptera larvae, some are hyperparasitoids, usually via braconids or ichneumonids, and others have been reared from the larvae of polistine wasps (Vespidae).

The taxonomy of the *Elasmus* species associated with African stem borers is far from clear, and further studies are necessary to clarify the situation. *Elasmus senegalensis* Risbec and *Elasmus* sp. were recorded as parasitoids of *Apanteles* (in the broad sense) by Risbec (1960) and Milner (1967), respectively. Other *Elasmus* species have been recorded, apparently as primary parasitoids of stem borers in Africa, by Risbec (1960: *Elasmus tolli* Risbec), Polaszek *et al.* (1994) and Vercambre (1977b). In each of these cases, the *Elasmus* sp. was recorded in association with rice borers in West Africa.

Encyrtidae

A single species of Encyrtidae has been recorded repeatedly in association with African cereal stem borers, *Exoristobia dipterae* (Risbec). Other (rare) records of Encyrtidae associated with borers, those of *Euzkadia integralis* Mercet (Risbec, 1960) and *Paraphaenodiscus risbeci* Ghesquière (Risbec, 1950a) and should be treated with suspicion until confirmed. The records of ?*Coccidencyrtus* and ?*Epiencyrtus* by Harris (1962) are misidentifications of *E. dipterae* (Risbec) (see below).

Exoristobia dipterae (Risbec) (Figs 516–518, 538, 539)

Diagnosis.

Female. Head and body entirely dark brown to black with slight green or purple metallic reflections. Wing (Fig. 518) with marginal vein relatively long for an encyrtid. Antennal club obliquely truncate apically (Fig. 516). Parasites of Diptera.

Male. As female, except genitalia and antennae (Fig. 517).

Distribution. Benin, Ghana, Kenya, Malawi, Nigeria, Senegal, Tanzania, Uganda.

Hosts. Hymenoptera: Braconidae: *Stenobracon* (*Euvipio*) sp. Diptera: Muscidae: *Atherigona pulla* Wiedemann, *Atherigona* sp. Anthomyiidae: *Delia arambourgy* Séguy, *Delia* sp. Tachinidae: *S. parasitica*. Reared from the following stem-borer hosts: Noctuidae: *B. fusca*, *Sesamia* sp. Crambidae: *C. partellus*, *C. ignefusalis* (Hampson). Pyralidae: *E. saccharina* Walker. The record from *Stenobracon* (*Euvipio*) sp. (Braconidae) is doubtful, given that all other records are from dipterous hosts, although Subba Rao (1970) records *Exoristobia philippinensis* Ashmead from an *Apanteles* and G.R. Prinsloo (personal communication) regards this as a probable synonym of *E. dipterae*.

Remarks. A widespread and very common species. Annecke and Mynhardt (1973) have drawn attention to the difficulty of separating *E. dipterae* from *E. philippinensis* Ashmead, an Asian species. Harris (1962) and Milner (1967) recorded ?*Coccidencyrtus* and ?*Epiencyrtus* from stem borers, records repeated and perpetuated by Mohyuddin and Greathead (1970). The material on which these records were based has been examined by Dr J.S. Noyes (NHM) and identified as *E. dipterae*.

Eulophidae

Five genera of Eulophidae have been recorded in association with African cereal stem borers, both as primary and secondary (hyper)parasitoids, sometimes facultatively. These genera can be separated as follows.

Key to the Genera of Eulophidae Associated with African Stem Borers

1. Scutellum with a single pair of setae (Fig. 529). Propodeum with distinct plica, and paired median carinae, which diverge posteriorly (Fig. 532). Scutellum without a pair of longitudinal, submedial grooves .... *Pediobius*

– Scutellum with two pairs of setae (Figs 527, 528). Propodeum without plica, and with a simple median carina or no median carina (Figs 530, 531). Scutellum with or without a pair of longitudinal, submedial grooves (Fig. 531) 2

2. Antenna with two funicular segments (Fig. 525); antenna inserted close to mouth margin. Fore wing with two or three dark spots or tufts of strong, black setae (Fig. 525). Coloration yellow to brown, non-metallic .... *Trichospilus*

– Antenna with three (female) or four (male) funicular segments; inserted in middle of face or only slightly closer to mouth margin. Fore wing hyaline. Coloration variable, often metallic or dark metallic 3

3. Submarginal vein of fore wing with three or more setae on dorsal surface. Scutellum without submedian lines, with deep sublateral lines (Fig. 526). Female antenna with third anellus larger than preceding two anelli and with setae .... *Tetrastichomyia*

– Submarginal vein of fore wing with one or two setae on dorsal surface. Scutellum with submedian lines. Female antenna without enlarged third anellus 4

4. Propodeum with a distinct, posteriorly bifurcate, paraspiracular carina (i.e. a carina which is more or less in the shape of an inverted Y just medial to the spiracle) (Fig. 531) .... *Tetrastichus*

– Propodeum without a carina medial to the spiracle (Fig. 530), or with a simple carina .... *Aprostocetus*

Pediobius Walker (Figs 529, 532)

Pediobius furvus Gahan is one of the most commonly reared primary parasitoids of stem borers in Africa. It is a gregarious larval–pupal parasitoid. *Pediobius furvus* has been used on several occasions for the attempted biological control of cereal stem borers, for example, in Madagascar (Appert and Ranaivosoa, 1971), the Comoros, the Mascarene Islands and the USA (Overholt and Smith, 1989). The biology has been described by Mohyuddin (1970).

Pediobius homoeus was described as a parasitoid of *B. fusca* by Waterston (1915). We have examined a series of specimens from Ivory Coast, which might be either primary parasitoids of *Sesamia calamistis* or hyperparasitoids of *Cotesia sesamiae*. Label data for these specimens are as follows: Ivory Coast, Bouaké, 'ex ?*Sesamia calamistis* ou *Apanteles sesamiae*'. Further rearing will be necessary to clarify this host record.

Studies on the taxonomy of the various *Pediobius* species associated with stem borers are currently in progress.

Tetrastichus Haliday

The genus *Tetrastichus* has been traditionally used in a very broad sense, and included a wide range of wasps. However, recent classifications of the Tetrastichinae have treated this genus in a much more restricted sense (Graham, 1987, 1991; Bouček, 1988; LaSalle, 1994). The most important character for distinguishing *Tetrastichus* from other tetrastichines is the presence of a posteriorly bifurcate carina on the propodeum, just medial to the spiracle, in the shape of an inverted Y (Fig. 531). Additionally, most species of *Tetrastichus* only have a single seta on the dorsal surface of the submarginal vein (Fig. 524); however *Tetrastichus howardi* and some of its Asian relatives have two (Fig. 523).

Although other hosts are known, most members of the genus *Tetrastichus* are parasites of Diptera, Coleoptera, Lepidoptera (sometimes as hyperparasitoids) and sawflies.

The species of *Tetrastichus* which are hyperparasitoids on African stem borers are all closely related. Graham (1991: 207) has already mentioned that the African *Tetrastichus atriclavus* and *Tetrastichus sculpturatus* hardly differ from the Asiatic *T. howardi* and *Tetrastichus inferens*, except in having only one dorsal seta on the submarginal vein.

Key to *Tetrastichus* Species Reared from African Cereal Stem Borers

1. Submarginal vein with a single seta on dorsal surface (Fig. 524) .... *Tetrastichus atriclavus* complex

– Submarginal vein with two or more setae on dorsal surface (Fig. 523) .... *Tetrastichus howardi*

Tetrastichus atriclavus complex (Figs 521, 522, 524, 528, 531)

The members of this species complex are being investigated in a separate paper. These species are very closely related, and some synonymy will be involved. The following names belong to this species complex:

Tetrastichus atriclavus Waterston (1915: 232–235)
Tetrastichus sculpturatus Waterston (1915: 238–241)
Tetrastichus dispar Masi (1917: 221)
Tetrastichus sesamiae Risbec (1951: 76–77)

Other species which have been considered as related to this complex, but which do not actually belong here are: *Tetrastichus mauripennis* Waterston; *Tetrastichus sesamiae* var. *angolensis* Risbec; *Tetrastichus procerae* Risbec. This latter species is now known as *Aprostocetus procerae* (Risbec) and is a senior synonym of *T. pachydisplosisae* Risbec (LaSalle and Delvare, 1994).

Diagnosis. Propodeum with a distinct carina in the shape of an inverted Y just medial to the spiracle (Fig. 531). Submarginal vein with a single seta on the dorsal surface (Fig. 524). Propodeum relatively rugose. Male antenna: club usually black or very dark brown in distinct contrast to pale funicle; scape with a row of long setae on the ventral margin (Fig. 522). Female antenna: funicle and club dark, in contrast to pale scape and pedicel; each successive funicular segment shorter and wider than preceding one (Fig. 521).

Host(s). Recorded hosts for *T. atriclavus* and *T. sesamiae* are (as parasitoids and hyperparasitoids): Noctuidae: *B. fusca, S. calamistis, Sesamia* sp.; Crambidae: *C.*

partellus, Chilo sacchariphagus, Chilo suppressalis (laboratory), *C. ignefusalis*; Pyralidae: *E. saccharina*.

Distribution. Mainly Afrotropical in distribution, although one species which appears to belong to this complex is known from China (J. LaSalle and D.W. Huang, unpublished information).

Remarks. As stated above, this species complex is currently under investigation (J. LaSalle and A. Polaszek, in preparation). Initial indications are that there are at least three species involved (one of them new). It is not yet clear whether all species have wide host ranges, or whether there is some degree of partitioning among hosts between these species.

Tetrastichus howardi (Olliff)

Diagnosis. As with the *T. atriclavus* species complex, but with two setae on the dorsal surface of the submarginal vein.

Hosts. *Tetrastichus howardi* is a facultative hyperparasitoid with a very wide host range. References to Asian hosts for *T. howardi* can be found in Bouček (1988) and Husain and Khan (1986, as *Tetrastichus ayyari* and *Tetrastichus israeli*).

Kfir *et al.* (1993) found that it would attack a wide range of hosts in the laboratory, including: the stem borers *C. partellus* (Crambidae), *E. saccharina* (Pyralidae), *B. fusca, S. calamistis* (Noctuidae); other Lepidoptera, such as *Cactoblastis cactorum* (Pyralidae), *Spodoptera littoralis* (Noctuidae), *Holocerina smilax* (Saturniidae); other insects, such as meal-worm, *Tenebrio molitor* (Coleoptera: Tenebrionidae), and the African honey-bee, *Apis mellifera scutellata* (Apidae); the stem-borer parasitoids *Mallochia pyralidis, Xanthopimpla stemmator* (Ichneumonidae), *Cotesia kazak* (Braconidae) and *Palexorista laxa* (Diptera: Tachinidae).

Distribution. Mainly Asiatic; however, *T. howardi* has been imported into Africa for investigation in biological-control attempts (Kfir, 1991c; Kfir *et al.*, 1993). Vercambre (1977b) reported that *T. israeli* had been introduced into Senegal, although it is not clear whether it was ever established, and this species has also been recorded as having been introduced into Madagascar (Appert, 1973).

Remarks. Bouček (1988) synonymized both *T. ayyari* and *T. israeli* with *T. howardi*. However, there does seem to be considerable variation within this species, and future studies may show that more than a single species is present. If that is the case, one or both of the above names may need reinstating.

Aprostocetus Westwood

Aprostocetus is the largest genus of Eulophidae, and one of the largest of all chalcidoid genera. It has a remarkably wide host range, and contains many of the species which were until recently included in *Tetrastichus*. For more complete accounts of the complicated taxonomic history of this genus, as well as synonymy lists and biological information, see Graham (1987), Bouček (1988) and LaSalle

(1994). There is a single species of *Aprostocetus* which is associated with cereal stem borers in Africa, *Aprostocetus theioneurus* (Masi).

Aprostocetus contains six subgenera (Graham, 1987; LaSalle, 1994). *Aprostocetus theioneurus* belongs to the subgenus *Ootetrastichus* and is the first known member of this subgenus which does not parasitize insect eggs.

Aprostocetus theioneurus (Masi) (Figs 519, 520, 527, 530)

Diagnosis.

Female. Head, mesosoma and metasoma bright metallic green to blue-green, except tegulae and upper angle of mesopleuron (just below tegula) bright yellow. Coxae and legs bright yellow, except hind coxae dorsally metallic green to blue-green. Antenna with scape yellow (slightly darkened dorsoapically), pedicel and flagellum brown. Wings hyaline, veins light brown. Propodeum without a carina just medial to the spiracle (Fig. 530). Submarginal vein of fore wing with two (rarely three) setae (Fig. 519). Club (both male and female) with down-curved apical spine (Fig. 520).

Male. Similar to female except in antennal and metasomal characters. Antennal club concolorous with the flagellum and not distinctly wider than funicle.

Hosts. In Kenya, *A. theioneurus* is a hyperparasitoid of the cereal stem borer *C. partellus*, through the braconid parasitoid *C. sesamiae*.

Distribution. Known from Kenya, Madagascar and the Seychelles.

Remarks. For a more complete discussion of this species, see LaSalle (1993).

Tetrastichomyia sp. A (Fig. 526)

Diagnosis. Length 1.70 mm. Dorsum of mesosoma as in Fig. 526, dorsellum of mesosoma divided by a median longitudinal ridge.

Host(s). Presumably a primary parasitoid, known only from *M. separatella* (Polaszek *et al.*, 1994).

Distribution. Uganda.

Remarks. This species differs from *Tetrastichomyia clisiocampae*, a widespread species (Italy and USA), which is also known to attack cereal stem borers (Graham, 1991).

Eupelmidae

Macroneura sp. A

Diagnosis. Length 3.50 mm. Mesopleuron convex, wings reduced. Pronotal dorsum with two prominent tufts of thick dark setae. Brown/black with blue/purple iridescence.

Hosts. Known only from two females reared from *M. separatella* in Senegal. Although Harris (1962) recorded a *Eupelmus* sp. from stem borers, possibly as a hyperparasitoid of *T. atriclavus*, the rare records of Eupelmidae associated with African cereal stem borers require confirmation.

Distribution. Senegal.

Eurytomidae

Eurytomidae display a very varied range of biologies, and species can be phytophagous or primarily or secondarily parasitic on insects and arachnids. Two species have been recorded regularly from cereal stem borers, both belonging to the genus *Eurytoma*. One, *Eurytoma oryzivora* Delvare, is a primary parasitoid, while *Eurytoma braconidis* Ferrière is a secondary parasitoid via a braconid primary. These two species can be easily distinguished as follows.

Key to *Eurytoma* Species Associated with African Cereal Stem Borers

1. Face with very distinct longitudinal striae converging towards the mouth, the clypeus recessed (Fig. 536). Metasoma (gaster) of female without a long petiole .... *E. braconidis*

– Face without distinct longitudinal striae converging towards the mouth, the clypeus not recessed (Fig. 537). Metasoma (gaster) of female with a long petiole .... *E . oryzivora*

Eurytoma braconidis Ferrière (Figs 534–536, 551, 553; Plate 11:5)

Diagnosis.

Female. Face as in Fig. 536, with very distinct longitudinal striae converging towards the mouth, the clypeus recessed. Antennae with all segments either transverse or subquadrate. Funicle 1 hardly longer than the pedicel. Funicle five-segmented, club distinctly two-segmented (Fig. 551; the club is unsegmented in a very closely related species, Fig. 552), although it may appear three-segmented. Mesosoma with the typical extremely coarse punctations, the propodeum strongly reticulate, sometimes with a transverse element. All legs black, except the knees and tarsi. Metasoma compressed, the ovipositor hardly protruding.

Male. Antennomeres with long whorls of hairs, the club two-segmented. Mesosoma as in female, metasoma petiolate and truncate. Genitalia as in Fig. 553.

Hosts. *Bracon* sp., *Stenobracon* (*Euvipio*) *unifasciata*. Reared from the following stem-borer hosts: Crambidae: *C. partellus*, *C. ignefusalis*; Noctuidae: *B. fusca*, *Sesamia cretica*, *Sesamia* sp.

Distribution. Benin, Ethiopia, Kenya, Malawi, Niger, South Africa, Sudan, Tanzania, Uganda, Zaïre. *Eurytoma braconidis* was recorded additionally from Cameroon and Togo by Delvare (1988).

Remarks. The *E. braconidis* group contains several species that are difficult to distinguish from one another, and related species may occur sympatrically. For example, *Eurytoma syleptae* Ferrière is common on *Syllepte derogata*, and an undescribed species has been reared from *Heliocheilus albipunctella* on millet (Fig. 552), together with *E. braconidis*, the latter reared from *C. ignefusalis*. Risbec

(1950b, 1960) recorded *Eurytoma verbena* Ferrière from *Rhaconotus soudanensis* (probably a misidentification of *Rhaconotus scirpophagae*; see Polaszek *et al.* (1994)) on *C. ignefusalis* in Senegal. This single repeated record is very likely to be based on a misidentification of *E. braconidis*, although no voucher material from this record has been examined.

Eurytoma oryzivora Delvare (Fig. 537)

Diagnosis. Length 3.2–3.6 mm. Almost entirely black, with some areas reddish brown. Metasoma petiolate, petiole longer than wide. Clypeus without a deep medial depression; face with foveolate/striate sculpture (Fig. 537).

Host(s). Assumed to be a primary parasitoid of Crambidae: *Chilo* sp.; Pyralidae: *M. separatella*.

Distribution. Cameroon, Senegal, Sierra Leone, Tanzania.

Pteromalidae

Two genera of Pteromalidae are known so far from cereal stem borers in Africa, *Conomorium* and *Norbanus* (= *Picroscytoides*). These genera can be distinguished as follows.

1. Antennal clava (club) of female distinctly acute apically or bearing a narrow spicula (Fig. 132) .... *Norbanus*
2. Antennal clava (club) of female not distinctly acute apically, without a narrow spicula .... *Conomorium*

The taxonomy of all the *Conomorium* and *Norbanus* species attacking African borers is currently under investigation. The following records of Pteromalidae in association with African cereal stem borers have been published: *Norbanus* 'sp. A'/*M. separatella* (Polaszek *et al.*, 1994); *Norbanus* sp. (as *Picroscytoides* sp.)/*Sesamia* sp. (Harris, 1962); *Norbanus* sp./*C. partellus* (Kfir, 1990a). A *Norbanus* species was reared recently from *S. cretica* on sorghum in Sudan (Plate 11:6). Milner (1967) recorded two species of Pteromalidae (*Pachyneuron* sp. and *Platecrizotes sudanensis* Ferrière) as hyperparasitoids of *S. parasitica* (Curran) (Diptera, Tachinidae) and *C. sesamiae* (Cameron) (Braconidae), respectively. These associations have never been confirmed, and it seems more probable that at least *P. sudanensis* is a primary parasitoid of a dipterous borer (Polaszek and LaSalle, 1995).

Trichogrammatidae

Trichogrammatids are small to extremely small species (0.15–1.20 mm), found throughout the world. They are easy to recognize, despite their small size, by having only three tarsal segments. All known species are primary (solitary or gregarious) egg parasitoids, with most hosts in the Homoptera, Lepidoptera and Coleoptera, although many other host orders are also attacked. Four genera of Trichogrammatidae have been recorded so far from the eggs of African stem borers; these can be separated as follows.

Key to the Genera of Trichogrammatidae Recorded from African Stem-borer Eggs

1. Antenna of both male and female without any funicle segments, instead with a five-segmented club (clava) and two very small ring segments (Figs 541, 542). Female antenna with a narrow spicula at the apex (Fig. 542) .... *Lathromeris*
– Antenna of female with two funicle segments (Figs 544, 547, 550). Male antennae various (Figs 544, 546, 549) but never with a five-segmented club ... 2

2. Female antenna with club unsegmented (Figs 547, 550). Male antenna either with flagellum composed of a single segment (Fig. 546) or, if as in Fig. 549, then vein track rs-1 absent from fore wing (compare Figs 545 and 548) 3
– Female and male antenna with club three-segmented (Fig. 544). Vein track rs-1 present in fore wing (Fig. 543) .... *Paracentrobia*

3. Male antenna with a single, long flagellar segment (Fig. 546). Fore wing (males and females) with vein track rs-1 present (Fig. 545) .... *Trichogramma*
– Male antenna with five flagellar segments (Fig. 549). Fore wing (males and females) with vein track rs-1 absent (Fig. 548) .... *Trichogrammatoidea*

Trichogramma and *Trichogrammatoidea* (Figs 545–550; Plate 11:3)

Records of species belonging to these two common genera of egg parasitoids are abundant in the literature concerning cereal stem borers. Species belonging to both genera have also been introduced as potential biological control agents, especially into South Africa against *E. saccharina* (Carnegie, 1991). Although it is currently possible for specialists to identify some of the species (by examination of the male genitalia), a thorough taxonomic review of these species is necessary before they can be given appropriate treatment.

Lathromeris ovicida (Risbec) (Figs 540–542)

This is the only species belonging to this genus known so far from eggs of cereal stem borers. It is sometimes found with *Telenomus* spp. in the same egg mass.

Diagnosis. Wing venation as in Fig. 540. Antennal club five-segmented, terminating in females in an elongate projection (Fig. 542). Funicle absent, two ring segments present.

Hosts. Noctuidae: *B. fusca, S. calamistis*; Crambidae: *C. zacconius, Scirpophaga subumbrosa*. Pyralidae: *M. separatella*.

Distribution. Benin, Cameroon, Ghana, Ivory Coast, Nigeria, Uganda.

Paracentrobia (Figs 543–544)

Records of *Paracentrobia* from African cereal stem borers are rare. Milner (1967) recorded *Paracentrobia dimorpha* from *B. fusca* eggs in Uganda, and Viggiani (1972) recorded the same species from unidentified eggs on sorghum, in Israel. The first author has recently examined a *Paracentrobia* species from eggs of *B. fusca* in Cameroon. Further studies are in progress to elucidate the identities of these specimens.

20 Ichneumonidae

K.W.R. Zwart
Department of Entomology, Wageningen Agricultural University, PO Box 8031, 6700EH, Wageningen, the Netherlands

INTRODUCTION

The family Ichneumonidae is probably the largest family of parasitic Hymenoptera. Townes (1969) estimated that 60,000+ species might constitute the world fauna, the majority of these still undescribed. Most ichneumonids are solitary parasitoids, and lepidopterous larvae and pupae are their main hosts. Endo- and ectoparasitism both occur also with regard to stem borers. Although most ichneumonids are primary parasitoids, secondary parasitism occurs in various subfamilies. Some are true hyperparasitoids, ovipositing into the primary parasitoid larva when this is active in its host.

Oviposition is usually either in or on the host larva or into the host pupa. Oviposition into the host egg occurs, but the ichneumonid larva develops in the host larva or sometimes even in the host pupa.

The mature larva of larval parasitoids usually leaves its host and spins a cocoon, in which to pupate. If the host larva spins a cocoon before pupation, the ichneumonid cocoon is found inside the cocoon spun by its host. Pupal parasitoids and larval–pupal parasitoids will usually pupate inside the host pupa; in these cases, also, the larva will usually spin its cocoon inside the host pupa.

In the present study, the literature on Afrotropical cereal stem borers and their ichneumonid parasitoids was checked, and specimens reared from stem borers in Africa were studied where these were available. All generic names that have been found are treated in the main part of this chapter, whether material representing these genera has been studied or not.

Some genera reported from African stem borers are based on misidentifications. In a few cases, the genus concerned, such as *Scenocharops* and *Casinaria*, is known to parasitize stem borers in other parts of the world. These names were corrected by Polaszek *et al.* (1994); both have been used for a species of *Venturia*. Other names are simply misidentifications, a few of which could not be traced. *Chriodes* sp. has been used for another species of *Venturia*. Two genera reported, *Coelichneumon* and *Coleocentrus*, are not known to occur in tropical Africa; their identity remains unclear until voucher material is found. These genera are discussed at the end as 'unplaced species'. It is hoped that the generic names *Casinaria, Chriodes, Coelichneumon, Coleocentrus* and *Scenocharops* will disappear from future papers on cereal stem borers and their parasitoids in Africa.

The genera and species are discussed under the appropriate subfamilies in alphabetical order, and the subfamilies are also arranged in this way. Although the taxonomic relationships between the subfamilies and lower taxa are obscured by this method, it will be easier for the non-taxonomist to find information. From the check-list of genera discussed, these can easily be attributed to the proper subfamily. Genera placed in square brackets have been introduced into Africa, without success, as far as is known. The last column contains names that have been used in the literature on stem borers for species belonging to the preceding genus. There are also a couple of wrongly spelled names, for genera as well as for species; most of these are given under the proper names in the general treatment of the genera.

A key to the genera that have been found to parasitize stem borers in the Ethiopian region is provided.

Table 20.1. Check-list of ichneumenid parasitoids of African stem borers.

Subfamily	Genus	Also reported as
Banchinae	*Syzeuctus*	
Campopleginae (= Porizontinae *sensu* Townes)	*Charops*	
	Eriborus	?*Chriodes*
	Venturia	*Scenocharops, Casinaria, Chriodes*
Cremastinae	*Pristomerus*	*Trathala*
		Temelucha
		?*Trathala*
Cryptinae (= Gelinae *sensu* Townes) (= Phygadeuontinae)	*Amauromorpha*	
	Diracela	
	Gambroides (genus near)	*Isotima, Menaforia*
	Vadonina	
	[*Mallochia*]	
	?*Paraphylax*	
	Genus A	
Ichneumoninae	*Dentichasmias*	
	Ichneumon	*Coelichneumon* ??
	Ischnojoppa	
	Procerochasmias	
Mesochorinae	*Mesochorus*	
Ophioninae	*Enicospilus*	*Amesospilus*
		Stauropoctonus
		Ophion (*Stauropodoctonus*)
Pimplinae (= Ephialtinae *sensu* Townes)	*Holcopimpla*	
	[*Itoplectis*]	
	Pimpla	
	Xanthopimpla	

For *Coelichneumon* and *Coleocentrus*, see unplaced species.

MORPHOLOGICAL TERMINOLOGY

The morphological terminology used in this chapter is largely based upon Townes (1969); where appropriate, synonymic terms as used by authors of other chapters are listed below.

Thorax and propodeum (Figs 557, 559):	mesosoma
Gaster (Townes: abdomen) (Fig. 563):	metasoma
Subgenital plate:	hypopygium
First intercubital vein (Fig. 558) (front wing):	rm = 2rm = 2rs-m
Second intercubital vein (Fig. 558) (front wing):	3rm = 3rs-m
Second recurrent vein (Fig. 558) (front wing):	2m-cu
Areolet (Fig. 558, 'A') (front wing):	second cubital cell = second submarginal cell
Discocubital cell (Fig. 558) (front wing):	discosubmarginal cell (the fusion of first cubital and first discoidal cells)

Some other terms that might require some explanation are:

Petiolus (Fig. 559) = narrow basal part of first gastral segment (in front of the spiracle) when this is relatively slender

Postpetiolus (Fig. 559) = broader distal part of first gastral tergite (behind the spiracle) when the basal part is relatively slender. In some genera, the postpetiolus is well separated from the anterior part of the tergite; in other genera, the tergite widens gradually.

Key to the Genera of Ichneumonidae Parasitic upon African Cereal Stem Borers

This key is constructed for practical use by non-specialists and is generally not based on real taxonomic differences between subfamilies and genera. It does not reflect the taxonomic relationships. Also, it should only be used for parasitoids reared from stem borers in graminaceous crops in sub-Saharan Africa. In other parts of the world and from other host species, different genera or species may be obtained, which could key out to the wrong genus if identified with this key.

Since the inexperienced user may sometimes have problems in making the right decisions when working with the key, some genera may be traced by following either half of a couplet.

To prevent the occurrence of wrong names in the literature on parasitoids of stem borers, it will often be advisable to have identifications checked; the author of this chapter is willing to do such identifications of Ichneumonidae without charge.

The basic morphological terminology as used in this key is largely based upon Townes (1969, 1970a, b, 1971) and illustrated in Figs 554–563; where appropriate, synonymous terms that might be used by authors of other chapters are listed (see above). For more detailed keys to the genera of Ichneumonidae, the advanced student is referred to the above-mentioned works of Townes.

Key to genera of Ichneumonidae associated with African cereal stem borers

1. Gaster clearly compressed in its distal half, higher than wide from its 3rd segment onwards (fig. 1a). The spiracle of the 1st gastral tergite is placed beyond the midlength of the segment (fig. 1b). In females the dorsal valve of the ovipositor usually has a notch before the apex, except in the subfamily Mesochorinae. This subfamily can also cause problems in deciding whether the gaster is compressed or depressed; it can therefore be traced through both halves of the couplet . . . 2

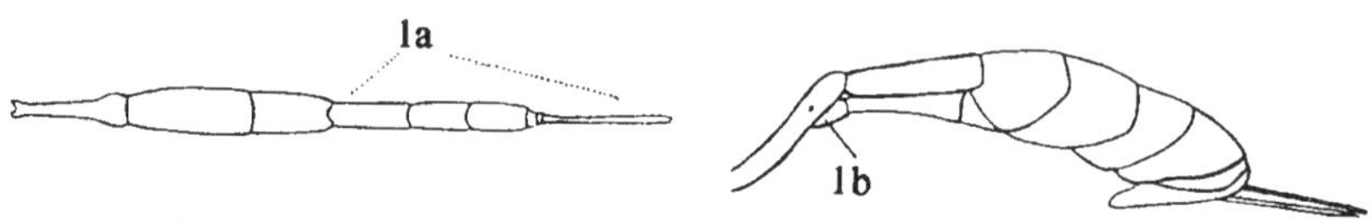

\- Gaster depressed; all segments wider than deep (fig. 1aa). Only the basal part of the 1st segment (= the petiolus) may be sometimes a little narrower than deep or as wide as deep. In females the dorsal valve of the ovipositor often does not have a notch before the apex, but a notch may sometimes be present 10

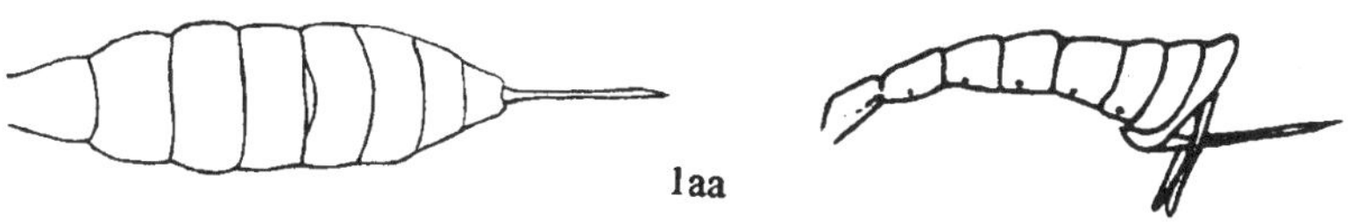

2. Front wing without areolet, only one intercubital vein (fig. 2a) present. The second recurrent (fig. 2b) vein conspicuously basad of the intercubital vein. The discocubital cell in front wing in most species with one or more sclerites in a hairless area (fig. 2c). Relatively large specimens (15 mm or more) of pale brown, yellowish colour . . . . *Enicospilus* (Subfam. Ophioninae)

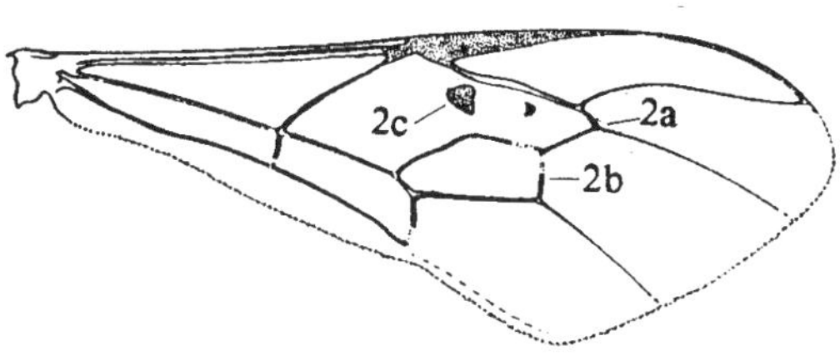

\- Front wing with or without areolet; if there is only one intercubital vein, then the second recurrent vein (fig. 2bb) is distad, opposite or just basad of the intercubital vein (fig. 2aa). Discocubital cell of front wing never with sclerites. Specimens of various sizes; the colour ranging from black to yellow .. 3

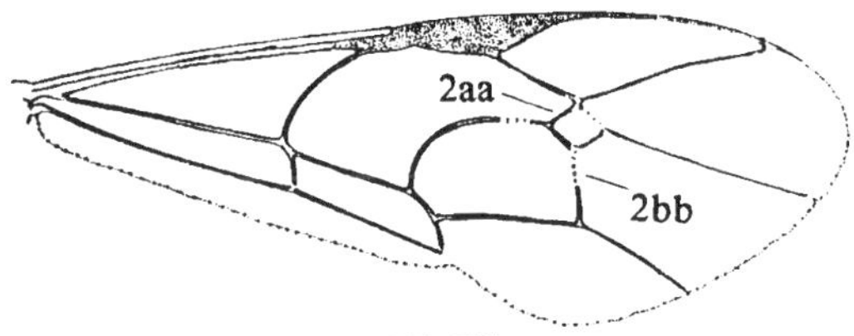

3. Clypeus not separated from the face either by a clear or a faint impression (fig. 3a) 4

\- Clypeus separated from face by an impression which is usually clear (fig. 3aa), but sometimes rather faint .... Subfamily **Cremastinae** 8

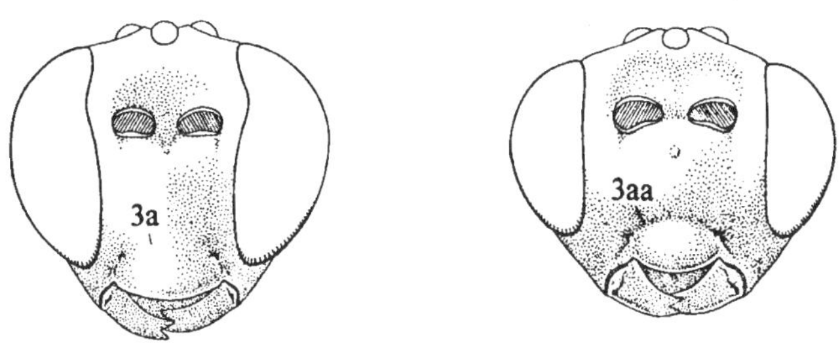

4. Areolet always present, relatively large and more or less diamond shaped (fig. 4a). Face and clypeus forming a broad and rather convex surface. Gaster polished from 2nd tergite onwards. Specimens rather small, usually not longer than about 5 mm, and usually yellowish or with brown markings on a yellow background. Claspers of male genitalia ending in long rods (fig. 4b). The female subgenital plate large; ovipositor without a dorsal notch .... *Mesochorus* (Subfam. Mesochorinae)

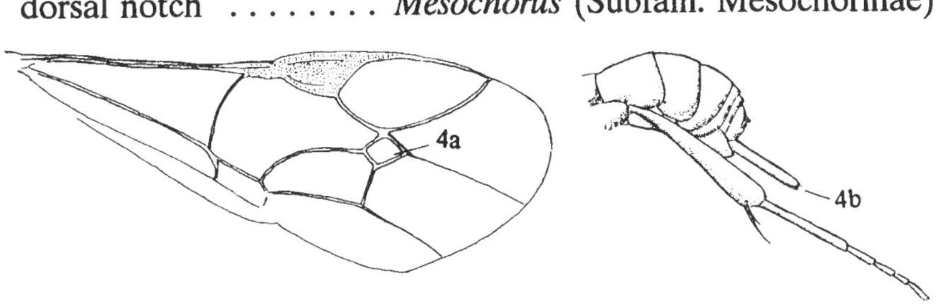

- Areolet present or absent; if present, much smaller, and not of the shape as in Mesochorinae. Gaster usually not polished; tergites often with conspicuous punctation. Specimens often considerably longer than 5 mm. Male claspers never ending in rod like structures. Female subgenital plate smaller and ovipositor with a preapical dorsal notch. General colour various, often black or very dark, but also yellow 5

5. Face and clypeus black, never yellow; without traces of a separating impression (see fig. 3a). Front wing with or without areolet . . . . Subfamily **Campopleginae** 6

- Face and clypeus usually yellow or brown, rarely black; always with traces of an impression separating them (see fig. 3aa). Front wing never with an areolet . . Subfamily **Cremastinae** 8

6. Petiolus long and slender, without a lateral pit or groove in front of the spiracle (fig. 6a). Cross section of petiolus near its basal third almost circular and rather smooth. Areolet present or absent; if absent then the head usually of lenticular shape (strongly transverse) 7

- Petiolus somewhat shorter, with a lateral pit or groove (= glymma) in front of its spiracle (fig. 6aa). Areolet always absent (see fig. 7bb) . . . . *Eriborus*

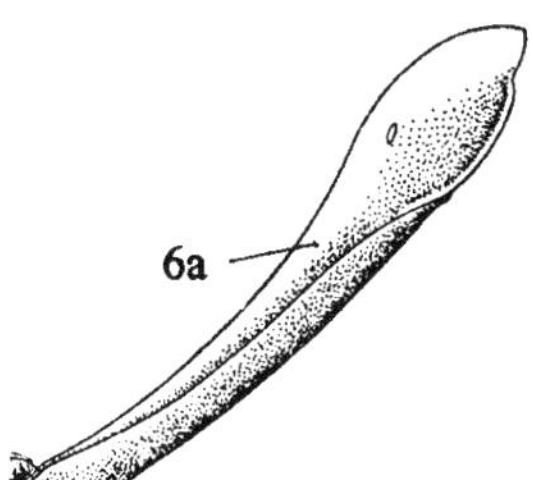

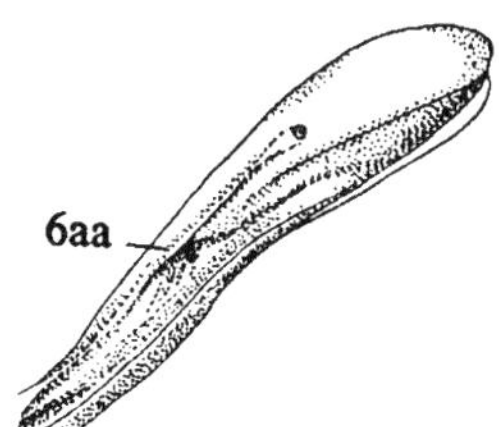

7. Inner margin of eye not or only very weakly emarginate near antennal socket (fig. 7a). Areolet present (fig. 7b). Ovipositor sheath of female comparatively long, more than twice as long as height of gaster near insertion of ovipositor. Head of 'normal' shape . . . . *Venturia*

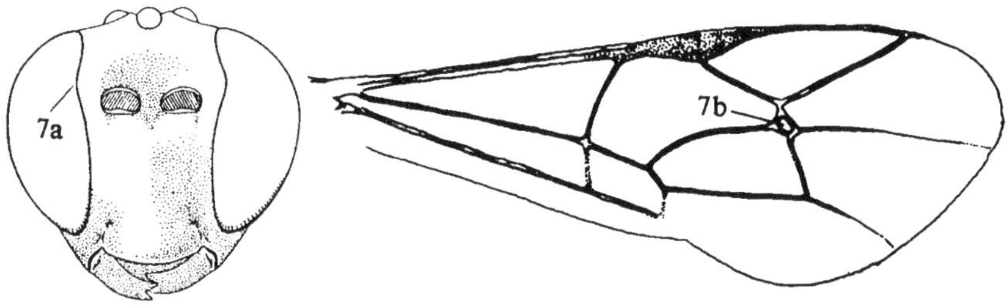

\- Inner margin of eye distinctly emarginate near antennal socket (fig. 7aa). Areolet absent (fig. 7bb). Ovipositor sheath of female less than twice as long as height of gaster near insertion of ovipositor. Head with temples very narrow, short and broad, of 'lenticular' shape . . . . *Charops*

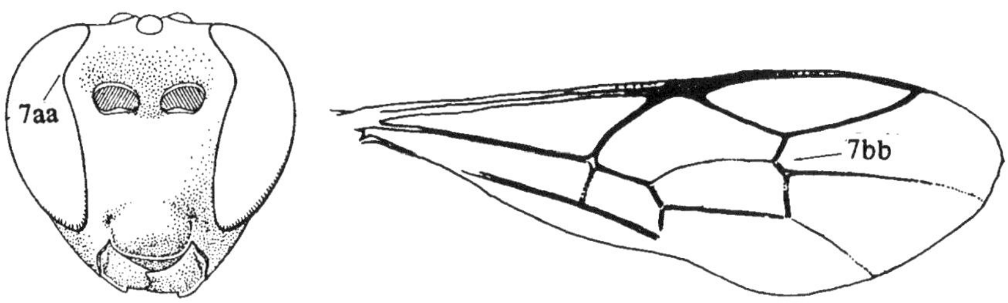

8\. Hind femur swollen, especially in male, with a strong tooth on its ventral surface; the tooth stronger in males than in females; usually also a row of smaller denticles distad of the tooth (fig. 8a). 2nd gastral tergite with thyridia . . . . *Pristomerus*

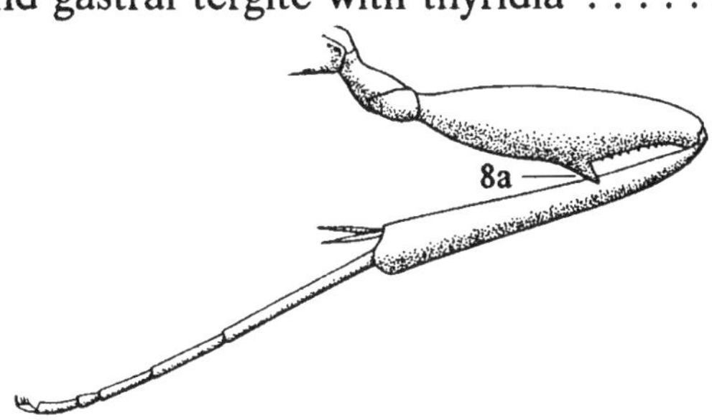

\- Hind femur not swollen, without a tooth or denticles on its ventral surface (fig. 8aa). 2nd gastral tergite without thyridia 9

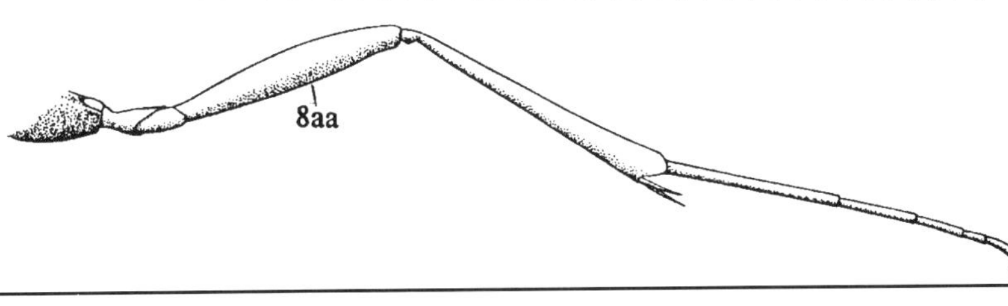

9. Ventral margins of 1st gastral tergite subparallel and clearly separate; the sternite well visible for all of its length (fig. 9a). Occipital carina of head almost complete above . . . *Trathala*

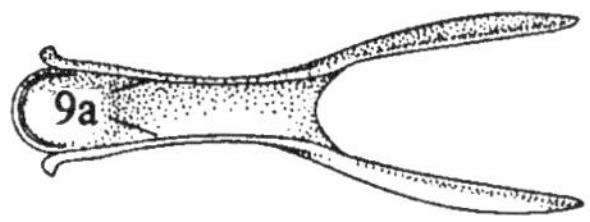

- Ventral margins of 1st gastral tergite bowed inward and touching or almost touching; the sternite therefore partly enclosed and (almost) invisible for part of its length (fig. 9aa). Occipital carina broadly interrupted dorsally . . . . *Temelucha*

9aa

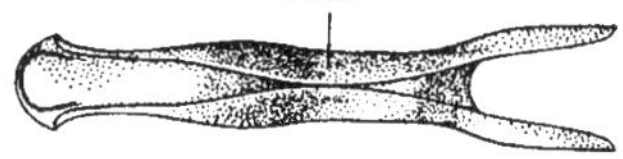

10. Areolet always present, relatively large and more or less 'diamond shaped' (fig. 4a). Face and clypeus forming a broad and rather convex surface. Gaster polished from 2nd tergite onwards. Specimens rather small, usually not longer than about 5 mm, and usually yellowish or with brown markings on a yellow background. Claspers of male genitalia ending in long rods (fig. 4b). The female subgenital plate large; ovipositor short, without a dorsal notch . . . . *Mesochorus* (Subfam. Mesochorinae)
- Areolet present or absent, if present, not as large and not of the shape as in Mesochorinae. Gaster usually not polished; tergites often with conspicuous punctation. Specimens often considerably longer than 5 mm. Male claspers never ending in rod like structures. Female subgenital plate various; ovipositor various, sometimes relatively long, with or without a preapical dorsal notch. General colour various, often black or dark, but also bright yellow 11

11. General colour of body and legs bright yellow, often with black markings on head, thorax and on gastral tergites. Or body pale brown 12
- General colour of body and legs usually a combination of red with black, very often with white markings; in some species with yellow markings, but never the general colour yellow or pale brown 15

12. First gastral tergite with petiolus considerably narrower at base than postpetiolus is broad, postpetiolus about 2.5 - 3 times as broad as petiolus at base (fig. 12a); spiracle of 1st tergite definitely behind the midlength (fig. 12b); head in dorsal view almost cubic, with the eyes relatively small as compared with the shape of the head (fig. 12c). Antenna of female with a white annulus, distinctly broadened beyond the annulus. Antenna of male often with a small white annulus, not broadened .... *Ischnojoppa* (Subfam. Ichneumoninae)

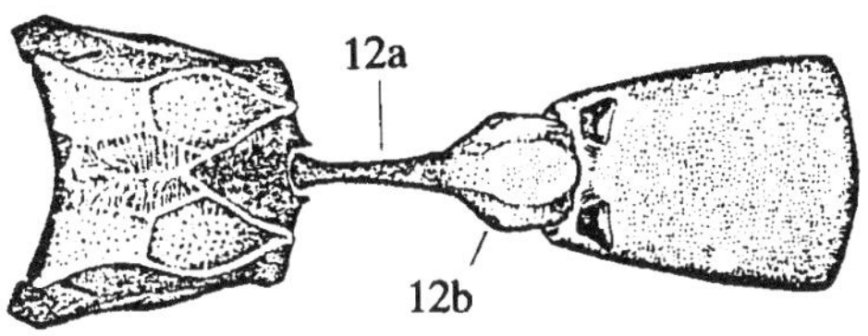

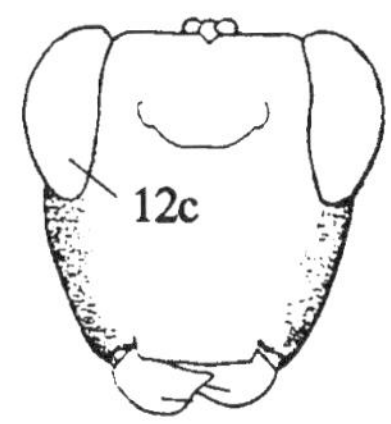

\- First gastral tergite not considerably narrower at base, the tergite at the apex about 1.5 times as broad as at base (fig. 12aa); spiracle of 1st tergite basad of the midlength (fig. 12bb) or approximately in the middle; head and eyes in dorsal view of 'normal' shape (fig. 12cc). Antenna of female not broadened distad of the middle, male and female antenna without white annulus 13

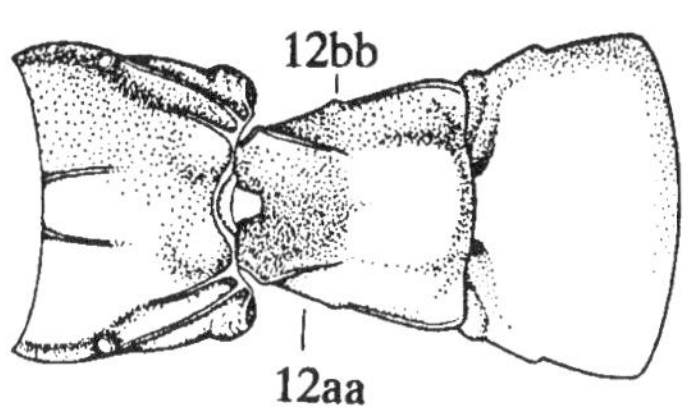

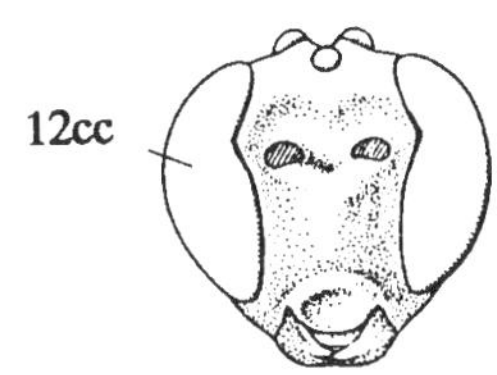

13. Mandible twisted to about 90° and considerably narrowed towards the apex, the lower tooth smaller than upper tooth, and almost invisible when mandibles are closed (fig. 13a). Body and legs yellow, often with black markings. Propodeum usually extensively areolate .... *Xanthopimpla* (Subfam. Pimplinae)

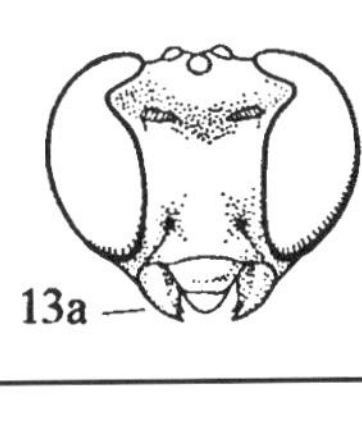

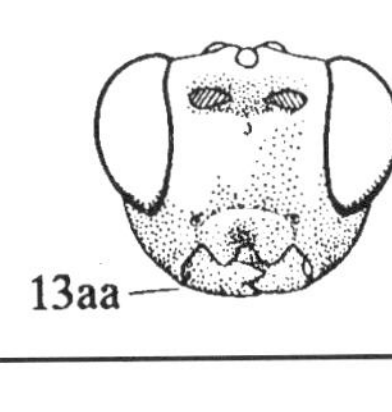

- Mandible not twisted and only slightly narrowed towards the apex; lower tooth not, or only slightly shorter than upper tooth (fig. 13aa) 14

14. Submetapleural carina forming a strong lobe just behind middle coxa (fig. 14a). Propodeum with at least basal transverse carina present; apical transverse carina present or absent. Female subgenital plate in lateral view large and conspicuous (fig. 14b). Ovipositor long and slender, with a subapical dorsal notch . . . . *Syzeuctus* [in part] (Subfam. Banchinae)

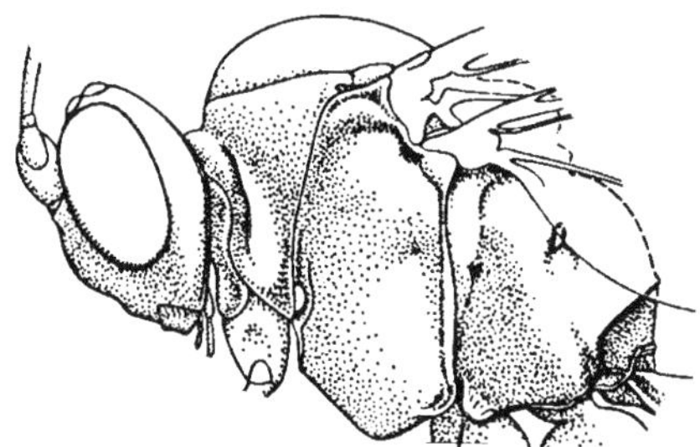

14a

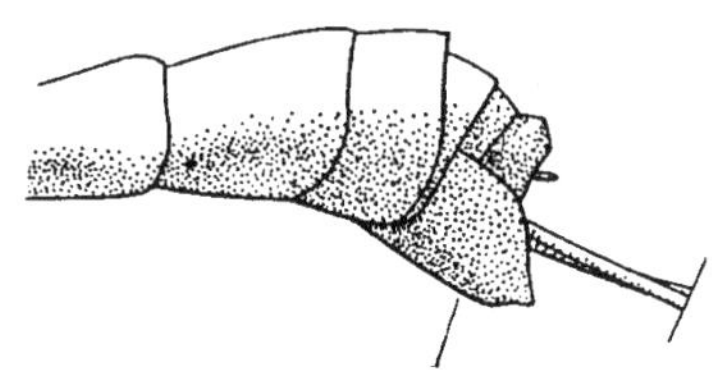

14b

- Submetapleural carina not forming a strong lobe just behind middle coxa (fig. 14aa). Propodeum without transverse carinae. Female subgenital plate in lateral view not conspicuous (fig. 14bb). Ovipositor rather long, without a subapical dorsal notch . . . . *Holcopimpla* (Subfam. Pimplinae)

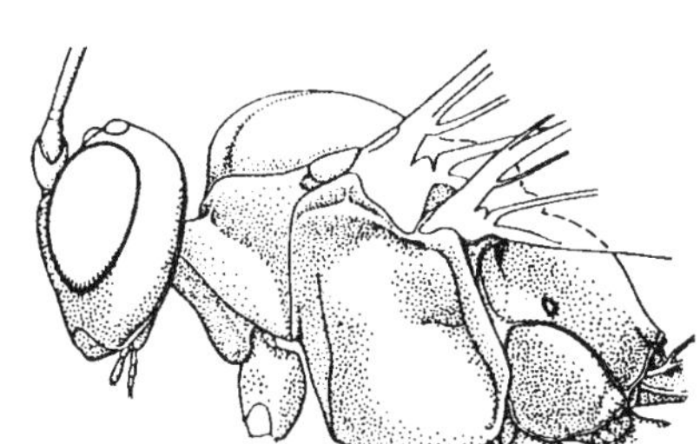

14aa

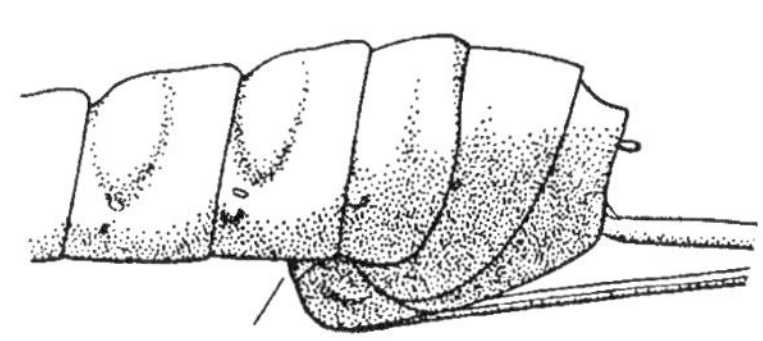

14bb

15. First gastral tergite at base not very much narrower than at the apex (fig. 15a); the tergite gradually broadened backward; spiracle of 1st gastral tergite definitely basad of the middle or at about the midlength (fig. 15b) 16

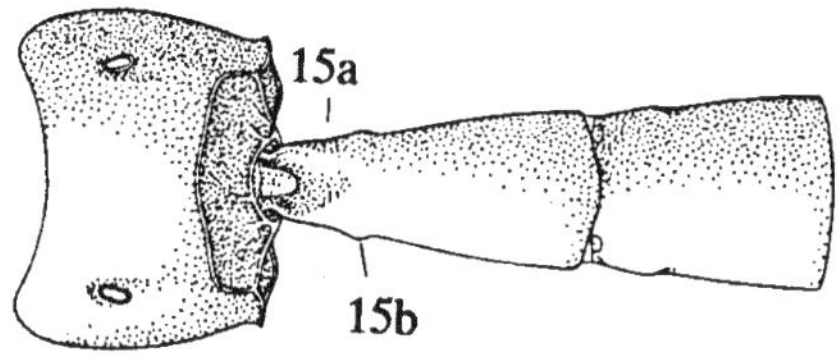

- First gastral tergite with petiolus at base considerably narrower than postpetiolus (fig. 15aa); spiracle of 1st gastral tergite definitely distad of the middle (fig. 15bb), or almost in the middle of the tergite 18

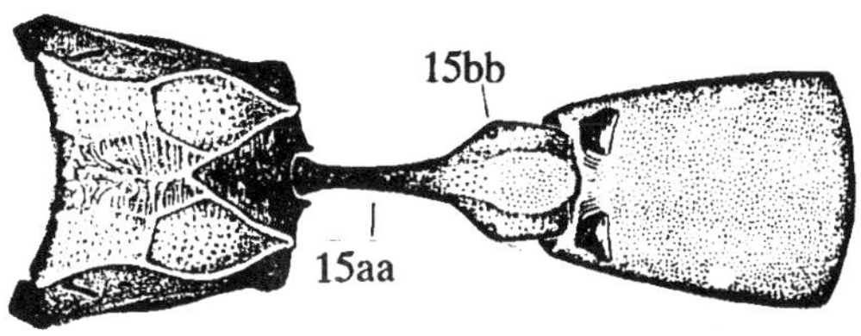

16. Areolet in front wing petiolate (intercubital veins fused at base) (fig. 16a). Submetapleural carina forming a strong lobe just behind middle coxa (see fig. 14a). First gastral segment at least twice as long as broad (fig. 16b). Propodeum without longitudinal carinae (fig. 16c). Female subgenital plate in lateral view large and conspicuous. Ovipositor long and slender, with a subapical dorsal notch . . . . *Syzeuctus* [part] (Subfam. Banchinae)

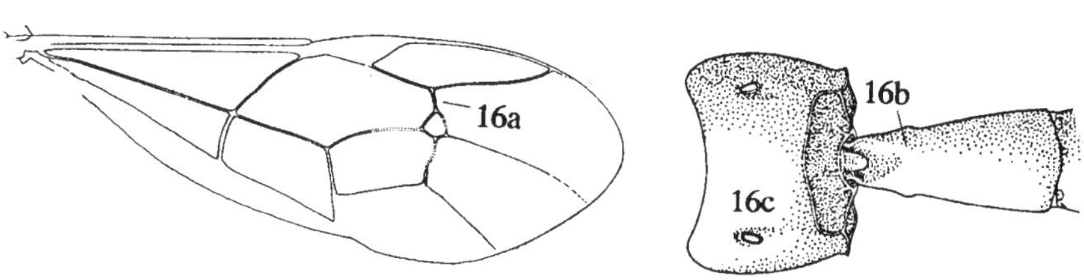

- Areolet not petiolate (intercubital veins not fused at base) (fig. 16aa). Submetapleural carina not forming a strong lobe just behind middle coxa (see fig. 14aa). First gastral tergite only little longer than broad (fig. 16bb). Propodeum with at least traces of longitudinal carinae (16cc). Female subgenital plate in lateral view not conspicuous. Ovipositor stout, without a subapical dorsal notch . . . . Subfamily **Pimplinae** [part] 17

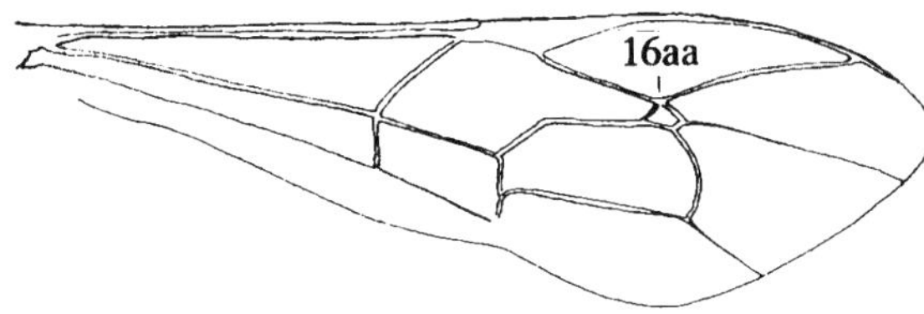

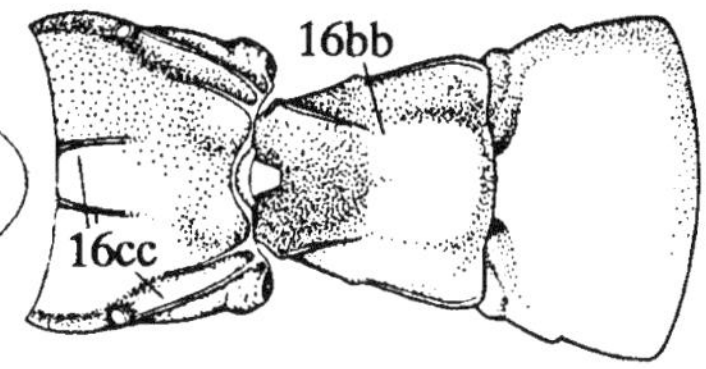

17. Hind tibia (except for base and apex), hind tarsus and all trochanters white or yellowish (hind tarsal segments black at their apex). Inner margin of eye strongly emarginate opposite antennal socket (fig. 17a) . . . . *Itoplectis*

\- Hind tibia, hind tarsus and all trochanters red or black, never partly white. Inner margin of eye, although clearly with a concavity just above antennal socket, not strongly emarginate (fig. 17aa) . . . . *Pimpla*

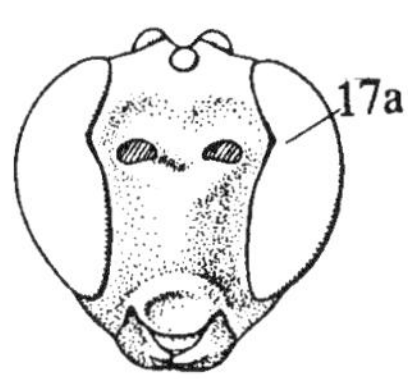

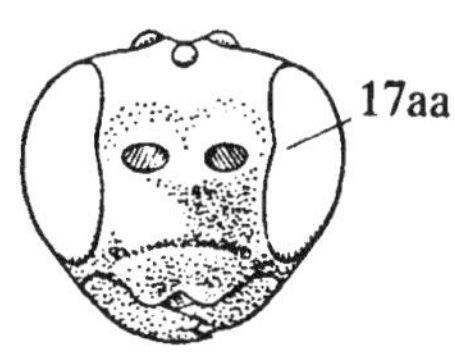

18. Propodeum always clearly areolate, with longitudinal carinae as well as transverse carinae present (fig. 18a). Areolet in front wing narrow in front, intercubital veins (almost) meeting at radial vein (fig. 18b), or in *Ichneumon* conspicuously convergent to front side; 2nd intercubital vein always present and pigmented. Ovipositor of female extending only comparatively little beyond apex of gaster (at most about 0.5 mm); in some species the antenna of female is considerably broader in the middle than at base and tip, in this area the flagellar segments are usually flattened ventrally . Subfamily **Ichneumoninae** 19

18a

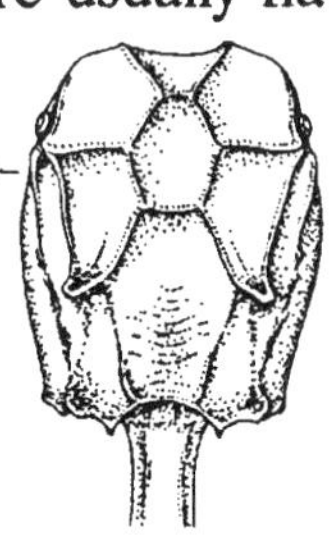

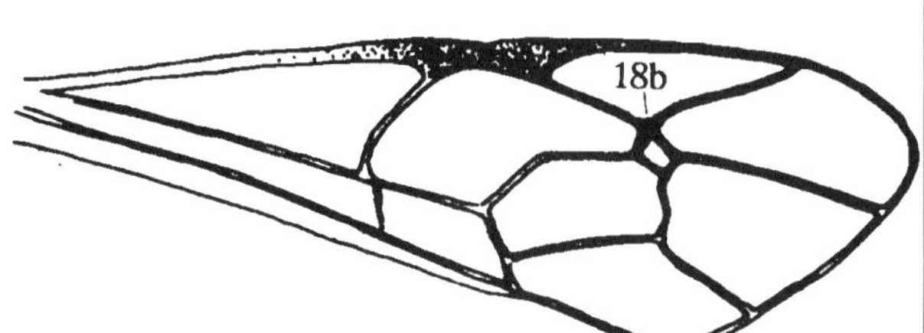

- Propodeum usually not areolate, median longitudinal carinae at most present at base to build the basal area; only transverse carinae well developed (fig. 18aa), (two genera have the propodeum areolate; they have no 2nd intercubital vein and can be reached through either half of the couplet). Areolet variable, but intercubital veins never meeting or almost meeting; if clearly convergent towards front side, then 2nd intercubital vein not pigmented (fig. 18bb); sometimes second intercubitus absent. Ovipositor of female always reaching evidently beyond apex of gaster, in dorsal view clearly visible; antennal segments in female never broadened and/or flattened in the middle of antenna . . . . Subfamily **Cryptinae** 23

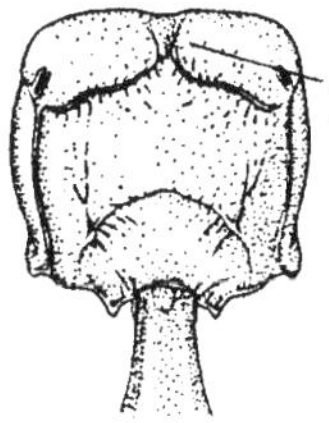

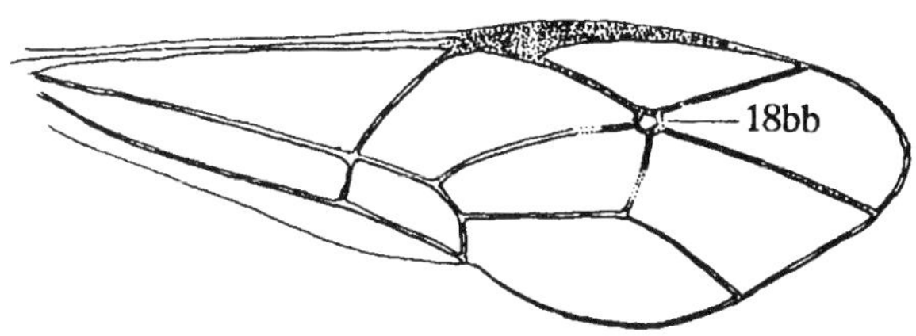

19. Front wing with one (fig. 19a) or two (fig. 19b) dark marks near stigma; 2nd intercubitus absent (areolet open) or rarely not pigmented; sternaulus well developed; rather small species with body length less than 7.5 mm. The only species which is positively reared in connection with cereal stem borers has very strong longitudinal sculpture on gastral tergites 2 and 3 (in males) or 2-4 (in females) (See *Diracela* and *Paraphylax* in Cryptinae) 23

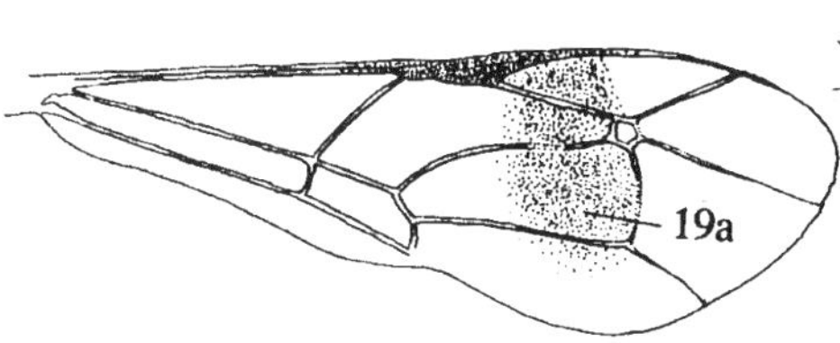

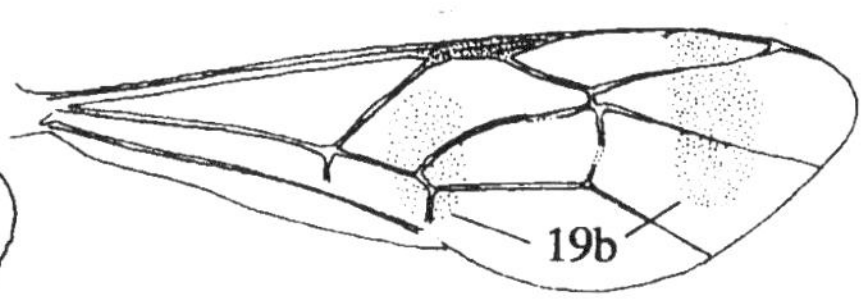

- Front wing without defined dark marks near stigma, except that in some species the whole wing or the outer margin may be darkened; 2nd intercubitus present and clearly pigmented (areolet closed); species usually considerably larger. - If body length less than 8 mm (as is sometimes observed in *Dentichasmias busseolae*) then gastral tergites 2 and 3 (in males) or 2-4 (in females) never with strong longitudinal sculpture 20

20. General colour of body yellow, with at most black apical mark on gaster and with black marks on legs; eyes relatively small as compared with the rather cubic shape of the head (see fig. 12c). Antenna of female strongly broadened distad of the annulus . . . . *Ischnojoppa*

\- Body largely black and red, in *Ichneumon* often with extensive orange and/or yellow markings, especially on gaster; eyes and head of 'normal' shape (see fig. 12cc) 21

21. Apex of clypeus with two small teeth (fig. 21a). Head usually red, ranging to black; thorax red with varying parts black, mesoscutum red; gaster largely red at base, black at apex, with tergites 6 and 7 always marked with white, but without yellow or white apical bands on other tergites. Specimens up to about 12 mm long, often less than 10 mm . . . . *Dentichasmias*

\- Apex of clypeus without small teeth. Colour pattern various; gaster whether without any white or yellow marks (and body length usually more than 12 mm), or at least a narrow yellow apical band on tergite 2 (usually also on other tergites) and usually with apical white marks on tergites 6 and 7 (and body length about 10-12.5 mm) 22

22. Intercubital veins meeting or almost so, forming a 'pointed' areolet (fig. 22a). Head, thorax and propodeum variously coloured, ranging from almost entirely black to almost entirely red; gaster often completely red, or with a black apex, no tergites marked with white or yellow. Body length usually 12-15 mm . . . . *Procerochasmias*

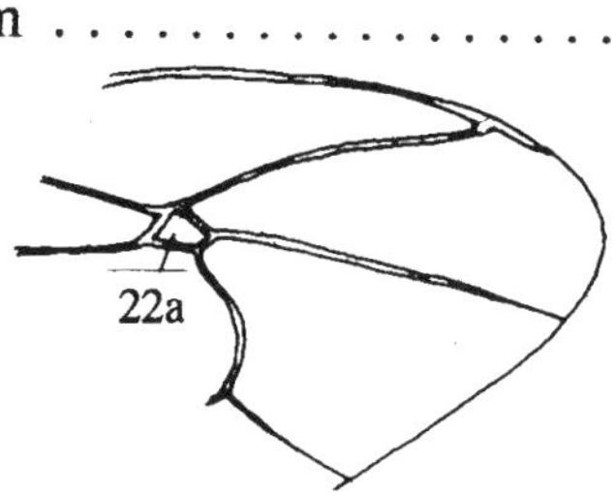

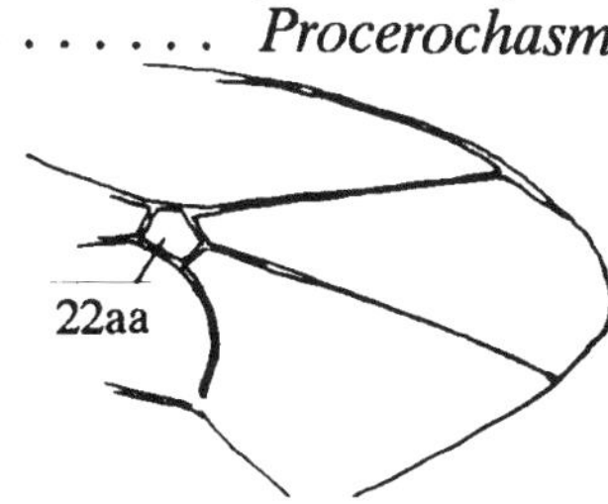

\- Intercubital veins, although clearly convergent towards the front, not meeting or almost so, areolet clearly pentagonal (fig 22 aa); at least 2nd gastral tergite with a narrow white or yellow apical band, but usually more tergites with yellow or white marks. Body length about 10-12.5 mm . . . . *Ichneumon*

23\. Propodeal spiracle very narrow and elongate, more than 4 times as long as wide (fig. 23a) (sometimes rather hard to recognize because of its narrowness). Head and thorax covered with very dense short hairs (these only visible at a high magnification). Areolet open at outer side and shape of areolet rather obscure (2nd intercubitus absent and not just unpigmented) (fig. 23b). Apex of clypeus without a well developed median tooth. Propodeum with basal transverse carina strong and apical transverse carina rather weakly developed medially. No longitudinal carinae present to build an 'areolate' propodeum (fig. 23c). Female orange-brown with dorsal side of head black; gastral tergites 4-6 darker, tergite 7 with a white band. Front wing without dark clouded mark. Male unknown, but probably darker than female . . . . *Amauromorpha*

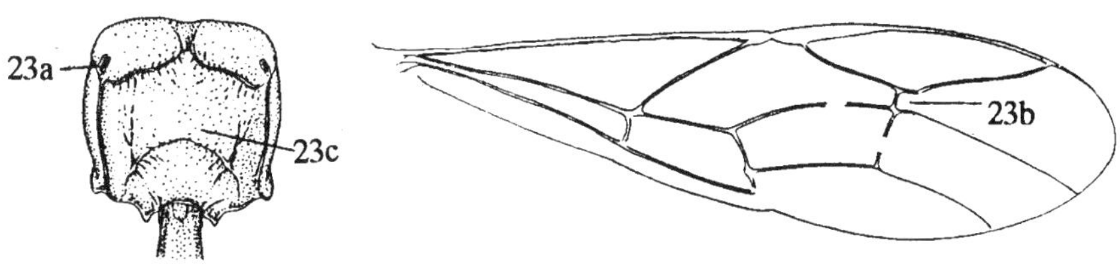

\- Propodeal spiracle circular or oval, less then twice as long as broad. Hairs on head and thorax less dense. Areolet closed or open, its shape at least clearly indicated; 2nd intercubitus usually present although sometimes not pigmented, rarely absent. Apex of clypeus various, often with a median pair of small tubercles (sometimes absent or hard to trace) or with a well developed median tooth. Propodeum various, whether with only basal transverse carina well developed and apical transverse carina at least absent medially, or completely areolate. Colour various, often (but not always) dark red with black; with or without white apical mark on gaster 24

24. Gastral tergites 1-3 (in males) or 1-4 (in females) with strong longitudinal sculpture (fig. 24a); in male segments 4-7 protruding very little beyond apex of tergite 3; head in dorsal view with a very large median notch at level of occipital carina; propodeum completely areolate (fig. 24b); 2nd intercubitus absent (areolet open) (fig. 24c); smaller sized species (body length up to about 7 mm) . . . . *Diracela*

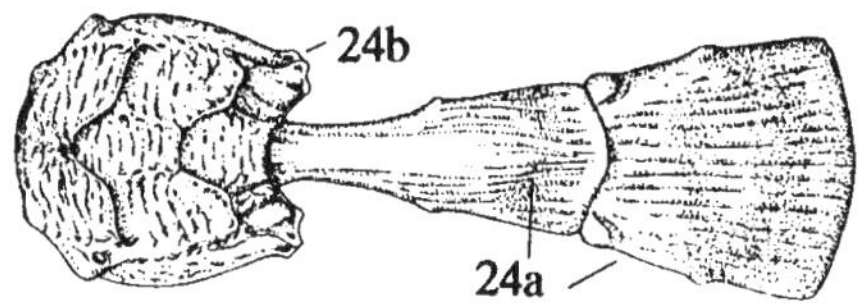

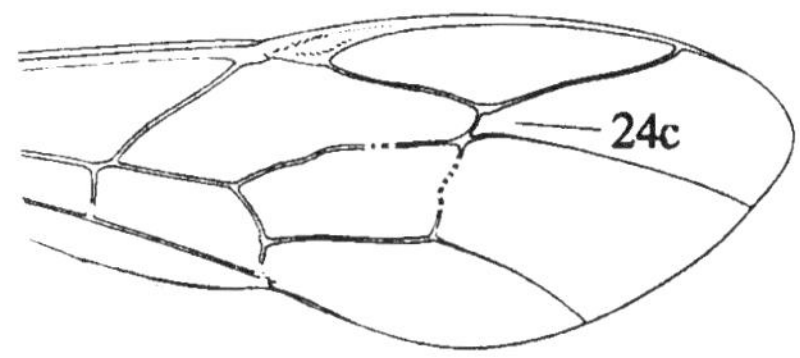

\- Gastral tergites 1-4 more or less strongly punctate, not with strong longitudinal sculpture (fig. 24aa), rarely with fine longitudinal striae; in male tergites 4-7 of normal shape; upper part of hind aspect of head without an incision at level of occipital carina; most species with propodeum not completely areolate (only transverse carinae present) (fig. 24bb) and body size usually more than 7 mm 25

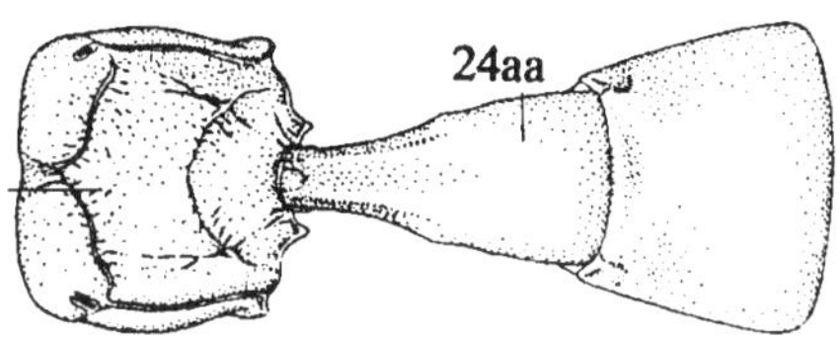

25. Propodeum almost completely areolate (fig. 25a); mesoscutum with a very fine median longitudinal groove on its median lobe; smaller sized species up to about 7.5 mm. (A few species of this genus show fine striation on tergites 2 and 3) . . . *Paraphylax*

25a
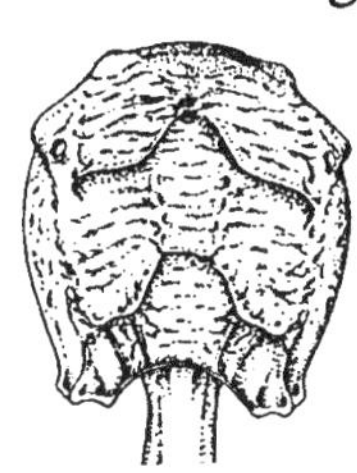

\- Propodeum with only transverse carinae present (fig. 25aa); usually only the basal transverse carina well developed; median lobe of mesoscutum without a trace of a fine median longitudinal groove; body size in most specimens more than 7.5 mm 26

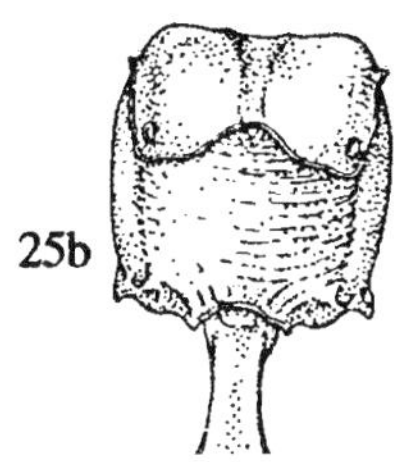

26. Females 27
- Males 29

27. Apex of clypeus with a single well developed median tooth. Front wing without dark cloud or bands in the middle, although apical margin of wing is darkened .... Genus A
- Apex of clypeus usually with a median pair of small teeth or tubercles, which are sometimes hard to see or even lacking, or rarely with a very poorly developed small median tooth. Front wing with one (sometimes faint) cloudy spot or band basad of the areolet (see fig. 28a), or with two well developed dark bands (fig. 28aa) 28

28. Front wing with only one (sometimes faint) clouded spot or band just basad of the areolet (see fig. 28a). Gaster dark red at base and largely black towards the apex, with a white band on tergite 7. - (In *G. javensis*, which is probably not established, the gastral tergites 1, 2, 5, 6 and 7 are usually marked with white)

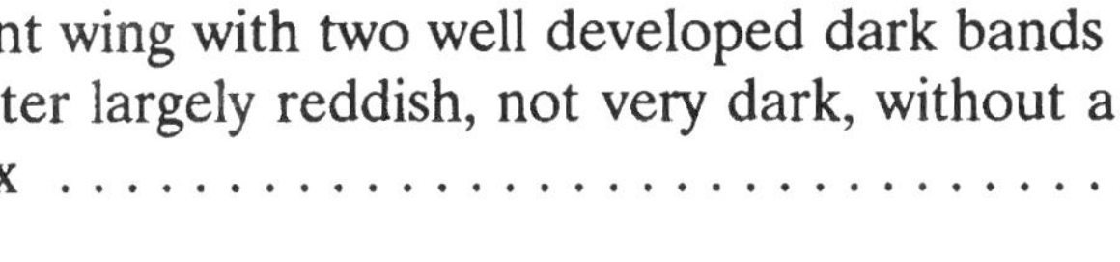

- Front wing with two well developed dark bands (see fig. 28aa). Gaster largely reddish, not very dark, without a white mark at apex .... *Mallochia*

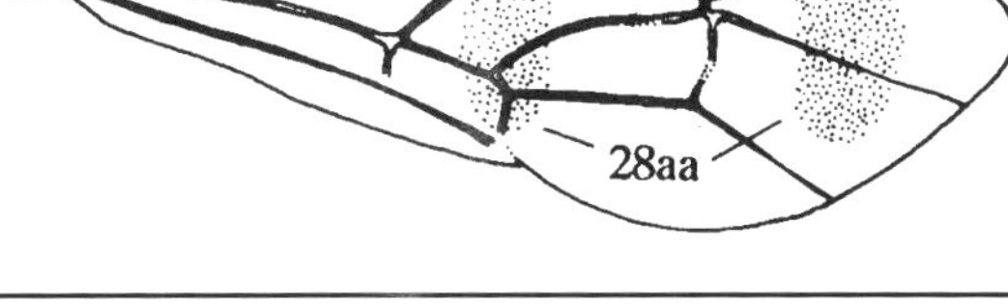

29. Apex of clypeus with a single well developed median tooth. Hind tarsus with segments 2-4 white; (thorax and propodeum not marked with white) . . . . Genus A
- Apex of clypeus usually with a median pair of small teeth or tubercles, which are sometimes hard to see or lacking, or rarely with a very poorly developed small median tooth. Hind tarsus pale brown to black. - If hind tarsus with white segments, in *Mallochia*, then also thorax and propodeum with extensive white markings 30

30. Thorax and propodeum red (in *G. javensis* - not established in Africa? - anterior part of thorax partly black); face black; gaster with a white mark at apex; hind tarsus pale brown to black . . . . *Gambroides*
- Thorax and propodeum black with extensive white contrasting marks; face white; gaster without a white mark at apex; hind tarsus with segments 2 and 3 white . . . . *Mallochia*

BANCHINAE

Banchinae are koinobiont endoparasitoids of lepidopterous larvae, and almost all are solitary. The majority of the species parasitize larvae of microlepidoptera that live in concealed situations, as stem borers do. Townes and Townes (1973) list 12 genera of this rather large subfamily occurring in sub-Saharan Africa. Only one of these, *Syzeuctus*, is known to parasitize stem borers.

Syzeuctus Foerster

Diagnosis. Females of *Syzeuctus* have a large subgenital plate; the body length is more than 5 mm, the ovipositor is relatively long, with a preapical dorsal notch, and the areolet is relatively small.

Biology. Oviposition is usually into the young host larva, and the parasitoid larva emerges from its host shortly before the host pupates. The parasitoid larva spins a longish cocoon in which to pupate. All cocoons that have been seen are brown.

There are a few records where *Syzeuctus* is mentioned as being a pupal parasitoid; these are assumed to be based on misinterpretations, where the cocoon has been mistaken for the host's pupa. In some cases, this could be checked.

All but one of the records of *Syzeuctus* parasitizing stem borers are from microlepidoptera. Milner (1967) reared two specimens of *Syzeuctus* spp. from larvae of 'probably *Busseola*'. Both specimens have been studied; their identification is correct, but I have some doubt about the determination of the host. Unfortunately, there are no host remains pinned with the specimens.

Distribution. *Syzeuctus* is a large genus of worldwide distribution. Townes and Townes (1973) list 46 species from the Ethiopian region, and many undescribed species have been seen.

Remarks. Several species have been studied which have been reared from stem borers, two of which are undescribed. A revision of the African species of *Syzeuctus* is planned, and preliminary research has started.

Key to the Species of *Syzeuctus* Reared from Stem Borers

1. Thorax (including propodeum) black with various contrasting white or yellow marks; head also black with white or yellow with black pattern 2

– Thorax never black with contrasting white or yellow pattern, but is either almost uniformly coloured bright yellowish red or brownish, or mesoscutum and propodeum darker; main colour of head in continental species yellowish red with or without some black on top and backside; head only black and white in Madagascan species 3

2. Gaster black with contrasting white apical bands or marks on tergites 1–4(5); thorax black, extensively marked with white (West Africa) .... *Syzeuctus* sp. A

– Gaster completely red, or tergites 1–3 sometimes darker and remaining tergites red; at least tergites 1 and 2, and sometimes more tergites with apical yellow band, tergite 1 always with two lateral yellow stripes at base (East and Southern Africa) .... *Syzeuctus tonganus*

3. Body in general aspect dark, especially gaster almost uniformly coloured blackish or dark brown-red; mesoscutum and upper part of propodeum dark brown, lower parts of thorax often orange-red (Madagascar and Comoros) .... *Syzeuctus gaullei*
– Body in general aspect bright yellowish red; with only few black marks on head, thorax and/or gaster, or without any black marks 4
4. Females 5
– Males 6
5. Gastral tergites 1, 2 and 6–8 with dark (brown/blackish) marks, not of the same red colour as other tergites .... *Syzeuctus* sp. B
– All gastral tergites more of less of same yellowish/red colour 7
6. Trochanter of hindleg black (coxa and trochantellus of hindleg yellow) .... *Syzeuctus* sp. B
– Trochanter of hindleg same yellow/red colour as coxa, trochantellus and femur III 7
7. Head without any black or dark markings at upper and backside (vertex and occiput bright yellowish red, like adjacent parts of head) (West Africa) .... *Syzeuctus senegalensis*
– Head red with a median dark (dark red to black) zone from just behind antennae backwards to postocciput, contrasting with rest of head. Mesoscutum in some specimens with dark median band (East Africa and Zaire) .... *Syzeuctus ruberrimus*

Syzeuctus gaullei Seyrig

This species is known from Madagascar, from where it is reported as a parasitoid of *Chilo sacchariphagus* in sugar cane by Appert (1967b, 1971, 1973), Appert *et al.* (1969) and Bordat *et al.* (1977) and of *Chilo orichalcociliellus* in sorghum and maize by Appert (1973) and Delobel (1975). From the Comoros, it is reported by Appert (1973) from *Chilo partellus* in maize and sorghum.

Diagnosis. The most striking character of this species is the almost unicoloured gaster, which is black to dark reddish, sometimes with some red on the sides. The thorax is usually dark brown above and orange-red on the underside, but other colour patterns occur. Also, the colour of the head is quite variable, the face being mainly white to brown.

Biology. Delobel (1975) gives data on the biology; oviposition is into first- and second-instar larvae, which have not yet entered the stem of sorghum and maize. The parasitoid larva leaves its host in the prepupal stage and spins a reddish brown cocoon in the gallery.

Hosts. *Chilo orichalcociliellus, C. partellus* and *Chilo saccharalis.*

Distribution. Comoros, Madagascar.

Syzeuctus ruberrimus Benoit

Besides the nominate subspecies from Zaïre, subspecies *S. ruberrimus mossambicanus* from Mozambique has been described. The latter is based on its smaller size, paler colour and coarser punctation on the propodeum. Specimens at hand cannot always be attributed clearly to one or other subspecies. Further research should show whether the subspecies should be retained or synonymized. This species has not been reported since its description.

Diagnosis. This species is very similar to *S. senegalensis*; it can be separated from that species by the black mark extending from just behind antennae backwards to the postocciput of the head. The gaster is uniformly coloured, without dark markings on tergites 1, 2 or on the apex in females. As in other species, the gaster is, however, slightly darkened towards the tip. In males, the gastral tergite 8 (which is often more or less intruded) is usually almost black. Males can be separated from *Syzeuctus* sp. B by the yellow trochanters III.

Biology. Like other species of the genus, this is a larval parasitoid. A specimen labelled as being reared from a pupa of *Chilo zonellus* (= *partellus*), by Robertson in Tanzania, is mounted together with a *Syzeuctus* cocoon, so it must have left its host larva.

Hosts. Specimens have been reared from stem borers in maize, rice and wild grasses in East Africa; the borer species are sometimes not specified. The following data are from labels: 'ex ?*Chilo* sp. in maize', 'ex stem borer in rice', 'ex stem borer of *Chloris gayana*, probably *Chilo phaeosema*', 'ex pupa [*sic*!] *Chilo zonellus*' (all from Tanzania); 'ex *Chilo partellus* on *Pennisetum purpureum*', 'ex *Chilo partellus* on *Zea mays*', 'ex *Chilo orichalcociliellus* in maize' from Kenya; 'ex striped caterpillar in rice' from Malawi.

Distribution. Kenya, Malawi, Mozambique (Benoit, 1959), Tanzania (including Zanzibar and Pemba), Zaïre (Benoit, 1959).

Syzeuctus senegalensis Benoit

This species seems to be a common parasitoid of *Coniesta ignefusalis* in West Africa. It is reported by various authors as '*Syzeuctus* sp.' reared from this host species. It was probably first reported from Senegal by Risbec (1950a) as a parasitoid of '*Chilo pyrocaustalis*' (= *C. ignefusalis*), although the figure (Risbec, 1950a: 75) does not represent *Syzeuctus*, but a braconid species. The holotype was reared by Risbec from *C. ignefusalis* in Senegal. Two specimens in The Natural History Museum were reared by Risbec in Bambey, Senegal, labelled 'ex *Chilo* on millet' (without a date). In a later paper, Risbec (1960) mentions *Syzeuctus* sp. from *Coniesta*. Other authors report the species from the same host in West Africa.

Diagnosis. This species can be separated from all other *Syzeuctus* reared from stem borers by its yellowish–reddish colour, without any black or dark brown marks on the head, thorax or gaster. Males are usually paler than females.

Biology. All authors state it is a larval parasitoid, except Harris (1961), who reports the species from a pupa of *Coniesta*. I have seen a female labelled 'Nigeria, Samaru, 9–7–58, ex pupa of *Coniesta ignefusalis*, KMH'. It is pinned together with the cocoon from which it apparently emerged, so it must have been reared from a host larva.

The species is reported as being the most important parasitoid of *Coniesta* in Nigeria by Harris (1962) and the most abundant in Niger (Markham, 1985). Bhatnagar (1987) states having collected large numbers in malaise traps in Senegal. Van Miltenburg (1968) reports that 8% of 50 diapausing larvae were parasitized by *Syzeuctus*.

Ceraphron sp. (Ceraphronidae) is reported by Harris (1962) as a secondary parasitoid of *Coniesta* in millet in Nigeria. Both *Aphanogmus fijiensis* and *Aphanogmus reticulatus* are known from *C. ignefusalis*, and the probability is that the Harris record refers to one of these, especially the former (A. Polaszek, personal communication). Bhatnagar (1984, 1987) reared *Mesochorus* sp. together with *Syzeuctus* from *Coniesta* in Senegal. In this case, *Mesochorus* was probably acting as a hyperparasitoid.

Host. So far only known from *C. ignefusalis*, almost all from millet.

Distribution. Benin, Burkina Faso, Ghana, Mali, Niger, Nigeria, Senegal, Togo.

Syzeuctus tonganus (Kriechbaumer)

Diagnosis. This species is characterized by its black thorax (including propodeum) with a pattern of yellow marks, the head being yellow with some black markings on face, frons, vertex and occiput, and the gaster being mainly red, or sometimes tergites 1–3 dark and following tergites red. First tergite with two longish yellow lateral marks at base and apical yellow mark. At least tergites 2–4 with yellow apical transverse band; in South African specimens all tergites apically yellow. All 'yellow' pattern is clearly yellow, and not whitish as in the next species.

Hosts. *Chilo orichalcociliellus* from maize, *C. partellus* from sorghum.

Distribution. Tanzania, South Africa.

Syzeuctus sp. A

This species was reared in large numbers by Dr Moyal from *Eldana saccharina* in maize in Ivory Coast. It was first reported by Pollet *et al.* (1974, 1978) as an ichneumonid parasitoid of *Eldana* in Ivory Coast. Although they do not use a generic name, in both publications a photograph is given, from which the species can be recognized. Bordat *et al.* (1977) mention *Syzeuctus* sp. from the same host from Ghana, and Dabiré-Binso (1980) published on the same species in Ivory Coast, but used the name *Syzeuctus cribrosus*, which is now known to be a misidentification.

Diagnosis. This species is characterized by the black and white pattern on the head, thorax and gaster, the 'white' being really white, and not 'yellow' as in *S. tonganus*.

Biology. *Syzeuctus* sp. A seems to be restricted to *E. saccharina* in maize in West Africa. The host species, however, is also an important pest in sugar cane and occurs in that crop in the same area. It is striking, therefore, that the species has not been reared from *Eldana* in sugar cane so far. There is a single specimen reared from a stem borer in sorghum; all other records of host plants indicate that maize is the crop from which borer were collected.

There are a few labels stating that a specimen is reared 'ex pupa *Eldana saccharina*', but, since in most cases the parasitoid cocoon is pinned or glued with these specimens, they will have been obtained from larvae. The cocoon has apparently been mistaken for the host's pupa.

Distribution. Ghana, Ivory Coast, Nigeria, Sierra Leone.

Remarks. This species will be officially described in a separate paper by K.W.R. Zwart and A. Delobel (in preparation). It has been widely misidentified by various authors as *S. cribrosus* (e.g. Benoit, 1959).

Syzeuctus sp. B

This is almost certainly the species reported by Jordan (1966) as a parasitoid of *Chilo zacconius* (as *Chilo zaleukos*) from Sierra Leone. Jordan reports having reared 12 or 11 specimens of a *Syzeuctus* species in 1964/65. From the NHM a single female has been studied, with a cocoon but without any host mentioned, which is labelled 'Lungi, SL, FJJ'. The specimen was identified by Kerrich in 1965 as *Syzeuctus* sp.

According to Khan *et al.* (1991), Akinsola (1979) mentions a *Syzeuctus* sp. reared from *Maliarpha separatella* in Nigeria. This paper could not be consulted, but his record almost certainly concerns this species. A male from the NHM is labelled 'Nigeria, Ibadan, 3.vii.1978, No. 7', 'ex larva E.A. Akinsola'. The host is not specified, but Akinsola worked on *Maliarpha.*

The species shows some variability, and also some sexual dimorphism. The available specimens are tentatively united as being one species, but it should be decided by further research whether this is correct. The fact that a male and a female are both reared from the same host and locality is sufficient evidence that the two sexes belong together.

Two small males from Senegal, ex *Chilo* sp. in rice, reared by Etienne (1977) in 1979, differ from other males in having the mesoscutum and propodeum more matt, and the mesosternum, mesopleurum, pronotum and prosternum very pale. They are provisionally included.

Diagnosis. The female is generally red, with black markings posteriorly on head and dark markings on the two basal and three apical gastral segments. It differs by these characters from other species reared from stem borers. The male is on average paler and lacks the dark marks on the gaster; the first trochanter of the hindleg, however, is black in contrast to the adjacent parts of the hindleg.

Hosts. *Chilo diffusilineus* in maize, *C. zacconius*, *M. separatella* (?), 'rice stem borer' from Nigeria and from Ivory Coast.

Distribution. Burkina Faso, Ivory Coast, Nigeria, Senegal (?), Sierra Leone.

CAMPOPLEGINAE

This is a large subfamily. Most species are solitary larval endoparasitoids of Lepidoptera. The species are usually koinobiont, and oviposition will probably usually be in the younger larval instars; the parasitoid larva often kills its host when the latter is mature. Where the host species is relatively large, the parasitoid larva might leave its host when this is still actively feeding.

Charops Holmgren

This genus has been mentioned often as parasitizing larvae of stem borers in Africa, but, upon analysing these references, it appears that almost all are based on very few original records. Names of host species mentioned include *C. zacconius* (as *Proceras africana*) and/or *Chilo* sp. near *C. phaeosema,* and possibly also *C. ignefusalis* (as *C. pyrocaustalis*) from Senegal, *C. partellus* (country not given), and *Busseola fusca* from Uganda. Although the available material includes specimens reared from Lepidoptera in rice fields, none of the specimens at hand can be attributed positively as having been reared from stem borers.

Risbec (1950a, b) was the first to report having reared *Charops* sp. from a larva of *C. zacconius* (as *P. africana*) in rice in Senegal. In a note on *Calliceras braconiphaga* Ghesquière (= *A. fijiensis* (Ferrière) (Ceraphronidae)) being a hyper-parasitoid via *Charops*, he also mentions rearing *Charops* sp. from *C. ignefusalis* (as *C. pyrocaustalis*), although this is not explicitly stated under that species.

Also Appert (1953) lists *Charops* sp. from *C. zacconius* (as *P. africana*) in rice in Senegal, and so does Descamps (1956). The latter author adds sorghum as a crop. Neither author states that he is quoting Risbec, although they probably both are. Risbec (1956b, 1960) and Appert (1964) both cite their earlier records, while Risbec (1960) states that the host species is *Chilo* near *C. phaeosema.* These records and those of subsequent authors (Nickel, 1967; Brenière, 1969; Grist and Lever, 1969; Rao, 1970; Bordat *et al.*, 1977) are probably all based on the record of Risbec (1950a, b). Whether this is based on a correct identification could not be checked. *Charops* is often mentioned in connection with a species referred to as '*Coleocentrus* sp.', which is almost certainly a misidentification.

A second, independent record of *Charops* is that of Milner (1967) who reared *Charops* spp. from *B. fusca* in Uganda. This record is apparently quoted by Mohyuddin and Greathead (1970) and Harris and Nwanze (1992). Greathead (1990) mentions *Charops* as being a widespread parasitoid of pupae of *C. partellus*, but does not give the source of this information. *Charops* is a larval parasitoid rather than a pupal parasitoid. Unfortunately, no material of the genus could be traced which is reared from stem borers.

Concluding, it remains uncertain whether *Charops* has been reared from stem borers in Africa. Since the genus is relatively easy to recognize, and it is definitely known as a parasitoid of stem borers in the Oriental region, the few records of the parasitoid are probably correct. However, it is unlikely to be as common and wide-spread as is suggested by the large number of records in the literature.

Chriodes Foerster

This genus occurs in the Old World tropics, with only three described species in Africa; there are no host records.

An unidentified species of *Chriodes* is reported as a larval parasitoid of *E. saccharina* in *Cyperus papyrus* and *Cyperus dives* in South Africa (Natal) by Conlong (1986, 1994), Conlong and Graham (1988) and Carnegie (1991). These records are based on a misidentification. Voucher specimens have kindly been supplied by Dr Conlong; all except two belong to an unidentified species of *Venturia* and two specimens belong to a species of *Eriborus*.

Eriborus Foerster

Two species of *Eriborus* (*E. sinicus* and *E. terebrans*) have been reported as parasitoids of several species of stem borers in the Oriental and eastern Palaearctic regions (Chiu, 1937; Nickel, 1964, 1967; Grist and Lever, 1969; Kamran and Raros, 1969; Chiu *et al.*, 1984; and others). Until now, the genus has not been reported parasitizing stem borers in Africa.

Eriborus sp. A

Two specimens of a species of *Eriborus* have been seen which were reared by Dr D. Conlong as larval parasitoids of *E. saccharina* from *C. papyrus* (or *C. dives*) in South Africa (Natal). These specimens were misidentified as *Chriodes* sp. and reported as such by Conlong (1986, 1994), Conlong and Graham (1988) and Carnegie (1991). The specimens do not belong to either of the above-mentioned species which are known from stem borers, but their identity has not yet been determined.

Diagnosis. The genus is characterized as follows. Front wing without areolet; hind wing with nervellus not intercepted (discoidella not reaching nervellus); glymma on first gastral segment obviously present; apical margin of clypeus not reflexed. The hind basitarsus bears a continuous median ventral row of very small and very closely spaced short hairs, which form a 'line' on the tarsal segment; this row of hairs is difficult to see and should be examined at high magnification.

Host. *Eldana saccharina* ex *Cyperus* sp.

Distribution. South Africa.

Venturia Schrotky

Diagnosis. All three species treated here have basically the same colours: head, thorax and propodeum black; mandibles and scape ventrally pale yellowish or white. The gaster in two species is black with red, in one species uniformly coloured black or dark brown; legs with or without various white parts. The species can be separated on the colour of gaster and legs, and in females on the relative length of the ovipositor sheath.

Biology. Most species of *Venturia* are solitary larval endoparasitoids of lepidopterous larvae, so oviposition is into the host larva. *Venturia canescens* (L.) accepts early-instar larvae as well as last-instar larvae of its host species for oviposition. The host species are various species of Phycitinae in stored products. One species, *Venturia ovivenans*, is known to oviposit into the eggs of its host, *Rupela albinella* (Cramer), a pyralid stem borer in rice in the neotropics; the parasitoid larva emerges from the mature host larva (Zwart, 1973). The exact biology of the African species attacking stem borers is not known.

Key to the Species of *Venturia* Reared from Stem Borers

1. Gaster black at base, from tergites 4 or 5 onwards red with more or less developed mediodorsal dark brown marks. Head rather 'buccate', not strongly

narrowed directly behind the eyes. Ovipositor various, either about as long as hind tibia or more than twice as long as hind tibia. In male, all coxae blackish, or coxa I reddish, never completely white 2
– Gaster uniformly black or dark brown, apical tergites without red colour. Head in dorsal view 'normally' narrowed behind the eyes. Ovipositor sheath in female slightly longer than hind tibia (1.2–1.4 times as long). Male with coxae and trochanters I + II white .... *Venturia* sp. B
2. Females 3
– Males 4
3. Ovipositor sheath about as long as hind tibia .... *Venturia jordanae*
– Ovipositor sheath more than twice as long as hind tibia .... *Venturia* sp. A
4. Front leg almost unicoloured reddish, with femur often somewhat darker, tibia dorsally not white; second trochanters (= trochantelli) pale red (not partly white) .... *Venturia jordanae*
– Front leg with coxa black, second trochanter pale whitish, tibia white .... *Venturia* sp. A

Venturia jordanae Fitton (Figs 580, 587)

Polaszek *et al.* (1994) suggest that this species might be a monophagous parasitoid of *M. separatella.* Jordan (1966) reports it (as *Scenocharops* sp.) to be the most common parasitoid of *Maliarpha* in Sierra Leone, having reared 455 specimens. Girling (1972), quoted by Ingram (1983), found a high rate of parasitization of the same host species in Kenya by '*Casinaria* sp.'. According to Polaszek *et al.* (1994), the species reported by Njokah and Okhoba (1985) as '*Cassarina*' (a misspelling of *Casinaria*) is also *V. jordanae.* Also misidentified as *Casinaria* sp. (Girling, 1972; Ingram, 1983) and as *Venturia crassicaput* (Morley) (Agyen-Sampong, 1980).

Diagnosis. Length 7–10 mm; front wing 4.3–5.4 mm; ovipositor sheath 1.5–2.0 mm; hind tibia 1.6–2.0 mm. On average, this is slightly smaller than both other *Venturia* species treated here. The ovipositor sheath relatively short, about as long as hind tibia (a good character to separate females). Front leg in both sexes almost uniformly coloured, pale red.

Host. *Maliarpha separatella* in rice.

Distribution. Ivory Coast, Kenya, Sierra Leone.

Venturia sp. A

This species is reported by Jordan (1966, as *Scenocharops*) to parasitize *Sesamia* spp. in rice in Sierra Leone. It is very closely related to the preceding species, from which the males are difficult to separate. On average, specimens are slightly larger than in *V. jordanae.*

Diagnosis. Length 9.5–11 mm; front wing 5.6–6.5 mm; ovipositor sheath 5.2–5.5 mm; hind tibia 2.4–2.5 mm. The ovipositor sheath long, more than twice as long as hind tibia (a good character to separate females). Front leg with coxa brown and second trochanter (= trochantellus) pale (whitish in male); tibia I in male almost white, in female dorsally white; tibia II in male dorsally white, in female brown.

Hosts. *Sesamia* spp. in rice.

Distribution. Sierra Leone.

Venturia sp. B

Conlong supplied material of a *Venturia* species reared from larvae of *E. saccharina* in *C. papyrus* and *C. dives* in South Africa (Natal). This species was misidentified and reported by him (Conlong, 1986, 1994; Conlong and Graham, 1988) and by Carnegie (1991) as ?*Chriodes* sp. The African species of the genus *Venturia* are in need of revision and there seem to be a rather large number of undescribed species. Whether this species is described or not has not yet been established.

Diagnosis. Length 8–9.2 mm; front wing 4.5–4.7 mm; ovipositor sheath 2.7–3.0 mm; hind tibia 2.2 mm. Gaster almost completely black or dark brown, without red on apical tergites. In female, legs rather uniformly brown to black, front leg somewhat paler, especially tibia. Ovipositor sheath relatively short, about 1.2–1.4 times as long as hind tibia. In male, legs with contrasting white parts: coxa I and usually coxa II, trochanters I and usually II; mandible and scape ventrally white.

Hosts. *Eldana saccharina* in *C. papyrus* and *C. dives*.

Distribution. South Africa, Uganda.

CREMASTINAE

Cremastine species are koinobiont solitary endoparasitoids. Their usual hosts are lepidopterous larvae living in concealed situations, e.g. in tunnels, pods, leaf rolls, etc., but host records also include free-living larvae.

Two genera have been reported as parasitoids of African stem borers, and a third – which is known to parasitize stem borers in the Oriental region – has been doubtfully reported from Senegal.

Pristomerus Curtis

The genus *Pristomerus* has a worldwide distribution; at present it contains nine described species in the Ethiopian region. Aubert and Shaumar (1978) and Fitton (in Polaszek *et al*., 1994) described one and two new species, respectively, from *M. separatella*. Another undescribed species has been studied, which is reared from other African stem borers. A revision of the African species of *Pristomerus* is planned.

Species of *Pristomerus* are solitary endoparasitoids of lepidopterous larvae. The host will usually live in a concealed way, such as inside stems, pods, etc.

The species show sexual dimorphism, the males having much stronger teeth on a more swollen hind femur than the females have.

Key to the Species of *Pristomerus* Reared from Stem Borers

1. Head with black parts: postocciput, occiput, gena dorsally except for outer orbits, vertex (including stemmaticum); frons and face with a large dark median mark, connected with black vertex; gastral tergites 1 and 2 for the greater part or completely blackish (Madagascar) .... *Pristomerus caris*

– Head orange and yellow, with some darker parts, but never black 2

2. Gastral tergites 1 and 2 dark brown, apically yellow; tergite 3 basal half dark brown; mesoscutum rather sparsely punctured (South Africa) ... *Pristomerus* sp. A

– Gastral tergite 1 red, tergites 2 and 3 often (in females) with a dark brown basal mark, in males only dark on tergite 2 3

3. Female hind femur somewhat swollen, with a rather small, but clearly visible tooth (Fig. 576) and a series of very small teeth on ventral surface; ovipositor sheath comparatively small, about 1.3 times as long as hind tibia. Male with tergite 2 not darker at base than at apex (East Africa) .... *Pristomerus bullis*

– Female hind femur more slender, on ventral surface only with a very small elevation representing the tooth; ovipositor sheath 1.7 times as long as hind tibia. In male, tergite 2 with a more or less developed dark mark at base (apical half of tergite 2 paler than at base) (West Africa) .... *Pristomerus africator*

Pristomerus africator Aubert & Shaumar (Figs 566, 571, 572, 578, 579, 584)

This species was described by Aubert and Shaumar as a subspecies of *Pristomerus pallidus*, from specimens reared from *M. separatella* in rice by Cocherau in Ivory Coast and a few specimens collected in Egypt. Horstmann (1990) studied the holotype and a paratype, both from Ivory Coast, and raised it to species level. According to Polaszek *et al.* (1994), this is the species reported by Jordan (1966) as *Pristomerus* sp. from *Maliarpha* in Sierra Leone.

Appert (1953) reports '*Pristomerus* sp. nr. *cunctator*' from an unknown host very numerous in rice in Senegal. This is probably *P. africator*, which, according to Polaszek *et al.* (1994), belongs to the same group of species. Brenière (1969) reports a '*Pristoderus*' [*sic*] sp. from *Maliarpha* in Sierra Leone. It is not clear whether this is an original record or a quotation of earlier authors.

In an unpublished report, Bianchi *et al.* (1991) mention a '*Trathala* sp.' reared from *Maliarpha* in Senegal and Ivory Coast. This record is based on a misidentification of *P. africator* (A. Polaszek, personal communication).

Diagnosis. Mesoscutum and scutellum closely punctate; hind femur in female slightly swollen, with a discernible tooth on ventral surface. Overall colour orange-brown with paler parts; backside of head red; gastral tergites 2 and 3 in female usually with basal half considerably darker than apical half, in male only tergite 2 somewhat darker at base. In male, lateral ocellus distant from eye by about its diameter. Female ovipositor sheath about 1.7 times as long as hind tibia.

Hosts. *Maliarpha separatella, Chilo* sp. in rice.

Distribution. Ivory Coast, Senegal, Sierra Leone.

Remarks. Closely related to *P. bullis*, the males especially hard to distinguish. Although Fitton (in Polaszek *et al.*, 1994) states that the visible portion of the ovipositor (which equals the length of the ovipositor sheath) is 1.4 times as long as hind tibia, I have measured it to be about 1.7 times as long in all three available females.

Pristomerus bullis Fitton (Figs 565, 569, 570, 574, 576, 577, 581)

Closely related to *P. africator* and very similar. The species is described from material reared from *Maliarpha* in rice from Tanzania (Morogoro region and Zanzibar). No other hosts are known. This is the same species as that reported by Bianchi *et al.* (1991) in an internal report as '*P. ?cunctator?*' ex *Maliarpha* from Tanzania.

Diagnosis. Mesoscutum and scutellum closely punctate; hind femur in female relatively slender, with a small tooth on ventral surface and a short series of smaller teeth. Overall colour orange-brown with paler parts; backside of head red; base of gastral tergites 2 and 3 not darker than apical half, in male lateral ocellus distant from eye by about 0.6 times its diameter (only one male available). Female ovipositor sheath 1.3 times as long as hind tibia (according to the description by Fitton; ovipositor not normally developed in the only available female).

Host. *Maliarpha separatella.*

Distribution. Tanzania.

Remarks. See under *P. africator.*

Pristomerus caris Fitton (Figs 564, 567, 568, 573, 575, 582, 583)

A third species described from *M. separatella* and which is not known from any other species of stem borers. It is described from a few males and females in the NHM, which are, according Polaszek *et al.* (1994), 'almost certainly reared from *M. separatella*'. The species is known from Madagascar only. This is the species reported by Bianchi *et al.* (1991) as '*Pristomerus* n.s.' from *Maliarpha* in Madagascar.

Diagnosis. Mesoscutum and scutellum closely punctate; hind femur in female rather slender, with a small elevation on ventral surface instead of a tooth. Overall colour orange-brown with paler parts; head for the greater part black, including postocciput, occiput, gena dorsally except for outer orbits, vertex, stemmaticum, frons in the centre (frontal orbits white); face with a large dark median mark, connected with black marks on frons and vertex. Gastral tergites 1 and 2 for the greater part or completely blackish. In female, ovipositor sheath 1.5–1.8 times as long as hind tibia (according to description) (in one ♀ studied: 1.7).

Host. *Maliarpha separatella.*

Distribution. Madagascar.

Pristomerus sp. A

Kfir (1988, 1990a, 1992c) reported a *Pristomerus* species as being reared from *C. partellus* in sorghum and maize in South Africa (Transvaal). The species hibernates inside the host for 4–5 months and emerges from the fifth-instar larva. I have studied a male reared by Dr Kfir; it probably belongs to an undescribed species. A second male, almost certainly conspecific, was reared from a maize stem borer at Chibuto, Mozambique, by Mr Domingos Cugala (A. Polaszek, personal communication).

Diagnosis. Mesoscutum and scutellum very sparsely punctate, strongly coriaceous between the punctations; overall colour orange-brown, with darker marks around scutellum and in the middle of base of propodeum; gastral tergites 1–2 blackish except at apex, tergite 3 blackish in basal half; all tarsi dark brown/black; backside of head red, stemmaticum a little darker than rest of vertex; inner orbits very strongly convergent in male; ocelli large, lateral ocellus just touching eye. Hind femur strongly swollen with a large tooth. Female not known.

Host. *Chilo partellus.*

Distribution. Mozambique, South Africa.

Temelucha Foerster

Bianchi *et al.* (1991) report a single specimen of a *Temelucha* species probably reared from *Maliarpha* in Madagascar. This specimen is listed as *Temelucha* sp. A in Polaszek *et al.* (1994). Besides this, a few specimens have been studied, reared from, or possibly from, lepidopterous stem borers.

Temelucha species do not seem to be common parasitoids of stem borers in Africa. In the Oriental and eastern Palaearctic regions, the following species have been reported as parasitoids of stem borers: *Temelucha biguttula* (Matsumura) and *Temelucha stangeli* (Ashmead) by Momoi (1968), *Temelucha japonica* Ashmead (Cushman, 1933) and *Temelucha philippinensis* Ashmead (Chiu, 1937).

Key to the Species of *Temelucha* Reported from African Stem Borers

1. Mesoscutum pale rufotestaceous, without black markings; mesopleurum and mesosternum completely yellowish; scutellum and face yellow .... *Temelucha basiornata*

– Mesoscutum and mesopleurum with black design; mesosternum extensively or completely black; scutellum and face various 2

2. Mesoscutum with a median black band in anterior half, two lateral black bands in posterior *c.* 0.7, which are not connected with the sides of mesoscutum; scutellum bright yellow; propodeum more or less extensively black, at least medially in basal half, but usually much more (South Africa) ... *Temelucha picta*

– Mesoscutum with median black band continuous from anterior margin to scutellum, two short brown bands at sides; scutellum black (Madagascar) .... *Temelucha* sp. A

Temelucha basiornata (Cameron)

In the NHM, there are a male and a female of this species, identified by Dr Kerrich. Both specimens have been reared in Ghana. The male is labelled '*Syzeuctus* sp. ex rice borer' by Scheibelreiter; the female 'ex *Diopsis* rice borer, F.J. Simmonds'. The former could also have been reared from a *Diopsis* sp., but, since the two specimens are reared by different people from different localities, there is a possibility that one is from a lepidopterous rice borer.

Diagnosis. Length 11 mm (both sexes); front wing 5 mm; ovipositor sheath (4.3 mm) a little more than twice as long as hind tibia (2.0 mm). Head yellow, with stemmaticum and small area around it black, connected with black on occiput; thorax and propodeum pale yellowish brown; in female, area lateral of scutellum black, connected with a short median black band on anterior half of propodeum; in male, this black mark much less developed or not present; gastral tergite 3 with comparatively strong fine longitudinal sculpture.

Hosts. 'Rice borer'; probably *Diopsis* (Diptera, Diopsidae).

Distribution. Ghana, South Africa (type).

Remarks. The specimens agree with the description by Cameron, except for some characters. Cameron states the female to be 7 mm and the male 5 mm. Despite the large gap in the distribution, the identification is assumed to be correct. Since the lectotype is in the NHM, Kerrich will have compared the material carefully with the type material.

Temelucha picta (Holmgren)

Kfir reared a specimen of *Temelucha* sp. from *B. fusca* in maize in South Africa. The specimen has been compared with material reared from other host species in the National Collection of Insects (NCI), Pretoria. The genus is in need of revision, but I assume that the specimen most probably belongs to *T. picta*, like some of the other material. The species seems to be quite variable in colour pattern.

Diagnosis. Length 5.5–8.5 mm; front wing 3.5–4.5 mm; ovipositor sheath about twice as long as hind tibia. Head yellow with black markings, black is usually a broad median band on face, running from (yellow) clypeus to (black) occiput; mesoscutum, scutellum and propodeum as mentioned in the key; non-black parts are pale to bright yellow; mesopleurum black with a broad yellow band from anterior margin to middle coxa, to almost completely yellow with some black marks.

Hosts. *Busseola fusca* (in maize); other hosts: *Helicoverpa armigera* (Hübner) (Noctuidae) and *Loxostege frustralis* (Zeller) (Pyralidae).

Distribution. South Africa.

Temelucha sp. A

One female specimen, probably representing an undescribed species, is reported by Bianchi *et al.* (1991) and by Polaszek *et al.* (1994) as a parasitoid of *M. separatella* in Madagascar.

Diagnosis. This is a very small species, only 4.3 mm long. The whole dorsal surface, from head to gaster, is black; mesosternum dark with some faint lighter marks; mesopleurum and sides of propodeum yellowish; legs whitish, but femur, tibia and tarsus III with dark clouds.

Host. *Maliarpha separatella.*

Distribution. Madagascar.

Trathala Cameron

In the literature on African stem borers, there is one record of *Trathala* sp.; in an internal report, Bianchi *et al.* (1991) mention it from Senegal and Ivory Coast. According to A. Polaszek (personal communication), this record is based on a misidentification; the specimens belong to *P. africator.* A species of *Trathala* is known as a parasitoid of cereal stem borers in the Oriental region. Since it has been recorded from Senegal, it is treated below.

Trathala flavoorbitalis (Cameron)

Trathala flavoorbitalis is a common and widely distributed, polyphagous species which is known to parasitize *Chilo suppressalis, Scirpophaga incertulas* and *Ostrinia nubilalis* in the Oriental region.

Gahukar (1981) lists the species as a parasitoid of *Mythimna loreyi* in Senegal, and Gahukar (1990), quoting his 1981 paper, mentions *Amsacta moloneyi* (both Noctuidae) as the host species. I have found no record that *T. flavoorbitalis* has been introduced into West Africa, but, if it has, the species could probably also establish on stem borers. I doubt, however, the correctness of the determination.

CRYPTINAE

Cryptinae (= Gelinae *sensu* Townes, = Phygadeuontinae of several modern authors) is by far the largest subfamily of Ichneumonidae. Several genera are known to parasitize stem borers. As is generally the case in species of this subfamily, most are known to be idiobiont ectoparasitoids of mature larvae or prepupae. This is true at least for the stem borer parasitoids *Amauromorpha accepta* (Tosquinet) and *G. javensis* (Rohwer) in the Oriental region, and in *Mallochia pyralidis* Wharton and *Strabotes rupelae* Zwart in the neotropical region. Ahmad and Mathur (1946) published a detailed study on the biology of *G. javensis.* This solitary ectoparasitoid accepts its host, *Scirpophaga nivella,* only as a prepupa (this is the mature larva which has ceased feeding and is preparing for pupation). The prepupa is paralysed and thus there is no pupation of the host.

Little is known about the biology of African species. *Gambroides nimbipennis* is reported (as genus near *Isotima* sp.) by Jordan (1966) as being a larva–pupal parasitoid of *M. separatella* in Sierra Leone. Other authors, however, record it as a larval parasitoid (Agyen-Sampong, 1980; Akinsola and Agyen-Sampong, 1984; Agyen-Sampong and Fannah, 1987). At least one species definitely oviposits in host pupae. Conlong reared a female of this species from a pupa of *E. saccharina* from sugar cane in South Africa (see Genus A).

Amauromorpha Ashmead

In the Oriental region, the genus contains two species, one of which, *A. accepta* (Tosquinet), is a widely distributed and well-known parasitoid of various species of stem borers; three subspecies are recognized. The recorded hosts are: *Scirpophaga excerptalis, S. incertulas, Scirpophaga innotata, S. nivella, C. sacchariphagus, C. suppressalis* and *Sesamia calamistis.*

Polaszek *et al.* (1994) report the genus for the first time from Africa.

Amauromorpha sp. A (Fig. 585)

Two reared and one non-reared female specimens of an apparently undescribed species are reported by Polaszek *et al.* (1994). One was reared from *M. separatella* in Togo, the second from a rice stem in Nigeria. As these authors stated, there is some difference between the former and both other specimens. Due to the lack of sufficient material, the species from *Maliarpha* was not described.

Diagnosis. The African specimens (all females) differ from females of *A. accepta* ssp. *accepta* from Java as follows:

1. Prothorax and mesothorax bright red; gastral segments 1–3 (4, 5) red, remaining black or blackish (with white mark on tergite 7); antenna with white annulus on flagellar segments 6–10 .... *Amauromorpha* sp. A

– Prothorax and mesothorax black (except for scutellum sometimes dark red); gastral tergite 2 dark or with a dark transverse band, tergites 1 and 3 red, 4 usually black, remaining tergites black (with white mark on seventh tergite in ♀); antenna with only small white mark(s) on flagellar segments 8 and 9a .... *Amauromorpha accepta*

Biology. Almost certainly, like other species of the genus, a larval ectoparasitoid.

Host. *Maliarpha separatella.*

Distribution. Malawi, Nigeria, Togo.

Diracela Townes

The genus *Diracela* is placed by Townes and Townes (1973) in the subtribe Chiroticina. Townes (1970a) states for the subtribe: 'The known hosts are small cocoons'. There are only a few host records for non-African species of this subtribe, among which are cocoons of other ichneumonids and braconids and spider egg cocoons. Some species of *Paraphylax* and particularly of *Chirotica* are also known to parasitize Psychidae. Only one host record of an African species of this group has been found (see *Paraphylax*).

Diracela is closely related to *Astomaspis*, but differs by the occipital carina having a deep median notch, and in the male particularly in lacking the lateroapical spines of the third gastral tergite. Only one species is described in the genus; there are no other host records.

Diracela latifasciata (Cameron)

A male of this species has been reared from *C. partellus* in Kenya. Unfortunately, there are no host remains mounted together with the specimen. The specimen was probably reared as a secondary parasitoid of *C. partellus*, from an ichneumonid or braconid primary parasitoid cocoon.

Diagnosis. Length 5–7 mm; front wing 4–5 mm; ovipositor sheath *c.* 0.6 mm; hind tibia 2 mm. Head at level of occipital carina with a deep median notch; pronotum medially with a high lamella; mesoscutum with coarse transverse punctation, notauli meeting beyond centre. In male, gastral tergites 1–3 strongly longitudinally striated, segments 4–7 protruding only very little beyond apex of tergite 3; latero-apical corners of third gastral tergite rounded, without spines; tergites 4–6 with or without apical white band. In female, tergites 1–4 strongly longitudinally sculptured; tergites 5–7 smooth and shining, with conspicuous hairs in apical half; tergites 5 and 6 with white apical band.

Head, thorax and propodeum dark red, or head brown to black; gaster dark red to dark brown or almost black; basal 0.2–0.3 of tibia III white; front wing with a small, well-defined brown mark against front margin at level of stigma.

Biology. Probably a secondary parasitoid.

Host. Probably a braconid or ichneumonid primary parasitoid of *C. partellus*?

Distribution. Kenya and South Africa.

Gambroides Betrem

Townes and Townes (1973) consider *Gambroides*, *Vadonina* and two other genera synonymous with *Isotima*, and list them under the latter name. Jonathan (1980) studying the Oriental species of this '*Isotima* complex' raised four of them again to generic rank, and treats *Vadonina* as a synonym of *Gambroides*. Polaszek *et al.* (1994) considered *Vadonina* a valid genus, and stated that it is different from *Isotima*.

Specimens mentioned in the literature on African stem borers as 'species near *Isotima*' constitute a puzzling group. As also stated by Polaszek *et al.* (1994), species limits are not clear and even generic limits are not clear in African material. The group is in need of revision and, as long as the species are not intensively studied, it is not clear whether there is only one or more genera and species. Considering the present poor knowledge of the complex, it seems advisable not to try to distinguish 'species' in the African material, although there are certainly differences between groups of specimens.

So far only one species is officially described from the Ethiopian region. Attempts have been made to introduce a second species from the Oriental region into Mauritius and Uganda against stem borers. These two taxa can be easily separated as follows.

Key to the Species of *Gambroides*

1. Gastral tergites 1 and 2 (and 3 in male) apically not white, only tergite 7 with a white apical mark (Africa and Madagascar) .... *G. nimbipennis*

– In female gastral tergite 2 (and often also tergite 1), in male tergites 1–3, apically white; apical white mark in female on tergites 6 and 7 (Oriental; presumably not established in Africa) .... *G. javensis*

Gambroides javensis (Rohwer)

This species is a well-known parasitoid of *S. excerptalis, S. incertulas* and *S. nivella* in the Oriental region. Since attempts have been made to introduce it into African countries, it is listed here. Introduction into Mauritius is reported by Rao (1969); according to Greathead *et al.* (1971), it was released in 1960 but did not establish. Ingram (1983) reports that the species was also unsuccessfully introduced from India into Uganda.

Diagnosis. Head black; mesoscutum usually black, sometimes red; propodeum red; first gastral tergite red, in male and often also in female, white at apex; apex of tergite 2 in both sexes white, in male also tergite 3 at apex white; white apical mark on tergites 6 and 7. Gastral tergites 2–4 densely punctate, punctation less dense on tergite 1 and apical tergites.

Biology. Larval ectoparasitoid (of prepupae).

Distribution. As far as known not established in Ethiopian region; widely distributed in Oriental region (Bangladesh, China, India, Java).

Gambroides nimbipennis Seyrig (Fig. 586)

This species was described from Madagascar. In the literature on stem borers, it is generally referred to as 'genus near *Isotima* sp.'. Jordan (1966) first reported it as a larva–pupal parasitoid of *M. separatella* in rice in Sierra Leone, and Milner (1967) reared it from pupae of *C. partellus, B. fusca* and *Maliarpha* in Uganda. Brenière (1969) reported it from *Maliarpha* in Swaziland and Appert (1971) from the same host in Madagascar. Mathez (1972) reared 'genus nr. *Isotima*' from pupae of Pyralidae in Kenya (*C. partellus and C. orichalcociliellus*). Agyen-Sampong (1977) listed '*Isotima* or "genus nr."' as an 'egg parasitoid of *Maliarpha*', which is definitely an error. '*Menaforia* spp.' was used for the same species (Agyen-Sampong, 1980; Akinsola and Agyen-Sampong, 1984; Agyen-Sampong and Fannah, 1987). Jonathan (1980) studied specimens of *G. nimbipennis* reared from *Maliarpha* in sugar cane from Ghana and Madagascar. (NB. Jonathan erroneously mentions Ghana to be in 'South Africa'.) Polaszek *et al.* (1994) report *Vadonina ?nimbipennis* from *Maliarpha* from most countries listed here.

Specimens are quite variable in a number of characters and, as stated above, it is uncertain whether all belong to one species. The fact that almost all have been reared from one host species could support the opinion that at least these might belong to *G. nimbipennis*.

Diagnosis.

Female. Head black; semicircular carina behind antennae usually very poorly developed or absent; prosternum usually black; pronotum usually dark red; mesothorax and propodeum usually red; gastral tergite 1 red to black; tergites 2–6 usually black, sometimes brown; tergite 7 with broad apical white mark. Legs red to black, tibiae and tarsi blackish. Apical carina of propodeum often present as two crests which are not connected in the middle, sometimes not traceable; basal carina usually present. Sternaulus usually visible and short (up to about 0.3 the distance to middle coxa), sometimes absent or poorly developed. Antenna with a white annulus on flagellar segments 6–9 (often also on apex segment 5). Gaster densely punctate from tergite 2 onwards, punctation less pronounced towards apex. Front wing with a dark mark, which can vary in size and intensity, but is always present.

Male. Some males have the same rather dark colour pattern as described for females; other males have the body very pale brown with the head dorsally black and apical tergites of gaster black (at least tergites 6 and 7); tergite 7 always with an apical white mark; legs various, in pale brown males legs also pale brown. Pale brown males have the whole of face, clypeus and lower gena white; dark males have at least the facial orbits white, and also white on outer orbits. Males tend to have the semicircular carina behind the antennae more developed than females; pale males especially seem to have a stronger developed carina.

Biology. Reported as a larval parasitoid, as a larva–pupal parasitoid and as being reared from pupae. Some specimens have been pinned together with the cocoons from which they emerged. These cocoons have apparently been removed from stems. This suggests that the species is at least not a pupal endoparasitoid, ovipositing into the host pupa. In that case, the adult wasp would almost certainly develop inside the host pupa and leave the pupal case of its host by making an exit hole. Most cocoons have some remnants of a lepidopterous pupa on their outside, which suggests that the lepidopteran pupated shortly before the ichneumonid larva matured and started to spin its (too large) cocoon inside the empty pupal case of its host. This then bursts and parts of the pupal case are present on the outside of the cocoon. One specimen left from a host that was about to pupate. The host pupa from which it left is still partly covered by the larval skin. It is very likely that *G. nimbipennis* is a larva–pupal endoparasitoid, which, as a mature larva, usually leaves its host very shortly after its pupation. Some specimens might possibly leave shortly before pupation of the host.

Hosts. Almost all records concern *M.* separatella; other hosts include *C. partellus* (and ? *C. orichalcociliellus*) and *B. fusca*.

Distribution. Ghana, Kenya, Madagascar, Senegal, Sierra Leone, Swaziland, Tanzania (including Zanzibar), Togo, Uganda.

Mallochia Viereck

This is a New World genus, with a small number of described species; most species occurring in the neotropical region and a few in North America. *Mallochia pyralidis*, which is discussed below, is the only species for which a host is known. The genus is related to *Strabotes*, two species of which, *Strabotes abdominalis* (Cresson) and *S. rupelae* Zwart, parasitize the crambid stem borer *Rupela albinella* (Cramer) in rice in South America and the Caribbean.

Mallochia pyralidis Wharton

Mallochia pyralidis is a parasitoid of the Mexican stem borer *Eoreuma loftini* (Dyar) in sugar cane. It was discovered in Mexico in 1983 and has been introduced into Texas (USA) (Browning *et al.*, 1985; Smith *et al.*, 1990), where it has established. Shortly thereafter, attempts were made to introduce it from the USA into South Africa against *C. partellus* in sorghum (Kfir, 1989a, 1991b, 1992d) and against *E. saccharina* in sugar cane (Carnegie, 1991; Conlong, 1994). Although some recoveries were made after releases, the species seems not to have established so far. Releases have now stopped (D.E. Conlong and R. Kfir, personal communication).

Biology. Larval ectoparasitoid.

Host. Natural host (not in Africa): *E. loftini* (Dyar) (Pyralidae).

Distribution. Almost certainly not established in South Africa.

Paraphylax Foerster

This large genus is known from the Old World tropics and Japan. According to Townes (1970a), 'The species parasitize small cocoons, including cocoons of other ichneumonids and spider egg cocoons. Several species attack Psychidae'.

Jordan (1966) reports having reared a specimen of 'Genus nr. *Strepsimallus* sp.' [= *Paraphylax*] from a braconid cocoon found in rice in Sierra Leone. There is no explicit indication that this was in connection with stem borers. The genus is related to *Diracela*.

Genus A

D.E. Conlong (personal communication) reared a single female of a cryptine species from a pupa of *E. saccharina* in sugar cane in Natal, South Africa. This female readily accepted *Eldana* pupae for oviposition: naked pupae as well as pupae put in paper straws, with some preference for the latter. In this way, he was able to obtain a series of male progeny.

The species does not match well with any of the genera described from Africa or from the Oriental region. It probably belongs to an undescribed genus, and is almost certainly also an undescribed species. The species (and genus?) is characterized by a sharp median tooth on the apex of the clypeus. Although the available material is sufficient to describe it, it seems prudent to be more certain about its generic relationship.

Diagnosis. Length 9–11.5 mm; front wing 5.5–7.5 mm. Head black, clypeus with a sharp preapical median tooth; thorax and propodeum red; first gastral tergite red or mainly red, tergites 2–5 black, apical half of tergite 6 and tergite 7 white; segments 2 (apical 0.6)–4 of hind tarsus white.

Biology. A pupal endoparasitoid accepting, under artificial conditions, naked as well as concealed pupae.

Host. *Eldana saccharina.*

Distribution. South Africa.

ICHNEUMONINAE

Most species of this subfamily are solitary endoparasitoids of lepidopterous pupae. Some species or groups, however, use mature larvae for oviposition. In this case, the adult wasp always emerges from the host pupa. Concerning the genera reared from cereal stem borers, there is some doubt about the host stage parasitized by *Ischnojoppa*. The other genera are certainly pupal parasitoids; all are endoparasitic.

Dentichasmias Heinrich

The genus was described by Heinrich (1968) as containing *D. busseolae*, which is the only species described so far. The genus is similar to *Procerochasmias*, from which it differs by the clypeus bearing a pair of small median teeth on its apical margin and by the scutellum being carinated until the apex.

Dentichasmias busseolae Heinrich (Plate 10:5)

Although the species is reported as being rather common in several countries, comparatively few specimens have been seen from most localities. Large samples could be studied from South Africa (Transvaal), where it was reared by Kfir in large numbers from pupae of *C. partellus*, and from a series which was malaise-trapped by Ebregt in maize in Mozambique. Smaller series were studied from Ivory Coast ex *C. diffusilineus* and from Kenya, where it was reared from *C. partellus* by Overholt and coworkers, and from Malawi, where Phiri reared it also from *C. partellus* in maize.

In the literature on stem borers, the species was generally reported until 1967 as 'Genus near *Chasmias* spec.'.

Diagnosis. Length 7–11.5 mm; front wing 5.5–8.5 mm; ovipositor slightly but clearly protruding beyond apex of gaster (*c.* 0.5 mm). The species is best characterized by the structure of the clypeus, which bears two small median teeth on the apical margin. The general colours are red and black, with the tergites 6 and 7 of gaster with a white mark and antenna with a white annulus (in females on segments 8–15, in males on segments 13–20 or less). The species is rather variable in colour pattern, in that several parts of the body (head, parts of thorax and parts of gaster) vary from red to black. Particular colour forms have not been noticed to occur in particular areas or countries. So there do not seem to be geographically separated races or subspecies.

Dr S.W. Kimani-Njogu (International Centre of Insect Physiology and Ecology (ICIPE), Nairobi) kindly sent me an aberrant specimen, in which both teeth on the clypeus are lacking (Kenya, Kwale District, *C. partellus*).

Hosts. The main host is *C. partellus*, but it is also reported from some other stemborer species. When describing the species, Heinrich had material available reared from *C. partellus*, *C. zacconius*, *B. fusca*, and from ?*Sesamia* sp. Although the species is named after *Busseola*, it seems not to be the usual host.

Besides the above-mentioned hosts, it is also reported from *C. diffusilineus*, *C. ignefusalis* and *S. calamistis*. Mohyuddin (1972) mistakenly reported that it will normally only parasitize *Chilo* species, but will accept noctuids and *Eldana* pupae if these are placed among *Chilo* frass.

Distribution. Burkina Faso, Cameroon, Kenya, Malawi, Mozambique, Nigeria, Sierra Leone, South Africa, Uganda. This species has not been recorded before from Burkina Faso, Malawi or Mozambique.

Remarks. The first record of this species as a parasitoid of stem borers is probably that of Ingram (1958), who states: 'An ichneumonid belonging to a genus near *Chasmias* has been recorded from *Chilo zonellus* [= *partellus*]; the adult parasitoid emerges from the host pupa'. This record is apparently quoted by Nye (1960). Harris (1962) mentions a 'Species nr. *Chasmias*' as being reared from the pupa of *C. ignefusalis* in millet in Nigeria. Jordan (1966) very definitely refers to the species, by stating:

> Gen. nr. *Chasmias* sp. is known as a parasite of stalk-boring Lepidoptera in Kenya, Uganda, Tanzania and Cameroon. It is being described in a monograph on the African Ichneumonidae which is being prepared in the U.S.A. Two specimens emerged from *C. zacconius* (as *zaleukos*) pupae ... one ... collected in November 1964 and one ... collected in September 1965 [on rice in Sierra Leone].

The material from Sierra Leone is not in the NHM, but there are three specimens from Cameroon, reared by Descamps in 1954 from pupae of *C. zacconius* (as *P. africana*) and identified by Kerrich in 1958 as 'Gen. nr. *Chasmias* sp.'. Milner (1967) reports 'Genus near *Chasmias* sp.' to be found as a parasitoid of *C. partellus* at 38 localities in East Africa. The species is said to be 'particularly successful in dry parts of northern Uganda'. Also, La Croix (1967) reports the species as a parasitoid of *C. partellus* from the Coast Region in Kenya.

Bordat (1979) mentions *Dentichasmias* as a parasitoid of *C. diffusilineus* in Ivory Coast (voucher material studied from CIRAD, Montpellier); this probably concerns the material described by Aubert (1979) as *Afrogenes clypeodontor*. The photograph in Bordat's paper, however, does not show *Dentichasmias*, but probably a cryptine species.

Ichneumon Linnaeus

This genus has an almost cosmopolitan distribution; in the Ethiopian region, it contains only four described species. Two of these have been reported in the literature to parasitize *Sesamia* and *Busseola*; only one specimen has been studied which quite probably represents voucher material of one of these records.

The gaster usually has white or yellow bands (at least in the species treated here), but a uniformly coloured gaster also occurs. The species show sexual dichroism. They will almost certainly parasitize the pupae of their host species.

Key to the Species of *Ichneumon* Reared from Stem Borers

1. Females 2
– Males 3
2. Gaster without apical white mark on tergites 6 or 7, only tergite 2 with an apical white band; thorax uniformly black; legs extensively black (Madagascar, Mauritius, Réunion) .... *Ichneumon unicinctus*
– Gastral tergite 6, and sometimes also 7, with an apical white mark, tergites 2 and 3 almost uniformly dark red (African continent) .... *Ichneumon rubriornatus*

3. Gastral tergites 1 + 2 black with apical white bands, never partly red (Madagascar, Mauritius, Réunion) .... *I. unicinctus*
– Gastral tergites 1 and (usually more than basal half of) 2 red with apical yellowish white bands (African continent) .... *I. rubriornatus*

Ichneumon rubriornatus Cameron

This is probably the most widely distributed species of the genus. It has been recorded from South Africa and Ethiopia so far, and is here reported from West Africa (Niger) for the first time.

Van Rensburg *et al.* (1988b) report the species as a parasitoid of *B. fusca* from maize in South Africa (Transvaal). This record is quoted by Harris and Nwanze (1992). Dr D.E. Conlong supplied a female specimen, which was reared by him in 1993 from a pupa of *S. calamistis* from sugar cane in South Africa (Natal). Contrary to Heinrich's description of the female, the wings are clearly infuscated in this specimen, and the seventh gastral tergite is not marked with white.

Other material of this species has been studied which was reared by Harouna from *M. loreyi* (Dup.) (Noctuidae) in Niger. These females are much lighter than the South African specimens, and the thorax is largely red instead of black.

Hosts. *Busseola fusca* (van Rensburg *et al.*, 1988b), *S. calamistis* (Agricultural University of Wageningen (WAU)). Other hosts: *M. loreyi* (Noctuidae) (WAU).

Distribution. Ethiopia, Niger, South Africa (Heinrich, 1967; van Rensburg *et al.*, 1988b).

Remarks. South African males of this species seem to be very variable in the colour pattern of the propodeum and gaster. Most specimens have the propodeum largely yellow and gastral tergites 1–3 with broad yellow apical bands. A few have the propodeum completely black and only tergites 2 and 3 with a narrow yellowish apical band. I assume here that the two colour types are conspecific.

Ichneumon unicinctus Brullé

Hosts. *Sesamia calamistis* (Williams, 1969, 1978, 1980, 1983; Appert 1971, 1973), *Sciomesa biluma* (Appert *et al.*, 1969; Appert, 1971, 1973). Other hosts: *Charadrina* sp. (Heinrich, 1967) (Noctuidae).

Appert *et al.* (1969) list *Ichneumon tananarive* from Madagascar as a parasitoid of *S. biluma* Nye, an endemic noctuid stem borer in maize. Under the same name, Appert (1971, 1973) reports the species to have been reared from *S. calamistis* and *S. biluma* in Madagascar. A single female specimen without data, which is only labelled '*I. tananarive*', has been studied from the CIRAD collection. It might be a voucher specimen of the material reported by Appert; the identification is correct, but the valid name is *I. unicinctus.*

The species is also reported as *I. unicinctus* by Appert (1973) as being reared once from *S. calamistis* in Madagascar. Williams (1969, 1978, 1980) mentions a single specimen of *I. unicinctus* reared by him from *S. calamistis* on sugar cane in

Mauritius in 1969. He reports the species as attacking the host larva and emerging from the pupa. These two specimens have not been seen.

Distribution. Mauritius (Heinrich, 1967; Williams, 1958, 1969); Madagascar (Heinrich, 1967; Appert, 1971, 1973; Appert *et al.*, 1969); Réunion (Heinrich, 1967).

Remarks. The species is not known from the Seychelles, as erroneously mentioned by Heinrich (1967) (*lapsus* for Mascarenes) and apparently quoted by Williams (1969).

Ischnojoppa Kriechbaumer

Diagnosis. This genus is represented in tropical Africa and in the Oriental region. It is characterized by a slender and elongate gaster, combined with the structure of the face and clypeus, which form, together with the malar space, a single, slightly convex plane, without sutures and elevations or depressions. The head is more 'swollen' than in other ichneumonid parasitoids of stem borers. The general basic colour is bright yellowish. In all but one species, the legs are also yellowish, with black markings on the femur and tibia. The tip of the gaster (segments 6 and 7) is usually darker (often black) with, in most species, a white anal mark on segment 7 (which is absent in *Ischnojoppa geniculata*). The dark antenna often bears a white annulus in females, which is flattened and broader beyond the middle. Most species are about 15 mm long.

The species are very similar and therefore rather difficult to separate without experience. Heinrich (1967) provides a key to the species of the Ethiopian region. He recognizes seven species, two of which are considered by Townes and Townes (1973) to be subspecies of one species. In the Oriental region, a single species occurs, which is very often reported as a parasitoid of stem borers. It has been reported erroneously from Africa, and is therefore included below.

Biology. Species of *Ischnojoppa* are reported as larval as well as pupal parasitoids. Whether they may indeed be reared from both larvae and pupae of their hosts is not clear. I assume that oviposition may be in mature host larvae and not in pupae; the species are definitely endoparasitic and, according to various records, probably hatch from pupae or from prepupae of their host species.

Smith, J.W. *et al.* (1993b) list *Ischnojoppa* as a parasitoid of stem borers. They list the genus erroneously as belonging to the subfamily Phygadeuontinae (= Cryptinae) and state, incorrectly, that it is a larval ectoparasitoid.

Two species have been reported as parasitoids of stem borers in the Ethiopian region, one (*I. geniculata*) from the continent, the other (*Ischnojoppa madagascariensis*) endemic to Madagascar. They seem to be uncommon parasitoids of stem borers. A few other species have been collected in rice fields; whether their relation is with rice borers or with other lepidopterous species is not known. At least some are parasitoids of hesperid species in rice.

All records of African species of *Ischnojoppa* as parasitoids of stem borers or occurring in graminaceous crops are summarized below. Since the Oriental species has erroneously been reported from Africa, it is also mentioned.

Key to the African Species of *Ischnojoppa* that have been Reared from Stem Borers or have been Collected in Rice Fields

1. Tibia III and tarsus III completely black or blackish; gastral tergites 6 and 7 black or blackish, in female both with a large white anal mark, in male only seventh tergite with white mark (Madagascar) .... *I. madagascariensis*

– Tibia III only black at base and at apex; tarsus III usually only partly black 2

2. No white anal mark on apex of gaster; in female gastral tergites 6 and 7 black, in male tergite 7 and usually at least part of tergite 6 black (Ethiopia to South Africa) .... *I. geniculatus*

– White anal mark on gastral tergite 7 or on 6 + 7; this white mark is, however, often absent or poorly developed in male of *Ischnojoppa seyrigi* 3

3. Dorsal transverse carina on front of pronotum (just behind head, and sometimes obscured by head in pinned specimens) with a deep and very distinct median incision; flagellum in both sexes with a dorsal white annulus (East and West Africa, Angola and Zambia; not yet known from South Africa) .... *Ischnojoppa flavipennis*

– Dorsal transverse carina on pronotum without a deep median incision; tibia III at apex dorsally suddenly broadened and in lateral view forming a 'hump'. This 'hump' is very evident in female but much less developed in male. Flagellum of male only exceptionally with a dorsal white annulus (present in female) (Ethiopia, Tanzania, Malawi, Angola and Madagascar) .... *I. seyrigi*

Ischnojoppa flavipennis (Brullé)

Hosts. This species is recorded by Agyen-Sampong (1980) as being reared from *Pelopidas mathias* or *Borbo fanta* (both Hesperidae) in rice in Sierra Leone. A. Polaszek (personal communication) has seen material of the species reared from *P. mathias* in the Musée Royale de l'Afrique Centrale (MRAC) collection.

Risbec (1950a, 1960) reports *Ischnojoppa luteator* as being reared from *P. mathias* (as *Pelopidius mathias*) in Senegal. The ichneumonid species is definitely misidentified; it is likely that *I. flavipennis* is involved, not only because of the host, but also because this is the only species so far recorded from West Africa. In his 1950a paper, Risbec describes how the wasp leaves the host pupa at the anterior end, after the colour of the pupa has turned yellow; in his 1960 paper, however, he states that the species emerges from the host larva.

Although the species is not recorded from stem borers, it might also parasitize these; related species apparently do.

Distribution. According to Heinrich (1967, 1968), Angola, Guinea, Kenya, Senegal, Tanzania, Togo and Zambia. Additional material has been studied from Ivory Coast, Kenya, Niger, Nigeria, Tanzania, Uganda and Zanzibar. This is the most widespread species in the African continent, but seems to be absent from the extreme south.

Ischnojoppa geniculata (Tosquinet)

Hosts. The first record of a species of *Ischnojoppa* as a parasitoid of stem borers in Africa is reported by Milner (1967), who reared one male specimen of *I. geniculata* from a 'cocoon found with remains of larva of *Sesamia* sp. in a stem of *Eleusine* in Kigezi (Uganda)'. Mohyuddin and Greathead (1970) list the species for 'East Africa', based on Milner's record. The specimen is preserved in the NHM and has been studied.

A second record of the species is reported by Appert (1973), who reared a specimen from *S. calamistis* in Madagascar. This specimen has not been located; its identification cannot be confirmed.

Distribution. According to Heinrich (1967), South Africa and Tanzania. Specimens from Ethiopia, Kenya, South Africa, Uganda and Zimbabwe have been studied; only the above-mentioned specimen from Uganda was reared. Two specimens were malaise-trapped by D.E. Conlong in a sugar cane field in South Africa.

Ischnojoppa madagascariensis Heinrich

Hosts. This species is listed by Appert *et al.* (1969) from an 'unknown host, probably *Proceras sacchariphagus*' from Madagascar. No voucher material has been located, nor has any other specimen of this species.

Distribution. Restricted to Madagascar.

Ischnojoppa seyrigi Heinrich

Hosts. The Zoological Museum, Amsterdam (ZMA) collection contains a specimen reared from 'pupa groene blad rups' by Feijen from rice in Malawi, which means the host is a green caterpillar feeding on the leaves of rice (probably a hesperiid?).

Distribution. Material has been studied from Ethiopia, Madagascar and Malawi. Heinrich (1967) mentions Angola, Madagascar and Tanzania.

Ischnojoppa luteator (Fabricius)

This is the only species of the genus occurring in the Oriental region; it has been reported by many authors and from several countries as a parasitoid of the crambid stem borer *S. incertulas* (Walker) (also as *Tryporyza incertulas* and as *Schoenobius bipunctifera* (Walker)). Other host species that have been found in the literature are: *Chilo polychrysus* (Meyrick), *Chilo* sp., *Cnaphalocrocis* [*medinalis* Gn.?], *S. innotata* (Walker), *S. nivella* (F.), *Scirpophaga* spp. (all Crambidae), and *P. mathias* (F.) (also as *Parnara mathias*) and *Borbo cinnara* Wallace (Hesperidae).

***Procerochasmias* Heinrich**

The genus was described by Heinrich in 1938 as containing two Madagascan species. In his revision of the African Ichneumoninae, Heinrich (1967) includes a few more and describes some new species. The genus now contains six species from the African continent, and two that are endemic to Madagascar. The host is known for only one species, *Procerochasmias nigromaculatus* (Cameron). Heinrich (1967) presents a key to the species and gives descriptions of all species of the genus.

Procerochasmias nigromaculatus (Cameron)

This is a common pupal parasitoid of *B. fusca* in East Africa above 1300 m (Milner, 1967) and in maize in South Africa (Kfir and Bell, 1993). Other references indicate that *Busseola* is the principal host in all countries. There are only a few records indicating that other stem borers might sometimes be parasitized.

Milner (1967) reports one specimen reared from *C. partellus* at Kawanda in Uganda. The record on map 3 in Mohyuddin and Greathead (1970) is most probably based on this specimen. The specimen could not be found in the NHM or in Kawanda (A. Polaszek, personal communication).

Mohyuddin and Greathead (1970) state, without further specification, that the species is reared in East Africa from *S. calamistis.* This is apparently cited by Mohyuddin (1972) and Appert (1973). The NHM possesses a specimen reared by Walker in 1956 from *Sesamia* in Tanzania, identified by Kerrich in 1956; this may be the (or a?) specimen on which the record of Mohyuddin and Greathead (1970) is based. Conlong reared a specimen from a pupa of *E. saccharina* in sugar cane in South Africa (Natal).

Procerochasmias nigromaculatus was originally described by Cameron (1906a) as *Exephanes nigromaculatus* from one female from King Williamstown in South Africa. In the same year, Cameron (1906b) described the male as being a parasitoid of *B. fusca* (as *Sesamia fusca*) from Komgha in South Africa. According to Mally (1920), this specimen was from a small series reared by him.

Morley (1916) described *Chasmias glaucopterus* as a new species from a female from the same material reared by Mally and preserved in the South African Museum, but without mentioning the host species.

Mally (1920) reported the species as being a pupal parasitoid of *B. fusca* in South Africa (material from Komgha, Grahamstown and Cedara). He also synonymized *C. glaucopterus* Morley, 1916 with *E. nigromaculatus* Cameron, 1906. This synonymy has been overlooked for a long time. The name *C. glaucopterus* was therefore in use in the literature on biological control of *Busseola* until Heinrich (1967) published the synonymy again as being new. He also placed the species in the genus *Procerochasmias.*

The species is reported, under the name *C. glaucopterus* or *Procerochasmias glaucopterus,* as a (pupal) parasitoid of *B. fusca* by Ritchie (1932), Le Pelley (1959), Nye (1960), Milner (1967), Grist and Lever (1969), Mohyuddin and Greathead (1970), Rao and Nagarkatti (1971), Appert (1973), Nwanze (1981) and Harris and Nwanze (1992). Only Mohyuddin (1972) and Kfir and Bell (1993) use the correct name *P. nigromaculatus.*

In his taxonomic treatment of the genus, Heinrich (1967) describes a new subspecies, *P. nigromaculatus melanopygus* Heinrich, which is also parasitic upon *B. fusca.* According to Heinrich (1967), the distribution of both subspecies is as follows:

- *P. nigromaculatus nigromaculatus*: South Africa (Cape, Natal, Transvaal), south and east Tanzania.
- *P. nigromaculatus melanopygus*: west Kenya, north-west Tanzania, Uganda.

Heinrich (1967) gives characters to separate the two subspecies. Although I have been able to attribute most material to one of the two subspecies, there are a few specimens from countries from which Heinrich did not study material which are somewhat different from each of the subspecies in some way. This aberrant material is too scarce to decide whether it will belong to one of the described subspecies or to a new subspecies. For the time being, I would prefer to treat all as belonging to *P. nigromaculatus* without separating subspecies. In particular, material from Ethiopia does not really match the descriptions given by Heinrich.

Diagnosis. Length 14–16.5 mm; front wing 9–10 mm; ovipositor slightly protruding (*c.* 0.5 mm). Head usually red, but almost completely black in Ethiopian specimen; thorax usually red, with mesosternum, mesopleurum and metapleurum black (specimens from Tanzania have pronotum and mesoscutum dark brown; black in specimens from Ethiopia). Scutellum red in female, yellow in male.

Biology. Pupal endoparasitoid.

Hosts. *Busseola fusca* (main host), *C. partellus, E. saccharina, S. calamistis.*

Distribution. Cameroon, Ethiopia, Kenya, Mozambique, South Africa, Tanzania, Uganda, Zimbabwe.

MESOCHORINAE

Mesochorinae is a rather small subfamily worldwide. Since Wahl (1993) recently synonymized *Stictopisthus* with *Mesochorus*, only two genera now occur in Africa. Mesochorinae are usually true hyperparasitoids, which means that they parasitize primary parasitoids as long as these are living in their hosts. The female oviposits into the larva of the primary parasitoid, which is almost always an endoparasitic larva of a braconid or an ichneumonid species. The mesochorine larva kills its host (the primary parasitoid) after the latter has left its host (often a caterpillar) and spun a cocoon in which to pupate. The mesochorine adult therefore emerges from the cocoon of a braconid or another ichneumonid species; rarely, tachinid flies may also be attacked. Only the genus *Mesochorus* is reported in connection with stem borers.

Mesochorus Gravenhorst

There are only a few records of *Mesochorus* species reared from stem borers. Williams (1980) mentions *Mesochorus* sp. reared by him in Mauritius in 1968 from *Marasmia* (as *Neomarasmia*) *trapezalis* in sugar cane. The only record from the African continent is by Bhatnagar (1984, 1987), who reared *Mesochorus* sp., together with *Syzeuctus* sp., from *C. ignefusalis* in millet and sorghum in Senegal. Whether there is any relation between these ichneumonid species is not clear, because he also reared a braconid (*Stenobracon unifasciatus*, as *Liomorpha* sp.). The percentages of hosts parasitized by these three species, however, indicate that *Syzeuctus* sp. (= *S. senegalensis*) might be the (main) host of *Mesochorus* sp. in this case.

No *Mesochorus* species has been seen which is definitely reared from stem borers or from their primary parasitoids.

Diagnosis. Males have rod-shaped claspers; these do not occur in other genera.

OPHIONINAE

This cosmopolitan subfamily is very well represented in the tropics, especially in the humid areas. It contains a relatively small number of genera in Africa. In their revision of the African Ophioninae, Gauld and Mitchell (1978) list 11 genera with 191 species, most of which (149 species) belong to the genus *Enicospilus*. This is the only genus that is recorded from stem borers in the Ethiopian region.

Ophionines are koinobiont solitary endoparasitoids of lepidopterous larvae, the parasitoid larva usually leaving its host when the caterpillar is mature. In most species, the parasitoid larva spins a cocoon, which is usually in the ground. Most ophionines have relatively large sizes, and their hosts also belong to the larger species.

Enicospilus Stephens

Despite an excellent revision of the genus by Gauld and Mitchell (1978), the species of *Enicospilus* are rather difficult to identify without sufficient material for comparison. This is because of the large number of relatively uniform species. The genus itself, however, is quite easy to recognize. It can generally be distinguished from other ophionine genera by having:

1. The mandible twisted and conspicuously narrowed toward the apex;
2. The posterior transverse carina of the mesosternum complete before the mid-coxae;
3. The occipital carina usually complete or only obscure mediodorsally;
4. The distal margin of mid- and hind trochantellus without a sharp tooth;
5. No sharp carina along inner orbit, just above the antennal socket.

Biology. Species of *Enicospilus* are known to be endoparasitoids of larvae of several lepidopterous families, including Noctuidae, Crambidae and Pyralidae. The larva leaves its host when this is fully grown or has spun a cocoon to pupate. The *Enicospilus* larva spins a rather strong cocoon in which to pupate; cocoons are generally in the ground or in leaf litter. Those species that parasitize hosts which spin cocoons before pupation generally spin their own cocoon within the cocoon of the host. Moutia and Courtois (1952) give extensive notes on the biology of *Enicospilus sesamiae* (as *Enicospilus* sp.).

Several species are reported as having been reared from stem borers. Most of these records concern the islands Madagascar, Mauritius and Réunion. Some are based upon misidentifications.

From the literature, *Enicospilus mauritii* (Saussure) seems to be one of the main parasitoids of *C. sacchariphagus* and *Sesamia* species in Madagascar and the Mascarenes (papers by Moutia and coworkers). However, Gauld and Mitchell (1978) studied part of the material of Moutia and coworkers, which is kept in the NHM. They found eight species of *Enicospilus* to be represented in the material.

Other species reported from Madagascar are *Enicospilus ambositrensis* Delobel (= *Enicospilus transvaalensis* (Cameron)) ex *S. calamistis* (Delobel, 1974), and *Enicospilus antimena* (Saussure, 1892), *Enicospilus plagiatus* (Saussure, 1892) and *Enicospilus punctipinnis* (Saussure, 1892) ex *C. sacchariphagus* (Gauld and Mitchell, 1978).

Only a few species have been reported as parasitoids of cereal stem borers from the African continent: *Enicospilus antefurcalis, Enicospilus ruscus* and *E. sesamiae*; these are treated below. According to Betbeder-Matibet (1981), *Enicospilus vorax* is mentioned in the literature as a parasitoid of *Eldana* on the African continent. He does not state by whom, and I have not found any primary record of this species. Also, Gauld and Mitchell (1978) do not give any host record; it is therefore omitted here.

All remaining records from the African continent concern '*Enicospilus* sp.', most of them as parasitoids of *S. calamistis* or *Sesamia* sp. *Enicospilus* sp. is reported as a parasitoid of *Sesamia* from Ghana (Forsyth, 1966; Quartey, 1975), Nigeria (Jerath, 1968; Anon., 1971) and East Africa (Mohyuddin and Greathead, 1970). Milner (1967) reports two specimens of an *Enicospilus* species from *B. fusca* in Uganda.

Key to the Species of *Enicospilus* Reared from Cereal Stem Borers on the African Continent

1. Clypeus clearly separated from face by a furrow; clypeus in profile flat, with the margin acute and out-turned. (Antenna with 55–59 flagellar segments.) .... *E. sesamiae*

– Clypeus not separated from face by a furrow; clypeus in profile convex or at least weakly convex, with the margin inturned or impressed. (Antenna either with more than 59 or fewer than 55 flagellar segments.) 2

2. Antenna with 60–63 flagellar segments. Distal sclerite in front wing (this is the biggest sclerite, nearest to the wing base) strongly pigmented, not confluent with the proximal sclerite (this is the long and narrow sclerite, nearest to the apex) .... *E. ruscus*

– Antenna with 52–54 flagellar segments. Distal sclerite in front wing confluent with the proximal sclerite and sometimes weakly pigmented ...... *E. antefurcalis*

Enicospilus antefurcalis (Szépligeti)

Hosts. This species was reared only once, by Decelle from *E. saccharina* in Zaïre (Yongambi): the specimen has been studied and reported on by Gauld and Mitchell (1978) and Gauld (1982).

Distribution. This species is distributed over almost all of the African continent and Madagascar.

Enicospilus ruscus Gauld & Mitchell

Hosts. Gauld and Mitchell (1978) studied material of *E. ruscus* that was reared by Courtois in Mauritius from *Sesamia vuteria* (or *Speia vuteria*) in 1949, and also from *M. loreyi* in 1950. According to A. Polaszek (personal communication), '*E. aff. ruscus*' was also reared from *Sesamia botanephaga* in Ivory Coast by Dabiré-Binso.

Distribution. Gauld and Mitchell (1978) have seen material from Madagascar, Mauritius, Nigeria, Senegal, Sierra Leone, Tanzania, Uganda and Zaïre.

Enicospilus sesamiae Delobel

The assembled material of *Enicospilus* was identified by Dr I.D. Gauld as *E. sesamiae*, which is apparently the commonest species reared from *Sesamia* species, not only in Madagascar and Mauritius, but also on the African continent. Prior to 1974, it was usually referred to as *Enicospilus antankarus* (a misidentification), which is also misspelled as *E. antancarus* or *E. antakarus*.

Hosts. Reliable host records are: *S. calamistis, S. vuteria* from Mauritius and Réunion (Gauld and Mitchell, 1978), *B. fusca* from Tanzania and Uganda.

Distribution. Ghana, Madagascar, Mauritius, Nigeria, Réunion, Sierra Leone, Tanzania, Uganda.

PIMPLINAE

Pimplinae (= Ephialtinae *sensu* Townes) is a moderately large subfamily containing some rather large genera. Three genera are known to contain parasitoids of stem borers in Africa and attempts have been made to introduce a species of a fourth genus into Madagascar, Réunion, the Comoros and Senegal.

Holcopimpla **Cameron**

Townes and Townes (1973) listed *Holcopimpla* (with *Delaulax*), *Flavopimpla* Betrem, 1932 and *Afrephialtes* Benoit, 1953 as synonyms of *Xanthophenax* Saussure, 1892, and included 43 African species in the genus. Gupta and Tikar (1969) argued that *Flavopimpla* and *Afrephialtes* are valid genera and provided a key to all four taxa. Gupta and Tikar (1976) and Gupta (1987) also maintained *Flavopimpla* and *Afrephialtes* as valid genera. Although they did not express themselves about a possible synonymy of *Xanthophenax* and *Holcopimpla*, I follow their opinion expressed in 1969, and consider *Holcopimpla* different from *Xanthophenax*.

The genus *Holcopimpla* is restricted to the Ethiopian region; some species show a slight superficial resemblance to *Xanthopimpla*. The genus has not been recorded previously from stem borers.

Holcopimpla concolor ssp. *concolor* (Brullé)

A specimen of this species has been studied which was reared from a pupa of *C. partellus* by Overholt and coworkers in Kenya. The species has a rather wide distribution and has several times been described as a new species. The reared material has been compared with the type series of *Xanthopimpla olsoufieffi* Seyrig, and also with the type of *Poecilopimpla testacea* Morley.

Diagnosis. Length 9 mm (range observed in other specimens 8–13.5 mm); front wing 8.5 mm; ovipositor sheath 6.8 mm; tibia III 2.3 mm. The whole body is pale brownish yellow, with the gaster slightly darker. Stemmaticum almost black; two small black marks at apical margin and close to the sides of gastral tergite 2; wings infumate; tarsi II and III brown. Mandible not strongly narrowed towards the tip and not twisted, mandibular teeth black; clypeus strongly concave. Mesopleurum polished, impunctate; propodeum with rather superficial but coarse punctations, sides impunctate; gaster with coarse punctation, tergites 2–5 with a shallow transverse impression.

Host. There are no other host records besides the present one: *C. partellus* (Kenya).

Distribution. Kenya, Madagascar, Mozambique, Senegal, South Africa, Tanzania, Togo and Zaïre.

Itoplectis Foerster

This genus has an almost worldwide distribution. Most species are rather polyphagous solitary endoparasitoids, mainly of lepidopterous pupae. Sometimes species also parasitize cocoons of other parasitoids, thus acting as secondary parasitoids. One species is known as a rather important parasitoid of stem borers in East and South-East Asia; attempts have been made to introduce it into the Ethiopian region.

Itoplectis naranyae (Ashmead)

Itoplectis naranyae was originally described as a parasitoid of *Naranya* (now *Naranga*) *diffusa* (Walker) (Noctuidae) from Japan; the related *Naranga aenescens* Moore, however, is also often reported as a host species. The host's name, *Naranga*, is the reason that the name is often incorrectly spelled '*Itoplectis narangae*' in the literature on stem borers. Other misspellings are *I. naranguae* and *I. naganguae*.

Rao and Nagarkatti (1971) mention *Itoplectis narangae* (!) from Japan as being considered for introduction into India against stem borers. This must have happened shortly thereafter, since Appert (1973) reports the species as being introduced from India into Madagascar, Réunion and Comoros against *Chilo* species and *M. separatella*.

According to Vercambre (1977c), it was also introduced into Senegal during 1972, at least until 1977. The author does not give information as to where the stocks originated. Introduction into Senegal (or 'West Africa') is also mentioned by Bordat *et al.* (1977), Gahukar (1981) and Akinsola and Agyen-Sampong (1984) (all using a misspelled name) and by Polaszek *et al.* (1994). All of these records state or suggest that the introductions were not successful.

Diagnosis. Head, most of thorax and apex of gaster black; gastral tergites 1–4(5) red; coxa II + III and femur III red (femur black at apex), all trochanters + tibia III + tarsus III mainly white/yellow (tarsal segments black at apex, and tibia with a dark base and apex).

Hosts. A selection of host species, other than *Naranga*, from various countries found in the literature is: *Sesamia inferens* (Noctuidae), *C. suppressalis, S. incertulas, Hedylepta accepta* and *C. medinalis* Gn. (Crambidae), *Pelopidas* (or *Borbo*) *cinnara* and *Parnara guttata* B.& C. (Hesperiidae), *Adoxophyes privatana* (Wlk.) and *Homona coffearia* (Nietn.) (Tortricidae), and *Oulema* (or *Lema*) *oryzae* (Kuw.) and *Lema tristis* (Col., Chrysomelidae). Also the ichneumonids *Charops* and *Casinaria* are recorded as host species, so, like other species of *Itoplectis, I. naranyae* can also act as a secondary parasitoid.

Distribution. Introduced but not permanently established in Africa. Widely distributed in East and South-East Asia and in Hawaii.

Pimpla Fabricius

This is a large cosmopolitan genus; most species are supposed to be rather polyphagous solitary endoparasitoids of lepidopterous pupae and prepupae. More than 20 species have been described from the Ethiopian region. Two species have been found to parasitize *Sesamia* pupae; these can be separated as follows:

1. Gaster black; wings almost hyaline .... *Pimpla hova*
– Gaster bright red, almost orange; wings brownish .... *Pimpla* sp. A

Pimpla hova Seyrig

This species is recorded from *S. calamistis* and *S. biluma* Nye in Madagascar by Appert (1971, 1973). A female and a male specimen have been seen from the MRAC collection and have been identified by Dr Benoit. Both specimens were collected in Madagascar, but do not bear any label suggesting that they have been reared; one was collected by Seyrig in 1944, while the other specimen does not bear a label with a collecting date. *Pimpla hova* is not reported from other countries.

Diagnosis. Head, thorax and gaster black, except for scutellum and postscutellum yellow/orange; hind margins of gastral tergites 1, 2 and 3 reddish or black; legs bright orange with black or dark brown. Black (dark brown) parts of legs are: coxa I + II, trochanters I–III, tibia II + III, tarsus II + III; wings almost hyaline.

Hosts. *Sesamia calamistis* (Appert, 1971, 1973); other hosts: *S. biluma.*

Distribution. Madagascar.

Pimpla sp. A

A single male specimen belonging to an unidentified species of *Pimpla* was reared by Dr Conlong from a pupa of *S. calamistis* in *C. papyrus* in South Africa.

Diagnosis. Head and thorax black, except for scutellum, postscutellum and subalar prominence yellow; gaster bright red with the apical segments brown; legs same colour as gaster, except for tibia + tarsus III blackish brown (tibia lighter at base); front and hind wings with brownish tinge.

Hosts. Reared only once from *S. calamistis.*

Distribution. South Africa.

Xanthopimpla Saussure

This is a very large genus, which is especially abundant in the Old World tropics, but is also represented in the neotropical region. In their revision of the genus *Xanthopimpla* in the Indo-Australian area, Townes and Chiu (1970) list 164 species, and Townes and Townes (1973) list 44 species for the Ethiopian region. Several species attack stem borers, most of them in the Oriental region. Almost all species have a bright yellow colour, with a pattern of black spots on various parts of the body. The fore wing is sometimes a bit darkened at the tip, but the wing itself is usually hyaline.

There are a few records of *Xanthopimpla* reared from stem borers in continental Africa; in one case, material has been seen on which a record is based. Two species occur very often in the literature on African stem borers: *Xanthopimpla citrina*, which is indigenous to Mauritius and the African continent, and *Xanthopimpla stemmator*, which was successfully introduced into Mauritius and probably also into Madagascar. Attempts are being made to introduce the latter into South Africa also. A third species is included below, because it is reported to have been reared from a non-stem-boring pyralid.

Key to the Species of *Xanthopimpla* Recorded from African Stem Borers

1. Body completely yellow, without any black markings on thorax, gaster or legs. Ovipositor sheath and the area between the ocelli (= stemmaticum) black .... *X. citrina*

– Gaster with black markings; usually also mesoscutum and propodeum with some black spots 2

2. Females 3

– Males 4

3. Gastral tergites 1–7 usually with a pair of black spots (eighth tergite always without); those on sixth tergite often much smaller or missing; second tergite always with a pair of black marks; mesoscutum with a pair of black spots in front and a smaller median spot in front of scutellum (very rarely mesoscutum completely yellow) .... *X. stemmator*

– Black markings absent from second, sixth and eighth gastral tergite, fourth tergite sometimes with small black spots; mesoscutum with a transverse black band in front, which is rarely broken into three separate blackish spots .... *X. punctata*

4. All gastral tergites with black spots (including eighth); mesoscutum as in female; hind tibia not black at base .... *X. stemmator*

– Black spots on gastral tergites 1, 3, 5 and 7, and sometimes additional smaller spots on tergites 2, 4 and/or 6; mesoscutum as in female; hind tibia black at base .... *X. punctata*

Xanthopimpla citrina (Holmgren)

This species is indigenous to Mauritius and is also reported from Kenya, Mozambique and Tanzania. The Mauritian population was traditionally known as *X. citrina*, whereas the continental populations were considered to be *Xanthopimpla luteola* and *Xanthopimpla stictopleura*.

The first record of this species as a parasitoid of stem borers is by Moutia and Mamet (1945), who report it as a parasitoid of *C. sacchariphagus* from Mauritius, and state that it was probably introduced from Madagascar. Two years later Moutia and Mamet (1947) repeat that it was introduced from Madagascar as a parasitoid of

C. sacchariphagus. In fact, the species did not occur in Madagascar, and attempts were made to introduce it from Mauritius in later years. Moutia and Courtois (1952) report *X. citrina* from *C. sacchariphagus* and *S. calamistis* (as *S. vuteria*). Carèsche (1962) states that the species had been successfully introduced into Réunion against *C. sacchariphagus* in sugar cane, but he expresses his doubts about the effects on the population of the pest. According to Greathead *et al.* (1971), the species had established in Réunion after introductions in 1953 and 1960, but Appert (1973) states that it is not present in Madagascar and Réunion.

There are a few records of *X. luteola*, which is the same species, as a parasitoid of stem borers: La Croix (1967) reared it from *C. partellus* or *Chilo argyrolepia* in Kenya (Coast Province), and Bordat (1979) mentions *X.* cf. *luteola* from pupae of *S. calamistis* in Ivory Coast. Two specimens in the CIRAD collection are apparently from the material reported by Bordat. Although there is a slight difference with other material identified as *X. citrina*, I tentatively consider them conspecific.

Other material that is reared from stem borers comprises one ♀ reared from a pupa of *C. partellus* in Kenya (Kilifi), and one ♀ reared from a pupa of *B. fusca*, from millet in northern Nigeria.

Diagnosis. See key.

Hosts. *Busseola fusca, C. argyrolepia* or *C. partellus, C. partellus, C. sacchariphagus, S. calamistis.* The species is also reported from *Nacoleia praeternalis* by Moutia and Mamet (1947), who state that the pupae are heavily parasitized in Mauritius.

Distribution. Ivory Coast (Bordat, 1979), Kenya (Seyrig, 1935; La Croix, 1967), Madagascar? (Appert, 1973), Mauritius (type of *Pimpla citrina*; Moutia and Mamet, 1945, and subsequent authors), Mozambique (type of *Pimpla luteola*), Nigeria, Tanzania (Szépligeti, 1908; type of *X. stictopleura*).

Xanthopimpla punctata (Fabricius)

The first record of this species as a parasitoid of cereal stem borers is by Morley (1913), who reported it from *C. suppressalis* in India. Various authors have mentioned other stem-borer species in the Oriental region as hosts, such as *Chilo infuscatellus, C. partellus* and *C. sacchariphagus* (Chu and Hsia, 1937; Sharma *et al.*, 1966; Chiu *et al.*, 1984).

The species has not yet been reported from stem borers in the Ethiopian region; it is, however, recorded as a parasitoid of the pyralid species *Sylepta derogata* (F.) in Nigeria (Lamborn, 1914; Vayssière and Mimeur, 1925; Risbec, 1950a, 1960). Probably all of these records are based on that of Lamborn. Furthermore, it is reported from the same host species by Silvie (1991) from Chad. I have studied a voucher specimen of the material reared by Silvie in 1987, which is kept in the NHM. The specimen is identified as *X. punctata*, which I consider a misidentification.

Krieger (1914a, b) mentioned *X. punctata* from Mauritius (non-reared material), from where it is also reported as a parasitoid of *C. suppressalis* by Nishida and Torii (1970).

No African material has been seen of the species, but, if it really occurs in Nigeria and/or Mauritius, it could be found parasitizing stem-borer pupae. It is therefore included here.

Diagnosis. See key.

Hosts. Not reared from stem borers in sub-Saharan Africa. Other hosts: *S. derogata?*

Distribution. Nigeria (Lamborn, 1914, and subsequent authors), Mauritius (Krieger, 1914a, b).

Xanthopimpla stemmator (Thunberg) (Plate 10:6)

Xanthopimpla stemmator was successfully introduced by Vinson in 1939 from Sri Lanka into Mauritius to control *C. partellus* (Jepson, 1939, 1941; Vinson, 1942). From Mauritius, it was successfully introduced into Réunion against *C. sacchariphagus* in sugar cane (Carèsche, 1962). Attempts have also been made to introduce it from Mauritius into Madagascar, where it probably did not establish (Carèsche and Brenière, 1962; Greathead *et al.*, 1971; Appert, 1973). Brenière (1965), however, lists the species as a parasitoid of *C. sacchariphagus* in Madagascar.

In Mauritius, it is well established and is reported to parasitize *C. sacchariphagus* and *S. calamistis* by Moutia and Mamet (1945, 1947), Moutia and Courtois (1952), Williams and Mamet (1954) and subsequent authors, who cite them.

In more recent times, attempts have been made to introduce it from Mauritius into South Africa against *E. saccharina* in sugar cane (Carnegie, 1991; Conlong, 1994) and against *C. partellus* in maize and sorghum (Kfir, 1989a, b, 1991b, 1992d). Although the species has been recovered a few times after releases, it has so far not established permanently. In laboratory cultures, it readily accepts pupae of both of the above-mentioned hosts (D. E. Conlong and R. Kfir, personal communication).

In the Oriental region, the species parasitizes various host species, such as the stem borers *S. nivella, Chilotraea infuscatella, Proceras venosata, S. inferens, S. calamistis, C. partellus, C. sacchariphagus, C. suppressalis, C. zonellus, Chilotraea auricilia, S. incertulas* (Townes and Chiu, 1970).

Diagnosis. See key.

Hosts. In Mauritius: *C. sacchariphagus, S. calamistis.*

Distribution. Mauritius (introduced: Vinson, 1942), ?Madagascar (Brenière, 1965).

UNPLACED SPECIES

Coelichneumon

This genus is known from the Holarctic, Oriental and neotropical regions; it is not known from sub-Saharan Africa.

A species of '*Coclichneumon*' [*sic*!] is reported by Jerath (1968), who reared it from a pupa of *Sesamia* in sugar cane in Nigeria; the record is quoted by Anon. (Jerath?) (1971). No voucher material could be traced on which the record is based, but it is almost certainly based on a misidentification. Since Jerath also mentions in the same paper 'Genus nr. *Chasmias*', which is *Dentichasmias*, there will be no confusion with this genus. I think it is most likely that *Coelichneumon* is a misidentification of a species of *Ichneumon*, probably *I. rubriornatus*, which is reported in the present chapter as parasitizing *Sesamia* in sugar cane. It is the only genus of known stem-borer parasitoids that can really be mixed up with *Coelichneumon*. It is also possible that confusion arose by the names used by Townes, who uses the name *Ichneumon* for what is generally understood as *Coelichneumon* by other authors and how it is used in the present chapter. Townes uses *Pterocormus* for '*Ichneumon* of authors'.

Less likely is a confusion with *Procerochasmias*. In the NHM, however, there is a specimen of *P. nigromaculatus* from South Africa, reared from *Busseola*, which was identified as ?*Coelichneumon* sp. by Ferrière in 1933.

In summarizing records from the literature, Bordat *et al.* (1977) also mention *Coelichneumon* sp. from *Sesamia nonagrioides botanephaga* in Nigeria; their record is probably based on Jerath's.

Coleocentrus

The genus belongs in the subfamily Acaenitinae, which are often parasitoids of wood-boring hosts (mainly Coleoptera). The name is almost certainly based on a misidentification, but it is not clear which species is meant. The species is reported for the first time as 'aff. *Coleocentrus* sp.' by Risbec (1950a, b), who reports having reared one specimen from a larva of *C. zacconius* (as *P. africana*) in rice in Senegal. Descamps (1956) reports '*Coleocentrus*' as being reared from pupae of the same host in Cameroon. In 1960, Risbec states that the host of the specimen he reported earlier was *C. phaeosema*. Nickel (1964) mentions *Coleocentrus* from *Maliarpha* in Cameroon, and states in 1967 that *Calocentrus* [*sic*] sp. from Cameroons appears promising for introduction into the Philippines against rice borers. Grist and Lever (1969) list *Coleocentrus* from '*Parerupa africana*' (= *C. zacconius*) in Cameroon, from where it is only reported from *Maliarpha*. Brenière (1969), quoting Risbec (1950a, b), introduced the incorrect spelling '*Coelocentrus*', which was subsequently used by various authors.

Bordat *et al.* (1977), almost certainly quoting Descamps (1956), list '*Coelocentrus*' as a larval parasitoid of *C. zacconius* in Cameroon. They are subsequently quoted by Nwanze (1981) and Beassem (1982), who use the same incorrect name.

Finally, Gahukar (1990), quoting Risbec (1950a, b), mentions *Coleocentrus* ex larva of *C. zacconius*.

Although the name appears quite often in the literature on stem-borer parasitoids, all 'records' may be based on very few rearings. Only Nickel suggests in his 1967 paper, by stating that the species is promising for introduction into the Philippines, that he reared a number of specimens. Unfortunately, no voucher material of his rearings seems to be stored. The identity of the species remains, therefore, unclear.

21 Scelionidae

A. Polaszek[1] and S.W. Kimani-Njogu[2]
[1]*CABI Bioscience UK Centre, c/o Entomology Department, The Natural History Museum, London SW7 5BD, UK;* [2]*Department of Entomology, National Museums of Kenya, PO Box 40658, and ICIPE, Nairobi, Kenya*

INTRODUCTION

Ten species of *Telenomus* have been reliably recorded as primary parasitoids in the eggs of African cereal stem borers. These species have been treated in detail in two publications: Polaszek and Kimani (1990) and Polaszek *et al.* (1993). The former paper included other *Telenomus* spp. recorded from non-stem-boring lepidopteran pests, and the latter included extralimital (Asian) *Telenomus* spp. attacking stem borers. The following, therefore, is the first complete account treating all of the known African *Telenomus* spp. associated with cereal stem borers. The degree to which *Telenomus* spp. attacking African stem borers were misidentified prior to 1990 is remarkable (about 90% of records, other than original descriptions), serving as a case-study for the importance of proper taxonomy in the study of natural enemies of pest species. An account of these misidentifications is provided by Polaszek and Kimani (1990). Several species of *Telenomus* associated with cereal stem borers are strongly dorsoventrally flattened (Plate 12:6), an apparent adaptation for entering leaf whorls.

Berger (1981) recorded a species of *Lepidoscelio* as a pupal parasitoid of *Chilo partellus* on maize in Mozambique. Voucher specimens on which this record is based have been located and studied, and are actually a *Dicroscelio* species. It is improbable that these really emerged from a *C. partellus* pupa, as all known records of the group to which *Dicroscelio* belongs associate species with eggs of Acrididae (Orthoptera).

IDENTIFICATION

The general methods for specimen preparation are given in Chapter 14, and the structure of male *Telenomus* genitalia, examination of which is essential for species-level identification, is given in Fig. 103. Descriptive terms used both for the male genitalia and for the male and female morphology in later sections of this chapter follow Johnson (1984). In many cases, comparison with the figures, followed by careful checking against the full species accounts, will give the correct answer. However, the key is provided in order to draw attention to the most useful characters in each case. The following characters, in particular, are subject to variation within a species: position of digiti (either one or both digiti may be in either the raised or the lowered position (see Fig. 103)), degree of pigmentation, length of the basal ring. Also, the aedeagal lobe, which is often very weakly sclerotized, may undergo distortion during slide preparation.

Slides should preferably be examined with phase-contrast illumination at a magnification of 200–300 ×.

Following the key, the species are treated in alphabetical order.

Key to *Telenomus* Species Associated with African Cereal Stem Borers

1. Ends of laminae volsellares with a projection centrally (cp. Figs 588–591) 2
– Ends of laminae volsellares without such a projection centrally (Figs 592–596) 5
2. Digiti small relative to aedeagal lobe, each digitus 0.4–0.45 × the length of the aedeagal lobe (Fig. 588) .... *Telenomus nemesis*
– Digiti larger relative to aedeagal lobe, each digitus at least 0.5 × the length of the aedeagal lobe (Figs 589–591) 3
3. Aedeagal lobe almost entirely truncate, digiti very large in relation to it (when digiti are in the lowered position, they nearly reach the tip of the aedeagal lobe (Fig. 591)) .... *Telenomus busseolae*
– Aedeagal lobe either pointed (Fig. 590) or truncate only at the tip, which is slightly bilobed (Fig. 589). Digiti smaller in relation to aedeagal lobe 4
4. Aedeagal lobe pointed, more than 0.33 × the total length of the aedeagovolsellar shaft (Fig. 590) .... *Telenomus creusa*
– Aedeagal lobe truncate at the tip, about 0.33 × the total length of the aedeagovolsellar shaft (Fig. 589) .... *Telenomus bini*
5. Aedeagovolsellar shaft constricted at a point approximately halfway along its length (Figs 592, 593, arrowed) 6
– Aedeagovolsellar shaft *either* with sides parallel-sided (Figs 595, 596) *or*, if constricted, then *not* at a point approximately halfway along its length (Fig. 594, arrowed) 7
6. Digiti large, one-half or more as long as the aedeagal lobe, with three to five teeth (Fig. 592) .... *Telenomus thestor*
– Digiti small, rounded, a third as long as the aedeagal lobe, with three teeth (Fig. 593) .... *Telenomus soudanensis*
7. Aedeagovolsellar shaft with sides bowed, its distal half wider (maximum width) than its proximal half, digiti small in relation to aedeagal lobe (Fig. 594) 8
– Aedeagovolsellar shaft parallel-sided, digiti large in relation to aedeagal lobe (Figs 595, 596) 9
8. Aedeagal lobe *usually* broadly rounded (Fig. 594, dotted line). Head and mesosoma of male largely pale brown, contrasting with dark metasoma. Hind femur of female largely dark. Second metasomal tergite of female less than 1.5 × its maximum width, and with hind margin straight .... *Telenomus versicolor*
– Aedeagal lobe tapering, triangular in shape, but often with a flattened tip (Fig. 594). Head, mesosoma and metasoma of male black. Hind femur of female largely yellow with a slight brown infuscation. Second metasomal

tergite of female more than 1.7 × its maximum width, and with hind margin curved .... *Telenomus nephele*

9. Aedeagal lobe with a pointed tip (Fig. 590). Mesosoma not flattened dorsoventrally .... *Telenomus creusa*

– Aedeagal lobe truncate at the tip (Figs. 595, 596). Mesosoma dorsoventrally flattened 10

10. Digiti and digital teeth elongate. Aedeagal lobe clearly truncate (Fig. 596) .... *Telenomus applanatus*

– Digiti and digital teeth short. Aedeagal lobe less truncate, or rounded (Fig. 595) .... *Telenomus isis*

Telenomus applanatus Bin & Johnson (Fig. 596; Plate 11:4)

Diagnosis.

Male. Genitalia (Fig. 596): aedeagus parallel-sided, digiti large, with elongate digital teeth.

Female. Mesosoma compressed dorsoventrally, dorsellum almost smooth, with very little sculpture; otherwise without any distinctive characters.

Host. Pyralidae: *Eldana saccharina* (Walker).

Distribution. Gabon, Ghana, Ivory Coast, Nigeria.

Remarks. Bin and Johnson (1982) described *T. applanatus* as including the morphologically very similar parasitoid of *Maliarpha separatella*, later described as a separate species, *T. bini*, by Polaszek and Kimani (1990). So far as is known, *T. applanatus* is restricted to *E. saccharina*, and has not been recorded outside West Africa. An unsuccessful attempt has been made to introduce this species into South Africa for biological control of *E. saccharina* (Conlong, 1994).

Telenomus bini Polaszek & Kimani (Fig. 589)

Diagnosis.

Male. Genitalia (Fig. 589): central projection present; digiti large relative to aedeagal lobe; aedeagal lobe about a third the total length of the aedeagovolsellar shaft and truncate at the tip, which is bilobed. Mesosoma depressed dorsoventrally, but this character rather variable.

Female. Mesosoma depressed dorsoventrally (variable); otherwise without any distinctive characters.

Hosts. Pyralidae, Phycitinae: *M. separatella*; Crambidae, Schoenobiinae: *Scirpophaga ?subumbrosa* Meyrick; Crambinae: ?*Chilo* sp. (majority of records from *Maliarpha*).

Distribution. Ghana, Ivory Coast, Kenya, Madagascar, Malawi, Senegal, Tanzania.

Remarks. This species was originally included under *T. applanatus* (Bin and Johnson, 1982; see above). Although the majority of records are from the white rice borer, *M. separatella* (Pyralidae), additional host records from crambid rice borers appear reliable.

Telenomus busseolae Gahan (Fig. 591)

Diagnosis.

Male. Genitalia (Fig. 591): central projection present, digiti large, much longer than half the length of the aedeagal lobe, which is truncate. A distinctive species, being strongly flattened dorsoventrally, including the head.

Hosts. Noctuidae: *Busseola fusca* (Fuller), *Chrysodeixis chalcites* (Esper), *Sesamia botanephaga* Tams & Bowden, *Sesamia calamistis* Hampson, *Sesamia cretica* Lederer, *Sesamia inferens* (Walker), *Sesamia nonagrioides* (Lefevre), *Sesamia* sp. Crambidae, Crambinae: *Coniesta* (= *Donacoscaptes*) *ignefusalis* (Hampson).

Distribution. Africa: Cameroon, Egypt, Ghana, Kenya, Mauritius, Niger, Nigeria, Réunion, Senegal, South Africa, Uganda. Elsewhere: Greece, Iran, Iraq, Israel.

Remarks. *Telenomus busseolae* is the most widespread *Telenomus* species attacking African stem borers, extending its range into Europe and the Near East. Sétamou and Shulthess (1995) gave an account of this species, although their data were pooled with data for *T. isis*, a species occurring sympatrically, and attacking the same hosts as *T. busseolae*.

Telenomus creusa Polaszek & Kimani (Fig. 590)

Diagnosis.

Male. Genitalia (Fig. 590): aedeagal lobe large, length greater than one-third of the total length of the aedeagovolsellar shaft, broadly pointed. Digiti large, about half the maximum length of the aedeagal lobe, usually with three digital teeth per digitus. Central projection present between distal ends of laminae volsellares, although this is often indistinct or unpigmented (best viewed with phase-contrast illumination).

Female. Head, mesosoma and metasoma without any distinctive characters.

Host. Crambidae, Crambinae: *Chilo diffusilineus* (J. de Joannis).

Distribution. Malawi.

Telenomus isis Polaszek (Fig. 595; Plate 12:6)

Diagnosis.

Male. Genitalia (Fig. 595): central projection absent, digiti small, rounded, less than half the length of the aedeagal lobe. F1 transverse, and all flagellomeres similar in colour. Antennal radicle long, more than one-third the length of the scape.

Female. Hardly distinguishable from the other species in this study; the antennal radicle long, more than one-third the length of the scape. Interantennal process short and broad.

Hosts. Noctuidae: *S. botanephaga* Tams & Bowden, *S. calamistis* Hampson, *Sesamia* sp., ?*Busseola* sp.

Distribution. Africa: Benin, Cameroon, Ghana, Ivory Coast, Nigeria.

Remarks. See *T. busseolae*, above.

Telenomus nemesis Polaszek & Kimani (Fig. 588)

Diagnosis.

Male. Genitalia (Fig. 588): distal ends of laminae volsellares with a very characteristically shaped central projection extending towards the distal end of the aedeagal lobe between the digiti. Digiti small (less than half the length of the aedeagal lobe) equipped with large teeth; aedeagal lobe smoothly rounded at its distal end.

Female. Head, mesosoma and metasoma without any distinctive characters.

Hosts. Crambidae, Crambinae: *Chilo orichalcociliellus* (Strand), *Chilo* sp.

Distribution. Ghana, Kenya, Mozambique, Senegal.

Remarks. *Telenomus nemesis* is known from *C. orichalcociliellus* in East Africa; the species of *Chilo* attacked in West Africa (Ghana and Senegal) is unknown.

Telenomus nephele Nixon (Fig. 594)

Diagnosis.

Male. Genitalia (Fig. 594): aedeagal lobe tapering, often truncate at the tip. Black with yellow legs. Genitalia very similar to, and sometimes indistinguishable from, those of *T. versicolor*, with which it is often found in sympatry.

Female. Black with yellow legs, the hind femora slightly infuscated. Body elongate, length of second metasomal tergite more than 1.7 times its maximum width and with a distinctly curved hind margin.

Hosts. Crambidae, Schoenobiinae: *Scirpophaga melanoclista* Meyrick, *Scirpophaga occidentella* (Walker), *S. subumbrosa* Meyrick, *Scirpophaga* spp.

Distribution. Cameroon, Ghana, Ivory Coast, Malawi, Mali, Senegal.

Remarks. *Telenomus nephele* has been confused in the past with *Telenomus tolli* Risbec, and as a morph of *T. versicolor* Bin & Johnson (see detailed account by Polaszek and Kimani, 1990).

Telenomus soudanensis (Risbec) (Fig. 593)

Diagnosis.

Male. Head and body brownish yellow (?discoloured). Genitalia (Fig. 593): digital teeth large, laminae volsellares most strongly pigmented laterally, such that aedeagovolsellar shaft appears to have two rods converging towards the digiti before diverging for a short distance. Central projection absent.

Female. Material examined without any characters to distinguish the species from the related species *Telenomus remus* and *Telenomus nawai*. Further study may reveal that *T. nawai* Ashmead, *T. remus* Nixon, *T. soudanensis* Risbec and *Telenomus spodopterae* Dodd are all conspecific.

Hosts. Crambidae: Crambinae *Chilo zacconius* Bleszynski; Pyralidae: *Crocidolomia binotalis* (new host record: recorded as *Liophanurus* sp. by Risbec, 1950a: 235).

Distribution. Mali, Senegal. Polaszek and Kimani (1990) reported the lectotype of this species as coming from Kayo, Niger, having been recorded originally by Risbec (1950a: 556) from 'Soudan français, Kayo, (Office du Niger)', and only the word 'Kayo' appearing on the lectotype slide. According to the US Board on Geographic Names Official Standard Names Gazeteers for Mali and Niger (both of which countries would have been included in 'Soudan français'), there are three localities called 'Kayo' in Mali, while there is none in Niger (although there are another three Kayos in Chad).

Telenomus thestor Nixon (Fig. 592)

Diagnosis.

Male. Genitalia (Fig. 592): central projection absent, aedeagus with sides concave, digiti large, usually with four or five teeth.

Female. Head, mesosoma and metasoma black, legs and antennae also almost entirely black. Head strongly transverse and appearing very low in side-view, giving this species a 'humpbacked' appearance. Otherwise, without any distinctive characters.

Hosts. Arctiidae: *Amsacta moloneyi* (Druce); Crambidae: Crambinae: *C. orichalcociliellus* (Strand).

Distribution. Ivory Coast, Kenya, Senegal, Uganda, Zaïre.

Telenomus versicolor Bin & Johnson (Fig. 594, dotted line)

Diagnosis.

Male. Genitalia (Fig. 594, dotted line): almost indistinguishable from those of *T. nephele* (Fig. 594), but aedeagal lobe usually broadly rounded. Head and thorax largely pale brown, mesoscutum usually darker centrally than at the sides, metasoma black.

Female. Black with legs infuscated, length of second metasomal tergite of female less than 1.5 times its maximum width and with the hind margin straight.

Hosts. Crambidae, Schoenobiinae: *S. melanoclista* Meyrick, *S. ?subumbrosa* Meyrick, *Scirpophaga* spp. (for notes on hosts, see *T. nephele*).

Distribution. Ghana, Ivory Coast, Malawi, Senegal.

Remarks. See *T. nephele*, above.

22 Diptera

K.M. Harris
Research Associate, CABI BIOSCIENCE UK Centre, c/o Entomology Department, The Natural History Museum, London SW7 5BD, UK

INTRODUCTION

Dipterous parasitoids of cereal stem borers in Africa are generally less abundant and less well known than the hymenopterous parasitoids. Only about ten species of Tachinidae can be definitely accepted as indigenous parasitoids of these pests and only two of these, *Sturmiopsis parasitica* Curran and *Descampsina sesamiae* Mesnil, are relatively abundant, widespread and well known. Some non-indigenous species of Tachinidae have been introduced to Africa to attempt classical biological control (but do not seem to have established) and some indigenous species have been moved within Africa. All relevant indigenous and non-indigenous species are listed in Tables 22.1 and 22.2. The main purpose of this chapter is to facilitate identification of reared adults of the indigenous species of Tachinidae and to provide summaries of information on their geographical and host ranges and other aspects of their biology and ecology.

Members of other families of Diptera have also been recorded as parasitoids of stem borers, but most of these records are based on inadequate observation of species whose larvae are phytophagous (e.g. shootflies) or saprophagous, feeding on dead or moribund stem-borer larvae or pupae. The main families involved are Phoridae, Chloropidae, Muscidae and Sarcophagidae. Notes and keys to assist identification of these families are provided and cases where further study is needed to determine whether or not the species associated with stem borers are truly parasitic are indicated.

The Diptera are true flies, with a single pair of fore wings (except in apterous groups) and with the hind wings modified into a characteristic pair of knobbed halteres (Fig. 601), which are balancing organs, used to maintain stability in flight. It is therefore easy to recognize adult Diptera. Larvae of most species are also easily recognized, as they lack jointed thoracic legs, generally have a small head capsule and, in most of the species that are of main concern here, are typical maggots, tapered anteriorly and with a characteristic internal, chitinized cephalopharyngeal skeleton (Fig. 619). More than 125 distinct families of Diptera are recognized by taxonomists. At least 110,000 species have been described in more than 6000 genera, but many species remain to be described and studied. This is especially so in Africa, as indicated by the relatively recent description of a new species of

Table 22.1. Species of Tachinidae that are indigenous to Africa and have been recorded as parasitoids of lepidopterous stem borers of cereals. (Note. The record of *Carcelia evolans* (Wiedemann) (as *Zenillia evolans*) from Zimbabwe by Cuthbertson (1936) has not been confirmed by subsequent rearings, is probably erroneous and is therefore not listed here. It has been authoritatively recorded only from West Africa, where its host is *Diparopsis watersi* Rothschild.) The most important species are indicated by asterisks * and species that have been recorded once only are bracketed [ ].

Tachininae
Nemoraeini
    Nemoraea bequaerti Emden
    Nemoraea discoidalis Villeneuve

Tachinini
    Chromatophania picta (Wiedemann)
    [*Dejeania bombylans* Fabricius]

Goniinae
Siphonini
    Actia cuthbertsoni Curran
    [*Actia exsecta* Villeneuve]
    [*Actia hargreavesi* Curran (= *Actia comitata* Villeneuve)]
    [*Peribaea orbata* (Wiedemann)]
    Siphona murina (Mesnil)

Sturmiini
    Drino sp. (= Palexorista sp.)
    Paradrino halli Curran

Eryciini
    Descampsina sesamiae Mesnil*
    Lydella sesamiae (Mesnil) (= *Metoposisyrops sesamiae* Mesnil)
    [*Pseudoperichaeta laevis* Villeneuve]
    Schembria eldana Barraclough
    Sturmiopsis parasitica (Curran)*

stem-borer parasitoid, *Schembria eldana* (Barraclough, 1991). This manual is focused very narrowly on a few families, genera and species. A wider knowledge of Diptera can be gained, if needed, from the excellent, well-illustrated *Manual of Nearctic Diptera*, published in three volumes (McAlpine *et al.*, 1981; McAlpine, 1987, 1989). The terminology used here is taken mainly from that Nearctic manual, which introduced conformity into a confused situation, where different terminologies had been developed and used for different groups of Diptera.

COLLECTION AND CURATION

Adult Diptera are usually preserved dry on pins, as they can then be examined easily with a stereoscopic microscope. Because of their fragility, they must be handled very carefully to avoid breakages and loss of antennae, wings, legs and setae, which provide many of the diagnostic characters used in the process of identification. Adults reared from known hosts are most valuable, but they should be kept alive long enough for them to harden and develop their full coloration. This can usually be done by leaving them in cages or other containers for a few hours after emergence. They should then be killed (usually with fumes of ethyl

Table 22.2. Species introduced to Africa as potential biocontrol agents of lepidopterous stem borers.

Paratheresia claripalpis (Wulp)	Mauritius (Moutia and Courtois, 1952) South Africa (Kfir, 1991c)
Lixophaga diatraeae (Townsend)	Mauritius (Moutia and Courtois, 1952) Comoros (Brenière *et al.*, 1985) Madagascar and Réunion (Greathead, 1971)
Lydella minense (Townsend) (= *Metagonistylum minense*)	Mauritius (Moutia and Courtois, 1952)
Zenillia palpalis (Aldrich)	Mauritius (Greathead, 1971)
Lydella striatalis (Townsend) (= *Diatraeophaga striatalis*)	Mauritius, Madagascar and Réunion (Etienne, 1971; Greathead, 1971; Ingram, 1983)
Sturmiopsis inferens Townsend	Uganda (Ingram, 1983) Ghana (Scheibelreiter, 1980) Mauritius (Greathead, 1971)
Palpozenillia diatraeae Townsend	South Africa (Carnegie, 1991)
Palpozenillia palpalis (Aldrich)	Mauritius (Greathead, 1971)

acetate) and pinned. Pinning soon after they have been killed is ideal but, if that is not possible, specimens may be kept for a day or so in a moist, cool atmosphere before pinning. Any possible wetting through condensation or by direct contact with water or other fluids must be prevented, as it will permanently damage colours and other important characters. It is therefore generally inadvisable to preserve adult Diptera in alcohol or other preservatives, although that may be done if there is no alternative. Large or medium-sized flies should be pinned directly on to long stainless-steel pins (35–40 mm), with the pin passing through the right or left side of the thorax. Alternatively, specimens may be pinned with stainless-steel headless micropins (10–15 mm) or other small pins for later staging on long pins through strips of closed cell foam (Plastazote), polyporus or other staging material. Card points are sometimes used to stage adult Diptera by gluing the side of the thorax to the point. This method is effective, if done carefully, and may be the best way of mounting specimens that have been dried out after preservation in fluids, but it does tend to obscure some characters. Formal setting of wings, antennae and legs is not normally used on adult Diptera, but subsequent examination will be easier if these appendages are well separated before the specimens are dried. It may also be possible to draw male genitalia out of the abdomen with a hooked pin so that they may be more easily examined and, if necessary, removed for detailed study. Puparial cases and larval skins, with associated cephalopharyngeal skeletal parts, should be retained whenever possible and staged on the same pin as the reared adult. Larvae of most Diptera are soft-bodied and are best preserved in 80% ethanol, or other fluid preservatives. They should be killed by immersion in hot water (60–70°C) or by collecting directly into the preservative and, if possible, they should be transferred to fresh preservative after a few days to minimize the risk of tissue discoloration resulting from dilution of the original preservative by body fluids. This is particularly necessary when large larvae are collected into small tubes. All specimens should be accurately labelled with information on locality, host, date of collection, name of collector and collector's serial reference number. Labels should be legible and durable, either handwritten or computer-generated.

EXAMINATION OF SPECIMENS

Dry, pinned specimens are usually examined with a stereoscopic microscope at magnifications from × 20 to × 200. A hand-lens or eyeglass may be adequate for preliminary examination at lower magnifications but will not permit accurate observation of many critical diagnostic characters. A compound bright-field or phase-contrast microscope may be needed for detailed examination of genitalia preparations, wing venation or other detailed study.

High-quality optical systems should be used whenever possible, as they reduce distortion of shapes and colours. Good illumination must also be set up, preferably with fibre-optic systems for stereoscopic microscopes or critically adjusted integral light sources in compound microscopes. Resolution of the finest detail at magnifications up to × 10,000 can be achieved by scanning-electron microscopy, but this is used more for research than for routine examination of specimens.

RECOGNITION OF FAMILIES

Adult flies belonging to the five families of Diptera generally associated with stem borers can be recognized by diagnostic combinations of external characters, especially wing venation, structure of the antennae and presence or absence of particular groups of setae on the sides of the thorax, as indicated in the following diagnostic key. Additional corroborative characters are given in square brackets and more detailed descriptions are given in the following sections on each of the main families. Keys for the identification of larvae are not given. In most cases, their identity is best determined by rearing adults.

Key to Families of Diptera Associated with Stem Borers of Cereal Crops in Africa

1. Wing venation distinctive with radial veins (*R*) thickened and crowded into the anterior part of the wing towards the costal margin; costa (*C*) ending near the middle of the anterior margin and with four other weak and peculiarly aligned veins in the remainder of the wing (Figs 597, 598). [Small flies (wing length 1–3 mm) with a hunchbacked appearance (Fig. 597); antennae apparently reduced to a single segment as the large flagellum conceals the scape and pedicel; maxillary palps large] .... Phoridae

– Wing with branches of R not strongly thickened and not crowded into the anterior part of the wing; C extending at least to the wing apex and other veins not peculiarly aligned (Figs 599, 611, 612) 2

2. Small flies (wing length 1–4 mm), almost devoid of bristles; ocellar triangle relatively large and conspicuous (Fig. 600a); wing venation without a *cup* cell (Fig. 599) .... Chloropidae

– Larger flies (wing length 5–12 mm), with many conspicuous bristles on head, thorax and abdomen; ocellar triangle relatively small (Fig. 609b); wings with a *cup* cell (Figs 603b, 611) 3

3. Meron (hypopleuron) with a distinct row of bristles (Fig. 603b) 4

– Meron (hypopleuron) without a distinct row of bristles (Fig. 603a) ..... Muscidae

4. Subscutellum (postscutellum) strongly developed and evenly convex in profile (Fig. 601a); antennal arista usually not plumose. [Generally dark, bristly flies but varying considerably in size (wing length 4–12 mm) and form] .... Tachinidae
– Subscutellum absent or, if weakly developed, concave in profile (Fig. 601b); arista plumose in basal half (Fig. 604). [Mostly flesh flies with distinctive grey and black shifting chequered pattern on abdomen and with three longitudinal black stripes on the thorax] .... Sarcophagidae

Tachinidae

About 8000 species of Tachinidae have been described and almost 1000 are known from tropical Africa. Larvae of all known species are parasitic on arthropods, especially on Lepidoptera but also on Hemiptera–Heteroptera, Coleoptera, Orthoptera, Dermaptera and some Diptera. Greathead (1971) summarized information on all attempted biological control of pests in Africa south of the Sahara, including Madagascar and other islands in the western Indian Ocean, and detailed the many attempts made to establish non-indigenous tachinids in Mauritius during the period 1947–1968 to control *Chilo sacchariphagus* on sugar cane. Six species were shipped from Trinidad, one from New Britain and two from Java (Table 22.2), but in some cases no releases were made and in others the numbers released were small. This campaign culminated in the shipment of the tachinid *Lydella striatalis* (= *Diatraeophaga striatalis*) from Java and adult releases over three successive years of 22,167 (1966), 23,697 (1967) and 15,850 (1968), but the species failed to become established. Similar attempts in Madagascar (Brenière *et al.*, 1966; Appert, 1973), Réunion (Etienne, 1971) and the Comoros (Brenière *et al.*, 1985) with this species and with *Lixophaga diatraeae* also failed. Ingram (1983) also referred to the failure to establish *L. striatalis*, despite sustained efforts and large-scale releases, and also noted that the Indian species, *Sturmiopsis inferens* Townsend, was released in Uganda but failed to establish. In South Africa, the American tachinid *Paratheresia claripalpis* was released at a rate of 4000–5000 per month in 1990, but no establishment was recorded (Carnegie, 1991). Betbeder-Matibet (1990) commented on the possibilities of laboratory mass-rearing of *L. striatalis* and *L. diatraeae* on *Chilo*, but prospects for the effective use of exotic tachinids in classical biocontrol of lepidopterous stem borers in Africa do not seem promising. A better approach may be to conserve and enhance the action of indigenous species by environmental manipulation, as suggested by Greathead (1990).

The most recent detailed taxonomic account of the Afrotropical Tachinidae was published by Crosskey (1984). He provided a classification and diagnostic keys for the identification of the 188 genera then known from tropical Africa, including the genera dealt with here. His publication provides an essential foundation for any work on the taxonomy of the Tachinidae in Africa, as does another publication by the same author (Crosskey, 1976) for the Oriental region. Figures 623–637 are illustrations of morphological terminology used by Crosskey in those publications, and should be studied before attempting the key to Tachinidae included here. Crosskey's objective in dealing with the African genera was to provide a framework for the considerable amount of taxonomic study that is still required to describe and classify species and genera. For most of the larger genera, published keys to the included species are not available. The following key to the most important indigenous species reared from stem borers is therefore likely to be of use only in confirming the probable identity of adults reared from lepidopterous stem borers and will not guarantee positive identification of caught adults. Until the tachinid parasitoids of stem borers in Africa are better known, it would be best to refer all specimens to specialists for confirmation of identifications, and that is certainly desirable when new host and/or distribution records are involved.

Key for the Identification of the Main Tachinid Species Known to be Parasitoids of Cereal Stem Borers in Tropical Africa

This key includes all of the species that are definitely parasitoids of stem borers. The following species have not been included, either because they have been recorded on only a single occasion or because there is some doubt that they really are parasitoids of stem borer hosts: *Pseudoperichaeta laevis* Villeneuve, *Peribaea orbata* (Wiedemann), *Dejeania bombylans*, Fabricius, *Drino* sp. (= *Palexorista* sp.), *Paradrino halli* Curran. The characters used in this key are readily seen in both sexes, which should ensure that it is user-friendly. Additional confirmatory characters are given in square brackets and more detailed descriptions of the species are given in the main body of the chapter. It is always possible that species may be encountered that have not been previously recorded from stem borers, and any doubtful cases should therefore be referred to a specialist in tachinid identification.

1. Two katepisternal (sternopleural) setae (Fig. 602a) 2
– Three katepisternal setae (Fig. 602b) 5
2. Eyes quite bare or with inconspicuous short hairs (Figs 604, 606, 608, 609, 616, 617) 3
– Eyes conspicuously haired (Fig. 610) [Relatively large, heavy-bodied species (wing length about 10–12 mm; head profile distinctive, with relatively small eyes, deep gena and protruding epistome (Fig. 610)] .... *Nemoraea discoidalis* Villeneuve/*Nemoraea bequaerti* Emden
3. Head profile markedly triangular, profrons much longer than height of gena (Fig. 608); *R4+5* node with a single strong seta [Medium-sized species (wing length about 5 mm), with narrow abdomen] .... *Lydella sesamiae* (Mesnil)
– Head profile not markedly triangular, profrons not longer than height of gena (Figs 605, 606); vein *R4+5* node with two to three short setulae (Figs 611, 612) .. 4
4. Parafacials haired (Fig. 605). [Medium-sized (wing length about 5–7 mm); generally dark-bodied but with grey pollinosity dorsally on thorax and abdomen. Puparium with posterior spiracles small, flush with surface and with serpentine slits (Fig. 620)] .... *Sturmiopsis parasitica* (Curran)
– Parafacials bare (Fig. 606). [Medium-sized (wing length about 5–7 mm) and very similar to *S. parasitica* in general appearance (habitus, Fig. 613). Males can be distinguished by the presence of dense black hair completely covering the sides and undersides of the fourth and fifth abdominal segments (compare Figs 614, 615). Posterior spiracles of puparium very similar to those of *S. parasitica*, with serpentine slits (Fig. 621)] .... *Descampsina sesamiae* Mesnil
5. Wings with discrete wide, dark bands across anterior half near wing apex (Plate 11:1); humeral callus with five to six setae. [Relatively large species (wing length

about 10 mm), generally brown with golden pollinosity on head and abdomen] .... *Chromatophania picta* (Wiedemann)
- Wings without dark bands near apex; humeral callus with no more than three to four setae 6

6. Proboscis very long, slender and geniculate, total length greater than the height of the head (Fig. 616). [Small species (wing length about 4 mm)] .... *Siphona murina* (Mesnil)
- Proboscis short, often only partly visible below peristome, much shorter than height of head 7

7. Vein *R4+5* node with only three to four setulae; vein *R1* without setulae. [Similar in size and general appearance to *S. parasitica* and *D. sesamiae*. Posterior spiracles of puparium straight, not sinuous as in those two species] .... *Schembria eldana* Barraclough
- Veins *R4+5* and *R1* with many small black setulae along their length [Small species (wing length about 4 mm), with prominent proboscis (Fig. 617)] .... *Actia* spp.

In the following section the most frequently encountered and best-known species are dealt with in greatest detail. Brief notes are provided for other species that have been recorded infrequently and are probably of little or no importance in limiting stem-borer populations. The species are grouped in the tribes defined by Crosskey (1984).

ERYCIINI

Five known parasitoids of African cereal stem borers (*S. parasitica* Curran, *D. sesamiae* Mesnil, *L. sesamiae* (Mesnil), *P. laevis* Villeneuve and *S. eldana* Barraclough) were included in the Eryciini by Crosskey (1984). The tribe is a polyphyletic convenience group for a miscellany of genera not readily assignable elsewhere. All Afrotropical genera included in it by Crosskey have a submedian ventral seta on the midtibia and all except *L. sesamiae* have the first and second abdominal tergites (T1+2) excavate to the hind margin or virtually so.

Sturmiopsis parasitica (Curran) (Figs 605, 611, 614, 619, 620)

This is the most widespread and commonly recorded species of tachinid stem-borer parasitoid in Africa.

Taxonomic notes. *Sturmiopsis parasitica* was described from Zimbabwe (Rhodesia) by Curran (1939) as type species of his proposed new genus, *Rhodesina* Curran. That name was later found to be preoccupied by *Rhodesina* Malloch. Townsend (1941) therefore proposed *Curranomyia* as a replacement name for *Rhodesina* Curran, which has since been synonymized in *Sturmiopsis*, a genus that now contains two Afrotropical species (*Sturmiopsis angustifrons* Mesnil and *S. parasitica* Curran) and one Oriental species (*S. inferens* Townsend).

Diagnosis. This is a medium-sized tachinid, about 8 mm long, basically black but with lighter grey to white pollinosity, especially dorsally on the thorax and in wide bands across the abdominal tergites. (This pollinosity disappears in greasy specimens and in specimens preserved in or affected by fluids, so that the specimens then look completely black.) Females are easily recognized by the presence of a pair of proclinate orbital setae, which are absent in males. The following combination of characters distinguishes this species from all other known Afrotropical Tachinidae. Wing venation as illustrated (Fig. 611), with acute M1 bend, and cell R5 open at costa. Costa black; other veins light brown. Head (Fig. 605) with relatively small eyes so that genal depth exceeds the width of the antenna; eyes haired but hairing very short and sparse, usually only discernible by very careful examination at high magnification; ocellar setae proclinate; facial ridges with weak setulae on lower half; parafacials with hairing extending to genal hairing in males but with an intermediate bare area in females. Vibrissae inserted about level with the epistomal margin. Prosternum haired; four posterior dorsocentral setae; two katepisternal setae and scutellum with two pairs of lateral setae; prealar seta stronger than other mesonotal setae, much longer than first posterior interalar seta (post ia) and nearly always longer than the first posterior dorsocentral seta (post dc); midtibia with two anterodorsal setae and with a submedian ventral seta; abdominal T1+2 excavate to hind margin. Puparium with posterior spiracles small, flush with surface and with serpentine slits (Fig. 620).

Hosts. The natural host range of this species in Africa, as recorded by various authors (Curran, 1939; Nye, 1960; Smithers, 1960b; Harris, 1962; Milner, 1967; Jerath, 1968; Mohyuddin and Greathead, 1970; Nagarkatti and Rao, 1975; Ajayi, 1985a; Alam, 1992), includes *Busseola fusca, Chilo orichalcociliellus, Chilo partellus, Coniesta ignefusalis, Eldana saccharina, Sesamia calamistis* and *Sesamia nonagrioides*. In addition, Nagarkatti and Rao (1975) reported that it developed on five Indian species of stem borer (*Sesamia inferens, Chilo auricilius, C. partellus, Chilo indicus* and *Chilo infuscatellus*) in laboratory experiments at Bangalore, India.

Biology. Smithers (1960b) reported that *S. parasitica* was the commonest parasitoid of *B. fusca* in the area around Harare (Salisbury), Zimbabwe during 1956 and 1957, with 3–35% of the larvae parasitized in six samples. Of the parasitized larvae, 65% produced one adult tachinid each; 21% produced two flies; 10% yielded three and 4% yielded four flies. Milner (1967) presented detailed results of field surveys on farmers' fields in Uganda, Kenya and Tanzania and concluded that this species was the most successful parasitoid of *B. fusca* over much of its range in East Africa. He recorded it at 20 localities. At half of these, it was the only primary parasitoid found, while at another seven localities it was the most abundant primary parasitoid. He also recorded that ?*Epiencyrtus* sp. was the commonest of several hyperparasites reared and must have exercised significant control over the numbers of *S. parasitica* in some areas. (Editorial note. ?*Epiencyrtus* sp. of Milner (1967) has been shown to be a misidentification of *Exoristobia dipterae* (Risbec) (Encyrtidae). See Chapter 19, this volume.) He noted that, despite the presence of hyperparasites, *S. parasitica* was an important controlling influence on *B. fusca* in some places, notably in Songea, Iringa, Mbeya and Singida in Tanzania. The species occurred between 500 and 6000 ft (150 and 1800 m) and within an annual rainfall range of 20–60 in (500–1524 mm).

Nagarkatti and Rao (1975) published a detailed account of the biology of this species and of the rearing techniques that they developed. In Ghana, it was more common on maize than on sugar cane (where *D. sesamiae* was usually dominant), especially on the second crop growing in October–December. Field sampling of *Sesamia* spp. and *E. saccharina* on second-crop maize in 1973 indicated levels of tachinid parasitism ranging from 0.9 to 17.7% during a period when the average temperature was about 25.5°C, with rainfall about 110 mm per month and relative humidity 87–90%. They also reported that in laboratory rearings at Bangalore, in India, mating was readily obtained by placing freshly emerged females singly with two to three older males in 8 cm × 2 cm glass vials and shaking them vigorously. Mating lasted 3–15 min and was encouraged by bright sunlight, but mating pairs had to be removed immediately to shade, as the females appeared to be agitated by bright light. Females mate only once, but males may mate many times and are able to fertilize all the ova in four to five females during their 3–4-week lifespan. Mating pairs were carefully set aside until the flies separated, after which the female was transferred to an 11 cm × 11 cm × 8 cm cage, until gestation was completed, which took 18–19 days at 26°C. Each female produced 500–900 maggots, and larval development in suitable stem-borer hosts at 26°C generally took 12–14 days, but occasionally up to 35 days. Full-grown larvae emerged from host larvae and pupae (mainly from pupae) and, after a prepupal period of 12 h, pupal development was completed and adults emerged after 12–19 days.

Smith, J.W. *et al.* (1993b) listed *Sturmiopsis* as one of the tachinid genera using planidial ingress to locate hosts. In this strategy, the female uses cues from stem-borer frass or the stem-borer tunnel to locate recent host activity. Females then larviposit mobile, planidial first-instar maggots at the tunnel entrance. They then find the stem-borer larva in the tunnel and penetrate the integument to feed internally as solitary or gregarious endoparasites. At first, they feed freely inside the host body, but then attach themselves to a tracheal branch or main trunk for the remainder of larval development. Later-stage stem-borer larvae are the targets of the planidia and adult flies emerge either from the mature host larva or, less commonly, from host pupae.

Distribution. This species is widespread in West, East and southern Africa. Harare, Zimbabwe, is the type locality but *S. parasitica* has also been recorded from Burkina Faso, Cape Verde, Ghana, Ivory Coast, Kenya, Malawi, Niger, Nigeria, Senegal and Tanzania and almost certainly occurs in many African countries where it is yet to be formally recorded.

Utilization. Nagarkatti and Rao (1975) reported shipment of puparia of this species from Ghana to the Commonwealth (now International) Institute of Biological Control (CIBC) station at Bangalore, India, during 1970–1972 and described the techniques used for shipment and laboratory rearing. Laboratory tests established that five species of Indian stem borer (*C. auricilius, C. partellus, C. infuscatellus, C. sacchariphagus indicus* and *S. inferens*) were parasitized and appeared to be suitable hosts. They suggested that this parasitoid has high potential as a biocontrol agent on sugar cane, maize and sorghum in southern India. Carnegie *et al.* (1985) reported that some *S. parasitica* were received in South Africa in consignments of *D. sesamiae* from Ghana and Nigeria in 1981/1982, but there were too few to culture.

Descampsina sesamiae Mesnil (Fig. 606, 612, 613, 615, 621)

Taxonomic notes. This species was described by Mesnil (1956) as type species of his new genus, *Descampsina*, which he distinguished from *Sturmiopsis* and

Pseudoperichaeta by the presence of dense black hair completely covering the sides and undersides of the fourth and fifth abdominal tergites in males, the lack of discal setae on the fifth abdominal segment, the lack of black microchaetae on the occiput and the short anterior claws in both sexes. Crosskey (1984) retained this as a monotypic genus, containing this single African species. It resembles *Sturmiopsis* in many of its morphological characters, as well as in its host range and biology, and would seem to be very closely related, possibly congeneric.

Diagnosis. *Descampsina sesamiae* is very similar to *S. parasitica* and shares most of the characters given for that species. It is about the same size, with a length of about 8 mm, but is darker in general colour, with much less light-coloured pollinosity on the thorax and abdomen. It can be separated from *S. parasitica* by the following combination of characters: eyes quite bare, with no traces of hairing; parafacials bare, without any hairing; no discal setae on the fifth abdominal segment; no black microchaetae on the occiput; short anterior tarsal claws; males with long, fine, closely appressed black hair on the whole of the sides and ventral surfaces of abdominal segments 3 to 5 (Fig. 615).

Hosts. *Descampsina sesamiae* was originally recorded from *Sesamia* sp. by Mesnil (1956) and has since been recorded by various authors (Jerath, 1968; Ajayi, 1985a; Lobo and Viereck, 1986; Sampson and Kumar, 1986; Carnegie, 1991) from species of that genus, especially *S. calamistis*, *Sesamia botanephaga* and *Sesamia penniseti*. Low levels of parasitism on *E. saccharina* in Ghana have been reported by Scheibelreiter (1980) and Sampson and Kumar (1986). Attempted introduction to South Africa for biological control of that species was not successful (see below).

Biology. Jerath (1968) reported that *D. sesamiae* was the major parasite found on *Sesamia* spp. larvae on sugar cane at Bacita, Nigeria, in 1963–1966 and that parasitism in the field ranged from 12.2 to 16.8% during 1964–1966. In 1963, 1964 and 1965, *D. sesamiae* comprised 84.0, 80.5 and 73.9%, respectively, of the total parasite population, including the hymenopterous parasitoids, as well as other tachinids. He recorded that *D. sesamiae* attacked all larval instars and pupae of *Sesamia*. Females oviposited within 48 h after emergence and adults lived 4–8 days. On hatching, young larvae immediately entered the host and it was evident that more than one egg was deposited, as in many cases three or four parasitoid puparia were obtained from a single host larva. The host larva usually died before the parasitoid larva completed its development and left its host to pupate close to it. In the laboratory, the life cycle was completed in 25–38 days. Nagarkatti and Rao (1975) reported that this species was the dominant tachinid parasitoid of sugar-cane stem borers (*S. penniseti, S. nonagrioides botanephaga, S. calamistis* and *E. saccharina*) in Ghana in 1970 and 1971, but they did not provide details of its biology. Carnegie and Leslie (1979) reported that this species would not mate in captivity, even when they used a method that was successful with *S. parasitica*. Scheibelreiter (1980) reported investigations of sugar-cane stem borers (mainly *Sesamia* spp.) in Ghana during 1970–1978. Tachinids were the dominant larval parasitoids, with *D. sesamiae*, which is more common in the savannah zone, most abundant. In 1971, it comprised 96% of all parasitoids reared, compared with 4% for *S. parasitica*, which is an important parasitoid of *Sesamia* spp. in the forest zone. The rate of parasitism of *E. saccharina* was usually near to zero and reached its highest values of 5–10% in the long dry season between November and February.

Sampson and Kumar (1986) studied parasitism by *D. sesamiae* on *Sesamia* spp. and *E. saccharina* on sugar cane in southern Ghana during 1978–1980. *Descampsina sesamiae* was the dominant larval parasitoid, followed by *S. parasitica*. Populations were highest during the rainy season from August to November, reaching peaks of 25.23% in September 1978, 47.73% in August 1979 and 42.72% in July 1980. Humidity appeared to be a factor regulating the rate of parasitism, possibly because newly deposited larvae require a moist surface while searching for hosts and because pupae do not develop at low humidities. They also recorded that parasitism was higher on outgrowers' fields (which were not burned before harvest) than on estate fields (which were burned). Smith, J.W. *et al.* (1993b) noted that the biology of *D. sesamiae* is similar to that described above for *S. parasitica*.

Distribution. Cameroon is the type locality and the species has also been recorded from Cape Verde, Ghana, Nigeria and Togo. It has been introduced to South Africa from Ghana and Nigeria but may not have established there (Carnegie *et al.*, 1985).

Utilization. Carnegie *et al.* (1985) reported the attempted use of this parasitoid against *E. saccharina* in South Africa. It was first introduced from Ghana in 1975, but a satisfactory culture was not established. A second importation was made from Ghana in 1981 and, with considerable difficulty, a culture was established on *S. calamistis*, to which *D. sesamiae* takes more readily than to *Eldana*. Further importations were made from Nigeria and, in November 1982, the culture was strong enough to make small-scale field releases, mainly into large field cages but also into smaller cages and directly into open fields where a high population of *Eldana* was present. Between November 1982 and December 1984, 26 releases were made of from two to 20 gravid females. Parasitized *Eldana* were recovered on only three occasions and parasitized *Sesamia* were recovered on six occasions, but these recoveries were only from cages and there were no recoveries from open fields. In 1984, parasitized *Eldana* larvae were dissected and it was found that the parasitoid was being encapsulated. It was also realized that, even in West Africa, the natural hosts of *D. sesamiae* are *Sesamia* spp., rather than *E. saccharina*, and the project was terminated.

Schembria eldana Barraclough (Fig. 607; Plate 11:2)

Taxonomic notes. Barraclough (1991) described this species and assigned it to the genus *Schembria*, which was previously known only from Malta and Israel. The genus is currently included in the tribe Eryciini, but its relationships to other genera of Tachinidae have not been established. The African species is distinguished by the following combination of characters: eyes haired; facial ridges with setulae on lower half to three-fifths; two plus one katepisternal setae; four posterior dorsocentral setae; midtibia with one anterodorsal seta.

Diagnosis. An illustrated description is given by Barraclough (1991). Adults are 6.0–8.5 mm long, generally dark brown or black, with conspicuous silver pollinosity.

The eyes are generally conspicuously haired; parafrontals have relatively long black hairing in males; the profrons extends well forward of the vibrissal angle in profile and the peristomal setae are well developed (Fig. 607). Wing cell R5 is just closed at the margin or very short–petiolate. Males have large oval areas of apparently yellow/brown pollinosity ventrally on abdominal tergites 3 and 4. For other characters, see Barraclough's well-illustrated original description.

Host. This species is known only as a parasitoid of *E. saccharina* on *Cyperus papyrus* in South Africa.

Biology. Barraclough (1991) noted that *S. eldana* is probably closely associated with stands of *Cyperus* attacked by *E. saccharina.* He also reported that, under laboratory conditions, adult flies emerged from late-instar larvae and pupae of *E. saccharina* and that, although *S. eldana* is micro-oviparous, little is known about its reproductive biology.

Distribution. This species is known at present only from a few localities in north-eastern Natal, South Africa.

Lydella sesamiae (Mesnil) (= *Metoposisyrops sesamiae* Mesnil) (Fig. 608)

Taxonomic notes. This species was described by Mesnil (1968) in the genus *Metagonistylum*, based on two single females reared at Serere, Uganda, from a larva of *Sesamia oriaula* Tams & Bowden by W.R. Ingram in 1957. Crosskey (1980) transferred the species to the genus *Metoposisyrops* and recorded it from Nigeria, where it had been reared from *Sesamia* by Jerath (1968). It has also been reared from *Sesamia* at three other locations in Nigeria: Badeggi, Mokwa and Ibadan. Woodley (1994) transferred this species to *Lydella.* It is the only species of *Lydella* known from Africa and is distinguished by the following combination of characters: head profile strongly triangular (Fig. 608); first and second abdominal segments (T1+2) not excavate to hind margin; arista thickened on its whole length and with extremely long second segment; basal node of wing vein R4+5 with one strong, curved seta; bend of vein M very abrupt and usually with a short stub-like appendix; humeral callus with only two setae and with only two katepisternal setae.

Pseudoperichaeta laevis Villeneuve

Milner (1967) reported that a single specimen of this species was reared from a pupa found with a dead larva of *C. partellus* in a maize stem at Kawanda, Uganda, and this record was also referred to by Mohyuddin and Greathead (1970).

STURMIINI

Drino sp. (= *Palexorista*)

Robertson (1965) recorded *Drino (Prosturmia) imberbis* Wiedemann (= *Drino laxa* Curran) from *B. fusca* and *Helicoverpa armigera* (Hübner) at Ukiriguru, Uganda. *Palexorista laxa* is exclusively a parasitoid of *Helicoverpa* and this record from *Busseola* seems unlikely to be correct. Kfir (1992c) also recorded a *Palexorista* sp. parasitising *C. partellus* in South Africa.

Paradrino halli Curran

Robertson (1965) reported rearing one specimen from *B. fusca* in Tanzania and two specimens reared from that host in South Africa by Kfir have recently been identified (K.M. Harris det.1994). There is some possibility of confusion between *Drino* (= *Palexorista*) and *Paradrino*, which look superficially similar but are morphologically quite distinct.

NEMORAEINI

Crosskey (1984) noted that in the Afrotropical region the tribe Nemoraeini contains only the single genus, *Nemoraea*, identified by the following combination of characters: eyes hairy; fore coxae haired on almost the entire anterior surface; prosternum bare; wing with m-cu not strongly oblique or remote from wing edge; setulae on vein R4+5 seldom extending much beyond the basal node; lower calypter with fine hair on the upper surface and scutellum with four or five pairs of marginal setae. Sixteen species are known and an identification key was published by van Emden (1960) which may be used to identify *N. bequaerti* and *N. discoidalis*.

Nemoraea discoidalis Villeneuve

Milner (1967) recorded that this large, solitary, pupal parasitoid had been reared on a few occasions from *B. fusca* pupae collected from maize at Kawanda and Mubende in Uganda and at Kitale in Kenya and from a *Sesamia* pupa collected from a stem of *Pennisetum purpureum* at Toro, Uganda. Mohyuddin and Greathead (1970) referred to these records and noted that this species is a rare solitary larval or larval–pupal endoparasite of *B. fusca* and *Sesamia* sp.

Nemoraea bequaerti Emden

Jerath (1968) recorded this species from *Sesamia* sp. larvae on sugar cane at Bacita, Nigeria, in 1964.

TACHININI

Chromatophania is one of six Afrotropical tachinid genera included in the tribe Tachinini by Crosskey (1984) who characterized the tribe by the following combination of diagnostic characters: eyes bare; prosternum bare; hind coxa with soft hair on the posterodorsal surface; wing venation usually with a definite appendix vein or fold extending from the bend of vein M; setae on humeral callus numerous (usually at least six). The genus contains five African species and an identification key was published by van Emden (1960).

Chromatophania picta (Wiedemann) (Fig. 609, Plate 11:1)

Jerath (1968) recorded this species from *Sesamia* sp. larvae at Bacita, Nigeria, in 1963 and 1965.

Dejeania bombylans Fabricius

Carnegie (1991) recorded that *S. calamistis* is well controlled by various natural enemies, including this species. This statement was based on identification of a single specimen of *D. bombylans* reared from a fifth/sixth-instar *S. calamistis* larva collected from sugar cane at Mount Edgecombe, South Africa (A.J.M. Carnegie, personal communication). Most records of this tachinid are from *H. armigera*.

SIPHONINI

The following three genera (*Actia*, *Peribaea* and *Siphona*) all belong in the tribe Siphonini. Species included in this tribe by Crosskey (1984) are all relatively small flies (length about 5 mm) with short legs bearing short stiff setae and with short, broad wings in which the costal margin is conspicuously incised at the end of vein Sc, and both R1 and Cu1 are often setulose. The head is widely dichoptic, with proclinate orbital and strong outer vertical setae in both sexes, and members of this tribe share the following combination of characters: eyes bare; prosternum haired; prealar seta short and weak, much shorter than the first post dc seta; lower calypter not bent downwards on its outer edge; and supra-alar region of thorax almost always with three setae. Crosskey (1984) provided a key to the five genera recorded from the Afrotropical region and noted that virtually no keys are available for the 89 species assigned to those genera. O'Hara (1989) revised the genera and subgenera of Siphonini on a worldwide basis and provided keys and a summary of known hosts.

Siphona murina (Mesnil)

Nye (1960) recorded this species as a parasitoid of *B. fusca* at nine localities in Uganda, Kenya and northern Tanzania. Milner (1967) reported the results of later surveys on farmers' fields in these countries, during which it was reared from *B. fusca* and *C. partellus* at 15 scattered localities in all three countries at altitudes between 4000 and 6000 ft (1200 and 1800 m). He recorded that generally one to three (but up to five) individuals emerge from a single host larva and pupate in the stem-borer tunnel, near the dead host. The pupal period lasted 15–16 days and no hyperparasites were recorded. He concluded that, although this species was widespread, it never occurred in large numbers (except for one small sample from Embu, in which eight out of 22 *B. fusca* larvae were parasitized) and did not appear to influence stem-borer populations in East Africa. Mohyuddin and Greathead (1970) reported that this species was widely distributed in East Africa but not common; that it is a gregarious larval endoparasite of *B. fusca, C. partellus, S. calamistis* and *E. saccharina*, with up to five individuals reared from one host larva; that the larvae pupate in the host tunnel near the dead host; and that the pupal period is 15–16 days.

Actia cuthbertsoni Curran

This species was recorded from *Chilo 'zaleukos'* (= *Chilo zacconius*) in Sierra Leone by Jordan (1966).

Actia exsecta Villeneuve, *Actia hargreavesi* Curran (= *Actia comitata* Villeneuve]

Milner (1967) reported, during his survey of stem-borer parasitoids in East Africa, that one specimen identified as *Actia* sp. ?*comitata* was reared from a larva of *B. fusca* at Karagutu, Toro, Uganda, and one specimen identified as *Actia ?exsecta* was reared from a dead larva of *B. fusca* in a maize stem at Kayaje, Toro, Uganda. Mohyuddin and Greathead (1970) tentatively recorded these two species as rare larval parasites of *B. fusca, C. partellus* and *E. saccharina* in Uganda.

Actia ?antiqua Mesnil.

Nonveiller (1984) recorded *Actia ?antiqua* as a parasitoid of *B. fusca, S. calamistis* and *E. saccharina* on maize at Nkolbisson, Cameroon, in 1965, 1967 and 1969.

Peribaea orbata (Wiedemann)

Milner (1967) recorded that a single specimen of this species was reared from a puparium associated with the remains of a larva of *Chilo* sp., probably *C. partellus*, at Kilifi on the Kenyan coast. Mohyuddin and Greathead (1970) also referred to this record. *Peribaea orbata* has often been recorded as a parasitoid of other lepidopterous hosts (especially Noctuidae) in Africa, Asia, Australia and the Pacific, but this seems to be the only occasion on which it has been recorded from a cereal stem borer.

Sarcophagidae

Nye (1960) recorded *Heteronychia (Uroxanthisca) benefactor* Malloch (as *Sarcophaga benefactor*) from a *B. fusca* larva collected at Singida, Tanzania. Milner (1967) noted this record and reported that he had also reared sarcophagids from *B. fusca, C. partellus* and *Sesamia* sp. at a number of localities in Uganda, Kenya and Tanzania. Most of the specimens were females, and therefore not identifiable to species, but two males were identified as *Sarcophaga (Chaetophalla) villa* Curran. He noted that *Sarcophaga* spp. are known to be saprophagous or facultatively parasitic and that records of associations with stem-borer larvae must be assessed with caution. Mohyuddin and Greathead (1970) noted these records but did not add to them. Although some Sarcophagidae are obligate parasitoids of Orthoptera, Coleoptera, Lepidoptera and some other insects, larvae of most species are saprophagous. There seem to be no adequately documented instances of any species being an obligate parasitoid of lepidopterous stem borers.

Muscidae

The genus *Atherigona* (Fig. 618) is often associated with stem borers, as its larvae feed in and on the same host plants. Milner (1967) noted that an undescribed species of *Atherigona* was recorded as a parasite of a *Chilo* sp., probably *C. partellus*, in Kenya by La Croix (1967) and reported that, in his own surveys, several specimens of *Atherigona* sp. n. were seen to emerge from moribund larvae of *C. partellus* and were definitely parasitic. He also noted that other species of *Atherigona* are saprophagous or even phytophagous, as in the case of *Atherigona soccata* Rondani (= *Atherigona infuscata* Emden), the sorghum shoot-fly, which is a primary pest of sorghum. He suggested that the undescribed species may be in the process of evolving from a saprophagous to a parasitic mode of life and recommended that records of this genus as a parasite should be assessed with caution. There appear to be no subsequent records of this undescribed species. Skidmore (1985), in his review of the biology of Muscidae, noted that larvae of species of the subgenus *Atherigona* (*Atherigona*) are phytophagous or saprophagous in graminaceous hosts but species of the subgenus *Atherigona* (*Acritochaeta*) are less specialized, most being facultative carnivores in decaying plant or animal matter. *Atherigona* (*Acritochaeta*) *yorki* Deeming has been reared from sorghum, millet and maize shoots and stems, often associated with *Coniesta, Chilo, Sesamia* or other stem-borer larvae. It seems likely that its larvae are sometimes predators on stem borers but unlikely that they are truly parasitic, although there is an unpublished record of this species as an important parasite of *S. nonagrioides* on sugar cane in Morocco (Hesse and Mohamed, personal communication). Records of other species of Muscidae (*Coenosia pachypoda* Bigot, *Myospila lenticeps* (Thomson), *Phaonia* sp.) as parasitoids of stem borers are probably erroneous, since larvae of species in these genera are most likely to be predacious.

Chloropidae

Risbec (1960) listed *Epimadiza* sp., *Polyodaspis milii* Séguy (as *Oscinosoma milii*) and *Polyodaspis risbeci* Séguy (as *Oscinosoma risbeci*) as parasites of stem borers in West Africa, and other chloropids (*Anatrichus erinaceus* Loew, *Polyodaspis ?robusta* Lamb and *Scoliophthalamus trapezoides* Becker) have also been cited by various authors. These records are almost certainly erroneous. Many chloropids are phytophagous (including some important pests of cereals and grasses), others are saprophagous and a few are predacious. None are known to be parasitoids of Lepidoptera or other insects. Deeming (1972), working in northern Nigeria, observed that, although some chloropid larvae were phytophagous and caused primary damage to cereal seedlings, they were often reared from shoots that had been previously damaged by shootflies of the genus *Atherigona*. In some cases, they may possibly have been predacious, especially on other dipterous larvae, and in other cases they were secondarily feeding as scavengers on decaying tissues. There are no records of true parasitism.

Phoridae

Risbec (1950b, 1960) and Ingram (1958) recorded rearing phorids from stem borers, but it is most unlikely that they were parasitoids. One species, *Megaselia scalaris* Loew (Fig. 597), is often associated with laboratory rearings of stem borers, but the larvae are saprophagous, feeding on many different types of dead organic matter, including dead insects. There are some parasitic species of Phoridae, but there is as yet no clear evidence that any species is an obligate parasitoid of lepidopterous stem borers.

OTHER DIPTERA

Risbec (1950b, 1960) recorded *Pipunculus risbeci* Séguy from *C. zacconius* in West Africa, but this must be erroneous as pipunculids are exclusively endoparasites of Homoptera, especially leaf hoppers and plant hoppers. Risbec (1960) recorded the ceratopogonid *Forcipomyia castanea* (Walker) (= *Ceratopogon risbeci* Séguy) from larvae of *C. ignefusalis*. Larvae of Forcipomyiinae generally feed on algae, plant debris or fungi in moist situations and are most unlikely to be parasitic. Descamps (1956) recorded a bombyliid, *Geron* sp., parasitizing *Saluria* sp., a phycitine stem borer of rice in Cameroon. There seem to have been no subsequent records of bombyliids from stem borers. More recently, van Rensburg *et al.* (1988a) recorded the platystomatid, *Elassogaster arcuata* Hendel, as a parasitoid of *B. fusca* in South Africa. This seems unlikely, as platystomatid larvae are generally saprophagous, feeding on a great variety of dead organic matter.

23 Nematoda, Fungi, Protozoa, Bacteria and Viruses

G.O. Poinar, Jr[1] and A. Polaszek[2]

[1]*University of California, Berkeley, USA; present address: Department of Entomology, Oregon State University, Corvallis, Oregon 97331-2907, USA;* [2]*CABI* BIOSCIENCE *UK Centre, c/o Entomology Department, The Natural History Museum, London SW7 5BD, UK*

INTRODUCTION

Among the natural enemies of cereal stem borers in Africa, the insect parasitoids and predators are the most conspicuous and probably most commonly encountered. However, various other parasitic or pathogenic organisms may often be encountered in the field attacking borers, and several have been the subject of biological control studies. In general, nematodes and pathogens are not of great importance in regulating numbers of stem borers because the borer life history does not usually contain the predisposing factors for the development of epizootics (Harris and Nwanze, 1992). In particular, the borer larva is concealed within the protective stem and is not often in contact with other larvae. The following brief account is intended as a summary of current knowledge concerning these other natural enemies of African cereal stem borers.

IDENTIFICATION OF PARASITIZED OR DISEASED LARVAE

It is often possible to identify causes of parasitism or disease in borers by examining both internal and external symptoms and signs. A microscopic examination of the internal tissues can reveal the identity of most pathogens, with the exception of non-inclusion viruses.

The following key can be used to identify the major groups of parasites and pathogens of stem borers, namely bacteria, fungi, nematodes, protozoa and viruses.

1. Insects covered with mycelium, often bearing spores, or with string-like or horn-like structures emerging from their bodies (Figs 655–660); cadaver often mummified, hard or cheese-like in consistency. Tissues containing hyphae and/or spores .... fungi
– Insects not covered with mycelium or elongate structures 2
2. Insects containing metazoan (many-celled) parasites in their body cavity 3
– Insects containing unicellular (single-celled) particles in their body cavity 4

3. Parasites mostly elongate and worm-like, lacking distinct segmentation (Figs 642–644) .... nematodes
– Parasites cigar-shaped or pear-shaped, distinctly segmented .... insects
4. Particles spherical, stain reddish with Sudan III .... fat globules
– Particles variable in shape, do not stain reddish with Sudan III 5
5. Particles rod-shaped or, rarely, spherical, usually motile, without polar filaments. Multiplication occurs extracellularly in body cavity .... bacteria
– Particles rod-shaped to spherical, usually non-motile, may have polar filaments. Multiplication occurs intracellularly in various tissues 6
6. Small to minute spherical to rod-shaped particles (greatest diameter usually less than 10 μm). Polar body absent. Bodies dissolve in a weak solution of sodium hydroxide (NaOH) .... inclusion viruses (non-occluded viruses are too small to be seen with the light microscope (Figs 664–666))
– Small to minute rod-shaped particles containing a polar body. Bodies do not dissolve in a weak solution of NaOH (Figs 661–662) .... protozoa

Table 23.1. The main steps in identifying stem-borer pathogens.

Pathogen	Location of pathogen	Characteristic stages	Method of examination	Method of culture	Diagnostic features	Storage methods
Nematodes	Haemocoele	Juvenile and adult stages	Bright field	Artificial media for rhabditids	Morphological characters	Refrigeration, liquid nitrogen
Protozoa	Haemolymph, tissues	Spores, cysts	Bright field, phase contrast	None needed	Spore and cyst size and shape	Refrigeration, host transfer
Fungi	Haemolymph, cuticle	Spores, hyphae	Bright field, phase contrast	Artificial media	Spores, hyphae	Transfer lyophilization
Bacteria	Haemolymph	Spores and vegetative cells	Bright field, phase contrast	Artificial media	Size, shape, biochemical and cultural characteristics	Cool, dry lyophilization transfer
Viruses	Cells and tissues	Inclusion bodies	Bright field, phase contrast	None practical	Size, shape	Freeze lyophilization

TECHNIQUES FOR STUDY

The main steps in identifying pathogens are the determination of the following:

1. Location in the host.
2. Characteristic stages.
3. Methods of examination.
4. Culture methods.
5. Diagnostic features.

These items are summarized for the above-mentioned five pathogen groups in Table 23.1.

NEMATODA

Introduction. To date, at least 16 distinct species of nematodes, belonging to 12 genera in five families have been recorded from cereal stem borers. The genera *Hexamermis* (Mermethidae) and *Steinernema* (Steinernematidae) are most commonly recorded. The families and genera of nematodes attacking cereal stem borers in Africa and elsewhere can be identified using the following key.

1. Adult and juvenile nematodes developing inside dead or dying borers. Infective stage juveniles (under 2 mm long) leave the insect cadaver 2
– Adult and/or juvenile nematodes developing inside living borers. Mature juveniles (over 2 mm long) leave the living host, which dies soon afterwards ... Mermithidae
2. Nematodes associated with a single type of bacterium (large, Gram-negative, rod-shaped) in the host cadaver. Host cadaver usually cream- or reddish-coloured 3
– Nematode species associated with a variety of different bacterial species (long and short rods, as well as cocci). Host cadaver usually turns brown or black 4
3. Female nematodes with a pointed tail (Figs 638, 643), males with a bursa (Figs 639, 644). Host cadaver usually a reddish colour.. .... *Heterorhabditis* Poinar (Heterorhabditidae)
– Female nematodes with a rounded tail (Fig. 640), males lacking a bursa (Fig. 641). Host cadaver usually a cream colour .... *Steinernema* Travassos (= *Neoaplectana* Steiner) (Steinernematidae)
4. Adult female with a single ovary that is reflexed posteriorly past the vulvar opening (Fig. 642). Males lacking a bursa .... *Panagrolaimus* Fuchs (Panagrolaimidae)
– Adult female with paired opposite ovaries that may or may not be reflexed posteriorly past the vulvar opening (Fig. 643). Males possessing a bursa (Fig. 644) .... *Rhabditis* Oujardin (Rhabditidae)

The following key can be used to identify genera of Mermithidae that have been recorded from stem borers.

1. Paired lip papillae present on adults and postparasitic juveniles (Fig. 645). Post-parasitic juveniles lack a tail appendage or scar .... *Mermis* Dujardin
– Paired lip papillae absent. Postparasitic juveniles with a tail appendage or scar (Figs 647, 648) 2
2. Vagina with curve in the middle (Fig. 650). Amphidial openings pore-like and closely associated with the lateral cephalic papillae (Fig. 649) .... *Ovomermis* Rubtsov
– Vagina with curve at the end (J-shaped). Amphidial openings pore-like and located posterior to the lateral cephalic papillae (Fig. 651) 3
3. Spicules twisted around each other, usually long (more than twice tail diameter (Fig. 653) .... *Amphimermis* Kaburaki & Imamura
– Spicules straight and short (less than twice tail diameter (Fig. 654)) 4
4. Postparasitic juveniles with a distinct tail appendage (Fig. 647) .... *Hexamermis* Steiner
– Postparasitic juveniles with a blunt tail scar (Fig. 648) .... *Agamermis* Cobb, Steiner and Christie

Table 23.2. Records of Nematoda attacking cereal stem borers in Africa.

Family	Genus	Species	Host(s)	Distribution	Reference(s)
Heterorhabditidae	*Heterorhabditis*	*bacteriophora*	*Eldana saccharina*	South Africa	Anon., 1986; Spaull, 1991
	Heterorhabditis	sp.	*E. saccharina*	South Africa	Anon., 1988
Mermithidae	*Hexamermis*	sp.	*Chilo partellus*, *Sesamia calamistis*, *E. saccharina*, *Busseola fusca*	Kenya, Tanzania, Uganda	Milner, 1965; Mohyuddin and Greathead, 1970; A. Polaszek, unpublished
	Hexamermis	sp	*E. saccharina*	Côte d'Ivoire	A. Polaszek, unpublished
	Mermis	sp.	*Coniesta ignefusalis*	Senegal	Bhatnagar, 1984
	Mermis	sp.	*Chilo sacchariphagus*, *Sesamia* sp.	Mauritius	Moutia and Courtois, 1952
	Mermis	sp.	*S. calamistis*	?	Remillet, 1970 (in Poinar, 1975)
	Genus not identified		*B. fusca*, *E. saccharina*	Côte d'Ivoire	Dabiré-Binso, 1980
	Genus not identified		*Sesamia* sp., *Chilo aleniellus*, *Chilo zacconius*, *Maliarpha separatella*	Sierra Leone	Jordan, 1966; Poinar, 1975
	Genus not identified		*S. calamistis*	Côte d'Ivoire	Odjo, 1984
	Genus not identified		*S. calamistis*	Benin	Shanower *et al.*, 1991
	Genus not identified		*C. partellus*, *S. calamistis*, *E. saccharina*	Kenya	Odindo *et al.*, 1989
Panagrolaimidae	*Panagrolaimus*	sp.	*C. partellus*	Kenya	Otieno, 1986
Rhabditidae	Genus not identified	spp.	*C. partellus*, *S. calamistis*, *E. saccharina*	Kenya	Odindo *et al.*, 1989
Steinernematidae	*Steinernema* (= *Neoaplectana*)	*intermedia*	*E. saccharina*	South Africa	Anon., 1986
	Steinernema	*glaseri*	*E. saccharina*	South Africa	Anon., 1986; Spaull, 1991
		sp.	*E. saccharina*	South Africa	Anon., 1988

In Africa alone, all five families of nematodes have been recorded, detailed records of which are summarized in Table 23.2.

Biology. The three most common families of nematodes attacking stem borers are the Mermithidae, the Steinernematidae and the Heterorhabditidae. Members of the former family enter the larval stage of the stem borer by means of an infective-stage juvenile that penetrates directly through the cuticle. Inside the body cavity of the host, the parasite moults into the parasitic stage, which in turn undergoes another moult as it rapidly outgrows the length of the stem-borer larva. When mature, the parasite leaves the host and becomes a postparasite in the soil. The host usually dies at the time of emergence or shortly afterwards. The parasite now completes its final moults and becomes an adult. The adult no longer feeds, relying on the reserves stored during its parasitic life. These adults mate and the females oviposit in the soil. The time of hatching depends upon the temperature and soil moisture. The newly emerged infective stage immediately begins to search for a host larva. Under certain conditions, the parasite may be carried into the pupal and even the adult stage of the host insect. This behaviour may result in increasing the distribution of the parasite.

The steinernematids and the heterorhabditids belong to the Rhabditoidea and have a relatively similar biology. Infection is initiated by a third-stage infective juvenile, which normally enters the host larva through natural openings, such as the mouth, anus and spiracles. Once inside the body cavity, the infective stage develops into a parasitic juvenile. At this time, bacterial cells belonging to a species of symbiont bacteria carried in the intestinal tract of the infective juvenile are liberated into the host's blood. These cells (members of the genus *Xenorhabdus* in *Steinernema* and the genus *Photorhabdus* in *Heterorhabditis*) assist the development of the nematodes directly, by supplying a food source, and indirectly, by breaking down the tissues of the insect. The host dies within 48 h after the nematodes enter, and the nematodes undergo one or two complete generations in the host cadaver before emerging as infective-stage juveniles. Up to several thousand infective juveniles can develop from a single host. Since these nematodes can be grown on artificial media, they are being produced in various parts of the world as commercial biological insecticides.

Economic importance. Most studies of nematodes attacking stem borers in Africa have been simply isolated host and distribution records, the main exception being the study by Spaull (1991) using *Heterorhabditis bacteriophora*. Preliminary studies on this nematode suggested that it might have some potential for biological control of *Eldana saccharina* in South Africa (Spaull, 1990). However, more detailed studies, including the use of polymer additives, eventually showed that, at least in this particular case, biological control using nematodes would be uneconomical (Spaull, 1992).

Outside Africa, the use of nematodes for biological control of stem borers was investigated in Korea, where tests on *Chilo suppressalis* (Walker), a rice stem borer, gave very favourable results (Choo *et al.*, 1991), while results of a similar study on *Scirpophaga incertulas* (Walker) in India were inconclusive (Rao *et al.*, 1971). Torii's (1975) feasibility study using the so-called DD-136 strain of *Steinernema* (= *Neoaplectana*) *carpocapsae* concluded that this species was potentially an extremely effective control agent for rice stem borers. It seems probable that the aqueous environment associated with rice may be more favourable for the beneficial activity of nematodes than the drier conditions associated with other cereals.

FUNGI

Introduction. In Africa, the first reliable studies of entomopathogenic fungi reared from field-collected cereal stem borers recorded the following genera: *Aspergillus*, *Beauveria* and *Metarrhizum* (Harris, 1962). Subsequent studies have not revealed any additional genera attacking stem borers, although *Cephalosporium*, *Isaria* and *Paecilomyces* spp. have been recorded from Asian and European stem borers (Tateishi and Murata, 1955; Yen and Ooi, 1967; Riba *et al.*, 1983).

These genera can be recognized as follows.

1. Conidia borne in slime or mucus balls at tip of conidia (Fig. 655) .... *Cephalosporium* Corda
– Conidia not borne in slime or mucus balls 2
2. Conidia borne in synnematia that emerge from the insect cadaver (Fig. 656) .... *Isaria* Pers.
– Conidia not borne in synnematia 3
3. Conidia borne singly, not in chains (Fig. 657) .... *Beauveria* Vuill.
– Conidia borne in chains (catenulate) 4
4. Phialides borne in groups on the enlarged apex of single conidiophores (Fig. 658) .... *Aspergillus* Mich.
– Not as above 5
5. Conidiophores in compact columns or groups, conidia elongate or rod-shaped (Fig. 659), colony often a shade of green .... *Metarrhizium* Sorok.
– Conidiophores not in compact groups, conidia usually round to elliptical (Fig. 660), colonies usually white, rose or red .... *Paecilomyces* Bain.

Published records of fungi attacking cereal stem borer larvae are summarized in Table 23.3.

Biology. All of the pathogenic fungi encountered in stem borers have a similar life cycle. The conidium or spore penetrates the cuticle of the insect larva and the invading hyphae grow throughout the haemocoele, often forming separate hyphal bodies, which circulate in the haemolymph. After filling the body cavity of the host with mycelium, the fungi produce emergence hyphae, which break out through the cuticle and produce spores on the surface of the host cadaver. Spores spread from infected to healthy hosts by wind, rain and physical contact between diseased and non-diseased insects.

Economic importance. Recent studies in Kenya, using both indigenous and exotic fungi to control the indigenous borer *Busseda fusca* and the introduced *Chilo partellus* (Maniania, 1991), have demonstrated the susceptibility of both stem-borer species in laboratory trials. Preliminary field tests of *Beauveria bassiana* against *C. partellus* (Maniania, 1993a, b) have been shown to reduce larval populations, foliar damage and tunnel length, as well as significantly increasing yield.

Table 23.3. Records of fungi attacking cereal stem borers.

Genus and species	Stem-borer host(s)	Distribution	Reference(s)
Aspergillus flavus	*Busseola fusca*	Nigeria	Harris, 1962
Aspergillus sydowii	*B. fusca*	Nigeria	Harris, 1962
Beauveria bassiana	*Chilo partellus*	Kenya	Maniania, 1991
	Chilo suppressalis	?France	
	Eldana saccharina	South Africa	Conlong, 1990
	Chilo auricilius	India	Israel and Padmanabhan, 1978
	Ostrinia nubilalis	France	Riba *et al.*, 1983
	Scirpophaga incertulas	India	Israel and Padmanabhan, 1978
	Maliarpha separatella	Senegal	Etienne, 1987
Cephalosporium sp.	*C. suppressalis*	Taiwan	Yen and Ooi, 1967
Isaria farinosa	*C. suppressalis*	Japan	Tateishi and Murata, 1955
Metarrhizum anisopliae (Metschnikoff)	*Coniesta ignefusalis*	Nigeria	Harris, 1962
Metarrhizum sp.	*B. fusca* *C. partellus*	Kenya	Nyambo and Mwangi, 1992
Paecilomyces fumoroseus (Wize)	*O. nubilalis*	France	Riba *et al.*, 1983

Clearly, the economics of controlling stem borers with fungal insecticides needs to be carefully evaluated following these promising results. Similarly promising results were obtained in Europe on *Ostrinia nubilalis* in the laboratory (Riba *et al.*, 1983), but field-trial results have not been located. Lynch and Lewis (1978) found that several species of fungi found in eggs of *O. nubilalis* were associated with previous injury by predators.

PROTOZOA

Introduction. Only the genus *Nosema* (Microsporidia) has been recorded with any frequency from stem borers, both in Africa and elsewhere. Five described and several undescribed or unidentified *Nosema* species have been recorded from African cereal stem borers, some from laboratory cultures outside Africa. The taxonomy of these *Nosema* species is currently far from clear and merits further study. Outside Africa, *Nosema* species have been recorded from the stem borers *O. nubilalis* and *Chilo infuscatellus* in North America (Hill and Gary, 1979) and Asia (Oshima, 1935; Wen and Sun, 1988, 1989). All published records of *Nosema* spp. attacking African cereal stem borers are summarized in Table 23.4.

Biology. Infection of stem borers by *Nosema* spp. occurs when the larvae ingest the spores of the protozoan. Infection is initiated in the gut epithelium and spreads to the fat body, malpighian tubules, hypodermis and reproductive organs. The initial intracellular sporont eventually forms a single spore (in *Nosema*) and the spores are borne free, not within a cyst. The infection can prove fatal to the larval

Table 23.4. Records of *Nosema* spp. attacking African cereal stem borers.

Species	Stem-borer host(s)	Distribution	Reference(s)
Nosema bordati Goudegnon	*Chilo partellus*	Laboratory (France)	Goudegnon, 1985
Nosema manieri Toguebaye & Bouix	*Chilo zacconius*	Laboratory (France)	Bordat *et al.*, 1984; Toguebaye and Bouix, 1983
Nosema marucae Odindo & Jura	*C. partellus*	Laboratory (Kenya)	Odindo *et al.*, 1993
Nosema partelli Walters & Kfir	*C. partellus*	South Africa	Walters & Kfir, 1993
Nosema sp. A	*C. partellus*	Laboratory (France)	Bordat *et al.*, 1984
Nosema sp. C	*Sesamia calamistis*	Laboratory (France)	Bordat *et al.*, 1984
Nosema sp.	*Chilo sacchariphagus*	Réunion	Fournier and Etienne, 1981
Nosema sp.	*C. partellus* *Chilo* spp.	Kenya	Odindo, 1990

stage, or it can be carried on through the pupal stage into the adult insect. In the latter situation, the reproductive tract of the female can become infected, and the infection can pass into the developing eggs, resulting in transovarial transmission (Ogwang and Odindo, 1992). Adults with a heavy infection will be less virile and have a shorter lifespan than non-infected insects. The production of thousands of spores in each infected insect ensures that some will be blown or otherwise carried to new infection sites for further infections.

Economic importance. Although the several species recorded in Table 23.4 successfully attack and kill stem borers, in many cases the presence of the particular *Nosema* species was regarded as detrimental, at least initially. It appears that the most severe infections known have tended to occur in laboratory cultures of moth larvae. For example, cultures of moth larvae used for biological control of a cactus in South Africa were adversely affected by *Nosema* spp. (Walters and Kfir, 1993). Most of the records in Table 23.4 are from laboratory cultures of various borer species, which were established in order to study the biology of the particular stem-borer species, with a view to controlling it. Instead, cultures were badly affected by *Nosema*. However, recent trials at the International Centre of Insect Physiology and Ecology (ICIPE), Kenya, have shown that at least one species, *Nosema marucae*, has great potential as a both cheap and efficient control agent for the introduced stem borer *C. partellus* (Odindo *et al.*, 1993).

BACTERIA

Introduction. Among the bacteria that regularly infect and kill insects, it is only *Bacillus thuringiensis* which is of any importance in connection with stem borers, although a few records of other bacteria associated with dead stem borers in Kenya are listed by Odindo *et al.* (1989).

Biology. *Bacillus thuringiensis* invades the host through its alimentary tract. The infective agent is the spore, which contains one or more parasporal crystals or parasporal bodies. These crystals paralyse and destroy the gut cells of Lepidoptera larvae, allowing the vegetative cells to enter the haemocoele and multiply. The first signs of infection are a cessation of feeding and sluggish behaviour. This is followed by a general body paralysis and death. Often the infected larvae will assume a characteristic position on the leaf or stem of the plant. Infected larvae can be diagnosed by finding the characteristic spores and parasporal bodies in their haemolymph. Usually the cadaver turns brown or black in colour.

Economic importance. *Bacillus thuringiensis* has been used widely as a bacterial pesticide for the attempted control of many different pest-insect species, in particular lepidopterous defoliators of coniferous woodlands in the northern hemisphere. Several attempts at using *B. thuringiensis* against stem borers have also been reported. Particular attempts to control borers in Africa have been reported by Berger (1981) for *C. partellus* in Mozambique and by Brownbridge (1990, 1991) for the stem-borer complex present in Kenya. In Mozambique, various combinations of *B. thuringiensis* with chemical pesticides were used, but results were inconclusive while suggesting that this type of treatment was probably too expensive. In Kenya, many locally collected strains of the bacterium were collected to select those with the most toxic effects on stem borers. Initial trials showed very high levels of borer control, with significantly increased yields (Brownbridge, 1991).

More recently, attempts have been made at using genes from *B. thuringiensis* for genetically engineering plants to protect them against attack by stem borers. The delta-endotoxin gene which produces the toxins lethal to borers and other insects has been successfully transferred into rice plants in both China (Yang *et al.*, 1991) and Japan (Fujimoto *et al.*, 1993). In the latter case, such genetically engineered rice plants were shown to be more resistant to borer attack. Transfer of *B. thuringiensis* deoxyribonucleic acid (DNA) into sugar-cane-inhabiting epiphytic bacteria (*Erwinia* and pseudomonads) has also been attempted in order to control *E. saccharina* in South Africa (Allison *et al.*, 1992). Thus *B. thuringiensis* is currently the subject of both direct and indirect attempts at controlling cereal stem borers.

VIRUSES

Introduction. Very little is yet known about the interactions between the different kinds of viruses and their stem-borer hosts. Few records exist of entomopathogenic viruses recorded from African stem borers, and even fewer detailed studies have been carried out for the African species. Records of viruses from cereal stem borers, including those found outside Africa, include the following families as stem-borer pathogens: Baculoviridae (nuclear polyhedrosis and granulosis viruses), Iridoviridae and Reoviridae (cytoplasmic polyhedrosis viruses). So-called polydnaviruses are found in some parasitoid hymenoptera and, consequently, sometimes in their stem-borer hosts (e.g. in *Cotesia sesamiae* and *Sesamia calamistis* (Maazouz, 1991)). They are actually not true viruses at all, but a part of the parasitoid genome with a particular function, often involving overcoming the host's defences.

Identification. The following key will separate the viruses known to infect stem borers.

1. Insects showing a blue, green or brown iridescence as a result of the internal tissues being filled with virus particles, inclusion bodies absent (Fig. 663) .... iridoviruses
– Insects not demonstrating iridescence, inclusion bodies present 2
2. Inclusion bodies ovoid, ellipsoidal, small (0.2–0.5 μm in length (Fig. 664)). Host often turns a cream colour .... granulosis viruses
– Inclusion bodies irregular, polyhedral, usually larger (0.2–20 μm in diameter). Hosts turn white, yellow, brown or black in colour 3
3. Inclusion bodies usually from 0.2 to 2.5 μm in diameter, normally formed in the midgut epithelial cells, virions spherical (Fig. 665), infected larvae usually shrivel and turn white or yellow .... cytoplasmic polyhedrosis viruses
– Inclusion bodies usually from 0.5 to 15.0 μm in diameter, formed in various tissues (Fig. 666), virions rod-shaped, infected larvae usually turn brown or black and become distended .... nuclear polyhedrosis viruses

Biology. Except for instances of transovarial transmission, all of the viruses mentioned above enter the insect larva through the mouth and alimentary tract. The virions, either separately or when released from their inclusion bodies, infect the intestinal epithelium and then often pass into the body cavity and infect additional tissues. Most viruses produce overt infections, which can be determined by discovering polyhedral bodies in the haemolymph or noticing an iridescent colour. In some cases, occult viruses will produce an unapparent infection that is difficult to diagnose. Stressing the larvae can often bring out such latent infections. Each virus appears to have its own characteristics regarding the location of replication and the effect on the insect. Once recognized, these characters can be used to diagnose the type of virus infection encountered.

Economic importance. Although several viruses of various kinds have been isolated from cereal stem borers, their use as control agents of these pests has been little investigated in Africa. A survey in Kenya identified granulosis viruses and polyhedral inclusion bodies from *B. fusca* and *C. partellus* (Odindo *et al.*, 1989), and, in Réunion, a cytoplasmic polyhedrosis virus was isolated from *Chilo sacchariphagus* (Morel *et al.*, 1981). A recent study has been carried out on the interactions between *S. calamistis*, its parasitoid *C. sesamiae* and a nuclear and two cytoplasmic polyhedrosis viruses affecting the stem borer (Maazouz, 1991). A similar study in Egypt on the borer *Chilo agamemnon* (Abbas, 1987) showed that infection by a nuclear polyhedrosis virus had a detrimental effect on development of a larval parasitoid (*Habrobracon brevicornis* (Wesmael)) but did not affect development of the egg parasitoid *Trichogramma evanescens* (Westwood). The economic implication of the latter study was that use of viruses as control agents of the borer may adversely affect larval parasitoid activity. Current knowledge on the potential of viruses to control cereal stem borers in Africa is summarized by Odindo (1990). In India, granulosis viruses have been used for the attempted control of *C. infuscatellus* and *C. sacchariphagus*, with some success in the case of *C. infuscatellus* (summarized by David and Easwaramoorthy, 1990). Also in India, laboratory studies found a granulosis virus to be pathogenic to the rice borers *Sesamia inferens*, *S. incertulas* and *Chilo auricilius* (Nayak and Godse, 1986). Among the Iridoviridae, *Chilo* iridescent virus (CIV, or iridovirus type 6) has been described from *C. suppressalis* in Japan (Fukaya and Nasu, 1966), and several aspects of the virus are currently being studied.

SUMMARY

This brief account of the nematodes and microorganisms known to attack African stem borers reveals primarily the lack of in-depth studies on their relationships with these microorganisms and the biological-control potential of the latter. Current investigations into the *Nosema*, fungal and bacterial pathogens of stem borers in Kenya, being carried out at ICIPE, are to be strongly welcomed. In India and the Far East, more detailed studies have been undertaken, particularly in association with rice borers. The results of the latter studies may also have consequences for stem-borer biological control in Africa. In particular, the possibility of integrating local strains of *B. thuringiensis* (bacteria), *B. bassiana* (fungi) and *N. marucae* (protozoa) into developing integrated pest management (IPM) programmes needs investigation.

24 Predators

M.J. Bonhof
International Centre of Insect Physiology and Ecology, Nairobi, Kenya

INTRODUCTION

The importance of predaceous arthropods as regulators of insect populations has long been underestimated. The value of ladybird beetles (Coleoptera: Coccinellidae), green lacewings (Hemiptera: Chrysopidae), nabids (Hemiptera: Nabidae) and other predators as enemies of several major pest species has been recognized since the mid-19th century. However, it is only recently that the critical role of predators in preventing outbreaks of many pests has been recognized. For example, populations of many noctuid pests in the USA and Canada are controlled by predators, accounting for 98.9% mortality in some cases (Whitcomb, 1981). The interaction between arthropod predators and stem-borer populations in Africa is still poorly understood. The importance of predators has been postulated by many authors (Mohyuddin and Greathead, 1970; Leslie, 1982; Kfir, 1988; Oloo, 1989; Greathead, 1990), but only a few studies have been conducted to quantify their impact. These studies include predator exclusion experiments, monitoring the fate of eggs, laboratory studies using Petri dishes and a serological technique. Most studies focus on predation of a specific life stage, but there is little insight into the overall influence of predators on stem-borer populations. Most stage-specific studies have been conducted on eggs and young larvae, as these life stages are usually exposed and therefore more vulnerable to predation than are the subsequent larval and pupal stages, which are well protected by the plant tissue and by the pupal case. Information on predators of older larvae, pupae and adults is merely anecdotal.

PREDATION ON EGGS

Eggs are among the most vulnerable life stages of stem borers, because of their relatively small size, their immobility and their position. Most Pyralidae and Crambidae (e.g. *Chilo* spp.) deposit their eggs on leaves and stems, where they are highly exposed to natural enemies (Plate 12:1). Two important exceptions in Africa are *Coniesta ignefusalis* Hampson and *Eldana saccharina* Walker, which lay their eggs in the crevices of leaf folds and in leaf sheaths. Noctuid eggs (*Busseola* and *Sesamia* spp.) are deposited in more concealed places, such as leaf sheaths (Smith,

J.W. *et al.*, 1993b), and seem less vulnerable to predation. Predators such as earwigs and spiders, often hiding in leaf sheaths, may consume these noctuid eggs or neonate larvae.

Two approaches to the study of egg predation have been practised. In the first one, predation is studied on artificially placed egg batches. The egg batches are usually obtained by letting moths oviposit on wax paper, which is then pinned to the plant in the field. The batches are collected just before emergence and the number of eggs predated upon or removed is counted. Using this method, Oloo (1989) found that, in western Kenya, 93% *Chilo partellus* eggs on maize, 97% on sorghum and 85% in maize/sorghum intercrop disappeared. At the Kenyan coast, predation on *C. partellus* eggs in maize was 30–77% (M.J. Bonhof, unpublished). Experiments on *E. saccharina* eggs in sugar cane in South Africa also showed a considerable impact of predators. In cages where arthropods were excluded, approximately 10% of the artificially infested eggs were eaten by unidentified predators (Leslie, 1988). In the control cages (no exclusion treatment), between 55% and 78% of the eggs were consumed. The importance of predators in this area was confirmed in another experiment, in which predators were excluded by applying insecticide. In plots where predators were excluded, the percentage of eggs recovered was 77–88%, while, in untreated plots, it was only 24–62% (Leslie, 1982). Ants belonging to the genera *Paratrechina* and *Solenopsis*, an unidentified mite and an unidentified thrips were observed feeding on the eggs. Leslie suggested that ants may have a considerable impact on stem-borer populations in this region because of their abundance in the field. However, abundance of a particular predator does not always imply that it feeds on a particular target pest (Whitcomb, 1981).

The above-mentioned studies were all conducted with artificially placed eggs. Predation on artificially placed eggs may be higher than on those naturally oviposited, as the latter are usually more concealed (Girling, 1978; Leslie, 1982; Oloo, 1989). Hence, the results of the above-mentioned studies may not represent the actual predation rate in a field. In Uganda, Girling (1978) confirmed this by pinning egg batches of *E. saccharina* on maize plants and then introducing mated female moths in cages over these plants to oviposit. He found that predation on eggs in natural positions was very high (92% of the eggs), but that some egg batches remained untouched, presumably because of their concealed position. Predation on artificially placed eggs was 100%. Egg disappearance on plants where predators were excluded was 6%, indicating that disappearance was mainly caused by predators. A study in South Africa also showed a large difference in predation on artificially placed and naturally oviposited *E. saccharina* eggs in sugar cane (Leslie, 1988). Predation on artificially placed eggs was 55–78% and on naturally laid eggs only 8–12%. The amount of predation depended on the degree of exposure of an egg batch. Egg batches placed in concealed positions were less severely attacked than exposed batches (37% and 54%, respectively). Egg batches near the base of the stalks were more severely attacked than batches near the top (Leslie, 1988).

Besides studying predation on artificially placed egg batches, the fate of naturally occurring eggs in the field can be determined. This second method of studying egg predation may give a better representation of the actual egg predation in the field. However, merely collecting naturally laid egg batches and determining their fate will underestimate predation in the field, as completely removed egg batches will not be detected and those removed will not be subject to further predation. The fate of eggs in the field was followed in the coastal region of Kenya. Seven per cent of *Chilo* eggs showed signs of predation, while another 9–21% had disappeared (M.J. Bonhof, unpublished data). Mathez (1972) collected and reared naturally laid egg batches in the same region and found no predation at all on *Chilo* spp. or *Sesamia calamistis* eggs, while M.J. Bonhof (unpublished) frequently detected predation on eggs of both species in a recent study. Mohyuddin and Greathead (1970) reported a disappearance of 90% of naturally laid eggs in Uganda. The ant species *Tetramorium guineense* F., *Pheidole megacephala* F., *Cardiocondyla badonei* Arnold and *Cardiocondyla emeryi* Forel were observed to prey on *C. partellus* and *Busseola fusca* eggs. Dwumfour *et al.* (1991) observed the earwigs *Diaperasticus erythrocephala* Olivier and *Forficula auricularia* L., the coccinellids *Cheilomenes sulphurea* Olivier and *Cheilomenes propinqua* Mulsant and *Cheilomenes* sp., the ants *Pheidole* sp. and *Camponotus* sp. and many spiders feeding on *C. partellus* eggs in maize fields in western Kenya. The predators did not distinguish between non-parasitized eggs and eggs that were parasitized by *Trichogramma bournieri*, and may therefore have had an impact on egg parasitoids.

In South Africa, egg predators of *E. saccharina* have been identified, using a serological technique. The relative importance of the predators in sugar-cane fields could be estimated by combining the results of the serological tests with estimates of predator abundance. Three Acarina species, two species of cockroaches, adults of unidentified beetles, Collembola, four ant species (*Pheidole* sp., *Solenopsis* sp., *Aenictus* sp. and *Paratrechina* sp.), species of Isopoda and pseudoscorpions were identified as egg predators. Some are normally not considered to be predacious (such as Blattaria species and Isopoda species) and it is likely that these arthropods only consumed the stem-borer eggs by exception (Leslie, 1988). Combining results of the serological test and relative abundance showed that mites and ants (*Pheidole* sp. and *Solenopsis* sp.) were the most efficient egg predators of *E. saccharina* on sugar cane in South Africa. Predation was greatest where host numbers were high; thus predation on *E. saccharina* eggs seems to be density-dependent. The overall level of predation was low in this study, as only 0.7–1.3% of the suspected predators had consumed *E. saccharina* eggs. An insecticide exclusion trial conducted in the same region and period showed that 60% of the artificially placed eggs were destroyed in untreated plots, compared to 19% of the eggs in plots treated with insecticide (Leslie, 1982). The discrepancy between this high predation rate and the low percentage of positive insects in the serological test may have been due to the artificial position of the eggs in the insecticide trial, which might have enhanced predation.

Laboratory studies have identified earwigs (*D. erythrocephala* and *F. auricularia*) (Dwumfour, 1990; M.J. Bonhof, unpublished), a green lacewing (*Chrysopa* sp.) (Dwumfour, 1990), *Tibellus* sp. (Araneida: Philodromidae) and other spiders, ants (*Crematogaster* spp. and *Technomyrmex* sp.), several Coleoptera, a hemipteran and a cockroach as being predators of *C. partellus* eggs (Table 24.1) (M.J. Bonhof, unpublished). The number of consumed eggs was highly variable, but could be as high as 93 eggs per day for the earwig *D. erythrocephala* (M.J. Bonhof, unpublished).

A list of egg predators recorded from Africa is given in Table 24.1.

Table 24.1. Predators attacking eggs of African stem borers.

Order, family, genus and species	Host species	Distribution	Reference
Acarina			
Prostigmata species	Es	South Africa	Leslie, 1988
Type 1	Es	South Africa	Leslie, 1982
Type 2	Es	South Africa	Leslie, 1982
Type 3	Es	South Africa	Leslie, 1982
Unidentified species	Es	South Africa	Carnegie, 1991
Araneida			
Unidentified species	Cp	Kenya	Seshu Reddy, 1983b; Dwumfour *et al.*, 1991; M.J. Bonhof, unpublished.
Philodromidae			
Tibellus sp.	Cp	Kenya	M.J. Bonhof, unpublished
Blattaria			
Type 1	Es	South Africa	Leslie, 1982
Type 3	Es	South Africa	Leslie, 1982
Unidentified species	Cp	Kenya	M.J. Bonhof, unpublished
Pseudoscorpionida			
Unidentified species	Es	South Africa	Carnegie, 1991; Leslie, 1988
Coleoptera			
Unidentified species	Cp	Kenya	M.J. Bonhof, unpublished
Unidentified species	Es	South Africa	Leslie, 1988
Coccinellidae			
Cheilomenes sp.	?	Kenya	Seshu Reddy, 1983b; Dwumfour *et al.*, 1991
	Cp	Kenya	
Cheilomenes sulphurea	Cp	Kenya	Dwumfour *et al.*, 1991
Cheilomenes propinqua	Cp	Kenya	Dwumfour *et al.*, 1991
Staphylinidae			
Paederus sp.	Cp	Kenya	M.J. Bonhof, unpublished
Collembola			
Unidentified species	Es	South Africa	Leslie, 1982, 1988; Carnegie, 1991
Dermaptera			
Forficulidae			
Diaperasticus erythrocephala	Cp	Kenya	Dwumfour *et al.*, 1991; Oloo, 1989; M.J. Bonhof, unpublished
	?	Ethiopia	Gebre-Amlak, 1985
	?	Ghana	Botchey and Kwapong, 1991
Forficula auricularia	Cp	Kenya	Dwumfour *et al.*, 1991
Hemiptera			
Chrysopidae			
Chrysopa sp.	?	Kenya	Dwumfour, 1990
Unidentified species	Cp	Kenya	M.J. Bonhof, unpublished

Table 24.1. Predators attacking eggs of African stem borers (*continued*).

Order, family, genus and species	Host species	Distribution	Reference
Hymenoptera			
Formicidae			
Aenictus sp.	Es	South Africa	Leslie, 1982; Carnegie, 1991
Camponotus rufoglaucus (Jerdon)	?	Kenya	Seshu Reddy, 1983b
Camponotus sericeus (F.)	Es	Uganda	Girling, 1978
Camponotus sp.	?	Kenya	Dwumfour *et al.*, 1991
Cardiocondyla schuckardi Forel (as *Cardiocondyla badonei* Arnold)	Bf	Uganga	Mohyuddin and Greathead, 1970
Cardiocondyla emeryi Forel	Bf	Uganda	Mohyuddin and Greathead, 1970
Crematogaster sp. 1	Cp	Kenya	M.J. Bonhof, unpublished
Crematogaster sp. 2	Cp	Kenya	M.J. Bonhof, unpublished
Dorylus sp.	?	Kenya	Dwumfour, 1990
Pheidole megacephala F.	Cp/Bf	South Africa	Kfir, 1988
Pheidole guineense F. (as *Tetramorium guineense*)	Bf	Uganda	Mohyuddin and Greathead, 1970
Pheidole sp.	Cp	Uganda	Mohyuddin and Greathead, 1970
Pheidole sp.	Es	Kenya	Oloo, 1989; Dwumfour *et al.*, 1991
Pheidole sp.	Es	South Africa	Leslie, 1982, 1988; Carnegie, 1991
Pheidole sp.	Cs	Mauritius	Williams and Mamet, 1954
Paratrechina sp.	Es	South Africa	Leslie, 1982; Carnegie, 1991
Solenopsis sp.	Es	South Africa	Leslie, 1982; Carnegie, 1991
Tetramorium bicarinatum Nylander	Es	Uganda	Girling, 1978
Technomyrmex sp.	Cp	Kenya	M.J. Bonhof, unpublished
Isopoda			
Unidentified species	Es	South Africa	Carnegie, 1991
Unidentified species	Es	South Africa	Leslie, 1982, 1988
Orthoptera			
Tettigonidae			
Clonia vittata Thunberg	Bf	South Africa	Akerman, 1932

Cp, *Chilo partellus*; Bf, *Busseola fusca*; Es, *Eldana saccharina*; Cs, *Chilo sacchariphagus*; ?, unknown.

PREDATION ON YOUNG LARVAE

Neonate larvae are vulnerable to predation, especially while migrating from the egg batch to the leaf whorl. The fate of young *E. saccharina* larvae in Uganda was studied by Girling (1978). He found that, 1 week after infestation, 4% of the larvae were found in the original host plant, 3% in the surrounding plants and 93% had disappeared. This disappearance may be caused by predators, but cannibalism, rainfall and desiccation by the sun may also have played a role. At the Kenyan coast, 6–17% of first-instar larvae disappeared while migrating to the leaf whorl (P.

Lammers, unpublished; Plate 12:2). In Uganda, four ant species (*T. guineense*, *P. megacephala*, *C. badonei* and *C. emeryi*) destroyed over 90% of first-instar *C. partellus* and *B. fusca* larvae (Mohyuddin and Greathead, 1970). Dick (1945), Waiyaki (1974) and Betbeder-Matibet (1989) also observed ants taking small *E. saccharina* larvae.

In the laboratory, green lacewings (*Chrysopa* sp.) and earwigs (*D. erythrocephala* and *F. auricularia*) readily preyed on small *C. partellus* larvae (Dwumfour *et al.*, 1991). M.J. Bonhof (unpublished) observed spiders (Araneida), coccinellid adults and larvae, unidentified ants (Formicidae), crickets (Orthoptera) chrysopid larvae and adults and nymphs of the earwig *D. erythrocephala* preying on first-instar *C. partellus* larvae. Predation by the widespread and abundant earwig *D. erythrocephala* was quite variable. Individuals starved for 24 h usually preyed on all the offered larvae (max. 50) or did not feed at all. The mean predation rate was eight first-instar larvae per day. Other voracious predators included coccinellids chrysopid larvae and spiders.

Several larval predators were identified using a serological technique, crossover electrophoresis. Mites (*Prostigmata* sp.), 11 genera of spiders, cockroaches, beetles, seven ant genera, bugs, pseudoscorpions and gryllids were identified as predators of *E. saccharina* larvae (Table 24.2) (Leslie and Boreham, 1981; Carnegie, 1991). Many species were predators of both eggs and young larvae. Spiders in the families Agelenidae and Argiopidae and ants in the genus *Dorylus* were the most frequent larval predators; 67%, 33% and 39% of the individuals had preyed on *E. saccharina* larvae, respectively. However, their abundance was low, so the impact of these insect groups may not be high. The ant genera *Pheidole* and *Plagiolepis* may have been more important than the results of the crossover electrophoresis indicated, because they were abundant in the field. The level of predation in the fields that were sampled for the serological study was quite low. Only 15% of the tested 1124 insects had preyed on *E. saccharina* larvae (Leslie, 1988).

In another study, nearly hatched *E. saccharina* eggs were introduced in field cages treated with insecticide to exclude predators or in untreated cages where predators were still present. Survival of larvae was only 5%, irrespective of treatment. Factors such as the inability to find a suitable host, cannibalism, rainfall and desiccation may have been more important causes of mortality than predation (Leslie, 1988).

A list of all recorded larval predators of African stem borers is given in Table 24.2.

PREDATION ON OLDER LARVAE

Late-instar larvae feed in the protected environment of the stem and seem therefore less vulnerable to predation. However, Berger (1993) found that some older *C. partellus* larvae (third to sixth instar) left the plant and walked on the ground or on the leaves to other plants. During this migration, larvae were susceptible to predation. Ants have been observed to attack large *C. partellus* larvae walking on the ground (see Plate 12:1) (W.A. Overholt, personal communication). Large larvae are also known to be attacked inside the stem. Kfir (1988) observed the ant *Dorylus helvolus* (L.) consuming *B. fusca* larvae in maize and sorghum stems in South Africa.

Table 24.2. Records of predators attacking larval stages of African stem borers.

Order, family, genus and species	Host species	Distribution	Reference
Acarina			
Unidentified species	Es	South Africa	Carnegie, 1991
Prostigmata sp.	Es	South Africa	Leslie, 1988
Araneida			
Agelenidae	Es	South Africa	Leslie, 1988
Argiopidae	Es	South Africa	Leslie and Boreham, 1981; Leslie, 1988
Drassidae	Es	South Africa	Leslie and Boreham, 1981; Leslie, 1988
Lycosidae	Es	South Africa	Leslie and Boreham, 1981; Leslie, 1982, 1988
Salticidae	Es	South Africa	Leslie and Boreham, 1981
Sicariidae	Es	South Africa	Leslie and Boreham, 1981; Leslie, 1988
Sparassidae	Es	South Africa	Leslie and Boreham, 1981
Theridiidae	Es	South Africa	Leslie and Boreham, 1981; Leslie, 1988
Thomisidae	Es	South Africa	Leslie and Boreham, 1981; Leslie, 1988
Blattaria			
Unidentified species	Es	South Africa	Leslie and Boreham, 1981; Leslie, 1988; Carnegie, 1991
Unidentified species	Cp	Kenya	M.J. Bonhof, unpublished
Pseudoscorpionida			
Unidentified species	Es	South Africa	Leslie and Boreham, 1981; Leslie, 1988; Carnegie, 1991
Coleoptera			
Unidentified species	Cp	Kenya	M.J. Bonhof, unpublished
Unidentified species	Es	South Africa	Leslie, 1988
Coccinellidae			
Cheilomenes sulphurea Olivier	Ci	Nigeria	Ajayi, 1985a
Cheilomenes sp.	?	Kenya	Seshu Reddy, 1983b
Ganocephalus simplex	?	Kenya	Dwumfour, 1990
Melyridae			
Unidentified species	Cp	Kenya	M.J. Bonhof, unpublished
Staphylinidae			
Paederus sp.	Cp	Kenya	M.J. Bonhof, unpublished
Dermaptera			
Forficulidae			
Diaperasticus erythrocephala Olivier	Cp	Kenya	Oloo, 1989; Dwumfour, 1990
		Ethiopia	Gebre-Amlak, 1985
		Ghana	Botchey and Kwapong, 1991
Forficula auricularia	Cp	Kenya	Dwumfour *et al.*, 1991

Table 24.2. Records of predators attacking larval stages of African stem borers (*continued*).

Order, family, genus and species	Host species	Distribution	Reference
Hemiptera			
Unidentified species	Es	South Africa	Leslie and Boreham, 1981; Leslie, 1988
Anthocoridae			
Orius sp.	?	Kenya	Dwumfour, 1990
Chrysopidae			
Chrysopa sp.	?	Kenya	Dwumfour, 1990
Hymenoptera			
Formicidae			
Acantholepis sp.	Es	South Africa	Leslie and Boreham, 1981
Aenictus sp.	Es	South Africa	Leslie and Boreham, 1981; Leslie, 1988
Camponotus rufoglaucus (Jerdon)	?	South Africa	Leslie, 1988; Carnegie, 1991
Camponotus sp.	?	Kenya	Seshu Reddy, 1983b
Camponotus sp.	Es	Kenya	Dwumfour, 1990
Cardiocondyla shuckardi Forel (as *Cardiocondyla badonei* Arnold)	Bf	South Africa	Leslie and Boreham, 1981
C. emeryi Forel	Bf	Uganda	Mohyuddin and Greathead, 1970
Dorylus affinis Schuckard	Bf	Uganda	Mohyuddin and Greathead, 1970
Dorylus helvolus (L.)	?	Nigeria	Ajayi, 1985a
Dorylus helvolus (L.)	Bf	South Africa	Kfir, 1988
Dorylus sp.	Es	Kenya	Dwumfour, 1990
Leptogenys sp.	Bf	South Africa	Leslie, 1988; Carnegie, 1991
Pheidole guineense F. (as *Tetramorium guineense*)	Cp/Bf	South Africa	Leslie and Boreham, 1981
Pheidole megacephala F.	Cp	Uganda	Mohyuddin and Greathead, 1970
Pheidole sp.	Es	Uganda	Mohyuddin and Greathead, 1979; Kfir, 1988
Pheidole sp.	Cs	Kenya	Oloo, 1989; Dwumfour, 1990; M.J. Bonhof, unpublished
Pheidole sp.	Es	South Africa	Leslie, 1988; Carnegie, 1991;
Paratechina	Es	Mauritius	Williams and Mamet, 1954
Polyrachis sp.	Es	South Africa	Carnegie, 1991
Plagiolepis sp.	Es	South Africa	Leslie, 1988; Carnegie, 1991
Solenopsis sp.	Cp	South Africa	Leslie, 1988; Carnegie, 1991
Technomyrmex sp.	?	South Africa	Leslie, 1982; Carnegie, 1991
		Kenya	M.J. Bonhof, unpublished
Isopoda			
Unidentified species	Es	South Africa	Leslie and Boreham, 1981
Orthoptera			
Saltatoria			
Unidentified species	Es	South Africa	Leslie and Boreham, 1981
Tettigonidae			
Clonia vittata Thunberg	Bf	South Africa	Akerman, 1932
Gryllidae			
Unidentified species	Es	South Africa	Leslie, 1988; Carnegie, 1991

Cp, *Chilo partellus;* Bf, *Busseola fusca;* Es, *Eldana saccharina;* Cs, *Chilo sacchariphagus;* Ci, *Coniesta ignefusalis;* ?, unknown.

Numbers of stem-borer larvae (*B. fusca* and *C. partellus*) and *D. helvolus* peaked simultaneously, indicating that stem-borer larvae may have been an important prey of *D. helvolus* (Watmough and Kfir, 1995). Another *Dorylus* species, *D. affinis* Schuckard, attacked *B. fusca* larvae in millet stems in Nigeria (Harris, 1962; Ajayi, 1985a). The long-horned grasshopper, *Clonia vittata* Thunberg, was observed feeding upon large *B. fusca* larvae outside the stem in South Africa (Akerman, 1932).

Predation on diapausing larvae has been reported from Nigeria. The mite *Pyemotes ventricosus* Newp. (Acarina: Pyemotidae) attacked diapausing *C. ignefusalis* larvae inside millet stems (Ajayi, 1985a).

PREDATION ON PUPAE

Predation on pupae of stem borers has not been well studied in Africa. Pupae, like the later-instar larvae, are sheltered in the stem and probably less attacked by predators than are eggs or young larvae. Preyed-on *Chilo* pupae have been recovered at the Kenya coast (W.A. Overholt, unpublished), but the extent of predation has not been quantified. A report from South Africa suggested that a grasshopper, *C. vittata*, preyed on *B. fusca* pupae (Akerman, 1932).

PREDATION ON ADULTS

Information on predation of adult stem borers is scarce. Some anecdotal observations of spiders attacking adults are available (Leslie, 1988) and the long-horned grasshopper, *C. vittata*, has also been reported to prey on adult moths (Akerman, 1932). Leslie (1988) assumed that the mobility and short lifespan of the moths made their predation of minor importance. A study confirmed that predation of gravid *E. saccharina* females was low over 3 days after mating, which is the period over which most eggs are laid (Leslie, 1989).

PREDATORS ASSOCIATED WITH STEM BORERS

The predator guild in maize and sorghum fields has been studied by several authors. Their studies focused on the predators on the African bollworm, *Helicoverpa* (*Heliothis*) *armigera* Hübner (Lepidoptera: Noctuidae) (van den Berg, 1993; Watmough and Kfir, 1995), or on the impact of predators on the total pest complex (du Toit, 1995). The predators identified in these studies are potential predators of stem borers. Watmough and Kfir (1995) showed a correlation between predation by the ant *D. helvolus* on *H. armigera* pupae and stem-borer numbers in maize and sorghum, and concluded that stem borers were an important prey for *D. helvolus*. Other predators recovered were (in descending order of importance) larvae of Cardiophorinae (Coleoptera; Elateridae), the ant *Aenictus* sp., larvae of *Astylus astromaculatus* Bl. (Coleoptera: Melyridae), tenebrionid larvae, larvae and adults of *Heteronychus arator* F., carabid larvae and adults, staphylinids and earwigs. In another study in South African commercial maize and sorghum fields, 56

Table 24.3. Potential predators associated with African stem borers.

Order, family, genus and species	Crop	Distribution	Reference
Araneida			
Ammoxenidae			
Ammoxenus sp.	M	South Africa	du Toit, 1995
Gnaphosidae			
Unidentified sp.	M, S	South Africa	du Toit, 1995
Linyphiidae			
Unidentified sp. 1	M, S	South Africa	du Toit, 1995
Unidentified sp. 2	M, S	South Africa	du Toit, 1995
Unidentified sp. 3	M, S	South Africa	du Toit, 1995
Lycosidae			
Unidentified sp. 1	M, S	South Africa	du Toit, 1995
Unidentified sp. 2	M, S	South Africa	du Toit, 1995
Salticidae			
Unidentified sp. 1	M, S	South Africa	du Toit, 1995
Unidentified sp. 2	M, S	South Africa	du Toit, 1995
Tetragnathidae			
Tetragnatha sp.	M, S	South Africa	du Toit, 1995
Theridiidae			
Enoplognatha sp.	M, S	South Africa	du Toit, 1995
Lactrodectus geometricus Kock	M, S	South Africa	du Toit, 1995
Lactrodectus indistinctus Cambridge	M	South Africa	du Toit, 1995
Chilopoda			
Unidentified species	M, S	South Africa	du Toit, 1995
Coleoptera			
Anthicidae			
Anthicus biplagiatus Laferte	M, S	South Africa	du Toit, 1995
Anthicus crinitus Laferte	M, S	South Africa	du Toit, 1995
Anthicus styguis Laferte	M, S	South Africa	du Toit, 1995
Endomia sp.	M, S	South Africa	du Toit, 1995
Formicomus tropicalis Krekick	M, S	South Africa	du Toit, 1995
Notoxus cucullatus Laferte	M, S	South Africa	du Toit, 1995
Pseudonotoxus testaceus (Laferte)	M, S	South Africa	du Toit, 1995
Carabidae			
Anthia omoplata Breme	M, S	South Africa	du Toit, 1995
Brachinus sp.	M, S	South Africa	du Toit, 1995
Calosoma sp.	M, S	South Africa	du Toit, 1995
Campolita sp.	M, S	South Africa	du Toit, 1995
Chlaenius sp. 1	M, S	South Africa	du Toit, 1995
Chlaenius sp. 2	M, S	South Africa	du Toit, 1995
Graphipterus cordiger Dejean	M, S	South Africa	du Toit, 1995
Harpalus sp.	M, S	South Africa	du Toit, 1995
Mesolestinus sp.	M, S	South Africa	du Toit, 1995
?*Microlestis* sp. 1	M, S	South Africa	du Toit, 1995
?*Microlestis* sp. 2	M, S	South Africa	du Toit, 1995
?*Microlestis* sp. 3	M, S	South Africa	du Toit, 1995
Unidentified sp. (Lebiinae)	M, S	South Africa	du Toit, 1995
Cicindelidae			
Lophyra sp.	M, S	South Africa	du Toit, 1995

Table 24.3. Potential predators associated with African stem borers (*continued*).

Order, family, genus and species	Crop	Distribution	Reference
Coccinellidae			
Cheilomenes lunata (Fab.)	M, S	South Africa	du Toit, 1995
Hippodamia variegata (Goeze)	M, S	South Africa	du Toit, 1995
Scymnus moreletti Mulsant	M, S	South Africa	du Toit, 1995
Elateridae			
Cardiophorinae sp.	M, S	South Africa	du Toit, 1995
Melyridae			
Astylus atromaculatus	M, S	South Africa	du Toit, 1995
Staphylinidae			
Unidentified sp. 1	M, S	South Africa	du Toit, 1995
Unidentified sp. 2	M, S	South Africa	du Toit, 1995
Unidentified sp. 3	M, S	South Africa	du Toit, 1995
Unidentified sp.	M, S	South Africa	du Toit, 1995
Tenebrionidae			
Unidentified sp.	M, S	South Africa	du Toit, 1995
Dermaptera			
Labiduridae			
Labidura riparia (Pallas)	M, S	South Africa	du Toit, 1995
Nala lividipes (Dufour)	S	South Africa	du Toit, 1995
Diptera			
Syrphidae			
Syritta ?flaviventris Macquart	S	South Africa	du Toit, 1995
Eumerus sp.	M, S	South Africa	du Toit, 1995
Hemiptera			
Anthocoridae			
Orius ?thripoborus (Hesse)	M, S	South Africa	du Toit, 1995
Reduviidae			
Pirates conspurcatus (Distant)	M, S	South Africa	du Toit, 1995
Ectomocris sp.	M, S	South Africa	du Toit, 1995
Onocephalus sp.	M, S	South Africa	du Toit, 1995
Hymenoptera			
Formicidae			
Aenictus sp.	M, S	South Africa	Watmough and Kfir, 1995
Anoplolepis sp.	M, S	South Africa	du Toit, 1995
Camponotus sp.	M, S	South Africa	du Toit, 1995
Camponotus sp.	M	Kenya	Dwumfour, 1990
Dorylus sp.	M, S	South Africa	du Toit, 1995
Dorylus sp.	M	Kenya	Dwumfour, 1990
Dorylus sp.	M, S	South Africa	Watmough and Kfir, 1995
Pheidole sp. 1	M, S	South Africa	du Toit, 1995
Pheidole sp. 2	M, S	South Africa	du Toit, 1995
Pheidole sp.	M	Kenya	Dwumfour, 1990

Table 24.3. Potential predators associated with African stem borers (*continued*).

Order, family, genus and species	Crop	Distribution	Reference
Neuroptera			
Anthocoridae			
Orius sp.	M	Kenya	Dwumfour, 1990
Chrysopidae			
Chrysoperla pudica (Navas)	S	South Africa	du Toit, 1995
Chrysoperla congrua (Walker)	M, S	South Africa	du Toit, 1995
Chrysopa sp.	WG	Kenya	Dwumfour, 1990
Orthoptera			
Gryllidae			
Gryllus bimaculatus de Geer	M, S	South Africa	du Toit, 1995
Unidentified sp.	M, S	South Africa	du Toit, 1995
Solpugida			
Unidentified sp.	M	South Africa	du Toit, 1995
Thysanoptera			
Phlaeothripidae			
Unidentified sp.	M	South Africa	du Toit, 1995

M, maize; S, sorghum; WG, wild grasses.

and 51 (maize and sorghum, respectively) species of predators were discovered (Table 24.3) in pitfall traps and by plant observation (du Toit, 1995). The predator complex in sorghum fields was more or less stable, while in maize large differences occurred between seasons. The study also showed that maize and sorghum have different predator guilds, both in epigeal and epiphytic predators. In maize the most prominent epigeal predators were ants and ground beetles (carabids), while in sorghum crickets were very prominent. Epiphytic predators of maize were mainly ants, anthocorid bugs and coccinellids, while the very low numbers of epiphytic predators in sorghum mainly consisted of coccinellids and syrphid flies (du Toit, 1995).

A study on natural enemies of *H. armigera* in Kenya showed that Anthocoridae and Formicidae were the most common predators in maize and sorghum fields, each group represented by a rich variety of species (van den Berg, 1993). Dwumfour *et al.* (1991) studied the predator guild in maize fields in western Kenya. He found that the ant species *Pheidole* and *Camponotus*, the coccinellids *Cheilomenes* sp. and *Ganocephalus simplex*, the earwigs *D. erythrocephala* and *F. auricularia*, a minute pirate bug (*Orius* sp.) and spiders were all associated with stem borers. However, the relative abundance of these species was low, and their impact on stem borers may not have been high.

A list of predators that are associated with stem borers but have not actually been observed feeding on stem borers is given in Table 24.3.

CONCLUSION

Studies on predation of cereal stem borers in Africa suggest that, in some environments, predators cause heavy mortality on eggs and young larvae, but not on older life stages. Of the predators identified, ants were frequently considered to be the most important. In the quantitative studies that have been conducted, mortality due to predation was usually expressed as a proportion of a certain life stage that was predated. This method gives little insight into the overall influence of predators on stem-borer population dynamics. Additional studies which quantify the generational mortality due to predators seem warranted. Finally, the high variation in predation found between different environments suggests that there may be possibilities to enhance predation through environmental manipulation. This will hopefully provide subsistence farmers in Africa with an effective, safe, sustainable and economically sound management system for stem borers in the future.

IV Control

25 Integrated Pest Management

K.V. Seshu Reddy
International Centre of Insect Physiology and Ecology (ICIPE), PO Box 30772, Nairobi, Kenya

INTRODUCTION

Maize and sorghum are two of the most widely grown cereal food crops by resource-limited small-scale farmers in many countries of Africa. Grain yields of these crops are low and one of the factors that affects the yields is undoubtedly the lepidopterous stem borers.

At present, *Chilo partellus*, *Busseola fusca*, *Sesamia calamistis*, *Sesamia cretica*, *Eldana saccharina* and *Coniesta ignefusalis* are considered to be of major importance in maize and sorghum in Africa. However, changes in agronomic practices, climate, distribution or cultivars might bring out a shift in the status of minor stem borers.

BIOLOGICALLY INTENSIVE INTEGRATED PEST MANAGEMENT STRATEGIES FOR STEM BORERS

In considering the strategies for the management of maize and sorghum stem borers, it is important and essential to develop biologically intensive integrated pest management (BIPM) strategies that are environmentally safe, economically feasible and acceptable to the resource-limited farmers. Biologically intensive integrated pest management fits well into the economy of maize and sorghum cultivation under subsistence farming conditions. Some useful information has been generated on different methods of controlling the stem borers. The aim of this chapter, therefore, is to review different methods of stem-borer management on maize and sorghum, which include the direct tactics: (i) developing plant resistance to stem borers; (ii) cultural control; (iii) biological control; and (iv) botanical control. Supportive tactics include: (i) the application of pheromone biology; and (ii) crop-loss assessment.

DAMAGE AND CROP LOSSES

The species of stem borers infesting maize and sorghum are for the most part the same and produce similar symptoms of damage. Newly hatched larvae congregate inside the leaf whorls and feed actively on folded central leaves, causing typical pinhole symptoms. Later on, they feed on the growing stem of the young plants, resulting in deadhearts. If the growing point has already moved upwards, only stem tunnelling takes place. The damaged plants become weak and stunted and bear very small cobs/panicles. The stem-borer larvae also bore into the maize cobs and feed on the developing seeds.

Reports on grain-yield losses in maize and sorghum due to stem borers are very few. In Kenya, 18% yield losses were attributed to *C. partellus* and *Chilo orichalcociliellus* in maize (Warui and Kuria, 1983). Revington (1986) estimated the yield losses due to *C. partellus* in maize and sorghum in western Transvaal and north-western Orange Free State in South Africa to be more than 50%.

Yield losses in sorghum due to *C. partellus* have been recorded up to 88% in Kenya (Seshu Reddy, 1988). Sithole (1989b) in Zimbabwe, reported a yield loss of 50–60% in sorghum due to *C. partellus.* Empirically determined yield-loss values in sorghum due to *C. partellus* vary with ecological zones, cultivars, stage of the crop and infestation levels (Seshu Reddy *et al.,* 1989).

One of the crucial features of the implementation of the BIPM programmes is to select and time appropriate BIPM components on the basis of economic injury levels (EILs) (see also economic threshold level (ETL), Chapter 26). Despite its importance, the EIL concept is often an underutilized tool in integrated pest management (IPM); furthermore, very little research based on EIL has been done on cereal stem borers in Africa. However, Seshu Reddy and Sum (1992) determined the EILs of *C. partellus* in three maize cultivars, based on the linear infestation–yield relationship, protection costs and the prevailing market price of maize. Economic injury level values were lowest in the susceptible variety INB-A, followed by the tolerant V-37, while the resistant ICZ-2 had the highest EIL. The EIL values increased with progressive crop growth stages, irrespective of the cultivar tested. However, EIL values are flexible and vary with the crop cultivar, age of crop, density of target borer at the time of attack, nature of pest damage, protection costs, market price of the crop, rainfall, agroecological zone, labour inputs and farming practices.

HOST-PLANT RESISTANCE

This is one of the most important and promising ways of reducing stem-borer damage and subsequent yield losses in target crops, either by resistance to initial attack, antibiosis once attacked or tolerance to attack.

Initially, in Africa, cereal improvement programmes paid more attention to yield improvement. However, at the International Centre of Insect Physiology and Ecology (ICIPE), research in this area is aimed at the development of strategies for the efficient utilization of plant resistance in maize and sorghum against the stem-borer complex. The activities undertaken include elucidation of components, mechanisms and genetic basis of resistance and improvement of crop cultivars so as to combine pest resistance/tolerance with other desirable characters, including yield.

In addition, in collaboration with national agricultural research and extension systems (NARES), several international agricultural research centres (IARCs) including the International Crops Research Institute for the Semi-Arid Tropics (ICRISAT), the International Institute of Tropical Agriculture (IITA) and the International Maize and Wheat Improvement Center (CIMMYT) have been spearheading the evaluation and development of sorghum and maize resistance to stem borers in Africa.

Research at ICIPE on the evaluation of maize and sorghum lines has shown good levels of resistance/tolerance to stem-borer attack (Dabrowski and Kidiavai, 1983; Omolo, 1983; Ampofo *et al.*, 1986). Seshu Reddy (1983a, 1985b) identified IS Nos. 1044, 1151, 3902, 4213, 4405, 5613, 10364, 10370, 10711, 12447, 18326, 18427, 18479, 18517, 18676, S-178, Tx 2780 and A & B Tx 2756 as multiple sources of sorghum resistance to stem-borer complex in east Africa. The sorghum genotype IS 1044 showed highest resistance to three major types of damage by the stem borer, *B. fusca*: leaf lesions, deadheart and stem tunnelling. The hybrid HYD-8 (1441A × IS 1044) was as highly resistant to deadheart and stem tunnelling as IS 1044 (K.N. Saxena and A.M. Nour, unpublished). Ampofo *et al.* (1986) evaluated several local and exotic maize lines against *C. partellus* oviposition and establishment and found Inbred A to be most susceptible, MP 702 moderately resistant and ICZ1-CM and MP 704 highly resistant.

A wide range of mechanisms were involved in *C. partellus* and *B. fusca* resistance in maize and sorghum, including non-preference for oviposition, reduced feeding, reduced tunnelling, tolerance of plants to leaf damage, deadheart and stem tunnelling and antibiosis. In addition, morphological, physical, chemical and other plant characteristics and non-plant factors, including photo- and geotactic stimuli, were involved (Dabrowski and Kidiavai, 1983, Saxena, 1985, 1990, personal communication; Kumar and Saxena, 1992; van Rensburg and Malan, 1992; Kumar, 1993).

Studies on the genetic basis of resistance in maize to *C. partellus* showed that both additive and non-additive and epistasis gene effects are important in the inheritance of deadheart, while the gene effect for leaf feeding and stem-tunnel length was predominantly additive. A recurrent selection procedure was therefore suggested as appropriate to increase the level of resistance in the maize population and to exploit heterosis breeding in F_1 crosses showing increases in yield and resistance. In the case of sorghum, it was shown that the additive gene effect was important in the inheritance of damage parameters. Dominance and epistasis gene effects were important only for leaf feeding. Transgressive segregation for a higher resistance level may therefore result from crosses through a recurrent selection programme (Pathak and Olela, 1983; R.S. Pathak, unpublished).

At ICIPE, efforts have been made to incorporate resistance of maize and sorghum to *C. partellus* into a good agronomic background, and many genotypes are already in the Kenya national yield trials. Several resistant lines/hybrids with good general combining ability have been identified (S.O. Ajala and A.M. Nour, unpublished).

CULTURAL CONTROL

Several cultural practices have been implicated in maize and sorghum stem-borer control, mainly in disrupting or slowing down the population build-up. These include appropriate disposal of crop residues, time of planting, tillage and mulching, spacing, intercropping, removal and destruction of volunteer and alternative host plants, removal of borer-infested plants, fertilizer application and crop rotation.

Appropriate disposal of crop residues. Stalks and stubble of maize and sorghum left standing in the fields are an important source of initial populations of several stem-borer species. Therefore, appropriate destruction and disposal of stubble and stalks have been recommended.

To control *B. fusca* in Nigeria, one recommendation was to burn the stalks completely after the harvesting of grain to kill larvae or to spread them thinly in the field to expose the larvae to the full effects of adverse weather conditions (Harris, 1962; Ajayi, 1978). However, as the stalks are used for building, fencing and as fuel, Adesiyun and Ajayi (1980) suggested partial burning of stalks immediately after grain harvest. Destruction of crop residues and stubble has also been recommended by several workers to reduce borer infestations (Ingram, 1958; Nye, 1960; Seshu Reddy, 1985a; Unnithan and Seshu Reddy, 1989).

Time of planting. Cultural control of stem borers based on time of planting follows the principle of growing the target crop when the pest is not present or planting at a time that coincides with when the pest is least abundant.

While experimenting with *C. partellus* and *C. orichalcociliellus*, Warui and Kuria (1983) found that early-planted maize had a lower number of borers per plant. For both *B. fusca* and *S. calamistis*, early planting has been found to lower stem-borer infestation (Abu, 1986a). Whether it is early or late planting in any given area, it is important to generate information on the population dynamics of the target pests of crops for at least a period of 3 years.

Tillage and mulching. Soil tillage may reduce borer populations through mechanical damage, either by burying them deeply or by bringing them to the surface, where they may be exposed to adverse weather factors as well as birds and other natural enemies. Mohyuddin and Greathead (1970) observed that untreated crop residues were often used to mulch the next crop in Uganda. Under these conditions, the newly planted crops were constantly infested by borers in the old stalks and this particular kind of mulching should be avoided.

Spacing. This may affect the relative rate of development of a plant and its pest population, as well as the behaviour of the pest in searching for food or an oviposition site (Lawani, 1982). First-instar larvae of *C. partellus* are known to migrate from their hatching site to the funnel of the plant on which they hatched or to other plants within the vicinity. During this process, as high as 100% mortality occurs (Mathez, 1972).

Intercropping. This is the most widespread traditional agricultural system and has long been practised by subsistence farmers in many tropical countries.

Intercropping of maize and sorghum intensified *C. partellus* infestation compared with their monocrops (Amoako-Atta *et al.*, 1983; Ogwaro, 1983). Interplanting bulrush millet with sorghum in alternate stands within the same row reduces larval infestation of *B. fusca* on sorghum, since the adults do not effectively utilize millet for oviposition (Adesiyun, 1983). On the other hand, intercropping sorghum with non-host crops, such as cowpea (Omolo and Seshu Reddy, 1985) and simsim (Kato *et al.*, 1982), reduced *C. partellus* damage in sorghum. Minja (1990) reported that planting sorghum and cowpea simultaneously or planting sorghum 2 weeks after cowpea significantly delayed *C. partellus* larval population build-up, compared with that under monocrop sorghum. Sorghum sown 2 weeks before cowpea had a slightly lower (but not significantly different) larval population from that on sorghum monocrop. Ampong-Nyarko *et al.* (1994) found reduced numbers of stem borer *C. partellus* in sorghum intercropped with cowpea. In the monocropped, unprotected sorghum, yield was reduced by 28% compared with the protected monocrop, while reduction in the unprotected intercropped sorghum was 15% compared with the protected intercrop.

Strip cropping of two, three or four rows of sorghum alternating with an equal number of cowpea rows was found to be as effective as the traditional single-row intercropping in reducing stem-borer incidence. The strip cropping had the added advantage of improved yields and increased efficiency and ease of field operations.

Removal and destruction of volunteer and alternative host plants. Destruction of alternative host plants, such as weeds and volunteer crop plants, may often lead to reduced pest populations. Several of the cereal stem borers can develop in graminaceous wild hosts (Ingram, 1958; Nye, 1960; Seshu Reddy, 1989); a thorough knowledge of these host plants is therefore important in any study of stem-borer population dynamics and their natural enemies.

Crop rotation. Rotating maize or sorghum with leguminous or root crops will help reduce the stem-borer population build-up.

Fertilizer application. A lowering of the total level of soil minerals by erosion and overcropping may improve conditions for the abundance of insect pests. Siddiq (1972) reported that the application of nitrogenous fertilizers to sorghum in Sudan increased the infestation of *C. partellus.*

All these cultural practices are subject to the variability of the weather and sometimes are heavily dependent on the cooperation of neighbouring farmers. Therefore, for these cultural measures to be effective, the farmers should be trained adequately.

BIOLOGICAL CONTROL

Although a range of naturally occurring biocontrol agents, such as parasitoids, predators and diseases, have been reported for different growth stages of stem borers of maize and sorghum in different parts of Africa, essentially very few studies on their effectiveness as well as host/parasite relationships have been made.

Biocontrol agents of interest in many countries in Africa include the egg parasitoids, such as *Trichogramma* spp., larval parasitoids, including *Cotesia sesamiae* and pupal parasitoids, e.g. *Pediobius furvus*, *Dentichasmias busseolae* and *Psilochalcis soudanensis* (Ajayi, 1989; Seshu Reddy, 1989; Sithole, 1989b; Minja, 1990). However, it should be noted that the overall rate of parasitism of stem borers was low and seldom reached 10–14%.

In eastern Africa, releases of the exotic parasitoids *Cotesia flavipes*, *Amyosoma chinensis*, *Isotima javensis*, *Trichogramma australiacum* and *Sturmiopsis inferens* against *C. partellus*, *B. fusca*, *S. calamistis* and *E. saccharina* have not been successful (Ingram, 1983). On the other hand, in the Indian Ocean islands there has been successful biological control of *C. partellus* and *Chilo sacchariphagus* using introduced and indigenous parasitoids (Betbeder-Matibet, 1971; Williams, 1983).

The ICIPE and Department of Entomology, Wageningen Agricultural University, have been jointly implementing a collaborative project on classical biological control of the introduced stem borer, *C. partellus*. The exotic larval endoparasitoid *C. flavipes* has been introduced into Kenya for laboratory studies as well as field releases. Investigations indicate that *C. flavipes* could attack and successfully develop not only on the coevolved host *C. partellus*, but also on two indigenous stem borers that occur on the Kenya coast. Field releases of *C. flavipes* were made at three locations on the Kenya coast in 1993. Recoveries of the parasitoid at all three sites demonstrated that *C. flavipes* could locate and successfully parasitize stem borers in a natural field setting. Parasitism of the stem-borer population reached nearly 40% at one release site (W.A. Overholt, personal communication).

Although the role of predators is not easy to assess, earwigs, ants, spiders and ladybird beetles have been found to feed on the eggs, larvae and pupae of the stem borers of maize and sorghum.

Diseases, particularly *Bacillus thuringiensis* (Bt), *Nosema marucae* and *Beauveria bassiana*, effectively reduced *C. partellus* infestation on maize and sorghum in the field, while *Metarhizium anisopliae* proved effective in laboratory studies (Maniania, 1993b; M.O. Odindo, personal communication). *Bacillus thuringiensis* and *Aspergillus* spp. were found in or on the larvae and pupae of *B. fusca* in Nigeria (Ajayi, 1989).

Interaction of Bt and *Nosema* with other IPM components, especially host-plant resistance, was studied at ICIPE. No incompatibility of the IPM components was observed in maize and sorghum cultivars (M.O. Odindo, personal communication).

BOTANICAL CONTROL

In Africa, botanical pesticide extracts have been used traditionally by resource-limited small-scale farmers to protect crops from ravages of pests. In Tanzania, 4% leaf extracts of *Tephrosia vogelii*, *Neurautanenia mitis* and *Cassia didymobotrya* plant in water (the leaves of these plants are locally used to control stalk borers in maize by the farmers) were compared with commercial insecticides. The extracts from *T. vogelii* and *N. mitis* plants gave promising results (Mallya, 1985, 1986; Marandu *et al.*, 1987), but they are not yet recommended.

In Zambia, *T. vogelii* leaf extracts prevented oviposition by *C. partellus* on maize in addition to being a phagodeterrent. The stem-borer infestation was highest in unsprayed control plots. The percentage of plants damaged by stem borers in control plots was almost three times higher than those sprayed with a 15% *T. vogelii* extract (C. Mugoya, personal communication).

Studies on the use of neem seed (*Azadirachta indica*) for the control of maize and sorghum stem borers have been initiated at the ICIPE. In both maize and sorghum, neem reduced stem-borer attack to the same magnitude as insecticide control (R.C. Saxena, personal communication).

CHEMICAL CONTROL

In Africa, control of maize and sorghum stem borers exclusively by insecticides by resource-limited small-scale farmers is uneconomical and not practical. However, despite the limitations of chemical control, in some countries a number of large-scale farmers as well as government-controlled farms have been applying insecticides for stem-borer control based mainly on calender of operations.

Several insecticides have been screened for the control of maize and sorghum stem borers in different regions in Africa. Those insecticides which have been found effective as sprays or dust treatments include carbofuran, carbaryl, deltamethrin, endosulfan, trichlorfon and synthetic pyrethroids (Seshu Reddy, 1985a; Ajayi, 1989; Minja, 1990; Sithole, 1990; van Rensburg and van den Berg, 1992a). However, in the absence of economic thresholds for different stem borers, it is doubtful that insecticide application is justified by infestation levels.

PHEROMONAL CONTROL

The search for alternative control measures that would overcome the drawbacks of insecticides has included research into the female sex pheromones and how these might be used in stem-borer control. These pheromones can be used for population monitoring and control by mass trapping and mating disruption.

The pheromone trap would be very useful for detection of *C. partellus* for monitoring its first emergence, particularly in areas where larval diapause is prevalent during the off-season and for monitoring the flight phenology (Unnithan and Saxena, 1990). They also suggested that pheromonal trapping may not be effective for estimating *C. partellus* population density and for mass trapping of the males when the pest-population levels are high.

Studies on the effectiveness of different synthetic sex-pheromone formulations and trap designs for evaluating the seasonal activity of *B. fusca* in the maize field indicated that the possibility of using pheromone trapping systems to identify seasons of potentially severe infestations seems to be feasible (van Rensburg, 1992). The use of pheromones has shown promise against stem borers when used in limited areas; however, large-scale field trials should be carried out to assess the effectiveness of this method.

INTEGRATION OF BIOLOGICALLY INTENSIVE INTEGRATED PEST MANAGEMENT COMPONENTS FOR STEM-BORER MANAGEMENT

On-farm trials with selected BIPM menus for reducing grain-yield losses in maize and sorghum due to stem borers were conducted by 50 resource-limited small-scale farmers under their management in western Kenya from 1987 to 1990. The components of the BIPM menus included: (i) plant resistance to insect pests; (ii) intercropping; (iii) adjustment of planting time; and (iv) crop-residue disposal. The incidence of stem borers of cereals showed a progressive decline on the participating farmers' plots during successive years relative to the project's preimplementation level. In contrast, the incidence of stem borers on the non-participating farmers' fields remained high. These observations indicate that the BIPM menus tested on the farmers' fields were quite effective in reducing the borer attacks. The trials demonstrated that yield increases of maize and sorghum of up to 40% can be achieved (Saxena *et al.*, 1991).

CONCLUSIONS

In Africa, a range of stem borers are a major constraint to maize and sorghum production. There is a long history of interaction between various species of stem borers and their host crops. The individual BIPM strategies discussed above have their limitations and none is sufficient to adequately control the stem-borer infestations. When no single control option is sufficient, one may try to exploit the interactions of different control strategies integrated in a pest management system. Biologically intensive integrated pest management takes into account the interactions between biotic, abiotic and economic factors of crop production, and pest management itself becomes part of managing or producing a crop. The limitations of individual control methods indicate that host-plant resistance and various cultural approaches should be major components in the BIPM of maize and sorghum stem borers. Although a number of natural enemies of stem borers have been identified, there is a great need for further studies which can contribute to a more effective programme of introduction, mass rearing and large-scale releases of the natural enemies.

Initiatives for the rational development and promotion of BIPM strategies for maize and sorghum stem borers have been taken by ICIPE, in collaboration with NARES. However, there should be more research and development efforts on the determination of economic thresholds for different stem borers to realize maximum benefit and implementation of integrated and locale-specific stem-borer management programmes.

26 Chemical Control

J. van den Berg[1] and A.F. Nur[2]
[1]*Agricultural Research Council, Grain Crops Institute, Private Bag X1251, Potchefstroom 2520, South Africa;*
[2]*Former Director, Plant Protection Service, Mogadishu, Somalia*

INTRODUCTION

Losses caused by pests of agricultural crops are estimated at a world average of 35%, with losses in Africa being among the highest in the world (Youdeowei and Service, 1983). Without pesticides, an estimated two-thirds of all crops would be lost, depriving millions of people of food. Agricultural economists have estimated that, in the absence of pesticides, production costs would increase by 60–200%, depending on the nature of the crop (Thyagarajan, 1988). In Africa, insecticides are primarily used for the control of pests of cash crops and of vectors of human and animal diseases but quantities used are relatively small compared with those in the developed world. However, it is estimated that importation of pesticides into Africa has risen by at least 100% over a period of 15 years prior to 1989 (Waage and Schulthess, 1989).

Chemical control of stem borers in Africa is particularly important in that borers can cause loss of up to 100% of staple foods, such as maize, millet and sorghum, depending on planting date and seasonal intensity of infestation. The only tool for effective control of stem borers, especially when borer epidemics occur, is the use of insecticides.

Whether or not chemical control is practised depends on the farmers' perception of the problem and the potential result after control actions have been taken. In large-scale commercial farming systems or where grain crops, such as maize and sorghum, are planted as cash crops, farmers are more likely to make use of insecticides for crop protection. On the other hand, in farming systems where grain sales only comprise one of the sources of financial income, stem-borer damage is often tolerated by farmers (Goldman, 1996). In West Africa, the cost of pesticides is largely beyond the means of the small farmer (Ajayi, 1989) and therefore usually used on cash crops rather than on food crops (Youm *et al.*, 1990).

Insecticides are not widely used by small farmers and one of the major factors permitting the tolerance of crop losses by these farmers is the availability of land. In areas where land is available, farmers increase the area under cultivation to compensate for expected losses, while, in areas where land expansion is not an option, higher yields per unit area are required and farmers need to practise intensive pest control. An example of this is the highly populated areas of central

Kenya, where 76% of farmers use insecticides for control of stem borers. In contrast, only 20% of farmers in the lower populated areas of western Kenya use insecticides for borer control (Goldman, 1996). In Senegal, in 1981, only 0.27% of the total amount of $US 7.2 million spent on pesticides was spent on millet and sorghum, by far the most important cereals in Senegal (Anon., 1985).

In commercial farming systems in Kenya, Nigeria, South Africa and Zambia, insecticides are intensively used for stem-borer control (Ajayi, 1989; Okech *et al.*, 1994; van den Berg and van der Westhuizen, 1995; Goldman, 1996). In South Africa, the potential market for stem-borer insecticides in epidemic years is estimated at $US 7 million. However, in average years, stem-borer insecticides to the value of only $US 2.5 million are used over an area of 3.5 million ha. An increase in pest incidence and associated insecticide use is observed in years of above-average rainfall (van Rensburg *et al.*, 1985).

The decision whether or not to apply insecticides should be seen from the farmer's perspective before guidelines or recommendations are made by extension workers, researchers or the insecticide industry. Unavoidable losses always occur, since 100% efficacy can never be achieved with insecticidal control (van den Berg and van Rensburg, 1991a). Farmers should be made aware of the principle that a part of the expected yield loss should be regarded as unavoidable. The extent to which insecticides are used in the control of stem borers in various farming systems can vary from chemically intensive programmes, on the one hand, to systems in which insecticides are used either occasionally or not at all.

Control measures should be aimed at the most vulnerable stages in the life cycle of the pest. For stem borers, these stages are the period when young larvae feed in plant whorls and the period when larvae are in diapause in the stalks. This chapter deals with the former option.

IMPORTANCE OF BIOECOLOGY IN CHEMICAL CONTROL OF BORERS

Stem borers are generally considered to be difficult to control. However, based on knowledge regarding the ecology and larval behaviour of stem borers such as *Busseola fusca* and *Chilo* spp., which feed in plant whorls, insecticides can be used with great success in controlling these pests. The identification of the most susceptible stage in its life cycle is, however, necessary to ensure timely and effective chemical control. Van Rensburg and van den Berg (1992a) found that a large percentage of *Chilo partellus* larvae feed behind leaf sheaths of sorghum, where they are not reached by insecticide applications. Late infestations of *B. fusca* feed in maize ears, where they are protected from insecticide application by husk leaves. In sugar cane, young *Eldana saccharina* larvae disperse in search of shelter and penetration sites, which include buds, root primordia and cracks (Leslie, 1993). *Sesamia calamistis* and *Sesamia cretica* penetrate stems after feeding under leaf sheaths, where hatching and initial feeding takes place. If reports by Ingram (1958) are correct in stating that *Sesamia poephaga* enters the stem through the whorl (contradicted by Harris, 1962), this *Sesamia* sp. can be controlled by means of whorl application of insecticide, a practice which is largely ineffective against other *Sesamia* species. Initial crop damage caused by the millet stem borer, *Coniesta ignefusalis*, is usually observed as deadhearts of millet seedlings, attributed to the feeding activities of young larvae. Leaf-feeding symptoms are not observed with this pest (Harris, 1962). The behaviour of these stem-borer species necessitates using systemic insecticides for effective control.

Although stem-borer larvae of all species exhibit cryptic behaviour, the majority of species spend time feeding in the whorls of plants, where they are vulnerable to insecticides. Walker (1960) and van Rensburg and van den Berg (1992a) found that *B. fusca* and *C. partellus* larvae feed in the whorls of sorghum plants for up to 2 weeks after hatching of the eggs. According to earlier literature, larvae are vulnerable to insecticides during this period and *B. fusca* larvae can be killed by placement of insecticide in the whorl, up to 10 days after hatching of eggs (Walker, 1960). However, from the first author's experience with modern insecticides, larvae may be effectively controlled until tasselling, when most larvae of *B. fusca* would be in the third and fourth instars. Although the tendency of *C. partellus* larvae to feed behind leaf sheaths protects them from insecticides, their migratory behaviour during the 2 weeks following egg hatching increases the likelihood of larvae coming into contact with insecticides applied during this period.

STEM-BORER CONTROL IN COMMERCIAL FARMING SYSTEMS

Chemical control forms the basis of pest control in commercial farming systems. Various studies, all aimed at increasing the timeliness and effectiveness of insecticide applications and application methods, have been undertaken to develop more efficient and economic control programmes (van Rensburg and van den Berg, 1992c, d; van den Berg and van Rensburg, 1996). Knowledge of the following aspects are crucial in pest control programmes in which insecticides play a pivotal role: economic threshold levels (ETLs), the economics of pest control, pest monitoring systems, application methods, comparative efficacy and classes of insecticides.

Economic Threshold Levels

In the absence of ETLs, it is doubtful whether insecticide application is justified. The ETL is defined as that level of damage at which control measures should be implemented to prevent an increasing pest population from reaching the economic injury level (Dent, 1991). The application of insecticides according to the ETL, rather than on a fixed-schedule basis, may reduce the number of insecticide-spray applications, or at least ensure that the number of applications is economically viable (Dent, 1991). Economic threshold models are dynamic and are influenced by factors such as yield potential of the crop, cost of insecticide and market value of the yield (Chiang, 1982; van den Berg *et al.*, 1997). Seshu Reddy and Sum (1992) indicated large differences in the economic injury level of *C. partellus* on maize in Kenya, based on the price of various insecticides. An ETL based on larval counts after destructive sampling has no practical application for stem-borer control. Seshu Reddy and Sum (1991) found the ETL for *C. partellus* in maize in Kenya to be when the mean infestation level per plant is 3.2 and 3.9 larvae, at 20 and 40 days after seedling emergence, respectively. The ETL for stem borers can also be defined in terms of the incidence of plants with visible damage, since it is a non-destructive, practical method of determining infestation levels (van den Berg *et al.*, 1997).

The ETL for control of *B. fusca* in commercial maize-farming systems in South Africa is when 10% of plants in a field show whorl-damage symptoms (van Rensburg *et al.*, 1988c). This ETL is very similar to the 16% determined by Sithole (1994) for *B. fusca* in maize and sorghum in Zimbabwe. The ETL for control of *B. fusca* and *C. partellus* in commercial sorghum in South Africa is when 10% of plants in a field show whorl-damage symptoms. This threshold was developed from a study conducted to evaluate the effect of plant resistance on ETLs (van den Berg *et al.*, 1997). An ETL of 5% oviposition, based on the incidence of *B. fusca* egg batches on maize plants was developed by van Rensburg *et al.* (1988c) and a sequential sampling plan developed for sampling egg batches in maize (van Rensburg and Pringle, 1989). This ETL is, however, not commonly used, due to its being labour-intensive. The monitoring technique on which an ETL is based must not involve intensive sampling, especially if it is to be carried out at regular intervals by the farmer (Dent, 1991). Control of *C. partellus* in commercial maize in South Africa is recommended when 40% of plants show symptoms of larval feeding in the whorls. The reason for the difference between the threshold values for *B. fusca* and *C. partellus* is ascribed to the lower comparative injuriousness of *C. partellus* on maize in South Africa (Bate and Van Rensburg, 1992). The ETLs for sorghum and maize are commonly provided on pesticide labels and are provided in official guides to the use of pesticides in South Africa (Krause *et al.*, 1996). These ETLs are considered to be very low and are often reached, but they provide for the timely application of control measures in high potential plantings. These relatively low ETLs are therefore a compromise to exclude control measures during periods of low population levels but ensuring timely control measures during epidemics.

ECONOMIC THRESHOLDS AND THE SMALL FARMER

Whereas the commercial farmer annually reaps the benefit of employing the ETL principle, the applicability of this technology to the small farmer is questioned. Infestation levels higher than 10% are common in maize and sorghum fields of small farmers throughout Africa, and this ETL level does not apply to these farmers. It is only in commercial farming systems, where the cash value of the crop is high, that this low ETL is of value. The practical applicability of the ETL to small farmers involves, firstly, whether sufficient variation in farmer circumstances in the agricultural community exists to warrant the definition of several distinct threshold levels and, secondly, the question of spatial distribution of these circumstances among the farming community (Farrington, 1977). The inclusion of a socioeconomic factor in a general economic threshold model was suggested by Chiang (1982). This would partly provide for small farmer circumstances since it would result in higher threshold values.

Insecticide Application on the Basis of Plant Growth Stages

Further recommendations regarding the timeliness of insecticide applications are also provided for commercial maize and sorghum farmers in South Africa. If the economic threshold principle is not applied, due to the 'nuisance factor' or fields being too large, insecticides can be applied on the basis of plant growth stage.

The following recommendations are based on the ecology of *B. fusca* and *C. partellus* in South Africa, where only one rainy season occurs and larvae go into diapause during winter. The broad principles applied to timeliness of insecticide application in the case of mixed stem-borer populations in commercial sorghum are as follows.

In the case of early plantings, insecticides must be applied at late plant-growth stages. When only one insecticide application is planned or economically warranted, it is preferable to postpone application rather than to spray too early. If the crop is planted early (before 15 November) and infestation is not very high, insecticide application can be postponed until the flag-leaf stage. Insecticide application during the flag-leaf stage results in higher yield under these conditions than a single application at an earlier growth stage or two applications at a later stage. Insecticide application after the boot stage in sorghum is ineffective. Similar recommendations were made by Nzibande (1993), who observed that application of endosulfan sprays on sorghum 50 days after crop emergence (which coincides with the flag-leaf stage) was effective for borer control in Swaziland.

In the case of late plantings, an early application is needed with the possibility of a follow-up application. In late plantings (after 15 November) an early insecticide application (approximately 4 weeks after crop emergence) is needed, with the possibility of a follow-up application during the flag-leaf stage. The need for a follow-up application is due to reinfestation during late plant-growth stages and very high population levels late in the growing season.

In the evaluation of programme sprays for *B. fusca* control in the tasselling and post-tasselling stages of maize, van Rensburg (1990) found that timing of insecticide application was critical. Insecticide application at 10% tasselling provided better control than sprays at either the whorl stage (6 weeks after crop emergence) or 50% tasselling. Spraying at the flag-leaf stage reduced the number of larvae to a lesser extent, probably because larvae were well concealed in the emerging tassel at that stage. Efficient control of larvae and associated increase in yield can be obtained after tasselling (van Rensburg, 1990). Ear damage can be reduced by spraying maize plants after tasselling, but control measures should preferably be taken before the tassel is furled in the last leaf. In the case of *B. fusca*, van Rensburg *et al.* (1987) found that insecticide application in the whorl should be effective until shortly before tasselling, since larvae remain in the whorl for prolonged periods. *Busseola fusca* larvae are also more likely to feed in whorls for longer periods than *C. partellus* larvae (van Rensburg and van den Berg, 1992a).

PROPHYLACTIC INSECTICIDE APPLICATION

A scheduled, prophylactic application (calendar sprays) implies that insecticide is applied to the crop at regular intervals without information on the level of pest infestation (Dent, 1991). Before the development of the ETL for *B. fusca* control in South Africa, prophylactic applications were commonly recommended for maize and sorghum (Bot and Hollings, 1975). Since 1977, the use of threshold values for control of *B. fusca* in maize has been propagated on insecticide labels and in insecticide guides (Bot *et al.*, 1977). Where the crop value is high or damage thresholds are low, e.g. the ETL for borer control in South Africa (van Rensburg *et al.*, 1987; van den Berg *et al.*, 1997), prophylactic control appears attractive (Dent, 1991). In South Africa, this practice was abandoned when dichlorodiphenyltrichloroethane (DDT) was replaced by newer, more expensive insecticides. Insecticide application at predetermined times, without considering the level of infestation, is an uneconomic and unacceptable practice, which can also influence the relative importance of other pests. Unnecessary application of insecticides for stem-borer control in sorghum, for example, will not only kill natural enemies of stem borers but also those of other pests, such as the aphid, *Melanaphis sacchari* (Homoptera: Aphididae), which can be a serious pest of sorghum in southern Africa. Follow-up insecticide applications will then be necessary, since the natural-enemy complex will not be able to recover and suppress aphid numbers. Insecticide application should not be done when the first symptoms of borer damage are observed, since grain crops, such as sorghum, are able to tolerate some degree of damage before yield loss commences (Flattery, 1982; van Rensburg and van den Berg, 1992b). It is therefore important to determine infestation levels before insecticide applications are made.

Seed- and grain-production fields under overhead irrigation are, however, exceptions to these rules. Because of the very high value of such crops and the likelihood of high borer infestations in such fields, programme sprays are recommended.

Utilization of Sex Pheromones for Control of Stem Borers

Pheromone-baited traps can be used as monitoring devices to assess population levels of stem borers. The use of traps in monitoring systems for stem-borer populations can provide useful information on timing of insecticide applications. Campion and Nesbitt (1983) reported extensively on the utilization of sex pheromones for stem-borer control. Sex pheromones for a number of stem-borer species, including *B. fusca, C. partellus, Chilo sacchariphagus, S. calamistis, S. cretica, Sesamia nonagrioides* and *C. ignefusalis*, have been identified. Commercial formulations of some of these pheromones are available for pest monitoring. Pheromone monitoring systems for *B. fusca* are used in commercial maize and sorghum in South Africa and have been used to determine moth flight patterns of this pest in various small-farming areas in South Africa (van Rensburg *et al.*, 1985; van Rensburg, 1997). The action threshold for *B. fusca* control, based on pheromone-trap catches, is when the average catch from three traps per site exceeds two moths per week for 3 weeks (Kfir and Bell, 1993). Youm and Beevor (1995) evaluated the use of pheromone-baited traps for population monitoring of *C. ignefusalis* in Niger.

THE ECONOMICS OF STEM-BORER CONTROL

Even in commercial agriculture, under favourable crop conditions and high yield potential, the application of insecticides needs to be carefully considered. Economic

Table 26.1. Gain thresholds calculated at different control costs and yield potentials.

Yield potential ha^{-1} (t)	Market price t^{-1} ($)	Gross income ha^{-1} ($)	Control cost* ($)		Gain threshold† (% of yield ha^{-1})	
			Tractor	Air	Tractor	Air
1	88	88	6.60	11	7.5	12.5
2	88	176	6.60	11	3.8	6.3
3	88	264	6.60	11	2.5	4.2
5	88	440	6.60	11	1.5	2.5
7	88	616	6.60	11	1.1	1.8

* Total cost for tractor or aerial application per hectare.
† Although the gain threshold (%) changes, the grain mass equal to the cost of one application remains constant at 75 kg and 125 kg for tractor and aerial applications, respectively.

analysis of insecticide application against *C. partellus* on maize in South Africa indicated that pest-control expenditures have a considerable effect on the profitability of maize farmers. Control expenditures could amount to 55% of the profit for irrigated maize and 25% for dryland maize (van Hamburg, 1987). The cost of insecticide application against *C. partellus* can amount to 12% and 7% of the total production costs on irrigated and dryland maize, respectively.

By calculating a gain threshold (GT), producers can get an indication of whether or not insecticide application will result in economic benefits. The GT is expressed as a percentage of the grain yield per hectare which is equal to the cost of an insecticide application. Insecticide application is only economically warranted when the GT is greater than the cost of the insecticide application. This aspect is particularly important in cases where fields of thousands of hectares have to be sprayed.

For example, a ground or aerial insecticide spray on a maize or sorghum field with a yield potential of 1 t will require a percentage gain of 7.5% or 12.5% of the yield potential, respectively, to warrant the cost of one insecticide application (Table 26.1). The economic viability of insecticide application in low-potential fields is therefore questioned. On the other hand, the gain in grain yield needed to warrant application in high-potential fields can be as low as 1.1%.

The GT can be determined by using the following equation:

$$\text{Gain threshold (GT)} = \frac{\text{Cost of control}}{\text{Gross income per hectare}} \times 100$$
$$= \text{\% of yield per hectare}$$

Cost of control = insecticide cost, tractor running costs, wages, etc.
Gross income = (yield potential ha^{-1}) × (market price t^{-1})

The GT can further be expressed in terms of grain mass. Examples of the GT are provided in Table 26.1. For example, if the yield potential of a field is 2 t and the market value is \$88 t^{-1}, a 3.8% (or 75 kg) gain in yield compared with an unsprayed field is necessary before the cost of tractor application of an insecticide is warranted. In the case of an aerial application, the GT would be 6.3%. If the expected price is \$66 t^{-1}, the GT increases to 5.0% (or 166 kg) of the yield per hectare on a field with a 2 t yield potential, in the case of a ground application.

APPLICATION METHODS

Since differences in the biology of stem-borer species affect the efficacy of insecticides, application techniques need to be adapted to accommodate the biology of different species (Jotwani, 1983; van den Berg and van Rensburg, 1996). Kishore (1989) reviewed research and recommendations on various methods of insecticide application against stem borers. Placement of insecticides in the whorls of maize and sorghum plants is the most effective method for controlling *B. fusca* and *C. partellus* on maize and sorghum (Figs 667, 668).

Formulations and Placement of Insecticides

Insecticide-application methods can largely be grouped into the application of systemic insecticides as seed dressings or granules with planting, or contact and stomach insecticides applied as granules or sprays to the plant itself.

GRANULAR APPLICATION TO PLANT WHORLS

Granular application of insecticides in the whorls of maize and sorghum plants seems to be the most effective and economic method for control of stem-borer species which feed in whorls. In an analysis of the economic injury level with differently priced insecticides, Seshu Reddy and Sum (1992) found granular trichlofon to be the most economic insecticide for *C. partellus* control on maize in Kenya. Insecticide granules have several important advantages for borer control in maize, since they are easy and safe to handle, can be applied by hand and can be applied accurately and in a controlled manner (Walker, 1972; Walker and Hodson, 1976). Granules disintegrate in the water that is often found in plant funnels and are retained there under conditions of wind and rain (Walker, 1961; Jotwani, 1983).

Jotwani and Young (1972) and Kishore (1980) demonstrated that granules were superior to sprays against stem borers. Recommendations for control of other stem borers, such as *Sesamia* spp. and *Maliarpha separatella*, include granular application of BHC, carbofuran, chlordimeform, chlorfenvinphos, diazinon, fensulfothion, mephosfolan and salithion (COPR, 1976). Granular application of stem-borer insecticides has been found to be effective against *B. fusca* and *C. partellus* in South Africa (Andersen and Wessels, 1959; van den Berg and van Rensburg, 1993) and East Africa (Mathez, 1972). Ahmed and Young (1969) found that granular formulations of endrin, lindane and carbaryl resulted in economic yield increases in sorghum infested with *C. partellus*. While no data are available on chemical control of *C. sacchariphagus*, which occurs in Mauritius, Madagascar and Réunion, it was reported that 0.25% demeton granules gave 80% control of this pest on sorghum in China (Anon., 1977).

Granular formulations of beta-cyfluthrin at a very low concentration of 0.5 g.a.i. kg^{-1} was found to be highly effective against *C. partellus* in South Africa (van den Berg and van Rensburg, 1993). The latter insecticide is registered for *B. fusca* and *C. partellus* control on maize in South Africa and it is foreseen that it will play an important role in stem-borer control in small farming in future, since it is packed in small, ready-to-use hand applicators.

Recommendations to small farmers regarding the application of granular insecticides for stem-borer control in South Africa are that granules must be applied in the whorls of infested plants as well as the whorls of one or two adjacent plants on either side. The rationale of this method is that the insecticide will also control larvae that migrate from the primary infested plants to adjacent plants. Application must be done as soon as whorl-damage symptoms are observed.

Whorl application of insecticide granules can also be done by means of a tractor-mounted applicator. The registered application rate for the different whorl-applied granular insecticides in South Africa is 3–4 kg ha^{-1}. The concentration of active ingredients in the different formulations was adapted to provide for this single application rate.

FOLIAR SPRAYS

Foliar sprays by means of ground or aerial application are the most common method of stem-borer control in commercial farming systems. The most commonly recommended and applied insecticides are the pyrethroids, endosulfan and monocrotophos (van den Berg and van Rensburg, 1993; Krause *et al.*, 1996). Pyrethroids are especially important for control of stem borers, since they are known to be particularly effective against lepidopterous larvae (King and Saunders, 1984). However, restrictions in the availability of water for foliar application of insecticide sprays limit its acceptance and use by the majority of small farmers in Africa, especially in the Sahel region (Nwanze, 1991). Shortage of water was also listed as one of the constraints in insecticide usage for *C. partellus* control in Sri Lanka (Subasinghe and Amarasena, 1988).

Van den Berg and van Rensburg (1992, 1993) evaluated various foliar sprays for their efficacy in control of *C. partellus* in sorghum. The efficacies of the different pyrethroids were similar. It was also found that systemic activity and/or the addition of pyrethroids in an insecticide mixture is a prerequisite for effective control of larvae behind leaf sheaths of sorghum. Foliar spray of endosulfan was reported to be effective against *C. partellus* and *B. fusca* on finger millet in Zimbabwe (Leuschner, 1990).

Effective control of stem borers, such as *S. cretica*, which feed behind leaf sheaths requires that insecticide sprays be directed more towards the axils or loose leaf sheaths (Jotwani, 1983). Whorl and split application of insecticide sprays directed at both the whorl and the sides of maize plants was found to give effective control of *B. fusca* and *C. partellus* in maize. In sorghum, whorl application was the most effective method for application of insecticide sprays (van den Berg and van Rensburg, 1996). The use of drop-arm application, aimed at the sides of plants only, is not a viable practice for stem-borer control (van den Berg and van Rensburg, 1996). Application of carbaryl and cytrolane reduced the damage of *S. calamistis, B. fusca, E. saccharina, S. cretica* and *Chilo agamemnon* on maize in Nigeria (Adeyemi *et al.*, 1966; Saad *et al.*, 1971).

APPLICATION OF DUSTS

Application of dusts of endosulfan, phenthoate, carbaryl, malathion and BHC was reported to be effective and economically feasible for *C. partellus* control. Application of carbaryl was effective against *S. calamistis, B. fusca* and *E. saccharina* on maize in Nigeria (Adeyemi *et al.*, 1966).

APPLICATION OF SYSTEMIC INSECTICIDES

Seed treatment

In a summary of seed treatments used for control of stem borers, Kishore (1989) found that none provided effective control. The possibility of seed treatments being effective against stem borers in grain crops is very slight, the reason being the very small amount of insecticide that can physically adhere to a single seed.

In-furrow application with planting

Granular systemic insecticides applied with planting have proved to be of considerable value for the control of stem borers in a number of crops. Application of insecticide granules is, however, expensive, and the efficacy thereof may not last for the desired period. To provide effective control of stem borers on maize, systemic activity up to 7 weeks after crop emergence is needed (van Rensburg and Walters, 1978). Nzibande (1993) found carbofuran to be highly effective against *B. fusca* in maize in Swaziland. Its application limited borer damage and deadheart development and the efficacy lasted until approximately 7 weeks after crop emergence. Drinkwater *et al.* (1979) found carbofuran to be consistently superior to aldicarb, phorate and disulfoton for control of *B. fusca* in maize and that it exerts a considerable residual controlling effect on larval numbers up to 7 weeks after crop emergence. This residual effect should have a strong influence on the level of control obtained under normal farming conditions, as *B. fusca* shows a distinct preference for oviposition on plants between 4 and 6 weeks after emergence (van Rensburg *et al.*, 1989). Egwuatu and Ita (1982) found the application of carbofuran at planting of maize efficient in the control *B. fusca* and *S. calamistis* in Nigeria.

A number of systemic insecticides for application within the furrow at planting are registered for use in South Africa (Krause *et al.*, 1996). Many of these insecticides have also been evaluated for their efficacy against stem borers in other parts of Africa. Application of carbofuran granules resulted in significant control of *B. fusca* in maize (van Rensburg and Walters, 1978). Cloethocarb resulted in significant reduction in *B. fusca* numbers and increased yield in maize; however, no dosage response was observed and further experimentation was suggested (van Rensburg, 1988).

Side-dressing after crop emergence

Side-dressing with systemic insecticides is not commonly practised by farmers, if at all. Attempts at this were made in Nigeria by Egwuatu and Ita (1982), who reported that split application of carbofuran resulted in increased yields but that the increase in yield did not justify the cost of labour for side-application. Van Rensburg and Malan (1982) found that side-dressing with carbofuran provided improved control of *B. fusca* in sorghum when applied 3 weeks after crop emergence. This practice is, however, due to possibilities of misuse and the danger of insecticide residues in green maize, not registered for borer control in South Africa. *Eldana saccharina* was effectively controlled by carbofuran side-dressing at anthesis of maize in experimental plots (Bosque-Pérez and Mareck, 1991).

TECHNICAL COMPONENTS OF INSECTICIDES

The technical components of chemical insecticides can be divided into four chemical groups: organochlorines, organophosphates, carbamates and pyrethroids (Dent, 1991). Insecticides belonging to all of these classes have been used successfully against many species of stem borers in Africa.

Organochlorines

Well-known insecticides, such as DDT, endosulfan and gamma-BHC, are classified as organochlorines. The broad-spectrum activity of these compounds, their persistence and the hazard to the environment make their use in insect-pest management inappropriate (Dent, 1991). Many of these compounds have been banned, but endosulfan remains one of the most frequently used insecticides against stem borers.

The efficacious use of DDT against stem borers was extensively reported on by Walker (1957, 1960). In dust form, DDT was used successfully to control *B. fusca* and other borers in Tanzania and Kenya. Using a manual application method and avoiding liquid in the formulation meant relatively cheap and simple application. Similar work using organochlorine insecticides is summarized by Ingram (1960).

Organophosphates

The organophosphate class of insecticides contains the greatest diversity of compounds which have at one time or another been used against stem borers. These compounds are usually non-persistent and are therefore less of a threat to the environment than organochlorines. The rapid breakdown of organophosphate insecticides is an advantage as far as the principles of integrated pest management (IPM) are concerned, but it also means that the timing of application is critical to ensure a high level of efficiency (Dent, 1991).

Pyrethroids

In commercial farming systems, pyrethroids are the most commonly recommended and applied insecticides (Krause *et al.*, 1996). Synthetic pyrethroids are known for their extremely high contact activity and are particularly effective against lepidopterous larvae (King and Saunders, 1984). Pyrethroids are also recommended in combination with other insecticides for stem-borer control (van den Berg and van Rensburg, 1993; Krause *et al.*, 1996).

Carbamates

The carbamate insecticides, of which carbofuran is the best known, have a broad spectrum of activity and few possess systemic activity (King and Saunders, 1984). However, those that do have systemic activity (e.g. carbofuran) are commonly used as soil-applied systemic insecticides. The use of carbaryl in a granular formulation has largely superseded the use of its more toxic relative carbofuran, and it is currently being promoted against stem borers, particularly for small farmers.

Botanical Pesticides

The use of compounds such as neem (*Azadirachta indica*) extract or extracts from *Tephrosia vogelii* has been attempted in stem-borer control in Africa and elsewhere. Neem is indigenous to India and is now widespread in Asia, Africa, Central and South America, Australia and many other countries where it has been introduced. It has been reported that, several centuries before synthetic insecticides became available, farmers in India protected their crops with natural insecticides found in neem fruits and leaves.

Specific examples of the use of botanical pesticides to suppress cereal stem borers in Africa are few. Hassanali *et al.* (1986) fed larvae of *E. saccharina* with maize leaves deliberately contaminated with harrisonin andobacunone, limonoid extracts of *Harrisonia abyssinica* roots. Initial results showed relatively high activity. The same research group conducted similar experiments with extracts of *Tephrosia elata* and found high levels of activity of isopongaflavone against *E. saccharina* (Bentley *et al.*, 1987). Mugoya and Chinsembu (1995) observed that aqueous fresh-leaf extracts of *T. vogelii* effectively reduced the incidence of *C. partellus* damage in maize in Zambia. Piperine, from *Piper guineense*, has also been shown to have high antifeedant activities against *C. partellus* (Torto *et al.*, 1992). Particularly encouraging results were found by Melamed-Madjar *et al.* (1989) using neem extract against larvae of *S. nonagrioides*. These authors found that no pupation occurred with 0.5% and 1.0% neem extract, and only 10% of the larvae pupated at 0.1%. In Togo, Dreyer (1987) observed that whorl treatment of maize plants with pure neem kernel powder, or a 50 : 50 mixture of neem powder and sawdust, reduced populations of *S. calamistis* to virtually zero.

In Somalia, neem is grown mainly in the coastal areas. The initial aim was to grow this tree for shade, as an ornamental and for firewood and toothbrushes. The government mobilized people to plant these trees, which easily adapted to the area. In early 1980, the Somali plant-protection services and agricultural research institutes began to do a number of trials to control various insect pests, using azadirachtin. Results obtained with neem derivatives used against *C. partellus* in maize were promising. The establishment of small industries to generate neem products is particularly required in rural areas where subsistence farming is the rule.

PROBLEMS ASSOCIATED WITH INSECTICIDE USE

Although insecticide use can be of benefit to farmers in the short term, the use thereof has not been without problems. In an overview of problems associated with pesticide use, Deedat (1994) listed the following: resistance to insecticides, adverse effects on non-target species, hazards of pesticide residues and direct hazards from pesticides.

The effect of insecticides on natural enemies of stem borers in Africa has so far received little attention. Du Toit (1995), however, evaluated the effect of foliar application of pyrethroids, endosulfan and monocrotophos on the predator-species complex occurring on maize and sorghum in commercial farming systems in South Africa. Epiphytic predators were found to be more susceptible to insecticides than epigeal predators. The predator complex was most adversely affected by monocrotophos, whereas endosulfan was the least disruptive (du Toit, 1995). While granular formulations of insecticides applied in plant whorls can generally be considered not to have a deleterious effect on natural enemies of stem borers, du Toit (1995) found that it had a significant influence on the predator complex of aphids which commonly occur in whorls of maize and sorghum plants.

Despite the proved efficacy of insecticidal control of stem borers under certain circumstances, a balance must be drawn between the pros and cons of insecticide use, and many less favourable reports have also been published. In a study of *M. separatella* control on rice in Nigeria, it was found that applications of carbofuran had no significant effect on total yield or individual yield components, despite significantly reducing damage to the crop (Ukwungwu, 1985a, b). Ho and Kibuka (1983) concluded that chemical control of stem borers in upland rice in Kenya was too expensive. In two studies at the International Crops Research Institute for the Semi-Arid Tropics (ICRISAT) Sahelian centre in Niger, results showed that, under low levels of stem-borer infestation (caused by *C. ignefusalis*), an unprotected pearl-millet crop gave higher yields than one that was protected by an insecticide (Taneja and Nwanze, 1989). Similar levels of compensation for borer damage on sorghum were also observed by Ingram (1958) and Harris (1962). The Food and Agriculture Organization (FAO) programmes involving stem-borer control on millet in the Gambia have adopted an integrated approach in an attempt to overcome the problems of overuse of pesticides (Zelazny *et al.*, 1985). Leslie (1982) found that the incidence of the stem borer *E. saccharina* increased with applications of pesticides in the soil, due to the effect on the natural enemies. Toxicity studies on deltamethrin, used against rice stem borers in Ivory Coast and other borers species in Africa, indicate that it can be dangerous to non-target organisms, such as fish and aquatic invertebrates (Dejoux, 1983). Chemical control of stem borers outside Africa has also met with a number of problems; for example, the rice stem borer *Chilo suppressalis* has developed resistance to organophosphorus insecticides in Asia (Konno *et al.*, 1986).

From these examples, it is apparent that the success of insecticide applications is not always guaranteed and some level of know-how is required for their effective use. The concept that the pesticide tactic requires less extension input, including the 'simplified' package, has clearly been a pathway chosen in error. Pesticide use is obviously not a simple approach but a highly complex decision-making strategy, which does, in fact, require high levels of extension if it is to be used appropriately in developing IPM programmes (Schaefers, 1994).

The development of insecticides which can be shown to be environmentally safe, which do not need water or any specialized application equipment and which can be made available to small-scale farmers as an economically viable proposition must be taken seriously as one of many alternatives for, or components of, a stem-borer control programme.

INTEGRATED PEST MANAGEMENT

While chemical control remains the most important tool against stem borers in commercial agriculture, current methods of stem-borer control in small-scale and subsistence farming in Africa are relying increasingly on cultural practices, development of resistant crop varieties, augmentation and encouragement of natural enemies and classical biological control.

It must be remembered that chemical control essentially involves the treatment of symptoms only. This treatment does not address the origin of the problem presented by high levels of stem-borer populations. A sustainable solution to stem-borer problems lies in long-term suppression of pest numbers. Van den Berg and van der Westhuizen (1998) and van den Berg (1998) demonstrated the impact of resistant sorghum plants on *C. partellus* populations in southern Africa. Kfir *et al.* (1989b) demonstrated the effect of cultural control through different tillage practices on *C. partellus* populations.

27 Cultural Control

J. van den Berg,[1] A.F. Nur[2] and A. Polaszek[3]
[1]*Agricultural Research Council Grain Crops Institute, Private Bag X1251, Potchefstroom 2520, South Africa;* [2]*Former Director, Plant Protection Service, Mogadishu, Somalia;* [3]*CABI Bioscience UK Centre, c/o Entomology Department, The Natural History Museum, London SW7 5BD, UK*

INTRODUCTION

Cultural control of stem borers implies the manipulation of the environment in such a way as to render it unfavourable to the pest. Cultural control is a prophylactic method and can rarely be used as a tactical means of control (Dent, 1991). Control through manipulating the crop and land is among the oldest traditional practices and, although on its own it is unlikely to reduce pest infestation to below economic-injury levels, it plays an important role in any pest management system. Cultural control is considered the first line of defence against pests and includes techniques such as crop rotation, tillage methods, intercropping and destruction of crop residues. Agronomic practices, of which some were developed over long periods of time, have built-in pest suppression mechanisms, apart from agronomical benefits. Manipulation of planting date, crop rotation and tillage practices are the three agronomic practices that can directly affect crop yield (Dent, 1991).

Cultural control is the most relevant and economic method of pest control for farmers in Africa. Many other control options exist but these are often of no use to the majority of farmers. Reasons for this are the cost of insecticidal control and other disadvantages associated with its use, lack of availability of stem-borer resistant maize, sorghum, millet and sugar-cane varieties, and the fact that biological control on its own will not reduce pest-infestation levels below economic-injury levels. Since some cultural-control measures have become routine practices employed by farmers, they can, after minor changes towards being more effective, be profitably integrated into the farming system. Furthermore, socioeconomic aspects may also influence the acceptability of management practices, since practical realities may have an effect on the adoption of otherwise effective cultural-control measures.

While it is true that many cultural-control practices are labour-intensive, they are readily available to the farmer and in most cases do not entail any extra investment in equipment to apply pest control. Cultural-control methods generally have little adverse effect on the environment unless the particular practice involves activities such as intensive cultivation or large-scale burning of crop residues which could otherwise have been used as organic input into the soil. Sometimes, even a slight reduction in stem-borer population levels caused by these practices may delay build-up of stem borers or reduce plant damage.

Cultural-control measures for stem borers could be aimed at various stages in the insects' life cycle. Although moth oviposition and larval survival during the growing season can be targeted by control measures such as intercropping and manipulation of planting date, larval survival during the off-season remains the most susceptible and therefore exploitable stage in the life cycle of stem borers.

Reviews of stem-borer cultural control have been published by Lawani (1982), Verma and Singh (1989), Seshu Reddy (1985a, 1990) and Minja (1990). These practices include the manipulation of wild host plants, management of crop residues and planting dates, crop rotation, intercropping, plant density, physical removal of infected plants, soil-fertility management, trap crops, variety selection and water management. Each of these could affect a specific stage during the insects' life cycle and the aim is to render the environment unfavourable during that stage.

IMPORTANCE OF BIOECOLOGY IN CULTURAL CONTROL

Understanding of adult and larval behaviour and of the relationships between stem borers and their respective crops is important for the development of efficient stem-borer management strategies. Changes in agronomic practices should not lead to other pest or disease problems. A cultural practice that effectively reduces a pest problem but creates or intensifies another pest or disease problem or proves to be agronomically unsound will be self-defeating.

Differences in the bioecology of *Eldana saccharina* in South and eastern Africa is a case in point on the importance of knowledge of pest behaviour in making control recommendations. In South Africa, sugar-cane farmers are recommended to cut stalks very low and to leave the tops, which are cut off at harvest, in the field. The reason for this is that *E. saccharina* larvae commonly occur in the lower parts of stalks and very few larvae are found in the tops of plants. In East Africa, however, *E. saccharina* larvae largely occur in upper plant parts and farmers are strongly recommended not to leave tops in fields since these residues would house the carry-over population.

Practically all stem-borer species enter a rest period towards the end of the cropping season, which they spend in crop residue left on fields after harvest. *Chilo partellus* go into diapause in the lower parts of plant stalks (Kfir, 1990b). In West Africa, larvae of *Coniesta ignefusalis* enter a facultative diapause in dry millet stems and become active again with the onset of the next season's rains (Youm and Gilstrap, 1994). In Nigeria, Kaufmann (1983) found that larvae of *Busseola fusca, E. saccharina* and *Sesamia calamistis* pupated in crop residues below the soil surface and observed a higher incidence of these borers and *Mythimna unipunctata* in no-tillage plots. In East Africa, *Chilo orichalcociliellus, C. partellus, E. saccharina* and *S. calamistis* can be found in stalks after harvest (Seshu Reddy, 1983a; Warui and Kuria, 1983). In Ethiopia, a considerable proportion of *B. fusca* larvae survives in stubble (Yitaferu and Gebre-Amlak, 1994), while, in South Africa, a large proportion of larvae survive the adverse winter conditions in the bases of maize stalks below soil level.

COMMON CULTURAL PRACTICES

Wild Host Plants

Although wild host plants of stem borers play an important role in the carry-over of pest populations from one season to another, the destruction thereof may not be as advantageous as reported in earlier studies (Ingram, 1958; Nye, 1960; Seshu Reddy, 1985a). Destruction of wild host plants of stem borers has been advocated by many researchers (Lawani, 1982; Seshu Reddy, 1983a, 1985a). Nye (1960) also suggested destroying crop residues and wild sorghum species. In Sudan, sorghum was observed to be a reservoir for *Sesamia cretica* infestation in sugar cane (El-Amin, 1984).

However, recent studies on the role of wild host plants in the ecology of stem borers in Kenya indicated that wild host plants, especially graminaceous hosts, can play an important role in the management of stem borers (Z.R. Khan, International Centre of Insect Physiology and Ecology (ICIPE), personal communication).

At ICIPE, an integrated technology for stem-borer management is being developed, using selected grasses of economic importance. It has been shown that several wild grasses act as trap plants and can provide natural control of stem borers. Although the stem borers oviposit heavily on some of these attractive grasses, only a few are able to complete their life cycles (Khan and Randriamananoro, 1995).

Napier grass and Sudan grass, commercial fodder grasses, can be used as trap plants for the management of stem borers. It was reported that stem-borer infestation on monocropped maize was several times more than on a maize crop surrounded by Sudan grass (Z.R. Khan, unpublished data). However, stem-borer infestation on Sudan grass was quite high, suggesting that Sudan grass can reduce stem-borer infestation by buffering the crop against stem-borer attack. Planting Sudan grass around the maize field also increased the efficiency of natural enemies. As compared with a maize field, where only 4.8% *C. partellus* and 0.5% *B. fusca* larvae were parasitized, 18.9% *C. partellus* and 6.17% *B. fusca* larvae were parasitized in a maize + Sudan-grass field. Planting Sudan grass around maize fields can reduce stem-borer infestation on maize and can thus increase crop yield. A significant increase in maize yield, from 3.44 t ha^{-1} in the control to 4.45 t ha^{-1} in maize + Sudan grass plots, was observed in field trials (Z.R. Khan, unpublished report).

Intercropping maize with non-traditional hosts and non-host companion plants can have a marked impact in reducing borer infestation on maize. In the 1995/96 short rainy season, planting maize into an already established crop of molasses grass (*Melinis minutiflora*) gave dramatic reductions in stem-borer incidence and damage (Z.R. Khan, unpublished data). More studies are under way to determine the optimum density of the non-host plants to protect the maize crop, as well as to avoid interspecific competitions.

It has also been established that several wild grasses, such as *Pennisetum purpureum, Echinochloa pyramidalis, Panicum* sp., *Cyperus* sp. and *Phragmites* sp., commonly growing near farmers' fields provide important refugia for natural enemies of stem borers, such as *Cotesia flavipes* and *Cotesia sesamiae*, after the cereal crops are harvested (ICIPE, 1996). Some species of borers, such as *Bactra stagnicolana*, *Phragmataecia boisduvalii* and *Poeonoma* sp., associated with these perennial wild hosts can be important hosts for *Cotesia* spp. and other natural enemies of crop pests. These host plants can therefore be very important in the ecology of natural enemies.

An increased understanding of the influence of plant- and associated arthropod-species diversity on pest populations will lead to the development of recommendations for utilizing grasses and their associated arthropod-diversity resources for pest management. At ICIPE, efforts are under way to formulate a push–pull strategy for diverting stem borers from crop plants to 'trap plants'. In this strategy, *M. minutiflora*, a grass that is repellent to ovipositing stem-borer females will be used along with Sudan grass or Napier grass, which is highly attractive to ovipositing females. Being highly attractive to natural enemies, Sudan grass can also help in attracting the natural enemies to a crop field.

This practice of using grass species that can also be utilized as animal fodder fits ideally into farming systems in Kenya and has additional advantages in that tall grasses can also serve as windbreaks in preventing lodging of maize.

Management of Crop Residues

Crop residues, in the form of stubble and stalks, form the primary source of borer infestation in the following season. These carry-over populations can be reduced through various agricultural practices, the most important being destruction of crop residues. However, the feasibility of managing crop residues will always have to be assessed before recommendations are made, since practices differ between communities and may also be influenced by the scale of farming.

In Nigeria, larvae of both *B. fusca* and *C. ignefusalis* spend the dry season in crop residues (Harris, 1962). In central Sudan, postharvest crop residues of maize and sorghum have been observed to be reservoirs of large populations of *S. cretica*, these infested stalks often being left standing adjacent to an off-season ('winter') crop (A. Polaszek, personal observation) (Fig. 669).

Youm and Gilstrap (1994) found that approximately 37% of *C. ignefusalis* larvae survived the dry period in millet stems, making residue destruction an attractive control option. However, farmers made use of residues and commonly stored more than 1000 millet stems in or near villages, thus creating a reservoir for infestation in the following season. The usual method of storing stems is to stack them in the open or under trees where they are left until commencement of the rains before being taken to villages. These overwintering populations are high enough to ensure the widespread reinfestation of new crops and, since they are concentrated in villages, early stem-borer attack is noticeably more severe in areas surrounding villages. This problem was solved through the development of practical measures, such as early cutting of stalks and laying them on the soil surface. This practice reduced the residual population of borers in millet from 16 per stem when stems were not cut to three per stem (Youm *et al.*, 1993b). Gebre-Amlak (1988b) obtained similar results with *B. fusca* in maize and sorghum in Ethiopia and indicated the duration of horizontal placement of stalks to be important in reducing the number of surviving larvae. Horizontal placement for a period of 4 weeks resulted in a 97% mortality of larvae in maize and 100% in

Table 27.1. Insect pests of maize which may be favoured by the introduction of reduced tillage (adapted from Walters, 1975).

Pest species	Stage regarded as most sensitive to cultivation
Lepidoptera	
Chilo partellus	Hibernating larvae
Busseola fusca	Hibernating larvae
Sesamia calamistis	Hibernating larvae
Euroa spp.	Larvae and pupae
Helicoverpa armigera	Hibernating pupae
Coleoptera	
Astylus atromaculatus	Larvae and pupae
Buphonella metallica	Larvae
Heteronychus arator	Pupae
Adoretus cribrosus	Larvae and pupae
Gonocephalum simplex	Larvae and pupae
Prostrophus spp.	Pupae
Elateridae	Pupae

sorghum. Päts (1996) ascribed the high level of mortality in horizontally placed stalks to the effect of temperature – more specifically, the reaching of the thermal threshold for survival. Destruction of crop residue therefore seems the most appropriate method of destroying off-season borer populations.

Destruction of crop residues, however, cannot always be recommended. The usefulness of sorghum and millet stems in the day-to-day lives of African farmers is illustrated by many examples. In Nigeria, Harris (1962) observed millet and sorghum stems being used for building houses and fences, while it was also used as fuel and occasionally as bedding for livestock. In South Africa, pearl-millet stems are used as fencing and to build roofs for huts (Fig. 670). In the Gambia, both sorghum and millet stalks are used for fencing purposes (Sagnia, 1983). In Kenya, cut maize stems are used as boundaries of terraces on slopes (Scheltes, 1978). Maize stems also play an important role in some tomato-farming systems in Kenya, where stems are used as stakes, holding up runners of tomato plants.

TILLAGE

Generally, tillage systems that bury crop residue act as a sanitation measure and reduce plant diseases and pests. Soil tillage may result in mechanical damage of overwintering larvae, bury them so deep that moths cannot emerge, or bring them to the surface, where they are exposed to natural enemies and adverse weather conditions. Destruction of crop residues was advocated as long ago as the 1920s, when Mally (1920) and Wahl (1926) recommended the use of railway metal or barbed wire, weighed down by scrap metal (probably leftovers from cannons and other war machinery which the Boers took from the British during the Anglo-Boer war), to be dragged across fields by means of oxen.

Reduced tillage not only results in increased survival of *B. fusca* larvae; many other insect pests of maize and sorghum are likely to be favoured by the introduction of reduced tillage (Table 27.1).

Table 27.2. Numbers of *B. fusca* and *C. partellus* moths emerging from diapause larvae in sorghum fields after different stubble treatments.

	Number of moths ha^{-1}				
Treatments	No till	Slash	Slash/remove	Slash/plough/disc	Slash/remove/plough
*C. partellus**	226,087	74,783	76,522	–	48,696
C. partellus†	194,783	53,913		12,174	
B. fusca†	90,435	26,087		5,217	

* Kfir *et al.* (1989b).
† Kfir (1990b).

Tillage practices in commercial agriculture

Destruction of crop residues can easily be done in highly mechanized, commercial farming systems. Tillage practices are viable options for *B. fusca* and *C. partellus* control in commercial agricultural systems, where large areas of monocrop maize or sorghum are planted and between 90,000 and 226,000 larvae overwinter per hectare (van Rensburg and van Hamburg, 1974; Kfir, 1990b). Kfir (1990b) and Kfir *et al.* (1989b) showed that conservation tillage is detrimental from a stem-borer control viewpoint (Table 27.2).

However, despite the proved success of various tillage methods in the destruction of crop residues and concomitant reduction in the survival rate of stem-borer larvae, minimum tillage is on the increase in South Africa and many other parts of the world. This can be ascribed to practical aspects which are considered when fields are to be cultivated after harvest. In mixed agricultural systems (animal husbandry together with crop production), which is common in South Africa, winter grazing of fields plays an important role, together with moisture conservation, while other factors, such as soil erosion and fuel and labour costs, are also considered before cultivation of fields during winter. The choice of cultivation method is therefore usually dictated by soil-moisture conservation measures. It is therefore not surprising that the benefits of not cultivating after harvest often outweigh the following season's possible stem-borer problem.

Minimum tillage, especially in sorghum fields, holds a further disadvantage in that volunteer growth, which is continuously produced during winter, serves as food for borer larvae and aphids. In South Africa, it may also provide food for a large proportion of the larvae from the first-generation flight of *C. partellus*, which occurs before any maize or sorghum has been planted, resulting in an increase in initial borer population numbers during the early crop-growth stages.

It is foreseen that changes from conventional tillage to conservation tillage in South Africa will result in increased stem-borer incidence in commercial farming systems (Walters, 1975). House and Alzugaray (1989) came to a similar conclusion regarding *Ostrinia nubilalis* on maize in Ohio, stating that no-tillage systems require more intensive pest management than conventional cropping systems because arthropod populations are enhanced at all trophic levels.

GRAZING

Gebre-Amlak (1988b) found cattle grazing to be a possible factor in reducing population numbers of overwintering *B. fusca* larvae. However, in a trial where cattle fed on maize plants after harvest, Duerden (1953) observed virtually no reduction in *B. fusca* larval numbers, since cattle preferred to feed on leaves and upper plant parts. The practicality and impact of using grazing as a major component of stem-borer management is therefore doubtful. In the Highveld region of South Africa, maize stubble forms an essential part of the fodder-flow system in the maize/cattle farming systems, and cattle feed on maize fields for up to 4 months after harvest. Judging by the high *B. fusca* infestation levels which annually occur in this area, the impact of grazing on borer numbers seems to be low.

BURNING OF CROP RESIDUES

Burning of crop residues is an effective method of controlling stem-borer numbers; however, many disadvantages are associated with this practice. In the majority of African farms where fertilizer is not applied, crop residues are the only input going back into soils. It is especially in areas where the organic content of soils is low that burning cannot be recommended, since, apart from no organic input into soils, soil erosion due to wind and rain is a serious problem. Furthermore, it takes considerable effort to collect and stack crop residues for burning, because fields often cannot be easily set alight, as is done with sugar cane, where plant population is very high and large quantities of dead leaf material are present.

Since crop residues are often valuable to subsistence farmers, Adesiyun and Ajayi (1980) have advocated partially burning sorghum stalks which kills up to 95% of larvae, while at the same time curing the stalks and making them more suitable as building materials or for use as firewood. In Botswana, the destruction by burning of all crop residues immediately after harvest is recommended for control of *C. partellus* and *B. fusca* (Ingram *et al.*, 1973). In Tanzania, almost complete eradication of *C. partellus* on maize and sorghum was obtained through burning of crop residues and stubble (Duerden, 1953).

Manipulation of Planting Date

Manipulation of planting date can be an effective measure to escape serious borer attack and should not merely be seen as purely of agronomic concern. In order to efficiently utilize planting date as a means of escaping borer damage, it is important to know the local seasonal patterns of borer life cycles. Planting dates can then be planned to ensure that the most susceptible stage of crop growth does not coincide with periods of peak moth activity. However, any effect which the date of planting might have on stem-borer incidence is clearly dependent on local or regional conditions. In the semiarid tropics, where rainfall is variable and unpredictable, the physiological and ecological advantages are nearly always with the early-sown and established crop. In such areas, delayed or late sowing is not desirable, since the yield of late-sown crops is low, even if free from stem borers (Seshu Reddy, 1990).

Manipulation of planting date as a possible control measure has been evaluated against many stem-borer species in Africa. Youm *et al.* (1993b) found manipulation of planting date a not very promising control measure to reduce *C. ignefusalis* infestation. Early planting to escape *C. partellus* damage is not a viable option in South Africa if planting is not done over a wide area. Isolated early plantings are often severely attacked by this pest, often necessitating programme sprays. It is recommended that sorghum in South Africa be planted after mid-October to avoid infestation from the first moth peak (van Hamburg, 1979). Manipulation of planting date is commonly recommended for *B. fusca* control in South Africa and Zimbabwe, since this borer has a distinct moth flight pattern with moths being virtually absent for a period of 2–4 weeks between the first- and second-generation moth flight (van Rensburg *et al.*, 1985; Sithole, 1989c). Similar observations were made in Ethiopia, where late planting of maize necessitates insecticide application for *B. fusca* control (Gebre-Amlak *et al.*, 1989). In Nigeria, Harris (1962) found early plantings of millet to suffer little damage, while *B. fusca* totally destroyed late plantings. In coastal Kenya, stalk-borer infestation accounted for 28% of yield loss in late-planted maize, compared with 9.7% in early-planted maize (Sithanantham *et al.*, 1995). In Tanzania, Swaine (1957) reported that plantings done 2–3 weeks after the first rains were heavily attacked by *B. fusca*, while those planted a few weeks later largely escaped damage. In Malawi, planting date was observed to influence infestation levels of *B. fusca* and *C. partellus* on sorghum; however, the choice of optimum sowing date depends on the variety of sorghum (Mchowa, 1990).

Despite the many benefits associated with manipulation of planting dates, farmers' choice of sowing dates is affected by many constraints and it is unlikely that a change in sowing date alone will result in higher sustainable yields (Nwanze and Mueller, 1989). In Africa, where rain is the most limiting factor to crop production, this practice is not always practical and farmers often plant with the first rains. It is also common practice for farmers to stagger planting dates, in order to spread the risk of total crop failure due to drought periods. Furthermore, factors, such as late sowing, drought, poor soil and intersowing of millet and sorghum, which affect the optimum development of the crop, will generally lead to more severe stem-borer damage.

Crop Rotation

Stem-borer damage may be considerably reduced by following the principle of crop rotation, in which planting of a host crop is followed by a non-host. A new generation of an insect pest that might have overwintered in the vicinity of the host crop will be faced with a different, non-host crop in a subsequent season (Dent, 1991). Crop rotation is generally most effective against pest species that have a narrow host range and a limited range of dispersal (Wright, 1984).

Although large-scale rotation of crops seems to be an effective method for stem-borer control, it is often not practical. In commercial farming systems, where monocropping allows this practice to be followed, it is unacceptable, due to reduced prices often paid for crops that could be used in a rotation programme. Furthermore, farm implements and infrastructure and the animal-husbandry factor in large-scale farming systems are of such a nature that large-scale changes in crop types are unacceptable. Not many examples of crop rotation as an effective method for borer control are mentioned in the literature. However, Sagnia (1983) reported that millet and sorghum in the Gambia are successfully rotated with groundnut. Where crop rotation is feasible, it will increase the distance from newly planted maize or sorghum to *C. partellus* eclosion sites and thereby diminish the immigration of mated females (Päts, 1992). In farming systems suited for rotation of sorghum, millet and maize with non-host crops, such as cowpea, cotton or fallow lands, the potential exists for suppression of borer populations. However, since the majority of farming systems in Africa consist of intercropping systems with a large number of crops, each playing an important role in food security, the impact of this method of control seems limited.

Intercropping

Intercropping systems are ideally suited to small farmers in Africa, since they provide a greater total land productivity and insurance against the failure or unsure market value of any single crop (Risch *et al.*, 1983). The majority of farmers in the semiarid tropics grow polycultures of grains, such as maize, pearl millet and sorghum, or mixtures of these crops with legumes. When two crops of similar plant type and host range for a particular insect pest are intercropped, the colonization, establishment and population build-up increase. On the other hand, intercropping of non-hosts brings about a considerable reduction in the incidence of most insect pests on the host species (Seshu Reddy, 1990). In general, crop mixtures reduce pest incidence but the individual components may not benefit equally (Nwanze and Mueller, 1989). Apart from reducing insect attack, intercropping also has the agronomically important effect of improving soil fertility.

Intercropping as a means of stem-borer control has received a great deal of attention. Most studies conclude that intercropping reduces the incidence of borers. In particular, studies at ICIPE, in Kenya, have concentrated both on the practice of intercropping itself and on the ways in which the developed systems can be adapted for use by small-scale farmers (e.g. Omolo and Seshu Reddy, 1985; Omolo, 1986; Seshu Reddy and Masyanga, 1988; Dissemond and Hindorf, 1990; Oloo and Ogeda, 1990). These studies in Kenya concentrated on maize and sorghum, usually intercropped with cowpea. In West Africa, maize, millet and sorghum intercropping were studied by Adesiyun (1983), Gahukar (1989) and Kwapong (1990). Päts (1992) found that 80% of *C. partellus* larvae on maize disappeared within 1 week from hatching, while 25% established on neighbouring plants. They concluded that intercropping might be an effective way of reducing damage caused by migrating larvae. This suggestion was confirmed by results obtained by Ampong-Nyarko *et al.* (1995), who observed that 30% of *C. partellus* oviposition in a maize/sorghum/cowpea intercropping system was on cowpea and that the number of larvae reaching host plants from the cowpea reduced with distance.

Maize/cassava intercropping systems in Nigeria were found to reduce larval numbers of *E. saccharina*, *B. fusca* and *S. calamistis* by approximately 50% compared with monocropped maize (Kaufmann, 1983). These results were ascribed to the more bulky cassava canopy, which conceals the maize plants, rendering them less accessible for oviposition. Similar observations were made by Mahadevan and Chelliah (1986), who found that canopy coverage of lablab and cowpea leaves in early crop-growth stages of sorghum reduced oviposition rate by *C. partellus*.

In 1976, agricultural research centre workers at Afgoi, Somalia, began advising subsistence farmers to intercrop their maize and sorghum with legumes in order to control stem borers and improve their diets, a practice with which many Somali farmers were already familiar (A.F. Nur, personal observation). In a study of the effects of maize/cowpea intercropping on *C. partellus*, *C. orichalcociliellus* and *S. calamistis* and their natural enemies in Kenya (Skovgard and Päts, 1996), it was observed that oviposition was not affected by intercropping, but significantly fewer larvae and pupae were found in the intercrop. Furthermore, egg parasitism by Hymenoptera was significantly higher in the intercrop, reaching as high as 80%. Larval and pupal parasitism remained between 3 and 20%, but neither of these were affected by intercropping or host density. They concluded that maize intercropped with cowpea has only limited potential for stem-borer control in maize.

The effect of a given intercropping system on pest incidence is influenced by many factors, such as location, variety of the crop, intercropping pattern and season. Recommendations will therefore have to be site-specific. Recommending changes in cropping patterns (and crops) without consulting local communities in the development thereof will result in non-adoption, as has been the case with many of the technologies developed on-station and reported in literature. Existing cropping practices and rural community structure may be subject to manipulation but rarely, if ever, to replacement (Amoako-Atta *et al.*, 1983). In a review of management options for farmers in the semiarid tropics, Nwanze and Mueller (1989) observed that, although there is scientific evidence that intercropping affects stem-borer ecology, there are no studies showing that farmers grow specific sorghum intercrops to exploit this effect.

Monocropping is practised in commercial farming systems, since it provides for conditions of intensive farming and highly mechanized inputs. In more intensive agricultural systems, the monocrop is maintained because with such high-input systems it remains economical even though crops grown in this way can be extremely susceptible to insect-pest attack (Dent, 1991). Strip intercropping of maize with sunflower or cowpea is occasionally done in commercial farming systems in South Africa, and pest control is done separately on strips.

Planting Density

In subsistence farming systems, it may be difficult to manage plant density as a tool for stem-borer control. Although sowing density could be manipulated to some extent, plant density after seedling emergence is influenced by many factors, including soil insect pests, seedling damping-off disease, seed-germination rate and seedling vigour, all factors which are largely beyond the control of the small farmer. In South Africa, the majority of small-scale sorghum and pearl-millet farmers sow seed mixtures of these grains, together with cowpea and various species of melons on unploughed fields, after which fields are then ploughed. This practice allows for only a limited degree of management of plant density.

Studies on *C. partellus* in maize planted at different densities (Firempong, 1988) and on *C. ignefusalis* in millet (Gahukar, 1989) showed no significant difference in oviposition or borer incidence. In sorghum infested by *B. fusca*, the lowest incidence of deadheart occurred at the lowest plant density, and initial plant density did not have a significant effect on either the degree of internal damage or the number of damaged stems (van den Berg and van Rensburg, 1991b). Van Rensburg *et al.* (1988d) found that a reduction in row width led to progressively more larvae being able to reach adjacent plant rows through migration and this in turn resulted in progressively more damaged plants. At a 0.9 m interrow spacing, which is common in commercial agriculture, migration of *B. fusca* larvae can take place from the hatching site over a distance of at least 3.6 m in one direction (over four rows). Harris (1962) found larvae to migrate up to 2.4 m. Studies in Nigeria found that increasing maize density resulted in higher borer incidence (Ogunwolo *et al.*, 1981). Moyal (1995) suggested that maize be grown with larger interrow spacings rather than with reduced plant number within the row to reduce *B. fusca* and *E. saccharina* damage.

Physical Control and Removal of Infested Plants

In cases where the size of fields allows for physical removal of infested plant whorls, this practice may result in control of borers on individual fields. In Nigeria, some farmers regularly practise removal of sorghum deadhearts caused by stem borers, but infestations are usually well established by the time deadhearts appear (Harris and Nwanze, 1992). Removal of deadhearts in sorghum and of shoots of sorghum or other cereals which show early pinhole damage (i.e. containing young larvae) has been found to be a successful practice in reducing infestations of *C. partellus* (Kishore and Jotwani, 1982; Seshu Reddy, 1985a). It is, however, important to remove infested whorls from the field, since larvae may leave rogued plants and reinfest other plants if they are not destroyed.

A study by Singh and Sandhu (1979) showed that most eggs of *C. partellus* are laid on the lower leaves of maize (up to fourth), and removing these reduced the damage. As this is not practicable on a large scale, they recommended the alternative practice of earthing-up of the crop to bury the lower leaves.

Volunteer Plants

Volunteer plants originate from seed that survives the off-season on fields. There is usually a large age difference between volunteer plants and the planted crop, since volunteer plants usually start growing before large-scale plantings are done. Since stem-borer moths prefer to lay eggs on plants older than 2–3 weeks, volunteer plants are preferred to seedlings for oviposition. If volunteer plants are hoed, increased incidence of stem-borer damage and especially deadheart symptoms can be observed in the newly planted crop, since larvae leave the hoed plants and infest young seedlings. Under normal conditions, a large proportion of larvae would have stayed on larger plants until pupation. The infestation of seedlings by large borer larvae results in deadheart development as well as large-scale formation of tillers, which may not produce any yield at all.

Hoeing of volunteer plants is not generally done by small-scale farmers. However, in commercial farming systems these plants are considered to be 'weeds' since they could also harbour cutworm (*Agrotis* spp.) larvae, which cause serious damage to the newly planted crop. Volunteer plants could be controlled through a number of methods. These include the use of harvesters which do not waste seed, grazing of sheep after harvest to rid fields of fallen grains, grazing of livestock on fields after germination of volunteer plants but before planting, or hoeing at least 1 week before planting. If plants are hoed after the new crop has been planted, they must be removed from the field and destroyed.

Fertilizer Use

Mineral nutrition is one of the major parameters influencing the health of plants as food for herbivores. This seems to be of particular importance in plant/stem-borer interactions, with the general tendency being an increase in borer infestation and survival with increased nitrogen content of plants. The use of fertilizers, especially nitrogen, can influence the response of insects, such as *C. partellus*, to plants and of plants to damage. Van den Berg and van Rensburg (1991c) indicated that sorghum plants which did not receive fertilizer or irrigation were less preferred by *C. partellus* for oviposition. However, while no differences were observed in *B. fusca* oviposition on fertilized and unfertilized maize plants, preference was shown for thick-stemmed plants (van Rensburg *et al.*, 1989). Sétamou *et al.* (1993) observed increased survival of *S. calamistis* larvae with increased nitrogen content of maize plants and suggested that the shorter larval period resulting from improved feeding status of plants may result in an increase in the number of stem-borer generations per year. More intensified agriculture in developing countries, accompanied by higher inputs, such as fertilizer, may therefore result in an increase in pest status of stem borers.

The possibility of exploiting nitrogen fertilization as a possible tool in integrated management of the millet stem borer *Acigona ignefusalis* was discussed by Ajayi (1990). He observed that time of nitrogen application influenced borer incidence on plants and suggested that the manipulation of time of nitrogen application may achieve a compromise between using low levels of nitrogen for low stem-borer infestation and using high levels for better yields. Ukwungwu (1985a) demonstrated that damage to rice in Nigeria by *Maliarpha separatella* increased with the application of fertilizer. Reduction in nitrogen (N) fertilization rate from 50 kg N ha^{-1} to 30 kg N ha^{-1} is recommended in areas where *E. saccharina* is a problem on sugar cane in South Africa (SASA, 1994).

Manipulation of fertilizer rate for pest control purposes in subsistence farming systems is largely impractical. Many farmers do not have access to fertilizers, but, if they have the means, fertilizers are used on cash crops. The lack of fertilizer use is a major constraint to crop production in Africa. To expect farmers who can afford fertilizers to use less than needed, just to reduce pest fecundity or damage, is unrealistic. The benefit of increased fertilizer use will largely result in yield increases above that of expected yield loss due to higher larval survival and fecundity.

Choice of Variety

The planting of a short-season variety may result in the crop escaping damage, due to it maturing before stem-borer infestation levels are very high. Early maturation also shortens the susceptible period of the crop. For example, the whorl stage (most susceptible stage) of a sorghum variety that flowers 70 days after planting is approximately 3 weeks shorter than that of a long-season variety flowering at 90 days after planting. The use of a short-season millet variety was recommended by Sagnia (1983) in the Gambia. Van Rensburg *et al.* (1988d) observed more pronounced yield losses in medium- and long-season maize cultivars than in short-season cultivars.

Recommending short-season varieties is not always practical, since farmers often prefer varieties of sorghum and pearl millet to mature late in the season. This is done to delay maturation and the harvesting process, which then provides time for other farming activities and harvesting early-maturing crops. Reduced stem-borer-inflicted yield loss in short-season varieties should not be the only criterion in recommendation of varieties to farmers. Yield potential of short-season varieties is lower than that of long-season varieties and the benefit of planting a long-season variety with a high yield potential generally outweighs the importance of escaping stem-borer damage. Van den Berg *et al.* (1993) observed that some short-season sorghum hybrids suffered negligible yield loss under *C. partellus* infestation but also indicated that the yield of high-potential, long-season hybrids suffering a high percentage yield loss was still above that of the short-season hybrids.

Trap Crops

Not many reports on effective use of traps crops occur in the literature. In Zimbabwe, Jack (1922, 1928) recommended the use of sorghum or maize as trap crops, while du Plessis (1936) reported it not to be effective in South Africa. In Tanzania, Duerden (1953) found that trap cropping was not effective and that the level of stem-borer infestation was too low to recommend trap cropping as a control measure for stem borers. Wahl (1926) suggested planting trap crops of maize and sorghum in rows of six or eight on the sides of fields 2 weeks before the main crop. It was further suggested that trap plants be removed and fed to cattle before larvae change into pupae. However, these suggestions were not based on data but on observations of selective oviposition of stem-borer moths on early and vigorously growing plants.

The use of trap crops seems not to be widely practised, if at all. In areas where arable land is scarce, this practice is not practical. Furthermore, the suggestion of planting sorghum as a trap crop and feeding it to cattle may result in animal deaths, due to the cyanide content of sorghum when plants are still green.

Water Management

Soil moisture influences crop damage by insect pests through its effect on plant vigour and growth. In Somalia it has been shown that sorghum grown under drought stress suffers greater damage from *C. partellus* than in the rain-fed areas (A.F. Nur, unpublished). It has also been reported that continuous high soil moisture in dryland agriculture, resulting from irrigation, favours the production of several generations of *Chilo agamemnon* in Israel and Egypt (Rivnay, 1967). Flooding of *C. agamemnon*-infested sugar-cane fields after harvest reduced damage to the subsequent crop (Ezzat and Atries, 1969). Moyal (1995) observed that more vigorously growing maize plants within water-stressed plantings were heavily attacked by *B. fusca* and *E. saccharina.*

INTERACTIONS BETWEEN HOST-PLANT RESISTANCE AND CULTURAL AND BIOLOGICAL CONTROL MEASURES

Some authors have suggested a relationship between tillage and biological control. Gahukar and Jotwani (1980) attributed reduced populations of *C. partellus* to off-season tillage, which they suggested exposes borers to their natural enemies, as well as to adverse climatic factors, such as high temperature and low humidity. This may certainly be true in the case of generalist predators, such as ants or birds, but is unlikely to increase parasitization, since most stem-borer parasitoids are adapted for finding concealed hosts. Kfir (1990b) also observed that, apart from tillage directly destroying larvae, they are more exposed to predators and adverse climatic conditions due to the breaking of protective plugs made by larvae from silk, frass and plant debris when they go into hibernation.

The effect of non-preference of stem-borer larvae for certain crop varieties could accentuate the effect of intercropping on migrating larvae. Apart from the increased distance between hatching sites and host plants in such a system, a non-preferred sorghum, maize or millet variety may further reduce larval establishment and density on the host crop.

RATOON CROPS

A number of studies have indicated that volunteer crops are disadvantageous to crops in the following season, due to the fact that these are responsible for carry-over of a number of pests. This practise is therefore not generally recommended. In South Africa, attempts have been made by commercial farmers to produce ratoon sorghum under overhead irrigation, but this practice was discontinued due to severe pest and weed infestation in the ratoon crop.

Aikins (1957) showed that *E. saccharina, B. fusca* and *Sesamia* spp. in Ghana survived dry periods on new growth produced by sorghum stubble in the ground. Ingram (1958) observed both *C. partellus* and *Sesamia* spp. in volunteer tillers of sorghum. Similar observations were made by van Hamburg (1979), who observed *C. partellus* to be present throughout the season, indicating the role of continuous volunteer growth of sorghum in maintaining borer populations. Volunteer growth of sorghum during the off-season also provides for survival of aphid populations. In East Africa, Mohyuddin and Greathead (1970) observed effective overwintering of large populations of *E. saccharina* in rootstocks of ratooned sorghum and concluded that ratooning is a serious source of infestation of stem borers in other crops. Milner (1967) noted that the practice of ratooning sorghum and sugar cane (leaving rootstock to tiller after harvest) was responsible for a significant carry-over borer population. Unnithan and Seshu Reddy (1989) found active breeding populations of *C. partellus* during the dry season in Kenya.

CONCLUSIONS

Although cultural-control options for stem-borer management appear promising, not many of the recommendations have been adopted by African farmers (Nwanze and Mueller, 1989). Field sanitation, crop rotation, specific intercropping patterns and manipulation of sowing dates each have their own constraints, making them unpractical or unattractive to farmers. Cultural control is therefore severely constrained by the limited management capabilities of farmers, especially in areas where farming communities lack the support of adequate advisory services (Harris, 1989b). Practical solutions of future problems have to take into account the preferences, skills, resources and constraints of the people whose problems are to be solved. A review of stem-borer management recommendations, done by Nwanze and Mueller (1989), indicated that the majority are impractical and have not been adopted by farmers in the semiarid tropics. Improved management practices will only become a reality once a farming-systems approach, taking into account the farmers' perspective of the problem, the practicality of recommended practices and the capacity to implement recommended practices, is applied to stem-borer management research.

28 Biological Control

W.A. Overholt
International Centre of Insect Physiology and Ecology, Nairobi, Kenya

INTRODUCTION

Lepidopteran stem borers are among the most important pests of graminaceous crops, such as maize, sorghum, pearl millet and sugar cane, in Africa. Estimates of losses vary greatly, but are typically about 20–40% of the potential yield (Youdeowei, 1989; Seshu Reddy and Walker, 1990). Several species of noctuid and pyraloid stem borers occur in mainland Africa and, with the exception of *Chilo partellus* (Swinhoe) (Lepidoptera: Crambidae), all are thought to be indigenous. A second exotic species, *Chilo sacchariphagus* (Bojer), has invaded the Indian Ocean islands of Madagascar, Mauritius, Réunion and the Comoros, but has not reached mainland Africa.

There have been several attempts over the past 50 years to introduce exotic parasitoids for biological control of stem borers in Africa, particularly for suppression of the exotic stem borers, *C. partellus* on mainland Africa and *C. sacchariphagus* on the Indian Ocean islands. A considerable amount of effort has also been made in South Africa against the indigenous stem borer, *Eldana saccharina* Walker. The emphasis on importation (= classical) biological control of stem borers may be due to the relatively low value of graminaceous crops, which makes the use of pesticides uneconomical. Furthermore, insecticidal control is difficult to achieve, due to the cryptic feeding behaviour of stem borers.

HOST PLANTS

Stem-borer larvae feed in communities of wild and cultivated grasses which have stems sufficiently large to accommodate stem-borer feeding behaviour (Harris, 1990). In addition to grasses, sedges (*Cyperaceae*) and cat's-tails (*Typhaceae*) are important wild host plants for some stem borers (Jepson, 1954; Conlong, 1990). Native grasses, sedges and cat's-tails are presumably the aboriginal host plants for the indigenous stem borers in Africa, and many species are attacked. However, stem-borer densities in wild hosts do not reach nearly the levels observed in cultivated crops (Mathez, 1972). Wild hosts may have certain antibiotic properties, be physically less suitable for oviposition or lack essential nutrients necessary for

Table 28.1. Development of *Chilo partellus* and *Sesamia calamistis* on selected diets.

Host plants	*Sesamia calamistis** (% pupation)	*Chilo partellus*† (% pupation)
Pennisetum purpureum (Napier grass)	1.0	1.0
Sorghum arundinaceum	6.5	13.0
Panicum maximum (Guinea grass)	0.3	0.0
Maize	30.0	46.0
Artificial diet	95.4	80.4

* Data from Shanower *et al.*, 1993b.
† Data from G.W. Oloo, unpublished..

optimal stem-borer growth. A high silica content may be one of the factors responsible for stem-borer resistance in wild grasses (McNaughton *et al.*, 1985; Sétamou *et al.*, 1993). Additionally, there is evidence that stem-borer natural enemies are more active in perennial native-plant communities (Conlong, 1994).

One of the most widely grown cultivated cereals in Africa is maize (CIMMYT, 1992). Maize is an introduced species, which has only been extensively cultivated during the last 100 years in East Africa and the last 400 years in West Africa (Purseglove, 1972). The introduction and widespread cultivation of maize has undoubtedly had an impact on the abundance of graminaceous stem borers by providing a highly nutritious and readily available food source with little inherent resistance to stem borers. The introduction of sugar cane from South-East Asia has probably had a similar effect. Studies by Bowden (1976) and more recent investigations conducted at the International Institute of Tropical Agriculture in Benin on the stem-borer *Sesamia calamistis* Hampson (Shanower *et al.*, 1993b) and at the International Centre of Insect Physiology and Ecology (ICIPE) in Kenya on *C. partellus* (G.W. Oloo, unpublished) demonstrated that larval survival and pupal weights were much higher when stem borers were reared on maize than when they were reared on selected wild grass hosts (Table 28.1).

STEM BORERS (Table 28.2)

A complex of lepidopteran stem borers feeds on cultivated grasses in Africa (Nye, 1960; Harris, 1962; Brenière, 1971; Bowden, 1976; Bosque-Pérez and Mareck, 1990). With two notable exceptions, the injurious species are native to Africa. Some of the more important native species are the pyraloids, *Chilo orichalcociliellus* Strand, *E. saccharina* Walker and *Coniesta ignefusalis* (Hampson), and the noctuids, *Busseola fusca* Fuller and *Sesamia* spp. *Chilo orichalcociliellus* occurs in coastal East Africa, Malawi and Madagascar at altitudes below 600 m (Nye, 1960; Mathez, 1972; Delobel, 1975), and has also been reported from Nigeria (Medler, 1980). The West African report of *C. orichalcociliellus* is questionable, as recent surveys have not reported its presence (Moyal and Tran, 1989; Bosque-Pérez and Mareck, 1990; Schulthess *et al.*, 1991; Shanower *et al.*, 1991; Gounou *et al.*, 1993). It is possible that a West African species, *Chilo aleniellus* (Stand), which is morphologically similar to *C. orichalcociliellus*, was misidentified as *C. orichalcociliellus*.

Table 28.2. Important stem borer species in Africa and the Indian Ocean islands.

Family	Species	Distribution	Host plants
Pyraloidea	*Chilo partellus*	Madagascar, East, southern, central Africa	Sorghum, maize, wild grasses
	Chilo sacchariphagus	Madagascar, Mascarene Islands	Sugar cane
	Chilo orichalcociliellus	Coastal East Africa, Madagascar, Malawi	Sorghum, maize, wild grasses
	Eldana saccharina	Sub-Saharan Africa	Sugar cane (South Africa) Maize (West, Central Africa) Wild grasses and sedges
	Coniesta ignefusalis	Sahelian Africa	Millet, sorghum, maize, wild grasses
Noctuidae	*Busseola fusca*	Sub-Saharan Africa	Maize, sorghum, wild grasses
	Sesamia spp.	Africa, Indian Ocean islands	Maize, sorghum, wild grasses

Eldana saccharina is widely distributed in Africa and has been reported from several grasses and sedges (Ingram, 1958; Girling, 1978; Atkinson, 1979; Conlong, 1990). In southern Africa, *E. saccharina* is a major pest of sugar cane but rarely causes damage in maize (Sithole, 1987a). In contrast, *E. saccharina* is considered to be a pest of both maize and sugar cane in West Africa (Scheibelreiter, 1980; Bosque-Pérez and Mareck, 1990; Gounou *et al.*, 1993; Shanower *et al.*, 1993b). *Coniesta ignefusalis* is the dominant stem borer of millet in the Sahelian zone of Africa, and also attacks sorghum, maize and wild grasses (Harris, 1989a).

Busseola fusca and *Sesamia* spp. occur throughout sub-Saharan Africa (Ingram, 1983). Of the *Sesamia* spp., *Sesamia calamistis* is probably the most widely distributed and economically important species, but several others, including *Sesamia cretica* Lederer, which occurs in Somalia, Sudan and Ethiopia, and *Sesamia botanephaga* Tams & Bowden, which is found in both East and West Africa, are also important. As with *E. saccharina*, the pest status of *B. fusca* varies by region. In East and southern Africa, it is a pest at higher altitudes (> 600 m) (Nye, 1960; Sithole, 1989b), but in West Africa *B. fusca* is a pest from sea level to > 2000 m (Tams and Bowden, 1953).

In addition to the native stem borers, two introduced species, *C. partellus* and *C. sacchariphagus*, are important pests. *Chilo sacchariphagus* is found in Madagascar and the Mascarene Islands, where it is an important pest of sugar cane but rarely attacks maize or sorghum (Betbeder-Matibet and Malinge, 1968; Williams, 1983). It is thought that *C. sacchariphagus*, which is widely distributed in Asia, was accidentally introduced into the Indian Ocean islands with the introduction of sugar cane around 1850 (Williams, 1983).

Chilo partellus is also considered to be indigenous to Asia (Bleszynski, 1970). This stem borer was first recorded in Africa from Malawi in 1932 (Tams, 1932), but was not reported again until some 20 years later in Tanzania (Duerden, 1953). Since that time, the geographical distribution of *C. partellus* has continued to expand, with reports from Kenya, Uganda, Rwanda, South Africa, Ethiopia, Sudan, Somalia,

Botswana, Malawi, Mozambique, Swaziland, Zimbabwe and Zaire (Nye, 1960; Ingram, 1983; Harris, 1990). In West Africa, *C. partellus* has been reported from Togo and Cameroon (IAPSC, 1985b). The veracity of the West African reports is questionable, because recent surveys in the region failed to recover *C. partellus* (Moyal and Tran, 1989; Bosque-Pérez and Mareck, 1990; Schulthess *et al.*, 1991; Shanower *et al.*, 1991; Gounou *et al.*, 1993). Regardless, if *C. partellus* is not already in West Africa, it is likely to eventually invade the region.

Displacement of Native Stem Borers

The immigrant stem borer, *C. partellus*, has proved to be a highly competitive colonizer in many of the areas it has invaded, often becoming the predominant and most economically important stem-borer species in maize and sorghum at elevations below 1800 m (Seshu Reddy, 1983b). There is evidence that *C. partellus* may be displacing indigenous stem-borer species in some areas. In the western Transvaal Highveld area of South Africa, stem-borer populations in maize have been sampled every year since 1985. Initially, *B. fusca* was the only stem borer found, but each year the abundance of *C. partellus* increased to the point where it is now the dominant stem borer in the area (R. Kfir, personal communication). In Madagascar, *C. partellus* was first recorded in 1972 and, by 1975, it was considered to be the most damaging stem borer in maize and sorghum, surpassing in importance the indigenous *C. orichalcociliellus* (Delobel, 1975). On the Kenya coast, we have evidence that *C. partellus* is displacing *C. orichalcociliellus* Strand in maize. Mathez (1972), working in the late 1960s, found that *C. orichalcociliellus* was the most common stem borer in maize. Sampling of maize and sorghum fields in the same area from 1978 to 1981 revealed that *C. partellus* and *C. orichalcociliellus* were more or less equally abundant (Warui and Kuria, 1983). Surveys in 1991/92 indicated that *C. partellus* was by far the predominant stem borer, typically accounting for > 80% of the total stem-borer population (Overholt *et al.*, 1994c). One could argue that, rather than displacing indigenous species, *C. partellus* populations are simply expanding into a partially empty ecological niche. If this were true, the number of *C. orichalcociliellus* would have remained more or less constant over time, but the total number of stem borers would have increased. However, this does not appear to be the case, as is evident from Table 28.3. The density of *C. orichalcociliellus* appears to be decreasing over time at the Kenya coast, while total stem-borer numbers have remained more or less constant.

The success of *C. partellus* in colonizing new areas and reaching economically important densities is not well understood. Delobel (1975) investigated the biologies of *C. partellus* and *C. orichalcociliellus* in Madagascar and concluded that the two species had approximately the same biotic potential, and was unable to explain the greater abundance of *C. partellus*. In comparative laboratory investigations on artificial diet, we found that *C. partellus* completed a generation in less time that *C. orichalcociliellus* (E.N. Kioko, unpublished data). We are now comparing development of the two stem borers on maize, sorghum, and several species of wild grasses. If the results we obtain on natural diet are similar to our findings on artificial diet, a relatively shorter developmental time may be one factor that has given *C. partellus* a competitive advantage over the indigenous *C. orichalcociliellus*. However, other, as

Table 28.3. Displacement of *Chilo orichalcociliellus* by *Chilo partellus* on the Kenya Coast.

Year	Stem borer density (mean borers plant^{-1})*	*Chilo orichalcociliellus* density (mean borers plant^{-1})	Proportion *Chilo orichalcociliellus*†	Reference
1965–1967	0.7–2.0	0.5–1.1	0.59	Mathez, 1972
1978–1981	0.7–1.4	0.2–0.7	0.43	Warui and Kuria, 1983
1991–1994	0.2–1.8	0.02–0.15	0.09	Overholt *et al.*, 1994c

* The range of means from different years and fields reported in the referenced paper.
† The mean of the range of *C. orichalcociliellus* divided by the mean of the range of *Chilo* spp.

yet unidentified, factors may be of equal or greater importance in the displacement of *C. orichalcociliellus*.

INDIGENOUS NATURAL ENEMIES

Predators

The most comprehensive work on predation of stem borers in Africa was done in South Africa on *E. saccharina*. Using serological techniques, several species of ants and spiders were shown to predate on *E. saccharina* eggs and larvae, although their impact on *E. saccharina* populations was generally low (Leslie, 1988). In contrast, Girling (1978) estimated that 91.6% of *E. saccharina* eggs and 93% of neonate larvae were consumed by predacious ants in Uganda. Also in Uganda, Mohyuddin and Greathead (1970) reported that four species of ants (*Tetramorium guineense*, *Pheidole megacephala*, *Cardiocondyla badonei* and *Cardiocondyla emeryi*) destroyed almost 90% of eggs and first-instar larvae of *C. partellus* and *B. fusca*. A study conducted at Mbita Point, in Kenya, revealed that two species of earwigs (*Diaperasticus erythrocephalus* and *Forficula* sp.), *Chrysopa* sp., ladybird beetles (*Cheilomenes* sp. and *Ganocephalus simplex*), ants (*Pheidole* sp., *Camponotus* sp. and *Dorylus* sp.), *Orius* sp. and several spiders commonly predated on *C. partellus* eggs (Dwumfour, 1990). Later instars, which feed in the more protected environment of the interior of the stem, are probably less vulnerable to predation.

Parasitoids

A complex of native parasitoid species attack stem borers in Africa, including species that attack eggs, larvae and pupae. The list of parasitoids recovered is impressive (see Chapters 15–23 in this volume); however, native parasitoids in most cases do not seem to be able to maintain stem-borer populations at economically acceptable levels (Williams, 1983; Oloo, 1989; Kfir, 1992c; Kfir and Bell, 1993; Overholt *et al.*, 1994c). Indigenous parasitoids may have a greater impact on stem-borer populations residing in wild-grass communities than on populations that periodically invade annual crops (Conlong, 1994). In perennial-grass habitats, as opposed to annual cultivated grasses, there is no need for natural enemies to migrate periodically to recolonize the habitat, and thus stem borers and their natural enemies may be able to approach stability. Additionally, because stem-borer survival is much lower in wild grasses than in maize, natural enemies may be better able to maintain stem-borer populations below the epidemic growth phase (Wiedenmann and Smith, 1993). There is also some evidence that stem-borer larval growth is slower in wild plants (Bowden, 1976), which would result in a wider temporal window of susceptibility to stage-specific parasitoids. Finally, natural enemies may be more effective in wild grasses due to physical or chemical constraints to host finding in cultivated grasses. Ectoparasitoids that attack concealed hosts by drilling through the stems of small-diameter wild grasses may be unable to parasitize stem borers in the relatively thicker-stemmed maize. Olfactory responses of parasitoids to volatiles emanating from wild plants and cultivated grasses may also differ.

Cotesia sesamiae

Cotesia sesamiae Cameron is the most commonly recovered larval parasitoid of several stem-borer species in maize and sorghum in many areas of sub-Saharan Africa (Ullyett, 1935; Ingram, 1958; Harris, 1962; Mohyuddin and Greathead, 1970; Mathez, 1972; Kfir, 1992c; Kfir and Bell, 1993). *Cotesia sesamiae* attacks medium- and large-instar larvae of several stem borers, including *Sesamia* spp., *C. partellus, C. orichalcociliellus* and *B. fusca* (Ullyett, 1935; Mohyuddin and Greathead, 1970; Kfir, 1992c; Kfir and Bell, 1993). Recent life-table studies conducted in the coastal zone of Kenya in maize have shown that, although *C. sesamiae* was the most common larval parasitoid, generational mortality of *C. partellus* was never greater than 3% and was typically < 0.5% (W.A. Overholt, unpublished data), although in other areas of Africa parasitism by *C. sesamiae* is reportedly much higher. In South Africa, Kfir and Bell (1993) found parasitism of *B. fusca* in sorghum as high as 75% at certain times of the year.

INTRODUCTION OF BIOLOGICAL CONTROL OF STEM BORERS IN AFRICA

There have been several attempts over the past 50 years to introduce exotic parasitoids for biological control of stem borers. In Madagascar and Mauritius alone, at least 76 introductions of exotic parasitoids have been made for the control of *C. sacchariphagus, S. calamistis* and *C. partellus* (Moutia and Courtois, 1952; Betbeder-Matibet and Malinge, 1968; Appert *et al.*, 1969; Greathead, 1971). There have also been serious efforts made in South Africa against *C. partellus* in maize (Kfir, 1994) and *E. saccharina* in sugar cane (Carnegie *et al.*, 1985; Conlong, 1994). In East Africa, the Commonwealth Institute of Biological Control (CIBC) (now International Institute of Biological Control) mounted a 5-year effort against *C. partellus* from 1968 to 1972 (CIBC, 1968–1972). In West Africa, introductions have been made for the control of stem borers in sugar cane and rice (Vercambre, 1977a; Bordat, 1983). With the notable exception of introductions into Madagascar and the Mascarene Islands (Mauritius, Réunion) against *C. sacchariphagus* in sugar cane and *S. calamistis* in maize, exotic parasitoids introduced into Africa have reportedly failed to establish. However, there is recent evidence that *Cotesia flavipes* is established in Kenya against *C. partellus*. Table 28.4 presents a list of the species that have been introduced in mainland Africa (as far as is known, only *C. flavipes* has established), and Table 28.5 lists the species successfully established in the Indian Ocean Islands.

Recent Evidence of the Establishment of *Cotesia flavipes* in East Africa

COASTAL KENYA

In the long rainy season of 1993 (March–July), *C. flavipes* was released at three locations in the southern coastal area of Kenya (3.3°–4.3°S, 39°–40°E) over a period of 6–8 weeks (Overholt *et al*., 1994). The number of female parasitoids liberated at each site was estimated to be between 18,100 and 24,200. *C. flavipes* successfully located and parasitized stem borers at the three sites during the season of release (Overholt *et al*., 1994), and all three stem borers found at the coast, *C. partellus*, *C. orichalcociliellus* and *S. calamistis*, were parasitized (Overholt *et al*., 1997), confirming the results of previous host suitability studies (Ngi-Song *et al*., 1995).

C. flavipes was released at a fourth site during the dry season of 1994 in an area where the vegetation was dominated by the wild grass, *Sorghum arundinaceum*. This release was conducted to determine whether *C. flavipes* could colonize the wild sorghum habitat and then move into a maize agroecosystem during the following rainy season. Approximately 6000 females were released over a period of 4 weeks. A maize field was planted adjacent to the wild sorghum once the long rains began and both plots were sampled throughout the rainy season. Approximately 2.2 and 5.5% of *C. partellus* medium- and large-sized larvae were parasitized by *C. flavipes* in wild sorghum and maize, respectively (Overholt *et al*., 1997).

In addition to sampling at the wild sorghum site, seven other maize fields were sampled in the 1994 long rains. Only one parasitized stem borer was recovered (at one of the 1993 release site). During the following short rains, *C. flavipes* was found at seven of 21 fields sampled. Seasonal parasitism at the recovery sites was low and ranged from 0.3% to 3.0%. In the 1995 long rains, 29 sites were sampled and *C. flavipes* was found at 4, but parasitism remained low, ranging between 0.05% and 1.0% (Overholt *et al*., 1997). Data from the 1996 long rains cropping season revealed a significant increase in the abundance of the *C. flavipes* as it was found to be the dominant parasitoid at five sites with average parasitism at one site of 7.1%. In 1997, the number of recoveries increased again and average parasitism at 15 sites in Kilifi District of coastal Kenya was 15%. The increase in recoveries from 1993 to 1997 is clear evidence of the establishment of *C. flavipes* in the coastal zone of Kenya.

OTHER AREAS IN EAST AFRICA

In May 1994, *C. flavipes* was collected from unidentified stem borers in the South Nyanza District of south-western Kenya. This was quite surprising as no releases had been made in this area. A more thorough survey was conducted in the area in June/July 1994, and *C. flavipes* emerged from *C. partellus* collected at seven locations in the same area (Omwega *et al*., 1995). In July 1995, a survey was conducted in northern and central Tanzania. *C. flavipes* was found at two locations, Tarime and Magu, which are both near Lake Victoria in the area bordering south-western Kenya. Parasitism of *C. partellus* larvae was 44.4 and 62.7% at Tarime and Magu, respectively (Omwega *et al*., 1997). Three possibilities could explain the establishment in the Lake Victoria area: (i) *C. flavipes* established from releases made by CIBC in Uganda and Tanzania from 1968–1972; (ii) *C. flavipes* moved from the coastal area where it was released in 1993 and 1994; or (iii) *C. flavipes* escaped from a laboratory colony that was maintained by ICIPE at Mbita Point Field Station in south-western Kenya in 1991. Based on evidence from surveys conducted prior to 1994 and on electrophoretic evidence, it was concluded that the most likely possibility was that *C. flavipes* escaped from a colony maintained at Mbita Point in 1991 (Omwega *et al*., 1995).

An extensive survey in south-eastern Kenya along a transect from Mombasa to Nairobi was conducted in 1997. At Voi, which is approximately 160 km from the nearest 1993 release site at the coast, parasitism of *C. partellus* by *C. flavipes* was 70%, and at Kibwezi (260 km from the coast), parasitism of 20% was found (W.A. Overholt, unpublished data). These results suggest that *C. flavipes* in increasing in abundance and has spread throughout much of southern Kenya from the coast to the shores of Lake Victoria.

In addition to the releases in Kenya, *C. flavipes* was released in Mozambique in 1996, and in Uganda and Somalia in 1997. Preliminary results from Mozambique indicate that the parasitoid colonized maize fields, but it is too early to determine whether it is established. Releases in several other countries are planned in 1998–1999.

Implications of the Results of Previous Classical Biological Programmes

ISLANDS VERSUS MAINLAND AFRICA

It is readily apparent from examining Tables 28.4 and 28.5 that there have been many more successful establishments on the Indian Ocean islands than on mainland Africa. Greathead (1971) discussed this differential rate of establishment, and suggested two factors that may be involved: (i) according to the theory of island biogeography, species are not packed as tightly on islands (fewer species per unit area with greater niche breadths) and therefore it may be easier for exotic specialists to establish (less interspecific competition); and (ii) for similar reasons, exotic pests are more likely to establish on islands and therefore obvious targets for classical biological control may be more plentiful.

ECOSYSTEM STABILITY

The geographical factor of islands versus the mainland is confounded with other potentially influential factors. For example, the major emphasis of the stem-borer biological-control activities on the Indian Ocean islands was against *C. sacchariphagus* in the sugar-cane ecosystem, whereas in mainland Africa the work has been divided between maize, rice and sugar cane. The only successes in maize were the introductions of *Pediopus furvus* and *C. sesamiae* from mainland Africa to Indian Ocean islands against *S. calamistis*, and the recently recognized establishment of *C. flavipes* in Kenya. There may be ecological differences between sugar-cane and maize ecosystems that influence the success rates. Sugar cane is grown throughout the year and thus provides a fairly stable habitat. Once an exotic natural enemy is established, it can remain in the system and reach an equilibrium density along with the stem-borer hosts. In contrast, the maize ecosystem is short-lived, providing a suitable habitat for stem borers and their natural enemies for only 2 or 3 months. During the non-cropping season, the stem borers must either survive in a state of rest in dry stalks or stubble, or migrate to alternative wild habitats and then reinvade the maize ecosystem during the next cropping season. Parasitoids introduced into such an ephemeral system would either have to remain in the crop residue and aestivate along with their hosts, be able to locate and parasitize diapausing hosts, or migrate to alternative habitats where non-diapausing hosts are found and reinvade the crop during the next growing season. Brenière *et al.* (1985) suggested that a lack of habitat stability on the island of Ngazidja (Comores) was the reason for several failures to establish parasitoids against *C. partellus* in maize. It may be inherently more difficult to identify parasitoids that

Table 28.4. Exotic parasitoids released in mainland Africa against graminaceous stem borers.

Host stage	Species	Target host	Origin	Country of introduction	Association (old/new)	Years	Reference
Egg	*Trichogramma fasciatum*	Complex of species	Barbados	Uganda Ghana	New	1971 1973/74	CIBC Annual Report Scheibelreiter, 1980
	Trichogramma flandersi	Complex of species	India	Uganda	New/old	1971	CIBC Annual Report
	Trichogramma chilotraeae	Complex of species	India	Uganda Ghana	New/old New	1971 1974	CIBC Annual Reports Scheibelreiter, 1980
	Trichogramma semifumatum	Complex of species	India	Uganda	New/old	1971	CIBC Annual Report
	Trichogramma californicum hybrid	Complex of species	India	Ghana	New	1974	Scheibelreiter, 1980
	Trichogramma brasiliensis	Complex of species	India	Ghana	New	1974	Scheibelreiter, 1980
	Trichogramma japonicum	Complex of species	India	Ghana	New	1974	Scheibelreiter, 1980
	Trichogramma perkinsi	Complex of species *Eldana saccharina*	India Columbia	Ghana South Africa	New	1973/74 1980	Scheibelreiter, 1980 D.E. Conlong, personal communication
	Trichogramma australicum	Complex of species *E. saccharina*	Taiwan	Uganda South Africa	New New	1971 1980–1983	CIBC Annual Report Carnegie *et al.*, 1985
	Trichogramma pretiosum	*E. saccharina*	USA	South Africa	New	1971	D.E. Conlong, personal communication
	Trichogramma australicum	*E. saccharina*	Taiwan	South Africa	New	1980–1983	Carnegie *et al.*, 1985
	Trichogramma evanescens	*E. saccharina*	Germany, Switzerland	South Africa	New	1984	D.E. Conlong, personal communication
	Trichogramma sp.	*E. saccharina*	Ivory Coast	South Africa	Old	1983	Carnegie *et al.*, 1985
	Trichogramma chilonis	*Chilo partellus*	Taiwan	South Africa	New	1988	Kfir, 1994
	Trichogramma ostrinia	*C. partellus*	Taiwan	South Africa	New	1989	Kfir, 1994
	Trichogrammatoidea eldanae	*E. saccharina*	Ivory Coast	South Africa	Old	1980–1983	Carnegie *et al.*, 1985
	Telenomus applanatus	*E. saccharina*	Ivory Coast	South Africa	Old	1981–1984	Carnegie *et al.*, 1985
Larva	*Bracon chinensis*	Complex of species	India	Uganda Kenya Ghana	New/old New/old New	1971/72 1971 1976/77	CIBC Annual Reports Scheibelreiter, 1980
	Allorhogas pyralophagus	*E. saccharina* *C. partellus*	Mexico	South Africa	New	1984 1988	Carnegie *et al.*, 1985 Kfir, 1994
	Xanthopimpla stemmator	*E. saccharina*	Mauritius	South Africa	New	1984	D.E. Conlong, personal communication
		C. partellus			Old	1988	Kfir, 1994
	Mallochia pyralidis	*C. partellus*	Mexico	South Africa	New	1988	Kfir, 1994
	Meloboris sinicus	*C. partellus*	Taiwan	South Africa	New	1993	Kfir, 1994

Table 28.4. Exotic parasitoids released in mainland Africa against graminaceous stem borers (*continued*).

Host stage	Species	Target host	Origin	Country of introduction	Association (old/new)	Years	Reference
Larva	*Isotima javensis*	*C. partellus*	India	Uganda	New	1971	CIBC Annual Report, 1971
	Sturmiopsis inferens	Complex of species	India	Ghana	New	1974–1977	Scheibelreiter, 1980
		C. partellus		South Africa	Old	1977	Kfir, 1994
	Descampsina sesamiae	*E. saccharina*	Ghana	South Africa	Old	1982–1984	Carnegie *et al.*, 1985
		C. partellus	Nigeria/ Ghana		New	1981	Kfir, 1994
	Paratheresia claripalpis	*E. saccharina*	Brazil, Columbia, West Indies	South Africa	New	1978 1985–1987 1991	D.E. Conlong, personal communication
		C. partellus	Brazil		New	1988, 1991	Kfir, 1994
	Metagonistylum minense	*E. saccharina*	Brazil, Columbia, West Indies	South Africa	New	1978, 1985, 1986, 1987	D.E. Conlong, personal communication
		C. partellus			New		Kfir, 1994
	Cotesia chilonis	*Chilo zacconius*	Japan	Senegal	New	1977	Bordat, 1983
				Côte d'Ivoire	New	1981	
		C. partellus		South Africa	New	1993	Kfir, 1994
	Cotesia flavipes	*C. partellus*	Pakistan/ India	Uganda	Old	1968–1972	CIBC Annual Reports
				Tanzania	Old		
				Kenya	Old		
				Kenya	Old	1991	Overholt *et al.*, 1994b
		Complex of species		Ghana	New	1974 1975	Scheibelreiter, 1980
		C. partellus		South Africa	Old	1984/85	Skoroszewski and van Hamburg, 1987
		C. zacconius		Cote d'Ivoire	New	1981	Bordat, 1983
				Senegal	New	1981	Bordat, 1983
Pupa	*Tetrastichus inferens*	Complex of species	India	Ghana	New	1973/74	Scheibelreiter, 1980
	Tetrastichus ayyari	Complex of species	India	Ghana	New	1974	Scheibelreiter, 1980
	Tetrastichus israeli	Complex of species	India	Ghana	New	1973/74	Scheibelreiter, 1980
				Senegal	New	1972	Vercambre, 1977a
	Tetrastichus spp.	Complex of species	India	Ghana	New	1973/74	Scheibelreiter, 1980
	Tetrastichus howardi	*C. partellus*	Philippines	South Africa	Old	1988	Kfir, 1994
	Trichospilus diatraea	*C. zacconius*	India	Senegal	New	1972	Vercambre, 1977a
	Itoplectis naranguae	*C. zacconius*	?	Senegal	New	1972	Vercambre, 1977a

Table 28.5. Exotic parasitoids successfully established in Mauritius, Réunion and Madagascar.

Island	Host stage attacked	Species	Target host	Origin	Association (old/new)	Years	Remarks	Reference
Mauritius	Larva	*Cotesia flavipes*	*Chilo sacchariphagus*	Unknown*	Old	Unknown	Important mortality factor	Williams, 1983
		Cotesia sesamiae	*Sesamia calamistis*	Kenya	Old	1951/52	Useful control agent	Greathead, 1971
		Bracon chinensis	*C. sacchariphagus*, *S. calamistis*	Sri Lanka	Old?† New	1939		Greathead, 1971
	Pupa	*Xanthopimpla stemmator*	*C. sacchariphagus*, *S. calamistis*	Sri Lanka	Old?† New	1939		Greathead, 1971
		Trichospilus diatraeae	*C. sacchariphagus*	India	Old?†	1963		Greathead, 1971
Réunion	Larva	*Cotesia flavipes*	*C. sacchariphagus*	Unknown*	Old?†	Unknown		Greathead, 1971
		Cotesia sesamiae	*S. calamistis*	Mauritius	Old	1953–1955		Greathead, 1971
	Pupa	*Xanthopimpla stemmator*	*C. sacchariphagus*, *S. calamistis*	Mauritius	Old New	1953, 1960		Greathead, 1971
Comores	Pupa	*Pediobius furvus*	*Sesamia calamistis*	Madagascar	Old	1969–71		Brenière *et al.*, 1985
			Chilo partellus		Old			
Madagascar	Larva	*Cotesia flavipes*	*C. sacchariphagus*	Mauritius	Old?†	1961	Good control	Appert *et al.*, 1969
		Cotesia sesamiae	*S. calamistis*	Uganda	Old	?		Brenière *et al.*, 1985
			C. partellus		Old			
	Pupa	*Pediobius furvus*	*S. calamistis*	Uganda	Old	1968	Suppresses *S. calamistis* populations	Appert, 1973
			C. partellus		Old			

* There is confusion regarding the origin of the *C. flavipes* populations in Mauritius and Réunion. Appert (1973) indicates that *C. flavipes* was introduced into Mauritius in 1917 and then later into Réunion, but provides no details or reference. Greathead (1971) indicates that there is no record of the introduction, but that it is assumed to be exotic. One possibility is that it arrived on the island along with its host, *C. sacchariphagus*, around 1850.

† The determination of old vs. new associations with *C. sacchariphagus* is difficult because the origin and taxonomic status of the population of *C. sacchariphagus* in Mauritius is unclear. It is speculated that the borer either arrived in Mauritius from Sri Lanka in 1848 or from Java in 1850 (Williams, 1983).

can adapt to unstable annual cereal crops than those that can adapt to more perennial ecosystems. Reviews of success rates in biological-control programmes in different categories of habitat stability support the hypothesis that establishment is more likely in stable ecosystems (Hall and Ehler, 1979).

SUCCESS OF *COTESIA* SPECIES PARASITOIDS

Cotesia flavipes has been introduced into more than 40 countries in the tropics and subtropics for biological control of pyralid stem borers in the genera *Chilo* and *Diatraea* (Polaszek and Walker, 1991). In Mauritius, where *C. flavipes* may have been accidentally introduced, parasitism of 4–50% of larvae of the introduced stem borer, *C. sacchariphagus*, has been reported (Rajabalee and Governdasamy, 1988). In Madagascar, where *C. flavipes* was introduced in 1960, parasitism of 60% of *C. sacchariphagus* larvae has been reported (Betbeder-Matibet and Malinge, 1968). *Cotesia flavipes* has also been introduced into several countries in the neotropics for biological control of *Diatraea saccharalis* (F.) in sugar cane, and substantial control has been reported in many areas (Alam *et al.*, 1971; Fuchs *et al.*, 1979; Macedo *et al.*, 1984). The success of *C. sesamiae* is limited to its establishment on Mauritius, Réunion and Madagascar against *S. calamistis.*

Several factors could be responsible for the success of the two *Cotesia* spp. In their areas of endemism, the *Cotesia* spp. parasitoids attack several species of crambid, pyralid and noctuid stem borers (Mohyuddin, 1971; Nagarkatti and Nair, 1973; Beg and Inayatullah, 1980; Mohyuddin *et al.*, 1981; Goraya *et al.*, 1982; Shami and Mohyuddin, 1987). The relatively wide taxonomic range of suitable hosts, coupled with a narrow habitat specificity, may favour its establishment. Stem borers often occur as species complexes, and a parasitoid that can exploit more than one of the hosts may be better able to colonize a new area than a monophagous parasitoid, due to a more constant availability of hosts. Another factor may be the high reproductive potential of *C. flavipes* and *C. sesamiae* in relation to most stem borers. Both *C. flavipes* and *C. sesamiae* have short generation times (*c.* 16–18 days) in comparison with their stem-borer hosts (30–50 days) and a fairly high fecundity (*c.* 30–40 progeny per oviposition with a female-biased sex ratio). A high host-searching ability may also be involved. Wiedenmann and Smith (1993) demonstrated that, even at low host densities, *C. flavipes* was able to successfully locate stem-borer hosts. The high host-searching ability of *C. flavipes* may, in part, be due to its behaviour of entering tunnels in plant stems to attack stem-borer larvae. Many other larval parasitoids of stem borers remain on the outside of the stem and parasitize larvae by drilling or locating breaches through the stem with their ovipositor. This strategy may be effective for attacking stem borers in wild grasses, such as *Panicum maximum*, but, in relatively larger-stemmed cultivated grasses, the length of the ovipositor may limit the number of hosts susceptible to attack.

OLD VERSUS NEW ASSOCIATIONS

Finally, it appears that old parasitoid–host associations are responsible for a disproportionate share of successful establishments when compared with new associations. All of the establishments on the Indian Ocean islands and the only known establishment on mainland Africa are probably old associations, although this is not certain (see footnotes to Table 28.5). The behavioural and physiological compatibility of old-association parasitoids and their hosts is implicit, whereas in new associations compatibility cannot be assumed. The necessary cues for host finding in new associations may be missing or the host may not be suitable for parasitoid development. Host suitability is particularly important for endoparasitoids, which have a more intimate relationship with their hosts than ectoparasitoids. The greater apparent success of old associations supports the contention of several authors (Hall and Ehler, 1979; Hall *et al.*, 1980) that old associations are more likely to result in establishment and suppression of the target host population. However, the greater likelihood of success with old associations should not preclude efforts against native pests. One of the greatest successes in stem-borer biological control was against the neotropical stem borer, *D. saccharalis*, with the new-association parasitoid, *C. flavipes*.

POSSIBLE REASONS FOR FAILURES TO ESTABLISH *COTESIA FLAVIPES* IN MAINLAND AFRICA

Several attempts have been made to establish *C. flavipes* in mainland Africa (Table 28.4) and, until recently, all were thought to be failures. In South Africa, *C. flavipes* was introduced against *C. partellus* and recovered several times during the season of release, but failed to make it through the austral winter (Skoroszewski and van Hamburg, 1987). In West Africa, *C. flavipes* was released in Ghana against a complex of sugar-cane stem borers (Scheibelreiter, 1980) and in Ivory Coast and Senegal against *Chilo zacconius* Bleszyinski in rice (Brenière and Bordat, 1982), but again failed to establish. In East Africa, *C. flavipes* was released against *C. partellus* in Tanzania, Kenya and Uganda, but reportedly failed to establish. There is little information on the probable reasons for the failures in Africa. The number of failures to establish *C. flavipes* against an old-association host, *C. partellus*, in mainland Africa is particularly surprising, especially in light of the success that has been achieved in the neotropics against the new-association host, *D. saccharalis*, in sugar cane.

One hypothesis that could explain the failures in East Africa is competition with the indigenous *C. sesamiae*. Both species have very similar ecological requirements and thus may not be able to coexist. Additionally, *C. flavipes* may have interbred with *C. sesamiae*. Interspecific mate competition could theoretically prevent establishment, particularly if outcrossing were common and the hybrid progeny not viable. Prior to releasing *C. flavipes* in Kenya in 1993, studies were conducted to examine interspecific mating behaviour and hybridization between the two species. In a no-choice situation, *C. flavipes* and *C. sesamiae* copulated, sperm was transferred, but no female progeny were produced (S.W. Kimani, unpublished data). As *Cotesia* spp. are arrhenotokous, this is strong evidence of postzygotic reproductive isolation. Olfactometer studies suggest that opportunities for interbreeding in nature may be limited, due to a lack of interspecific attraction to pheromones between the opposite sexes of the two *Cotesia* spp. (Kimani and Overholt, 1995). Additionally, previous studies on *C. flavipes* have reported that individuals mate with their siblings soon after emergence (Arakaki and Gahana, 1986) and that females mate only once (Kajita and Drake, 1969). If competition over resources or mates precluded establishment, it would be difficult to explain how the two parasitoids are able to coexist in Mauritius, Réunion, Madagascar and south-western Kenya.

A second possibility is that *C. flavipes* may have been released in areas where stem borers unsuitable for the development of *C. flavipes* were abundant. In work conducted at ICIPE, we have found *C. partellus, C. orichalcociliellus* and *S. calamistis* are suitable hosts for *C. flavipes*, but that *B. fusca* is totally unsuitable for parasitoid development (Ngi-Song *et al.*, 1995). However, *B. fusca* is attractive to the parasitoid and quite acceptable for oviposition. If *C. flavipes* had been released in an area where *B. fusca* occurred, the female parasitoids would have searched for *B. fusca* and oviposited, thereby impairing their reproductive success and decreasing the chance for establishment.

A third possibility is that *C. flavipes* did actually become established, but the establishment was not recognized. During the initial period following a colonization of *C. flavipes*, the density may be extremely low and difficult to detect as was shown in our surveys at the Kenya coast in 1994, one year after the release. In Barbados, where *C. flavipes* was established in sugar cane on *D. saccharalis*, the first recovery was not made until more than 1 year after releases (Alam *et al.*, 1971). Similarly in Madagascar, *C. flavipes* remained rare during the first year after releases (Greathead, 1971). Thus, surveys with a greater temporal separation from the time of release may be more reliable for assessing establishment than those conducted during the first year after releases. The rate of increase of population densities of *C. flavipes* may be related to the size of the geographical area with suitable habitat. On a small island, regardless of whether it is a true island or an ecological island, the parasitoid population may rapidly disperse to all suitable habitats and then begin to increase to detectable levels. In larger areas of suitable habitat, as may be encountered more frequently on mainlands, the dispersal to all suitable habitats and subsequent density increases may take much longer.

Failure to recognize the establishment of *C. flavipes* may also have been due to misidentifications. *Cotesia flavipes* and *C. sesamiae* are morphologically quite similar, and earlier identifications were based primarily on coloration, which is now known to be unreliable (Polaszek and Walker, 1991). *Cotesia flavipes* may have established, as it has in south-western Kenya, but was misidentified as *C. sesamiae*. In South Africa, *C. flavipes* was introduced from Pakistan into an area that included both *C. partellus* (the target species of the introduction) and *B. fusca* (Skoroszewski and van Hamburg, 1987). Earlier work had demonstrated that *C. flavipes* could not successfully develop in *B. fusca* (Mohyuddin, 1971), and this has recently been confirmed in laboratory exposures at ICIPE (Ngi-Song *et al.*, 1995). Nevertheless, Skoroszewski and van Hamburg (1987) reportedly recovered *C. flavipes* from *B. fusca*. Although this may have been true, it would seem more likely that the recoveries in *B. fusca* were based on misidentifications of *C. sesamiae*. Additional collection and identification of *Cotesia* spp. in the areas where *C. flavipes* has been introduced are warranted.

CONCLUSIONS

Over the past 40 years, there have been several attempts to introduce biological-control agents for suppression of exotic and native stem borers in Africa. Most attempts failed, but the reasons for failure are largely unknown. The success rate has been much higher on islands, and old parasitoid/host relationships are responsible for a disproportionate share of the establishments. Future biological-control programmes must place a greater emphasis on investigating ecological relationships within and between plants, stem borers and natural enemies as a means of increasing our knowledge of the colonization process. Only through understanding the results of past introductions can we hope to improve the chances for success in the future.

References

Abbas, M.S.T. (1987) Interactions between host, egg and larval parasitoids and nuclear polyhedrosis virus. *Bulletin of the Entomological Society of Egypt, Economic Series* 16, 133–141.

Abu, J.F. (1972) The bionomics of *Diopsis* (Diptera: Diopsidae) and *Epilachna similis* (Coleptera: Coccinellidae) on *Oryza sativa* L. in the Accra plains. M.Sc. thesis, University of Ghana, Legon.

Abu, J.F. (1986a) *Biology and Control of the Insect Pests of Sorghum in the Southern Guinea Savanna Zone of Nigeria.* Institute for Agricultural Research Samaru, Zaria, Nigeria, 23 pp.

Abu, J.F. (1986b) Dry season populations of *Sesamia calamistis* Hampson (Lepidoptera: Noctuidae) on sorghum at Mokwa in the southern Guinea savanna zone of Nigeria. Paper presented at the 16th Annual Conference of the Nigerian Society for Plant Protection, 16–20 March 1986, Ahmadu Bello University, Samaru, Zaria, Nigeria, 11 pp.

Acland, J.D. (1971) *East African Crops.* FAO and Longmans, London, 252 pp.

Adesiyun, A.A. (1983) Some effects of intercropping sorghum, millet and maize on infestation by lepidopterous stalk-borers, particularly *Busseola fusca*. *Insect Science and its Application* 4, 387–391.

Adesiyun, A.A. (1986) Control of the stem borer, *Busseola fusca*, on sorghum with granular insecticides. *Samaru Journal of Agricultural Research* 4, 35–43.

Adesiyun, A.A. (1988) Influence of sorghum and millet diets on the survival, development and oviposition of the stem borer, *Busseola fusca* (Fuller). *Nigerian Journal of Entomology* 9, 1–6.

Adesiyun, A.A. and Ajayi, O. (1980) Control of the sorghum stem borer, *Busseola fusca*, by partial burning of the stalks. *Tropical Pest Management* 26, 113–117.

Adeyemi, S.A.O., Donelly, J. and Odetoyinbo, J.A. (1966) Studies on chemical control of stem borers of maize. *Nigerian Agricultural Journal* 3, 61–66.

Adkinson, P.L., Vanderzant, E.S., Bull, D.L. and W.E. Allison (1960) A wheat germ medium for rearing the pink bollworm. *Journal of Economic Entomology* 53, 759–762.

Agyen-Sampong, M. (1977) Insect pests of rice in Ghana and their control. In: *WARDA Seminar Proceedings No. 4, Plant Protection for the Rice Crop.* WARDA, Monrovia, Liberia, March 1977, pp. 51–58.

Agyen-Sampong, M. (1980) Parasites of rice pests of mangrove swamps of northern Sierra Leone. *WARDA/ADRAO Technical Newsletter* 2, 1–3.

Agyen-Sampong, M. and Fannah, S.J. (1987) Seasonality of the white rice borer *Maliarpha separatella* Rag. (Lepidoptera: Pyralidae) in the mangrove swamp rice ecology of northwest Sierra Leone. *WARDA/ADRAO Technical Newsletter* 7, 4–7.

Ahmad, T. and Mathur, I.D. (1946) The biology and ecology of *Melcha ornatipennis* Cameron, a parasite of the top shoot borer of sugarcane, *Scirpophaga nivella* Fabr. *Indian Journal of Entomology* 7, 21–35.

Ahmed, S.M. and Young, W.R. (1969) Field studies on the chemical control of the stem borer *Chilo partellus* on hybrid sorghum in India. *Journal of Economic Entomology* 62, 478–482.

Aikins, J.S. (1957) Dry season investigation of the stemborers, Northern Region. *Ghana Farmer* 1, 190–191.

Ajayi, O. (1978) *Control of Sorghum, Millet and Wheat Insect Pests.* Samaru Conference Paper 24, Institute for Agricultural Research, Ahmadu Bello University, Zaria, Nigeria.

Ajayi, O. (1985a) *A Check-list of Millet Insect Pests and Their Natural Enemies in Nigeria.* Samaru Miscellaneous Papers 108, Ahmadu Bello University, Samaru, Zaria, Nigeria, 16 pp.

Ajayi, O. (1985b) The effects of planting date and varieties on stem borer infestation and damage of pearl millet. *Nigerian Journal of Entomology* 6, 71–77.

Ajayi, O. (1986) *The Incidence of Stem Borers on Irrigated Wheat in Nigeria.* Samaru Miscellaneous Papers 108. Ahmadu Bello University, Samaru, Zaria, Nigeria.

Ajayi, O. (1987) Insecticidal control of the sorghum stem borer. In: *Cereals Research Programme. Cropping Scheme Report, 1987.* Institute for Agricultural Research, Samaru, Zaria, Nigeria, pp. 35–38.

Ajayi, O. (1989) Stem borers of sorghum in west Africa with emphasis on Nigeria. In: *International Workshop on Sorghum Stem Borers, Patancheru, India, 17–20 November, 1987.* International Crops Research Institute for the Semi-Arid Tropics, Pantancheru, India, pp. 27–30.

Ajayi, O. (1990) Possibilities for integrated control of the millet stem borer, *Acigona ignefusalis* Hampson (Lepidoptera: Pyralidae), in Nigeria. *Insect Science and its Application* 11, 109–117.

Ajayi, O. (1991) Yield reduction in sorghum caused by *Busseola fusca* (Fuller) and *Sesamia calamistis* Hampson. Paper presented at the 9th Conference of the African Association of Insect Scientists, Accra, Ghana, 23–27 September, 1991.

Akerman, C. (1932) On the carnivorous habits of the long-horned grasshopper, *Clonia vittata* Thunberg. *Annals of the Natal Museum* 7(1), 143–144.

Akinsola, E.A. (1975) Present status of different rice stem borers in parts of Nigeria. *Rice Entomology Newsletter* 3, 28.

Akinsola, E.A. (1979) The biology and ecology of rice stem borers in Nigeria. PhD thesis, University of lbadan, Nigeria.

Akinsola, E.A. (1980a) Effets d'infestation de *Diopsis thoracica* (West) sur les plants de riz. *Bulletin Technique ADRAO* 2(2), 3–4.

Akinsola, E.A. (1980b) Notes on damage caused by *Maliarpha separatella* on deep flooded rice in Mali. *WARDA Technical Newsletter* 2(2), 1.

Akinsola, E.A. (1990) Management of *Chilo* spp. in rice in Africa. *Insect Science and its Application* 11(4/5), 815–823.

Akinsola, E.A. and Agyen-Sampong, M. (1982) The ecology, bionomics and control of rice stemborers in West Africa. Paper presented at the International Study Workshop on Crop Borers and Emergency Strategies for their Control. ICIPE, Mbita Point Field Station, Kenya.

Akinsola, E.A. and Agyen-Sampong, M. (1984) The ecology, bionomics and control of rice stem-borers in West Africa. *Insect Science and its Application* 5, 69–77.

Alam, M.M., Bennet F.D. and Carl K.P. (1971) Biological control of *Diatraea saccharalis* (F.) in Barbados by *Apanteles flavipes* Cam. and *Lixophaga diatraea* Tns. *Entomophaga* 16, 151–158.

Alam, M.S. (1988) Seasonal abundance of rice stem borer species in upland and irrigated rice in Nigeria. *Insect Science and its Application* 9, 191–195.

Alam, M.S. (1990) Final report to the USAID/IITA/Zaire Applied Research and Outreach Project, Lubumbashi, Zaire (unpublished).

Alam, M.S. (1992) A survey of rice insect pests in Nigeria. *Tropical Pest Management* 38, 115–118.

Alam, M.S. and Efron, Y. (1986) Resistance of rice to stalk-eyed fly *Diopsis macrophthalma* Dalman. *International Rice Commision Newsletter* 35(1), 40–45.

Alam, M.S., John, V.T. and Kaung Zan (1985) Insect pests and diseases of rice in Africa. In: *Rice Improvement in Eastern, Central and Southern Africa. Proceedings of the International Rice Workshop at Lusaka, Zambia, 9–19 April 1984*, IRRI, Los Baños, pp. 67–82 .

Alghali, A.M. (1979) Weed hosts of diopsid (Diptera) rice stem borers in Southern Nigeria. *International Rice Research Newsletter* 4(4), 21–22.

Alghali, A.M. (1981) Studies in the relative susceptibility of rice varieties to *Diopsis thoracica*, West. (Diptera: Diopsidae). PhD thesis, University of Ibadan, Nigeria.

Alghali, A.M. and Osisanya, E.O. (1981) Effects of rice plant age on diopsid oviposition and plant susceptibility. *International Rice Research Newsletter* 6(6), 17–18.

Alghali, A.M. and Osisanya, E.O. (1982) Oviposition by *Diopsis macrophthalma* Dalman (Diptera: Diopsidae) on different rice varieties and development of deadheart. *Bulletin of Entomological Research* 72, 583–588.

Alghali, A.M. and Osisanya, E.O. (1984) Effects of damage by stalk-eyed fly *(Diopsis thoracica)* on yield components of rice. *Experimental Agriculture* 20, 225–233.

Allison, J.C.S., Nuss, K.J. and Bond, R.S. (1992) How can genetic engineering contribute to sugarcane improvement? *Technical Communication, Department of Agricultural Development, South Africa* 232, 90–92.

Amoako-Atta, B., Omolo, E.O. and Kidega, E.K. (1983) Influence of maize, cowpea and sorghum intercropping systems on stem/pod-borer infestations. *Insect Science and its Application* 4, 47–57.

Ampofo, J.K.O., Saxena, K.N., Kibuka, J.G. and Nyangiri, E.O. (1986) Evaluation of some maize cultivars for resistance to the stem borer *Chilo partellus* (Swinhoe) (Lepidoptera: Pyralidae) in western Kenya. *Maydica* 31, 379–389.

Ampong-Nyarko, K., Seshu Reddy, K.V., Ruth A. Nyang'or and Saxena K.N. (1994) Reduction of insect pest attack on sorghum and cowpea by intercropping. *Entomologia Experimentalis et Applicata* 70, 179–184.

Ampong-Nyarko, K., Seshu Reddy, K.V. and Saxena, K.N. (1995) Mechanisms of low insect pest incidence in intercropping: *Chilo partellus* oviposition on non-host plant cowpea. In: *Proceedings of the 10th Meeting and Scientific Conference of the African Association of Insect Scientists, 5–10 September 1993*. Mombasa, Kenya, pp. 103–108.

Andersen, V.S. and Wessels, P.P.M. (1959) The maize stem borer and its control in South Africa. *Höfchen-Briefe* 12, 276–285.

Angladette, A. (1966) *Le Riz*. Editions Maisonneuve et Larose, Paris, 930 pp.

Annecke, D.P. and Mynhardt, M.J. (1973) New and little known African Encyrtidae, with descriptions of two new genera (Hymenoptera: Chalcidoidea). *Journal of the Entomological Society of Southern Africa* 36(2), 211–228.

Anon. (1971) Pests of sugar cane. In: *Annual Report of the Federal Department of Agricultural Research 1966–67*. Lagos, Nigeria, pp. 20–22.

Anon. (1977) Bionomics and control of striped stem borer *Proceras venosatus* Walker. (In Chinese, summary in English.) *Acta Entomologica Sinica* 20, 417–425.

Anon. (1984) Upland rice in West Africa. In: *An Overview of Upland Rice Research, Proceedings of the 1982 Bouaké, Ivory Coast Upland Rice Workshop*. IRRI 1–566, Los Baños, pp. 21–43.

Anon. (1985) Pesticides: du poison contre la faim? *Vivre Autrement (Dakar)* 2, 7–13.

Anon. (1986) Entomogenous nematodes as control agents of *Eldana*. In: *Annual Report, South African Sugar Association Experiment Station 1986*, Mount Edgecombe, Natal, p. 54.

Anon. (1988) Nematodes to control *Eldana*. In: *Annual Report, South African Sugar Association Experiment Station 1988*. Mount Edgecombe, Natal, p. 53.

Anon. (1990) *Final Technical Report on IITA/USAID/Zaire Applied Agricultural Research and Outreach Project*. IITA, Ibadan, Nigeria.

Anon. (1991) ICRISAT helps Namibia improve production of its staple crop. *SAT News (ICRISAT)* 7, 1.

Appert, J. (1953) L'entomofaune économique du casier rizicole de Richard Toll et des cultures de la basse vallée du Sénégal. *Agronomie Tropicale* 8, 129–154.

Appert, J. (1957) *Les Parasites animaux des plantes cultivées au Sénégal et au Soudan*. Gouvernement General de l'Afrique Occidentale Française, Paris, France, 292 pp.

Appert, J. (1964) Les chenilles mineuses des céréales en Afrique tropicale. *Agronomie Tropicale* 19, 60–74.

Appert, J. (1967) Notes techniques sur les insectes nuisibles aux cultures Malagasy: I Lépidoptères. *Agronomie Tropicale* 22, 153–230.

Appert, J. (1970) *Insects Harmful to Maize in Africa and Madagascar*. Madagascar Institute of Agronomic Research, 71 pp.

Appert, J. (1971) Les lépidoptères foreurs des graminées à Madagascar, aux Comores et aux Mascareignes. *Agronomie Tropicale* 26, 500–508.

Appert, J. (1973) Entomofaune parasitaire des foreurs des graminées à Madagascar. *Entomophaga* 18, 77–94.

Appert, J. (1975) Exemples des méthodes d'élévage et de dissémination des parasites et prédateurs d'intérêt économique à Madagascar. In: *Rapport du premier cours en formation à la lutte contre les ennemies des cultures et plus spécialement la lutte contre le criquet pélérin et aux recherches sur cet acridien, 17 fev. 1976*. FAO Rome, pp. 239–242.

Appert, J., Betbeder-Matibet, M. and Ranaivosoa, H. (1969) Vingt années de lutte biologique à Madagascar. *Agronomie Tropicale* 26, 555–572.

Appert, J. and Ranaivosoa, H. (1971) Un nouveau succès de la lutte biologique à Madagascar: contrôle des foreurs de la tige de maïs par un parasite introduit *Pediobius furvus* Gahan (Hym. Eulophidae). *Agronomie Tropicale* 26(3), 327–331.

Arakaki, N. and Gahana, Y. (1986) Emergence pattern and mating behavior of *Apanteles flavipes* (Cameron) (Hymenoptera: Braconidae). *Applied Entomology and Zoology* 21, 382–388.

Arnold, H.C. (1928) Top dressing of maize against stalk borer. *Rhodesia Agricultural Journal* 25, 162–165.

Aroga, R. (1988) Entomologie des céréales. In: Rapport d'activities: années 1987 et 1988. Cameroun National Cereals Research and Extension Project (unpublished).

Artschwager, E. and Brandes, E.W. (1958) *Sugarcane* (Saccharum officinarum)*: Origin, Classification, Characteristics and Descriptions of Representative Clones*. United States Department of Agriculture, Agriculture Handbook 122, Washington, DC, 307 pp.

Atkinson, P.R. (1979) Distribution and natural hosts of *Eldana saccharina* Walker in Natal, its oviposition sites and feeding patterns. *Proceedings of the South African Sugar Technologists' Association* 53, 111–115.

Atkinson, P.R. (1980) On the biology, distribution and natural host-plants of *Eldana saccharina* Walker (Lepidoptera: Pyralidae). *Journal of the Entomological Society of Southern Africa* 43, 171–194.

Aubert, J.F. (1979) Ichneumonides pétiolées inédites avec quatre genres nouveaux. *Bulletin de la Societé Entomologique de Mulhouse* 1979, 1–8.

Aubert, J.F. and Shaumar, N. (1978) Supplément aux ichneumonides d'Egypte. *Bulletin de la Societé Entomologique de Mulhouse,* 1978, 13–20.

Bagdanow, E.A. (1908) Uber die Abhangigkeit des und Fermenten und über Variabilitat und Verebung bei den Fleishfliegen. *Archiv für Anatomie und Physiologie* 2 (Suppl.), 173–200.

Barraclough, D.A. (1991) A new species of Tachinidae (Diptera) parasitic on the sugarcane borer *Eldana saccharina* (Lepidoptera: Pyralidae) in Natal, South Africa. *Bulletin of Entomological Research* 81, 133–136.

Barrow, M.R. (1985) The effect of different maize genotypes on the maize stalk-borer, *Busseola fusca* (Fuller) (Lepidoptera: Noctuidae), feeding in whorl tissue. *Journal of the Entomological Society of Southern Africa* 48, 113–119.

Barrow, M.R. (1989) Screening and breeding for resistance to *Busseola fusca.* In: *Proceedings of the International Symposium on Methodologies for Developing Host Plant Resistance to Maize Insects, 9–14 March 1987.* CIMMYT, Mexico City, pp. 184–191.

Barry, B.D. (1980) Where are we, and where are we going with insect resistance in sorghum? *African Journal of Plant Protection* 2(2), 149–153.

Bartlett, A.C. (1985) Guidelines for genetic diversity in laboratory establishment and maintenance. In: Singh, P. and Moore, R.F. (eds) *Handbook of Insect Rearing.* Elsevier, Amsterdam, pp. 7–17.

Bate, R. and van Rensburg, J.B.J. (1992) Predictive estimation of maize yield loss caused by *Chilo partellus* (Swinhoe) (Lepidoptera: Pyralidae) in maize. *South African Journal of Plant and Soil* 9, 150–154.

Bate, R., van Rensburg, J.H. and Giliomee, J.H. (1991) Distribution and occurrence of the stem-borer *Chilo partellus* (Swinhoe) (Lepidoptera: Pyralidae) on maize in South Africa. *Journal of African Zoology* 105, 11–15.

Beassem, J. (1982) Etude en laboratoire de l'influence de facteurs éco-physiologiques sur la biologie d'*Apanteles flavipes* (Cam.) (Hym: Braconidae) élevé sur *Chilo zacconius* Blesz. (Lep: Pyralidae). Thèse (Dr Ingénieur), Montpellier, 131 pp.

Beg, M.N. and Inayatullah, C. (1980) Studies on *Apanteles flavipes,* a parasite of graminaceous borers. *Pakistan Journal of Agricultural Research* 1, 50–53.

Bennett, F.D. (1965) Tests with parasites of Asian graminaceous moth borers on *Diatraea* and allied genera in Trinidad. *Technical Bulletin, Commonwealth Institute of Biological Control* 5, 101–116.

Bennett, F.D., Cock, M.J.W. and Diaz, F.A. (1983) *Allorhogas* sp. n. (braconids), a potential biological control agent for graminaceous stem borers from Mexico. *Entomology Newsletter, International Society of Sugarcane Technologists* 14, 9–12

Benoit, P.L.G. (1959) Hymenoptera: Ichneumonidae; Banchinae, Bassinae, Pimplinae et Tryphoninae. In: Hanström, B., Brinck, P. and Rudebeck, G. (eds) *South African Animal Life. Results of the Lund University Expedition in 1950–1951,* Vol. 6, Lund University, Lund, pp. 441–482.

Bentley, M.D., Hassanali, A., Lwande, W., Njoroge, P.E.W. and Sitayo, O. (1987) Insect antifeedants from *Tephrosia elata* Deflers. *Insect Science and its Application* 8, 85–88.

Berger, A. (1981) *Biological Control of the Spotted Stalkborer,* Chilo partellus *Swinhoe in Maize by Using the Bacteria* Bacillus thuringiensis. Project UNDP/FAO MOZ/75/009, Report, Instituto Nacional de Investigacao Agronomica, Maputo, Mozambique.

Berger, A. (1993) Larval migration in the stem borer *Chilo partellus* (Lep., Pyralidae). Dissertation, Swedish University of Agricultural Sciences, Uppsala, 80 pp.

Berger, R.S. (1963) Laboratory techniques for rearing *Heliothis* species on artificial medium. *US Agricultural Research Service (ARS)* 33–84.

Betbeder-Matibet, M. (1971) La lutte biologique contre *Chilo sacchariphagus* le 'borer ponctué' de la canne à sucre à Madagascar. *Agronomie Tropicale* 26(3), 332–336.

Betbeder-Matibet, M. (1977) Activités du laboratoire central d'entomologie de l'IRAT en matière de foreurs des graminées. Bilans et objectifs. Réunion sur les foreurs des graminées, Bouaké, Côte d'Ivoire, 8–11 Novembre, pp. 1–12.

Betbeder-Matibet, M. (1981) *Eldana saccharina* Walker, foreur des tiges de la canne à sucre en Afrique. *Agronomie Tropicale* 36, 279–293.

Betbeder-Matibet, M. (1985) *Eldana saccharina* Walker, pyrale de la canne à sucre et des cereales. In: *Insectes nuisibles aux cultures vivrières d'Afrique, de Madagascar et des Mascareignes.* IRAT Fiches Techniques, IRAT–CIRAD, Montpellier, France, 2 pp.

Betbeder-Matibet, M. (1986) *Maliarpha separatella* Ragonot, Pyralidae, foreur du riz. In: *Insectes nuisibles aux cultures vivrières d'Afrique, de Madagascar et des Mascareignes.* IRAT Fiches Techniques, IRAT–CIRAD, Montpellier, France, 2 pp.

Betbeder-Matibet, M. (1987) *Acigona ignefusalis* (Hampson), foreur du mil et du sorgho. In: *Insectes nuisibles aux cultures vivrières d'Afrique, de Madagascar et des Mascareignes.* IRAT Fiche Techniques, IRAT–CIRAD, Montpellier, France, 2 pp.

Betbeder-Matibet, M. (1989) Biological control of sorghum stem borers. In: *International Workshop on Sorghum Stem Borers, India, 17–20 November 1987.* International Crops Research Institute for the Semi-Arid Tropics, India, pp. 89–94.

Betbeder-Matibet, M. (1990) Elévage de plusieurs espèces du genre *Chilo* et de certains de leurs parasites pour la lutte biologique contre les foreurs des graminées en Afrique. *Insect Science and its Application* 11(4/5), 617–623.

Betbeder-Matibet, M. and Malinge, P. (1968) Un succès de la lutte biologique: controle de *Proceras sacchariphagus* Boj. Borer ponctué de la canne à sucre à Madagascar par un parasite introduit: *Apanteles flavipes* Cam. *Agronomie Tropicale* 22, 1196–1220.

Bhatnagar, V.S. (1984) *Rapport d'activité (novembre 1982–octobre 1983): Programme de Lutte Biologique.* Project CILSS, Nioro du Rip, Sénégal, 78 pp.

Bhatnagar, V.S. (1987) *Rapport de synthèse (1981–86) et récommandations, sous-programme de lutte biologique, project CILSS de Lutte Intégrée, Sénégal.* FAO Rapport Technique no. 1, Rome, 168 pp.

Bianchi, G., Delucchi, V., Bousses, P. and Rahalivavololona, N. (1991) La lutte biologique contre le borer blanc africain du riz *Maliarpha separatella* Rag. à Madagascar. Unpublished internal report, Institut de Phytiatrie, EPF, Zurich, 71 pp.

Biggler, F. and Laing, J. (1991) The first draft of criteria for quality control (product control) of *Trichogramma.* In: Biggler, F. (ed.) *Proceedings of the 5th Workshop of the Global Working Group on Quality Control of Mass Reared Arthropods, 25–28 March.* Wageningen, the Netherlands, pp. 200–201.

Bin, F. and Johnson, N.F. (1982) Some new species of *Telenomus* egg parasitoids of Tropical Pyralid pests. *Redia* 65, 229–252.

Blackburn, F. (1984) *Sugar-cane.* Longman, London, 414 pp.

Bleszynski, S. (1962) Studies on the Crambidae (Lepidoptera) Part xxxvi: On some species of the genus *Chilo* Zincken. *Acta Zoologica Cracoviensia* 7, 107–133.

Bleszynski, S. (1963) Studies on the Crambidae (Lepidoptera). Part 40: Review of the genera of the family Crambidae with data on their synonyms and types. *Acta Zoologica Cracoviensia* 8, 91–132.

Bleszynski, S. (1970) A revision of the world species of *Chilo* Zincken (Lepidoptera, Pyralidae). *Bulletin of the British Museum (Natural History) (Entomology)* 25, 101–195.

Bojer, W. (1856) Report of the committee on the 'cane borer'. Port Louis, Mauritius, pp. 1–46.

Boller, E. (1972) Behavioural aspects of mass rearing of insects. *Entomophaga* 17, 9–25.

Bonzi, M. (1980) *Rapport annuel 1979.* Défense des cultures, IRAT, Ouagadougou, Haute-Volta, 50 pp.

Bonzi, M. (1982a) *Chilo diffusilineus* J. de Joannis (Lepidoptera Pyralidae) borer des tiges de céréales en culture irrigués et pluviales en Haute Volta. *Agronomie Tropicale* 37, 207–209.

Bonzi, M. (1982b) Note on sorghum insect pests in Upper Volta. In: *Sorghum in the Eighties: Proceedings of the International Symposium on Sorghum, 2–7 November 1981, ICRISAT Center, India*, Vol. 2. International Crops Research Institute for the Semi-Arid Tropics, Patancheru, India, p. 747 (abstract).

Bonzi, M., Doumbia, Y.O., Selvaraj, C.J. and Konate, A. (1986) Pest problems of sorghum in the Sahel. In: *Proceedings of an International Seminar of the CILSS Project on Integrated Pest Management, 6–13 Dec 1984, Niamey, Niger.* CILSS (Comite Permanent Inter-états de Lutte Contre la Sécheresse dans le Sahel), Ouagadougou, Burkina Faso, pp. 108–118.

Bordat, D. (1979) Sondages effectués en Côte d'Ivoire concernant les foreurs du riz et leurs parasites. IRAT, Montpellier, France, 40 pp. (unpublished report).

Bordat, D. (1983) Mis au point de l'elevage de masse d'*Apanteles chilonis* Matsumura, 1912 et d'*Apanteles flavipes* (Cameron, 1891) (Hymenoptères Braconidae) sur trois lepidoptères Pyralidae foreurs des graminées (*Chilo zacconius* Bleszynski, 1970; *Chilo partellus* (Swinhoe, 1884) et *Diatraea saccharalis* (Fabricius, 1794) dans un objectif de lutte biologique. Thèse, Université des Sciences et Techniques du Languedoc, Academie de Montpellier.

Bordat, D. and Pichot, M. (1978) *Chilo zacconius* Blez. Technique d'élevage en milieu artificiel et observations sur sa biologie en laboratoire. *Agronomie Tropicale* 33(4), 337–343.

Bordat, D. and Renand, M. (1987) *Mussidia nigrivenella* Ragonot (Lepidoptera, Pyralidae): mise au point d'un élevage de masse sur milieux nutritifs artificiel. *Agronomie Tropicale* 42, 139–144.

Bordat, D., Brenière, J. and Coquard, J. (1977) Foreurs de graminées africaines: parasitisme et techniques d'élévage. *Agronomie Tropicale* 32, 391–399.

Bordat, D., Coquard, J. and Renand, M. (1984). Quelques moyens de lutte pour enrayer les nosémoses de trois foreurs des graminées élevés en laboratoire sur milieu nutritif artificiel. *Agronomie Tropicale* 39(3), 275–285.

Borror, D.J. Triplehorn, C.A. and Johnson, N.F. (1989) *An Introduction to the Study of Insects*, 6th edn. Saunders College Publishing, Philadelphia, 875 pp.

Bosque-Pérez, N.A. and Dabrowski, Z.T. (1989) Mass rearing of the maize stem borers *Sesamia calamistis* and *Eldana saccharina* at IITA. In: *Toward Insect Resistant Maize for the Third World. Proceedings of the International Symposium on Methodologies for Developing Resistance to Maize Insects.* CIMMYT, Mexico City, pp. 22–26.

Bosque-Pérez, N.A. and Mareck, J.H. (1989) Maize entomology research and insect resistance breeding at IITA. In: *Towards Production Technologies for Maize and Cowpea in Semi-arid West and Central Africa. SAFGRAD Maize–Cowpea Joint Workshop, Lomé, Togo.* Semi-Arid Food Grain Research and Development Project, Burkina Faso, pp. 227–235.

Bosque-Pérez, N.A. and Mareck, J.H. (1990) Distribution and species composition of lepidopterous maize borers in southern Nigeria. *Bulletin of Entomological Research* 80, 363–368.

Bosque-Pérez, N.A. and Mareck, J.H. (1991) Effect of the stem borer *Eldana saccharina* (Lepidoptera: Pyralidae) on the yield of maize. *Bulletin of Entomological Research* 81, 243–247.

Bosque-Pérez, N.A., Mareck, J.H., Dabrowski, Z.T., Everett, L., Kim, S.K. and Efron, Y. (1989) Screening and breeding for resistance to *Sesamia calamistis* and *Eldana saccharina*. In: *Toward Insect Resistant Maize for the Third World. Proceedings International Symposium on Methodologies for Developing Resistance to Maize Insects.* CIMMYT, Mexico City, pp. 163–169.

Bosque-Pérez, N.A., Ubeku, J.A. and Polaszek, A. (1995) Survey for parasites of *Sesamia calamistis* (Lepidoptera: Noctuidae) and *Eldana saccharina* (Lepidoptera: Pyralidae) in southwestern Nigeria. *Entomophaga* 39(1994), 367–376.

Bot, J. and Hollings, N. (1975) *A Guide to the Use of Pesticides and Fungicides in South Africa.* Plant Protection Research Institute, Government Printer, Pretoria.

Bot, J., Vermeulen, J.B. and Hollings, N. (1977) *A Guide to the Use of Pesticides and Fungicides in South Africa.* Plant Protection Research Institute, Government Printer, Pretoria.

Botchey, M. and Kwapong, P. (1991) The population fluctuation of *Chilo aleniella* (Lepidoptera: Pyralidae) in relation to the earwig *Diaperasticus erythrocephalus* (Dermaptera: Forficulidae) in the maize ecosystem in Ghana. Unpublished.

Bottger, G.T. (1942) Development of synthetic food media for use in nutritional studies of the European corn borer. *Journal of Agricultural Research* 65, 493–500.

Bouček, Z. (1988) *Australasian Chalcidoidea.* CAB International, Wallingford, 832 pp.

Bowden, J. (1954) The stem-borer problem in tropical cereal crops. In: *Report 6th Commonwealth Entomological Conference.* Commonwealth Institute of Entomology, London, pp. 104–107.

Bowden, J. (1956) New species of African stem-boring Agrotidae [Lepidoptera]. *Bulletin of Entomological Research* 47, 415–428.

Bowden, J. (1976) Stem borer ecology and strategy for control. *Annals of Applied Biology* 84, 107–134.

Box, H.E. (1953) List of sugar-cane insects: a synonymic catalogue of the sugar-cane insects and mites of the world, and of their insect parasites and predators, arranged systematically. *Commonwealth Institute of Entomology, London* 1–101.

Brenière, J. (1965) Liste des parasites et prédateurs des principaux insectes nuisibles aux cultures à Madagascar. *Agronomie Tropicale* 20, 344–349.

Brenière, J. (1969) Importance des problèmes entomologiques dans le développement de la riziculture de l'Afrique de l'Ouest. *Agronomie Tropicale* 24, 906–927.

Brenière, J. (1970) Entomological research carried out in French speaking West Africa in the fields of sorghum and millet. *African Soils/Sols Africains* 15, 93–99.

Brenière, J. (1971) Les problèmes des lepidoptères foreurs des graminées en Afrique de l'Ouest. *Annales de Zoologie et Ecologie Animale* 3, 287–296.

Brenière, J. (1976) Reconnaissance des principaux lépidoptères du riz de l'Afrique de l'Ouest. *Agronomie Tropicale* 31(3), 212–231.

Brenière, J. (1982) Lepidopterous borers of rice in West Africa: biology, damage and control. In: *Integrated Pest Management in Rice in West Africa. Proceedings of a Course, Concepts, Techniques, and Applications of Integrated Pest Management in Rice in West Africa, 10–28 January 1982.* West Africa Rice Development Association, Fendall, Liberia, pp. 20–32.

Brenière, J. (1983) *The Principal Insect Pests of Rice in West Africa and Their Control.* West Africa Rice Development Association, Monrovia, Liberia, 87 pp.

Brenière, J. and Bordat, D. (1982) Intérêt et difficultés de la lutte biologique par introduction d'entomophages – cas des foreurs du riz en Afrique. *Agronomie Tropicale* 37, 203–206.

Brenière, J., Rodriguez, H. and Ranaisova, H. (1962) Un ennemi de riz à Madagascar, *Maliarpha separatella* Rag. ou borer blanc. *Agronomie Tropicale* 17, 223–302.

Brenière, J., Pfeffer, P., Betbeder-Matibet, M. and Etienne, J. (1966) Tentative d'introduction à Madagascar et à la Réunion de *Diatraeophaga striatalis* parasite de *Proceras sacchariphagus* – boreur ponctué de la canne à sucre. *Entomophaga* 11, 231–238.

Brenière, J., Bordat, D., Vercambre, B., Hamza, H. and Renard, M. (1985) Les opérations de lutte biologique contre le foreur du maïs *Chilo partellus* (Swinhoe), Lepidoptera, dans l'île de Ngazidja. *Agronomie Tropicale* 40, 157–166.

Bridge, J., Terry, P.J. and Tunstall, J.M. (1978) *A Survey of Crop Pests, Diseases and Weeds of the Gambia.* Ministry of Overseas Development, London, 45 pp.

Brownbridge, M. (1990) The role of bacteria in the management of *Chilo* spp. *Insect Science and its Application* 11, 779–783.

Brownbridge, M. (1991) Native *Bacillus thuringiensis* isolates for the management of Lepidopteran cereal pests. *Insect Science and its Application* 12, 57–61.

Browning, H.W., Melton, C.W. and Saldana, R.R. (1985) *Release and Evaluation of Exotic Parasites for Biological Control of the Mexican Rice Borer on Sugar Cane in the Lower Rio Grande Valley, 1983–85.* Texas Agricultural Experiment Station, Progress Report No. 4344, 7 pp.

Brullé, A. (1846) Hyménoptères. In: Le Peletier de Saint Fargeau, A.L.M. (ed.) *Histoire naturelle des insects*, Vol. 4, Lepeletier, Paris, pp. 1–689.

Buddenhagen, I.W. and Persley, G.J. (1978) *Rice in Africa.* Academic Press, New York.

Cameron, P. (1906a) Descriptions of new species of parasitic Hymenoptera, chiefly in the collection of the South African Museum, Cape Town. *Annals of the South African Museum* 5, 17–186.

Cameron, P. (1906b) On the hymenopterous parasites of the mealie stalk borer (*Sesamia fusca*, Hampson). *Transactions of the South African Philosophical Society* 16, 334–336.

Campion, D.G. and Nesbitt, B.F. (1983) The utilization of sex pheromones for the control of stem-borers. *Insect Science and its Application* 4, 191–197.

Cardwell, K.F., Schulthess, F., Ndemah, R. and Ngoko, Z. (1997) A systems approach to assess crop health and maize yield losses due to pests and diseases in Cameroon. *Agriculture, Ecosystems and Environment* 33–47.

Carèsche, L. (1962) Les insectes nuisibles à la canne à sucre dans l'Ile de la Réunion. *Agronomie Tropicale* 17, 632–646.

Carèsche, L. and Brénière, J. (1962) Les insectes nuisibles à la canne à sucre à Madagascar; aspects actuels de la question. *Agronomie Tropicale* 17, 608–631.

Carnegie, A.G.M. and Leslie, G.W. (1979) Attempts at the biological control of *Eldana saccharina. Proceedings of the South African Sugar Technologists' Association* 53, 116–119.

Carnegie, A.J.M. (1991) Cane borers in Africa and recent work on *Eldana saccharina.* Discussion Document, International Society of Sugarcane Technologists' Workshop, West Palm Beach, Florida, 11–15 March 1991. 11 pp.

Carnegie, A.J.M., Conlong, D.E. and Graham, D.Y. (1985) Recent introductions of parasitoids against *Eldana saccharina* Walker (Lepidoptera: Pyralidae). *Proceedings of the South African Sugar Technologists' Association* June, 160–163.

Carter, A.O. (1985) An evaluation of the importance of stem borers and their control on maize grown in south eastern Nigeria. MSc thesis, University of Reading, UK, 57 pp.

Chacko, M.J. and Ogedah, K. (1991) Survey for other parasitoids of sorghum and maize stem borers. In: *(ICIPE) Eighteenth Annual Report, 1990.* ICIPE, Nairobi, p. 31.

Chatterji, S.M., Siddiqui, K.H., Panwar, V.P.S., Sharma, G.C. and Young, W.R. (1968) Rearing of the maize stem borer *Chilo zonellus* Swinhoe on artificial diet. *Indian Journal of Entomology* 30, 8–12.

Chiang, H.C. (1982) Factors to be considered in refining a general model of economic threshold. *Entomophaga* 27, 99–103.

Chiasson, H. and Hill, S.B. (1993) Population density, development and behaviour of *Diopsis longicornis and D. apicalis* (Diptera: Diopsidae) on rice in the Republic of Guinea. *Bulletin of Entomological Research* 83, 5–14.

Chiu, S.C., Chou, L.Y. and Chou, K.C. (1984) *A Check List of Ichneumonidae (Hymenoptera) of Taiwan.* Taiwan Agricultural Research Institute, Taipei, Special Publication no. 15.

Chiu, Shin-foon (1937) Notes on the natural enemies of the paddy borer, *Schoenobius incertellus* Walker in Canton with a list of its natural enemies in the world. [In Chinese; English summary.] *Entomology and Phytopathology* 5, 442–457.

Choo, H.Y., Kaya, H.K., Kim, J.B. and Park, Y.D. (1991) Evaluation of entomopathogenic nematodes, *Steinernema carpocapsae* (Steinernematidae) against rice stem borer *Chilo suppressalis* (Walker) (Lepidoptera: Pyralidae). *Korean Journal of Applied Entomology* 30(1), 50–53.

Chu, J.T. and Hsia, S.H. (1937) A list of the known hymenopterous parasites of the European corn borer (*Pyrausta nubilalis* Hübner). *Entomology and Phytopathology* 5, 136–147.

CIBC (1968–1972) Annual Reports of the Commonwealth Institute of Biological Control, Farnham Royal.

CIE (CAB International Institute of Entomology) (1988) Busseola fusca *(Fuller)*. Distribution Maps of Pests, Series A (Agricultural) no. 499, International Institute of Entomology, London, 2 pp.

CIE (CAB International Institute of Entomology) (1989) *Chilo partellus* (Swinh). Distribution Maps of Pests, Series A (Agriculture) no. 184, International Institute of Entomology, London.

CIMMYT (1992) *1991–92 CIMMYT World Maize Facts and Trends: Maize Research, Investment and Impacts in Developing Countries*. CIMMYT, Mexico City, 57 pp.

Cochereau, P. (1978) Fluctuation des population imaginales de *Diopsis thoracica* Westwood et *Diopsis apicalis* Westwood (Diptera, Diopsidae) en liaison avec la phénologie d'un riz de bas-fond à Bouaké (Côte d'Ivoire). *ORSTOM séries Biologie* 13(1), 45–58.

Cochereau, P. (1985) *Ecologie des populations, en Côte d'Ivoire, du foreur africain de la canne à sucre et du maïs* Eldana saccharina *Walker*. Côte d'Ivoire, Institut des Savanes, ORSTOM, 99 pp.

Conlong, D.E. (1986) Biological control. In: *Progress Report, March 1985–February 1986*. South African Sugar Association Experiment Station, Mt Edgecombe, pp. 123–128.

Conlong, D.E. (1989) Technique development for mass rearing. In: *Proceedings of Insect Rearing Workshop, Pieteramaritzburg, July 11, 1989*. Entomological Society of Southern Africa, Pretoria.

Conlong, D.E. (1990) A study of pest–parasitoid relationships in natural habitats: an aid towards the biological control of *Eldana saccharina* Walker (Lepidoptera: Pyralidae) in sugarcane. *Proceedings of the South African Sugar Technologists' Association* 64, 111–115.

Conlong, D.E. (1993) *Directory of South African Insect Rearers*. South African Sugar Association, Natal, pp. 1–19.

Conlong, D.E. (1994) A review and perspectives for the biological control of the African sugarcane stalkborer *Eldana saccharina* Walker (Lepidoptera: Pyralidae). *Agriculture, Ecosystems and Environment* 48, 9–17.

Conlong, D.E. and Graham, D.Y. (1988) A comparison of the known life histories of the bethylids *Parasierola* sp. and *Goniozus natalensis* Gordh collected from *Eldana saccharina* Walker (Lepidoptera: Pyralidae) in Uganda and southern Africa respectively. *Journal of the Entomological Society of Southern Africa* 51, 143–144.

Conlong, D.E., Graham, D.Y. and Hastings, H. (1988) Notes on the natural host surveys and laboratory rearing *Goniozus natalensis* Gordh (Hymenoptera: Bethylidae), a parasitoid of *Eldana saccharina* Walker (Lepidoptera: Pyralidae) larvae from *Cyperus papyrus*. *Journal of the Entomological Society of Southern Africa* 51, 115–127.

Cook, M. (1997) Revision of the genus *Maliarpha* Ragonot (Lepidoptera: Pyralidae), based on adult morphology with description of three new species. *Bulletin of Entomological Research* 87, 25–36.

COPR (Centre for Overseas Pest Research) (1976) *Pest Control in Rice*. PANS Manual no 3, COPR, London, UK.

Coquard, J. (1987) *Goniozus procerae* biologie et fécondité en laboratoire. *Mémoires et travaux de l'IRAT* 6, 1–44.

Crosskey, R.W. (1976) A taxonomic conspectus of the Tachinidae (Diptera) of Australia, including keys to the supraspecific taxa and taxonomic and host catalogues. *Bulletin of the British Museum (Natural History), Entomology Supplement* 21, 1–221.

Crosskey, R.W. (1980) Family Tachinidae. In: Crosskey, R.W. (ed.) *Catalogue of the Diptera of the Afrotropical Region*. British Museum (Natural History), London, pp. 822–882.

Crosskey, R.W. (1984) Annotated keys to the genera of Tachinidae (Diptera) found in tropical and southern Africa. *Annals of the Natal Museum* 26, 189–337.

Curran, C.H. (1939) African Tachinidae II. *American Museum Novitates* 1022, 1–5.

Cushman, R.A. (1933) The identity and synonymy of three oriental species of *Cremastus* (Hym., Ichneumonidae). *Proceedings of the Entomological Society of Washington* 35, 73–75.

Cuthbertson, A. (1936) Biological notes on some Diptera of southern Rhodesia. *Occasional Papers of the Rhodesia Museum* 5, 46–63.

Dabiré-Binso, L.M. (1980) Biologie et écologie d'*Eldana saccharina* Walker (Lepidoptera Pyralidae Galleriinae) foreur du maïs en Côte d'Ivoire et inventaire des autres Lépidoptères foreurs du maïs. Thèse Doctorale, 3e cycle, Université P. et M. Curie, Paris, 196 pp.

Dabrowski, Z.T. and Kidiavai, E.L. (1983) Resistance of some sorghum lines to the spotted stalk borer, *Chilo partellus* under western Kenya conditions. *Insect Science and its Application* 4(1–2), 119–126.

Dale, D. (1994) Insect pests of the rice plant – their biology and ecology. In: Heinrichs, E.A. (ed.) *Biology and Management of Rice Insects.* Wiley Eastern, New Delhi.

Dang, K., Anand, M. and Jotwani, M.G. (1970) A simple improved diet for mass rearing of sorghum stem borer *Chilo partellus* Swinhoe. *Indian Journal of Entomology* 32, 130–133.

David, H. and Easwaramoorthy, S. (1990) Biological control of *Chilo* spp. in sugarcane. *Insect Science and its Application* 11, 733–748.

Deedat, Y.D. (1994) Problems associated with the use of pesticides: an overview. *Insect Science and its Application* 15, 247–251.

Deeming, J.C. (1972) A review of the taxonomy of African shoot flies of sorghum. In: Jotwani, M.G. and Young, W.R. (eds) *Control of Sorghum Shoot Fly.* Oxford and IBH Publishing, New Delhi, India, pp. 3–26.

Dejoux, C. (1983) The toxicity of some new insecticides for the aquatic fauna. III – Deltamethrin. *Revue d'Hydrobiologie Tropicale* 16, 263–275.

Delobel, A. (1974) Deux *Enicospilus* nouveaux parasites de *Sesamia calamistis,* 'Borer rose' du maïs, à Madagascar (Hym. Ichneumonidae). *Bulletin de la Société Entomologique de France* 79, 98–104.

Delobel, A. (1975) *Chilo orichalcociliellus* Strand (Lepidoptera, Pyralidae), foreur des tiges du sorgho et du maïs à Madagascar. I. Charactéristiques morphologiques. *Cahiers d'ORSTOM, Série Biologie* 10(1), 3–9.

Delvare, G. (1988) Revision des *Eurytoma* (Hym.: Eurytomidae) d'Afrique Occidentale decrits par Risbec. *Annales de la Société Entomologique de France* 24, 117–149.

Delvare, G. and Aberlenc, H.P. (1989) *Les insectes d'Afrique et d'Amérique tropicale. Clés pour la reconnaissance des familles.* CIRAD, Montpellier, France, 298 pp.

Dent, D. (1991) *Insect Pest Management.* CAB International, Wallingford, Oxon., UK.

de Saeger, H. (1944) Microgasterinae (Hymenoptera Apocrita). *Exploration du Parc National Albert Mission de Witte* 47, 1–342.

de Saeger, H. (1946) Euphorinae (Hymenoptera Apocrita), fam. Braconides. *Exploration du Parc National Albert Mission de Witte* 50, 1–245.

de Saeger, H. (1948) Cardiochilinae et Sigalphinae (Hymenoptera, Apocrita). Fam. Braconidae. *Exploration du Parc National Albert Mission de Witte* 53, 1–272, figs 1–306.

Descamps, M. (1956) Insectes nuisibles au riz dans le nord Cameroun. *Agronomie Tropicale* 11(6), 732–755.

Dessart, P. (1989) *Aphanogmus maniboti* sp. n. espèce nouvelle d'Afrique (Hym. Ceraphronoidea Ceraphronidae). *Bulletin et Annales de la Société Royale Belge d'Entomologie* 125, 61–65.

De Wet, J.M.J. (1978) Systematics and evolution of sorghum Sect. Sorghum (Gramineae). *American Journal of Botany* 65, 477–484.

Dick, J. (1945) Some data on the biology of the sugarcane borer (*Eldana saccharina* Walker). *Proceedings of the South African Sugarcane Technologists' Association* 19, 75–79.

Diémé, E. (1982) *Connaisance de l'entomofaune de l'Agrosystème Maïs-Coton au Sahel.* Institut Sénégalaise de Recherches Agricoles (ISRA), pp. 1–48.

Diop, T. (1979) *Entomological Problems in the Rice Growing Areas Supervised by the SAED.* WARDA Report, February 1979, Monrovia, Liberia.

Dissemond, A. and Hindorf, H. (1990) Influence of sorghum/maize/cowpea intercropping on the insect situation at Mbita/Kenya. *Journal of Applied Entomology (ZAE)* 109, 144–150.

Doggett, H. (1988) *Sorghum.* 2nd edn. Longman, London, 512 pp.

Doumbia, Y.O. (1980) Entomologie du mil-sorgho. In: *Cellule de la défense des cultures résultats de la campagne 1979.* Document no. 6, Section de Recherches sur les Cultures Vivrières et Oléagineuses, Bamako, Mali, pp. 3–4.

Dreyer, M. (1987) Field and laboratory trials with simple neem products as protectants against pests of vegetables and field crops in Togo. Natural pesticides from the neem tree (*Azadirachta indica* A. Juss) and other tropical plants. In: Schmutterer, R. (ed.) *Proceedings of the 3rd International Neem Conference, Nairobi, Kenya, 10–15 July 1986.*

Drinkwater, T.W., Walters, M.C. and van Rensburg, J.B.J. (1979) The application of systemic insecticides to the soil for the control of the maize stalk borer, *Busseola fusca* (Lep.: Noctuidae), and *Cicadulina mbila* (Naudé) (Hem.: Cicadellidae), the vector of maize streak virus. *Phytophylactica* 11, 5–12.

Duerden, J.C. (1953) Stem borers of cereal crops at Kongwa, Tanganyika, 1950–52. + Corrigendum (1954). *East African Agricultural Journal* 19, 105–119 + 211 (1954).

du Plessis, C. (1936) The maize stalk borer. *Farming in South Africa* 14, 2 pp.

du Plessis, C. and Lea, H.A.F. (1943) The maize stalk borer, *Calamistis fusca* (Hmps). *Bulletin, Department of Agriculture and Forestry, Union of South Africa* 238, 51 pp.

du Toit, H. (1995) Die natuurlike predatorkompleks van somergraanplae en die invloed van chemiese plaagbeheer daarop. MSc thesis, University of Potchefstroom, South Africa, 140 pp.

Dwumfour, E.F. (1990) Predators of *Chilo partellus.* In: *ICIPE 18th Annual Report.* ICIPE, Nairobi, pp. 32–36.

Dwumfour, E.F., Owino, J. and Andere, M. (1991) Discovery capacity by parasitoids and predators of *Chilo partellus* eggs. In: *ICIPE 19th Annual Report,* ICIPE, Nairobi, pp. 23–24.

Egwuatu, R.I. and Ita, C.B. (1982) Some effects of single and split applications of carbofuran on the incidence of and damage by *Locris maculata, Busseola fusca* and *Sesamia calamistis* on maize. *Tropical Pest Management* 28, 277–283.

El-Amin, E.M. (1984) Relative susceptibility of seven sugarcane varieties to the stem borer, *Sesamia cretica* Led. under conditions of natural infestation at Sennar, Sudan. *Beitrage zür tropischen Landwirtschaft und Veterinärmedizin* 22, 72–77.

Elemo, K.E. and Ajayi, O. (1989) Effects of nitrogen on stem borer damage in sorghum/millet mixture. *Insect Science and its Application* 10(5), 601–605.

Endrody-Younga, S. (1968) The stem borer *Sesamia botanephaga* Tams and Bowden (Lep., Noctuidae) and the maize crop in central Ashanti, Ghana. *Ghana Journal of Agricultural Science* 1, 103–131.

Etienne, J. (1971) Tentative d'implantation de deux espèces de tachinaires contre *Chilo sacchariphagus* à la Réunion. *Agronomie Tropicale* 26(3), 337–340.

Etienne, J. (1977) Lutte contre les foreurs des graminées à la Réunion. Réunion sur les foreurs des graminées. Bouaké, Côte d'Ivoire, 8–11 Novembre, Annexe VI. IRAT, Réunion, 4 pp.

Etienne, J. (1987) Les problèmes entomologiques actuels en riziculture casamançaise. *Agronomie Tropicale* 42, 47–60.

Ezzat, Y.M. and Atries, I.E. (1969) Physical and mechanical measures for the control of sugar cane borers (Lepidoptera). *Bulletin of the Entomological Society of Egypt, Economic Series* 3, 167–177.

Fahringer, J. (1935) *Opuscula braconologica. 2. Aethiopische Region* 6–8, 385–635.

Fajemisin, J.M. (1992) Maize production and research in West and Central Africa: a perspective. In: Fajemisin, J.M. (ed.) *Maize Production in West and Central Africa: Trends and Research Orientation.* Semi-Arid Food Grain Research and Development Project (SAFGRAD), International Institute of Tropical Agriculture (IITA), Ouagadougou, Burkina Faso, pp. 1–5.

FAO (1991) *FAO Production Yearbook 44, 1990.* FAO Statistics Series No. 99, Rome, 283 pp.

FAO (1993) *FAO Production Yearbook 46, 1992.* FAO Statistics Series No. 112, Rome, 283 pp.

Farrington, J. (1977) Economic thresholds of insect pest infestation in peasant agriculture: a question of applicability. *PANS* 23, 143–148.

Feijen, H.R. (1979) Economic importance of rice stem borer *(Diopsis macrophthalma)* in Malawi. *Experimental Agriculture* 15, 177–186.

Feijen, H.R. (1985) The correct name of the African rice stem-boring Diopsidae (stalk-eyed fly). *International Rice Research Newsletter* 10(5), 21–22.

Ferrière, C. (1933) Chalcidoid and Proctotrupoid parasites of pests of the coconut palm. *Stylops* 2, 86–108.

Firempong, S. (1988) The effect of maize cultivar and planting density on *C. partellus* oviposition in the field. In: *ICIPE 15th Annual Report 1987*, ICIPE, Nairobi, pp. 11–12.

Flattery, K.E. (1982) An assessment of pest damage on grain sorghum in Botswana. *Experimental Agriculture* 18, 319–328.

Forsyth, J. (1966) *Agricultural Insects of Ghana.* Ghana Academy of Sciences, Accra, 163 pp.

Fournier, D. and Etienne, J. (1981) Effet du parasitisme par les Nosematidae sur la réproduction de *Chilo sacchariphagus* (Boj.) (Lep. Pyralidae). *Agronomie Tropicale* 36, 389–390.

Fuchs, T.W., Huffman, F.R. and Smith, J.W., Jr (1979) Introduction and establishment of *Apanteles flavipes* (Hym.: Braconidae) on *Diatraea saccharalis* (Lep.: Pyralidae) in Texas. *Entomophaga* 24, 109–114.

Fujimoto, H., Itoh, K., Yamamoto, M., Kyozuka, J. and Shimamoto, K. (1993) Insect resistant rice generated by introduction of a modified delta-endotoxin gene of *Bacillus thuringiensis. Bio/Technology* 11, 1151–1155.

Fukaya, M. and Nasu, S. (1966) A *Chilo* iridescent virus (CIV) from the rice stem borer *Chilo suppressalis* (Walker). *Applied Entomology and Zoology* 1, 69–72.

Gahan, A.B. (1928) Some reared parasitic Hymenoptera from Sudan. *Bulletin of Entomological Research* 19, 255–257.

Gahukar, R.T. (1981) Biological control of insect pests of sorghum and pearl millet in West Africa. In: *Proceedings of the International Conference on Biological Control of Pests: Its Potential in West Africa.* USAID, Dakar, Senegal, pp. 69–91.

Gahukar, R.T. (1984) Insect pest of pearl millet in West Africa: a review. *Tropical Pest Management* 30, 142–147.

Gahukar, R.T. (1989) Pest and disease incidence in pearl millet under different plant density and intercropping patterns. *Agriculture, Ecosystems and Environment* 26, 69–74.

Gahukar, R.T. (1990) Overview of insect pest management in cereal crops in sub-Saharan West Africa. *Indian Journal of Entomology* 52, 125–138.

Gahukar, R.T. and Jotwani, M.G. (1980) Present status of field pests of sorghum and millets in India. *Tropical Pest Management* 26, 138–151.

Gauld, I.D. (1982) A revised key to the *Enicospilus antefurcalis* (Szepligeti) (Hymenoptera:

Ichneumonidae) species group of the Afrotropical Region. *Bulletin of Entomological Research* 72, 33–38.

Gauld, I.D. and Bolton, B. (eds) (1988) *The Hymenoptera*. British Museum (Natural History)/ Oxford University Press, London, 332 pp.

Gauld, I.D. and Mitchell, P.A. (1978) *The Taxonomy, Distribution and Host Preferences of African Parasitic Wasps of the Subfamily Ophioninae (Hymenoptera: Ichneumonidae)*. CAB, Slough, 287 pp.

Gebre-Amlak, A. (1985) Survey of the lepidopterous stem borers attacking maize and sorghum in Ethiopia. *Ethiopian Journal of Agricultural Sciences* 7, 15–26.

Gebre-Amlak, A. (1988a) Development of maize stalk borer *Busseola fusca* (Fuller) in wild host plants in Ethiopia. *Journal of Applied Entomology* 106, 390–395.

Gebre-Amlak, A. (1988b) Survival of maize stalk borer, *Busseola fusca* (Fuller) in crop residues in Ethiopia. *Crop Protection* 7, 183–185.

Gebre-Amlak, A., Sigvald, R. and Petterson, J. (1989) The relationship between sowing date, infestation and damage by the maize stalkborer, *Busseola fusca* (Noctuidae), on maize in Awassa, Ethiopia. *Tropical Pest Management* 35, 143–145.

Ghani, M.A. and Williams, J.R. (1962) An attempt to establish the Javanese fly, *Diatraeophaga striatalis*, Towns in Mauritius for control of the cane moth-borer, *Proceras sacchariphagus* Boj., with notes upon parasites of cane moth-borers from Java. In: *Proceedings 11th [congress] International Society of Sugarcane Technologists, Mauritius, 1962*, pp. 626–641.

Gibson, G.A.P. (1985) Some pro- and mesothoracic structures important for phylogenetic analysis of *Hymenoptera*, with a review of terms used for the structures. *Canadian Entomologist* 117, 1395–1443.

Gillespie, D.Y. (1993a) Rearing the African sugarcane borer *Eldana saccharina*. In: *Proceedings of the 3rd National Insect Rearing Workshop: Effective Insect Rearing, Johannesburg, 1 July 1993*. South Africa, pp. 17–24.

Gillespie, D.Y. (1993b) Development of mass rearing methods for the sugar cane borer *Eldana saccharina* (Lepidoptera, Pyralidae) and its indigenous parasitoid *Goniozus natalensis* (Hymenoptera, Bethylidae). In: *Proceedings of the 3rd National Rearing Workshop: Effective Insect Rearing, Johannesburg, 1 July 1993*. South Africa, pp. 25–43.

Girling, D.J. (1972) Report on investigations on graminaceous stem-borers in East Africa (November 1969–March 1972). CIBC Report, 22 pp.

Girling, D.J. (1978) The distribution and biology of *Eldana saccharina* Walker (*Lepidoptera: Pyralidae*) and its relationship to other stem borers in Uganda. *Bulletin of Entomological Research* 68, 471–488.

Girling, D.J. (1980) *Eldana saccharina* as a crop pest in Ghana. *Tropical Pest Management* 26, 152–156.

Goldman, A. (1996) Pest and disease hazards and sustainability in African agriculture. *Experimental Agriculture* 32, 199–211.

Gonçalves, L. (1970) *A. broca* milho *Chilo partellus* Swinhoe (Lepidoptera: Crambidae) em Moçambique. Contribuicao para o sen estudo. *Agronomia Moçambicana* 4, 239–426.

Goraya, A.A., Mushtaque, M. and Attique, M.R. (1982) Suitability of *Apanteles flavipes* (Cameron) (Hym: Braconidae) as biocontrol agent for graminaceous stem-borers in Pakistan. In: *Proceedings of the Eighteenth Annual Convention of the Pakistan Society of Sugar Technology, Rawalpindi, 4–5 September*, pp. 177–182.

Gordh, G. and Moczar, L. (1990) A catalog of the world Bethylidae (Hymenoptera: Aculeata). *Memoirs of the American Entomological Institute* 46, 1–364.

Goudegnon, A.E. (1985). Le foreur ponctué de graminées, *Chilo partellus* (Swinhoe, 1884) (Lepidoptera = Pyralidae) et ses parasites, *Apanteles flavipes* (Cameron, 1891) (Hymenoptera = Braconidae) et *Nosema bordati* n. sp. (Microsporidia = Nosematidae):

cycles de développement et inter-relations. Thèse de Doctorat, Academie de Montpellier, Université Montpellier II Sciences et Techniques du Languedoc, 230 pp.

Goulet, H. and Huber, J.T. (eds) (1993) *Hymenoptera of the World: An Identification Guide to Families.* Agriculture Canada, Ottawa, 668 pp.

Gounou, S., Schulthess, F., Shanower, T., Hammond, W.N.O., Braima, H., Olaleye, I., Cudjoe, A.R., Adjakloe, R. and Antwi, K.K. (1993) *Stem and Ear Borers of Maize in Ghana.* Plant Health Management Division, Research Monograph, International Institute of Tropical Agriculture, Ibadan, Nigeria.

Graham, D.Y. and Conlong, D.E. (1988) Improved laboratory rearing of *Eldana saccharina* (Lepidoptera, Pyralidae) and its indigenous parasitoid *Goniozus natalensis* (Hymenoptera, Bethylidae). *Proceedings of the South African Sugarcane Technologists Association* 62, 116–119.

Graham, M.W.R. de V. (1987) A reclassification of the European Tetrastichinae (Hymenoptera: Eulophidae), with a revision of certain genera. *Bulletin of the British Museum (Natural History), Entomology Series* 55(1), 1–392.

Graham, M.W.R. de V. (1991) A reclassification of the European Tetrastichinae (Hymenoptera: Eulophidae): revision of the remaining genera. *Memoirs of the American Entomological Institute* 49, 1–322.

Granger, C. (1949) Braconides de Madagascar. *Mémoires de l'Institut Scientifique de Madagascar, Sér. A* 2, 1–428.

Greathead, D.J. (1971) *A Review of Biological Control in the Ethiopian Region.* Commonwealth Institute of Biological Control, Technical Communication No. 5, Commonwealth Agricultural Bureaux, Farnham Royal, UK, 162 pp.

Greathead, D.J. (1990) Utilization of natural enemies of *Chilo* spp. for management in Africa. *Insect Science and its Application* 11, 749–755.

Greathead, D.J., Lionnet, J.F.G., Lodos, N. and Whellan, J.A. (1971) A review of biological control in the Ethiopian Region. *Technical Communication* 5, CIBC, 162 pp.

Grist, D.H. (1983) *Rice.* 5th edn. Tropical Agriculture Series, Longman, London, 601 pp.

Grist, D.H. and Lever, R.J.A.W. (1969) *Pests of Rice.* Tropical Science Series, Longmans, London, 520 pp.

Gupta, V.K. (1987) The Ichneumonidae of the Indo-Australian area (Hymenoptera). A synonymic catalogue of the taxa described through 1985 together with a bibliography, 1960–1985. *Memoirs of the American Entomological Institute* 41, 2 parts, 1210 pp.

Gupta, V.K. and Tikar, D.T. (1969) Taxonomic identity of the Pimpline genera *Flavopimpla* Betrem and *Afrephialtes* Benoit (Hymenoptera: Ichneumonidae). *Oriental Insects* 3, 269–277.

Gupta, V.K. and Tikar, D.T. (1976) *Ichneumonologia orientalis or a Monographic Study of the Ichneumonidae of the Oriental Region, Part I. The Tribe Pimplini* (*Hymenoptera: Ichneumonidae: Pimplinae*). Oriental Insects Monograph No. 1, Association for the Study of Oriental Insects, Delhi, 313 pp.

Hall, R.W. and Ehler, L.E. (1979) Rate of establishement of natural enemies in classical biological control. *Bulletin of the Entomological Society of America* 25, 280–282.

Hall, R.W., Ehler, L.E. and Bisabri-Ershadi, B. (1980) Rate of success in classical biological control of arthropods. *Bulletin of the Entomological Society of America* 26, 111–114.

Hall, D.R., Beevor, P.S., Cork, A., Lester, R., Nesbitt, B.F., Nyirenda, G.K.C., Phiri Nota, D.D., Blair, B.W. and Tannock, J. (1981) The female sex pheromone of the maize stalk-borer, *Busseola fusca* (Fuller) (Lepidoptera: Noctuidae): identification and initial field trials. *Zimbabwe Journal of Agricultural Research* 19, 111–122.

Hampson, G.F. (1919) Descriptions of new Pyralidae of the subfamilies Crambinae and Siginae. *Annals and Magazine of Natural History* 3, 533–547.

Harris, K.M. (1961) Stem borers. In: *Annual Report, Department of Agricultural Research, Nigeria, 1959–60*, pp. 35–38.

Harris, K.M. (1962) Lepidopterous stem borers of cereals in Nigeria. *Bulletin of Entomological Research* 53, 139–171.

Harris, K.M. (1964) Annual variations of dry-season populations of larvae of *Busseola fusca* (Fuller) in northern Nigeria. *Bulletin of Entomological Research* 54(4), 643–647.

Harris, K.M. (1989a) Bioecology of sorghum stem borers. In *International Workshop on Sorghum Stem Borers, 17–20 November 1987*, International Crops Research Institute for the Semi Arid Tropics, Patancheru, India, pp. 63–71.

Harris, K.M. (1989b) Recent advances in sorghum and pearl millet stem borer research. In: *International Workshop on Sorghum Stem Borers, 17–20 November 1987.* ICRISAT, Patancheru, India, pp. 9–16.

Harris, K.M. (1990) Bioecology of *Chilo* species. *Insect Science and its Application* 11(4/5), 467–477.

Harris, K.M. and Nwanze, K.F. (1992) Busseola fusca *(Fuller), the African Maize Stem Borer: a Handbook of Information.* Information Bulletin 33, International Crops Research Institute for the Semi-Arid Tropics, Patancheru, India, 84 pp.

Harrison, M.N. (1981) *Maize in the Tropics.* Longman, London, 400 pp.

Hassanali, A., Bentley, M.D., Ole-Sitayo, E.N., Njoroge, P.E.W. and Yatagai, M. (1986) Studies on limonoid insect antifeedants. *Insect Science and its Application* 7, 495–499.

Heath, J. and Emmet, A.M. (1979) *The Moths and Butterflies of Great Britain and Ireland*, Vol. 9. Curwen Press, London.

Heinrich, G.H. (1967) *Synopsis and Reclassification of the Ichneumoninae stenopneusticae of Africa South of the Sahara (Hymenoptera).* Vol. 1, *Introduction; Key to Tribes and Subtribes of Ichneumoninae stenopneusticae; Synopsis of the Protichneumonini, Ceratojoppini, Ischnojoppini, Trogini.* State College Press, Farmington, pp. 1–250.

Heinrich, G.H. (1968) *Synopsis and Reclassification of the Ichneumoninae stenopneusticae of Africa South of the Sahara (Hymenoptera).* Vol. 3, *Synopsis of the Ichneumonini: Subtribes Ichneumonina and Amblytelina.* State College Press, Farmington, pp. 481–692.

Hill, R.E. and Gary, W.F. (1979) Effects of the microspridium *Nosema pyrausta* on field populations of European corn borers in Nebraska. *Environmental Entomology* 8, 91–95.

Hillis, D.M. and Moritz, C. (1990) *Molecular systematics.* Sinauer Associates, Sunderland, Massachusetts, pp. 45–126.

Hinton, H.E. (1946) On the homology and nomenclature of the setae of lepidopterous larvae. *Transactions of the Royal Entomological Society of London* 97, 1–37.

Ho, D.T. and Kibuka, J.G. (1983) Stem borers in various rice ecosystems in Kenya. *International Rice Research Newsletter* 8, 18.

Ho, D.T., Njokah, J.J. and Kibuka, J.G. (1983) Studies on rice stem borers in Kenya with emphasis on *Maliarpha separatella* Rag. *Insect Science and its Application* 4, 65–73.

Hoekstra, N. and Kfir, R. (1997) Microbial pathogens of the cereal stem borers *Busseola fusca* (Fuller) (Lepidoptera: Noctuidae) and *Chilo partellus* (Swinhoe) (Lepidoptera: Pyralidae) in South Africa. *African Entomology* 5, 161–163.

Holloway, J.D. (1985) The moths of Borneo: family Noctuidae: subfamilies Euteliinae, Stictopterinae, Plusiinae, Pantheinae. *Malayan Nature Journal* 38, 157–317.

Holloway, J.D. (1989) The moths of Borneo: family Noctuidae, trifine subfamilies: Noctuinae, Heliothinae, Hadeninae, Acronictinae, Amphipyrinae, Agaristinae. *Malayan Nature Journal* 42, 57–226.

Holloway, J.D., Bradley, J.D. and Carter, D.J. (1987) *IIE Guides to Insects of Importance to Man. 1. Lepidoptera* (edited by C.R. Betts). CAB International, Wallingford, UK, 262 pp.

Horstmann, K. (1990) Die westpalaearktischen Arten der Gattung Pristomerus Curtis, 1836 (Hymenoptera, Ichneumonidae). *Entomofauna* 11, 9–44.

House, G.J. and Alzugaray, M.D.R. (1989) Influence of cover cropping and no-tillage practices on community composition of soil arthropoda in a North Carolina agroecosystem. *Environmental Entomology* 18, 302–307.

Husain, T. and Khan, M.Y. (1986) Family Eulophidae. *Oriental Insects* 20, 211–245.

IAPSC (Inter-African Phytosanitary Council) (1985a) Busseola fusca *(Fuller)*. Distribution Maps of Major Crop Pests and Diseases in Africa, 193. Inter-African Phytosanitary Council, Yaoundé, 2 pp.

IAPSC (Inter-African Phytosanitary Council) (1985b) Chilo partellus *(Swinhoe)*. Distribution Maps of Major Crop Pests and Diseases in Africa, 199. Inter-African Phytosanitary Council, Yaoundé.

IAR (Institute for Agricultural Research) (1975) Entomology. In: *Cereals Improvement Programme: Report to the Board of Governors on the Institute's Work in 1974–75*. IAR, Samaru, Zaria, Nigeria, pp. 28–38.

IAR (Institute for Agricultural Research) (1976) Entomology. In: *Cereals Improvement Programme: Report to the Board of Governors on the Institute's Work in 1975–76*. IAR, Samaru, Zaria, Nigeria, pp. 17–38.

IAR (Institute for Agricultural Research) (1984) Sorghum. In: *Science and Technology Briefing, Lagos, December 1984*. IAR, Samaru, Zaria, Nigeria.

Ibekwe, G.O. (1986) *Le Maïs: Publications en langue française disponibles à la bibliothèque de l'IITA*. Library and Documentation Centre, Documentation, Information and Library Program, IITA, Ibadan, Nigeria, 244 pp.

ICIPE (1996) Understanding the role of the wild habitat in stemborer invasion of cereal crops. In: *1995 ICIPE Annual Report*. ICIPE, Nairobi, Kenya, pp. 22–25.

ICRISAT (1984a) *Sorghum Bibliography 1977–80*. ICRISAT Sorghum and Millets Information Centre, Patancheru, India.

ICRISAT (1984b) *Sorghum Bibliography 1981*. ICRISAT Sorghum and Millets Information Centre, Patancheru, India.

ICRISAT (1986a) *Sorghum Bibliography 1982*. ICRISAT Sorghum and Millets Information Centre, Patancheru, India.

ICRISAT (1986b) *Sorghum Bibliography 1983*. ICRISAT Sorghum and Millets Information Centre, Patancheru, India.

IITA (International Institute of Tropical Agriculture) (1984) Upland rice in Africa. In: *An Overview of Upland Rice Research, Proceedings of the 1982 Bouaké Ivory Coast Upland Rice Workshop*. IRRI 1–566, Los Baños, pp. 3–19.

IITA (International Institute of Tropical Agriculture) (1986) *Maize Program Annual Report for 1985*. Ibadan, Nigeria, 167 pp.

IITA (International Institute of Tropical Agriculture) (1993) *Plant Health Management Division Annual Report for 1992*. Cotonou, Republic of Benin, 149 pp.

Ingram, W.R. (1958) The Lepidopterous stalk borers associated with Graminae in Uganda. *Bulletin of Entomological Research* 49, 367–383.

Ingram, W.R. (1960) Experiments on the control of stalk borers on sorghum in Uganda. *East African Agriculture and Forestry Journal* 25, 184–189.

Ingram, W.R. (1983) Biological control of graminaceous stem borers and legume pod borers. *Insect Science and its Application* 4, 205–209.

Ingram, W.R., Irving, N.S. and Roome, R.E. (1973) *A Handbook on the Control of Agricultural Pests in Botswana*. Government Printer, Gaborone, 129 pp.

Isaac, P.V. and Rao, K.V. (1941) A key for the identification of the larvae of the known lepidopterous borers of sugarcane in India based on morphological characters. *Indian Journal of Agricultural Science* 11(5), 795–803.

ISRA (Institut Sénégalais de Recherches Agricoles) (1989) *Recherches pluridisciplinaires sur le riz pluvial et submergé: principaux résultats de la campagne 1988*. Centre de Recherche Agricoles de Djibelor, Ziguinchor, Sénégal.

Israel, P. and Padmanabhan, S.Y. (1978) *Biological Control of Stem Borers of Rice in India.* Report, Central Rice Research Institute, ICAR, Cuttack, India, 155 pp.

Jack, R.W. (1918) A note on the maize stalk borer. *Rhodesia Agricultural Journal* 15, 449–450.

Jack, R.W. (1922) Notes on the maize stalk borer or 'top worm'. *Rhodesia Agricultural Journal* 19, 87–88.

Jack, R.W. (1928) Trap cropping against maize pests *Rhodesia Agricultural Journal* 25, 1228–1231.

Jackai, L.E.N. and Raulston, J.R. (1982) Rearing two maize stem borers and a legume pod borer on artificial diet. *IITA Research Briefs* 3, 1–6.

Jacquot, M. and Courtois, B. (1983) Le riz pluviale. *Le Technicien d'Agriculture Tropical* 1, 1–134.

Jepson, W.F. (1939) Progress in parasite importation during 1938. *Revue Agricole Sucrière Maurice* 105, 82–84.

Jepson, W.F. (1941) Entomological Division. In: *Annual Report of the Department of Agriculture, Mauritius for 1939,* Port Lewis, pp. 16–19.

Jepson, W.F. (1954) *A Critical Review of the World Literature on the Lepidopterous Stalk Borers of Tropical Graminaceous Crops.* Commonwealth Institute of Entomology, London, UK, 127 pp.

Jerath, M.L. (1968) Parasites of sugarcane stem borers in Nigeria. *Journal of Economic Entomology* 61, 435–436.

Johnson, N.F. (1984) Systematics of Nearctic *Telenomus.* Classification and revisions of the *podisi* and *phymatae* groups. *Bulletin of the Ohio Biological Survey* 6(3), 113 pp.

Jonathan, J.K. (1980) *The Isotima-complex (Hymenoptera: Ichneumonidae).* Records of the Zoological Survey of India, Miscellaneous Publications, Occasional Paper 17, 146 pp.

Jordan, F.J. (1966) Report on an investigation into the presence and prevalence of rice stem-borers and their parasites in Sierra Leone, 1964–1965. West African Rice Research Station, Rokupr, Sierra Leone and International Rice Research Institute, Los Baños, Philippines, 47 pp. (unpublished report).

Joslyn, D.F. (1984) Maintenance of genetic variability in reared insects. In: King, E.G. and Leppla, N.C. (eds) *Advances and Challenges in Insect Rearing.* ARS, USDA, US Government Printing Office, Washington, DC, pp. 20–29.

Jotwani, M.G. (1983) Chemical control of stem borers. *Insect Science and its Application* 4, 185–189.

Jotwani, M.G. and Young, W.R. (1972) Recent development on chemical control of insect pests of sorghum. In: Rao, N.G.P. and House, L.R. (eds) *Sorghum in the Seventies.* Oxford and IBH, New Delhi.

Kajita, H. and Drake, E.F. (1969) Biology of *Apanteles chilonis* and *A. flavipes* (Hymenoptera: Braconidae), parasites of *Chilo suppressalis. Mushi* 42, 163–179.

Kamran, M.A. and Raros, E.S. (1969) Insect parasites in the natural control of species of rice stem borers on Luzon Island, Philippines. *Annals of the Entomological Society of America* 62, 797–801.

Kato, D., Karel, A.K. and Ndunguru, B.J. (1982) Effect of insecticide spray on insect pest and yield of sorghum and simsim in pure stand and in intercropping – summary. In: *Proceedings of the Second Symposium on Intercropping in Semi-Arid Areas, Morogoro, Tanzania, 4–6 August 1980.* IDRC-186e, Ottawa, Ontario, Canada, pp. 117–118.

Kaufmann, T. (1983) The behavioral biology, feeding habits and ecology of three species of maize stem borers, *Eldana saccharina, Sesamia calamistis* and *Busseola fusca* in Ibadan, Nigeria, West Africa. *Journal of the Georgia Entomological Society* 18, 259–272.

Keaster, A.J. and Harrendorf, K. (1965) Laboratory rearing of south-western corn borer *Zeadiatrea grandiosella* (Dyar) on a wheat germ medium. *Journal of Economic Entomology* 58, 923–924.

Kfir, R. (1988) Hibernation by the lepidopteran stalk borers, *Busseola fusca* and *Chilo partellus* on grain sorghum. *Entomologia Experimentalis et Applicata* 48, 31–36.

Kfir, R. (1989a) Biological control of stalk borers. *Bulletin of the Plant Protection Research Institute, South Africa* 16, 5.

Kfir, R. (1989b) Prospects of biological and cultural control of lepidopteran stalk borers in summer grain crops in South Africa. *Proceedings of the 7th Entomology Congress, Entomological Society of Southern Africa* 7, 79.

Kfir, R. (1990a) Parasites of the spotted stalk borer, *Chilo partellus* (Lepidoptera: Pyralidae) in South Africa. *Entomophaga* 35, 403–410.

Kfir, R. (1990b) Prospects for cultural control of the stalk borers, *Chilo partellus* (Swinhoe) and *Busseola fusca* (Fuller), in summer grain crops in South Africa. *Journal of the Entomological Society of Southern Africa* 53, 41–47.

Kfir, R. (1991a) Duration of diapause in the stem borers, *Busseola fusca* and *Chilo partellus*. *Entomologia Experimentalis et Applicata* 61, 265–270.

Kfir, R. (1991b) Non-chemical control of grain stalk borers. *Bulletin of the Plant Protection Research Institute, Pretoria* 23, 4–5.

Kfir, R. (1991c) Selecting parasites for biological control of lepidopterous stalk borers in summer grain crops in South Africa. *Redia* 74(3), 231–236.

Kfir, R. (1992a) Alternative, non-chemical control methods for the stalk borers, *Chilo partellus* (Swinhoe) and *Busseola fusca* (Fuller), in summer grain crops in South Africa. *Technical Communication, Department of Agricultural Development, Republic of South Africa* 232, 99–103.

Kfir, R. (1992b) A simple artificial diet for mass rearing of the stem borer *Chilo partellus* (Swinhoe) (Lepidoptera, Pyralidae). *Journal of the Entomological Society of South Africa* 25, 283–284.

Kfir, R. (1992c) Seasonal abundance of the stem borer *Chilo partellus* (Lepidoptera: Pyralidae) and its parasites on summer grain crops. *Journal of Economic Entomology* 85, 518–529.

Kfir, R. (1992d) Selecting parasites for biological control of lepidopterous stalk borers in summer grain crops in South Africa. *Redia* 74(3) (Appendix), 231–236.

Kfir, R. (1994) Attempts at biological control of the stem borer, *Chilo partellus* (Swinhoe) (Lepidoptera: Pyralidae), in South Africa. *African Entomology* 2, 67–68.

Kfir, R. (1995) Parasitoids of the African stem borer, *Busseola fusca* (Fuller) (Lepidoptera: Noctuidae) in South Africa. *Bulletin of Entomological Research* 85, 369–377.

Kfir, R. (1997) Competitve displacement of *Busseola fusca* (Fuller) (Lepidoptera: Noctuidae) by *Chilo partellus* (Lepidoptera: Pyralidae). *Annals of the Entomological Society of America* 90, 619–624.

Kfir, R. and Bell, R. (1993) Intraseasonal changes in populations of the African maize stem borer, *Busseola fusca* (Fuller) (Lepidoptera: Noctuidae), in Natal, South Africa. *Revue de Zoologie Africaine* 107, 543–553.

Kfir, R., Graham, D.Y. and Van Vuuren, R. (1989a) An improved method for mass rearing *Paratheresia claripalpis* for biological control of lepidopteran stalk borers. *Entomologia Experimentalis et Applicata* 51, 37–40.

Kfir, R., van Hamburg, H. and van Vuuren, R. (1989b) Effect of stubble treatment on the post-diapause emergence of the stalk borer *Chilo partellus* (Swinhoe) (Lepidoptera: Pyralidae). *Crop Protection* 8, 289–292.

Kfir, R., Gouws, J. and Moore, S.D. (1993) Biology of *Tetrastichus howardi* (Olliff), a facultative hyperparasite of stem borers. *Biocontrol Science and Technology* 3, 149–159.

Khan, Z.R. and Randriamananoro, J.J. (1995) Behavioual and physiological responses of *Busseola fusca* to selected wild host plants. Paper presented at 1995 Entomological Society of America Annual Meeting, Las Vegas, Nevada, USA, 17–21 December 1995.

Khan, Z.R., Litsinger, J.A., Barrion, A.T., Villanueva, F.F.D., Fernandez, N.J. and Tayo, L.D. (1991) *World Bibliography of Rice Stem Borers, 1794–1990.* International Rice Research Institute, Los Baños, Philippines, and International Centre of Insect Physiology and Ecology, Nairobi, Kenya, 415 pp.

Khan, Z.R., Ampong-Nyarko, K., Chiliswa, P., Hassanali, A., Kimani, S., Lwande, W., Overholt, W.A., Pickett, J.A., Smart, L.E., Wadhams, L.J. and Woodstock, C.M. (1997) Intercropping increases parasitism of pests. *Nature* 388, 631–639.

Kimani, S.W. and Overholt, W.A. (1995) Biosystematics of the *Cotesia flavipes* complex (Hymenoptera: Braconidae): interspecific hybridization, sex pheromone and mating behaviour studies. *Bulletin of Entomological Research* 85, 379–386.

Kimani-Njogu, S.W. and Overholt, W.A. (1997) Biosystematics of the *Cotesia flavipes* complex (Hymemoptera: Braconidae) parasitoids of gramineous stemborers. *Insect Science and its Application* 17 (in press).

Kimani-Njogu, S.W., Overholt, W.A., Woolley, J.B. and Walker, A.K. (1997) Biosystematics of the *Cotesia flavipes* Cameron complex (Hym: Bracondiae): Morphometrics and morphology of selected allopatric populations. *Bulletin of Entomological Research* 87, 61–66.

King, A.B.S. and Saunders, J.L. (1984) *The Invertebrate Pests of Annual Food Crops in Central America. A Guide to their Recognition and Control.* Overseas Development Administration, London.

Kioko, E.N., Overholt, W.A., Omwega C.O. and Mueke, J.M. (1995a) Taxonomic significance of isoenzymes in two stemborers (Lepidoptera: Pyralidae) of maize and sorghum in Kenya. *African Entomology* 3(2), 167–171.

Kioko, E.N., Overholt, W.A. and Mueke, J.M. (1995b) Larval development in *Chilo orichalcociliellus* and *Chilo partellus*; a comparative study in the laboratory. In: *Beneficial African Insects: A Renewable Resource.* Proceedings of the 10th meeting and scientific conference of the African Association of Insect Scientists, 5–10 September 1993, Mombasa, Kenya. ICIPE Science Press, Nairobi, 251 pp.

Kishore, P. (1980) Integrated approach for control of *Chilo partellus* (Swinhoe), a serious pest of sorghum. PhD thesis, Indian Agricultural Research Institute, New Delhi, India.

Kishore, P. (1989) Chemical control of stem borers. In: *International Workshop on Sorghum Stem Borers, 17–20 November 1987.* ICRISAT, Patancheru, India.

Kishore, P. and Jotwani, M.G (1982) Estimation of avoidable losses caused by the earhead caterpillars on sorghum and their control. *Entomon* 7, 65–69.

Königsmann, E. (1977) Das phylogenetische System der *Hymenoptera.* Teil 2 *Symphyta. Deutsche entomologische Zeitschrift* (*NS*) 24, 1–40.

Königsmann, E. (1978a) Das phylogenetische System der *Hymenoptera.* Teil 3 *Terebrantes* (Unterordung *Apocrita*). *Deutsche entomologische Zeitschrift* (*NS*) 25, 1–55.

Königsmann, E. (1978b) Das phylogenetische System der *Hymenoptera.* Teil 4 *Aculeata (Unterordung Apocrita*). *Deutsche entomologische Zeitschrift* (*NS*) 25, 365–435.

Konno, Y., Shishido, T. and Tanaka, F. (1986) Structure–resistance relationship in the organophosphorus-resistant rice stem borer, *Chilo suppressalis. Journal of Pesticide Science* 11, 393–399.

Krause, M., Nel, A. and Van Zyl, K. (1996) *A Guide to Pesticides and Fungicides in the Republic of South Africa.* Directorate of Livestock Improvement and Agricultural Production Resources, Department of Agriculture, Pretoria.

Krieger, R. (1914a) Ueber die Ichneumonidengattung *Xanthopimpla* Sauss. *Archiv für Naturgeschicte* (*A*) 80(6), 1–148.

Krieger, R. (1914b) Ueber die Ichneumonidengattung *Xanthopimpla* Sauss. (Fortsetzung). *Archiv für Naturgeschicte* (*A*) 80(7), 1–152.

Kumar, K.A. (1989) Pearl millet: current status and future potential. *Outlook on Agriculture* 18, 46–53.

Kumar, H. (1993) Responses of *Chilo partellus* (Lepidoptera: Pyralidae) and *Busseola fusca* (Lepidoptera: Noctuidae) to hybrids of a resistant and a susceptible maize. *Journal of Economical Entomology* 86(3), 962–968.

Kumar, H. and Saxena, K.N. (1992) Resistance in certain maize cultivars to first and third-instar *Chilo partellus* larvae. *Entomologia Experimentalis et Applicata* 65(1), 75–80.

Kwapong, P.K. (1990) The bionomics of *Eldana saccharina* and the effects of intercropping and time of planting on the stemborer complex on maize in the Central region of Ghana. MSc thesis, University of Cape Coast.

La Croix, E.A.S. (1967) Maize stalk borers in the coast province of Kenya. *East African Agriculture and Forestry Journal* 33, 49–54.

Lamborn, W.A. (1914) The agricultural pests of the Southern Provinces, Nigeria. *Bulletin of Entomological Research* 5, 197–214.

Laporte, B. (1984) Noctuidae (sauf les Plusiinae). In: Rougeot, P.C. (ed.) *Missions Entomologiques en Ethiopie 1976–1982.* Vol. II. *Mémoires du Muséum National d'Histoire Naturelle, N.S.A. Zoologie* 128, 14–41.

LaSalle, J. (1993) *Aprostocetus (Ootetrastichus) theioneurus* (Masi) (Hymenoptera: Eulophidae): a hyperparasitoid on the cereal stemborer *Chilo partellus* (Lepidoptera: Pyralidae) in Africa. *Zoologische Mededelingen* 67, 445–451.

LaSalle, J. (1994) North American genera of Tetrastichinae (Hymenoptera: Eulophidae). *Journal of Natural History* 28, 128 pp.

LaSalle, J. and Delvare, G. (1994) *Aprostocetus procerae* (Risbec): a senior synonym of *Tetrastichus pachydiplosisae* Risbec (Hymenoptera: Eulophidae). *African Entomology* 2, 133–135.

Lawani, S.M. (1982) A review of the effects of various agronomic practices on cereal stem borer populations. *Tropical Pest Management* 28, 266–276.

Le Pelley, R.H. (1959) *Agricultural Insects of East Africa*. East African High Commission, Nairobi, 307 pp.

Leppla, N.C. (1989) Laboratory colonization of fruit flies. In: *World Crop Pests, Fruit Flies, Their Biology, Natural Enemies and Control.* Elsevier Science Publishers, Amsterdam, the Netherlands, pp. 91–103.

Leppla, N.C. and Fisher, W.R. (1989) Total quality control in insect mass production for insect pest management. *Journal of Applied Entomology* 108, 452–461.

Leslie, G.W. (1982) A study of egg predators of *Eldana saccharina* Walker (Lepidoptera: Pyralidae). *Proceedings of the South African Sugar Technologists' Association* 56, 85–87.

Leslie, G.W. (1988) The identification and importance of predators of *Eldana saccharina* (Lepidopera: Pyralidae). *Proceedings of the South African Sugar Technologists' Association* 62, 124–128.

Leslie, G.W. (1989) Effect of pre-trashing sugarcane on oviposition, and on predation of eggs and moths of *Eldana saccharina* (Lepidoptera: Pyralidae). *Proceedings of the South African Sugar Technologists' Association* 63, 192–196.

Leslie, G.W. (1993) Dispersal and behaviour of neonate larvae of the pyralid borer *Eldana saccharina*. *Proceedings of the South African Sugar Technologists' Association* 67, 122–126.

Leslie, G.W. and Boreham, P.F.L. (1981) Identification of arthropod predators of *Eldana saccharina* Walker (Lepidoptera: Pyralidae) by cross-over electrophoresis. *Journal of the Entomological Society of Southern Africa* 44(2), 381–388.

Leuschner, K. (1990) Sorghum and millet entomology in the SADCC region 1988/89. In: *Proceedings of the Sixth Regional Workshop on Sorghum and Millets for Southern Africa, 18–22 September 1989.* ICRISAT, Bulawayo, Zimbabwe, pp. 105–117.

Lewvanich, A. (1981) A revision of the Old World species of Scirpophaga (Lepidoptera: Pyralidae). *Bulletin of the British Museum (Natural History), Entomology Series* 42, 185–298.

Leyenaar, P. and Hunter, R.B. (1977) Effect of stem borer damage on maize yield in the coastal savanna zone of Ghana. *Ghana Journal of Agricultural Science* 10, 67–70.

Libby, J.L. (1968) *Insect Pests of Nigerian Crops.* Research Bulletin no. 289, University of Wisconsin, Madison, Wisconsin.

Lobo, M.-L. and Viereck, A. (1986) Entomological problems of the main crops in Cape Verde. In: *Proceedings of an International Seminar of the CILSS Project on Integrated Pest Management, Niamey (Niger), 6–13 December 1984*, CILSS, Ouagadougou, pp. 55–66.

Ly, L.S. (1976) Etude morphologique et biologique de *Goniozus procerae* Risbec (Hymenoptera, Bethylidae) parasite de *Chilo suppressalis* (Walker) (Lepidoptera: Pyralidae). Thèse, Faculté des Sciences, Université de Paris VI. P & M Curie.

Lynch, R.E. and Lewis, L.C. (1978) Fungi associated with eggs and first-instar larvae of the European corn borer. *Journal of Invertebrate Pathology* 32, 6–11.

Maafo, L.K.A. (1975) Laboratory mass rearing of exotic *Tetrastichus* spp for the control of maize and sugarcane stemborers in Ghana. *Ghana Journal of Agricultural Sciences* 8(2), 89–93.

Maazouz, S. (1991) Etude des interactions entre l'hymenoptère braconide *Cotesia sesamiae* (Cameron, 1906), virus de polyhédroses nucléaires et virus de polyhédroses cytoplasmiques chez les chenilles de *Sesamia calamistis* Hampson, 1910, Lepidoptera, Noctuidae. Thèse de Doctorat, Académie de Montpellier, Université Montpellier II Sciences et Techniques du Languedoc.

McAlpine, J.F. (ed.) (1987) *Manual of Nearctic Diptera*, Vol. 2. Research Branch, Agriculture Canada, Monograph No. 28, Agriculture Canada, Ottawa, pp. 675–1332.

McAlpine, J.F. (ed.) (1989) *Manual of Nearctic Diptera*, Vol. 3. Research Branch, Agriculture Canada, Monograph No. 32, Agriculture Canada, Ottawa, pp. 1333–1581.

McAlpine, J.F., Peterson, B.V., Shewell, G.E., Teskey, H.J., Vockeroth, J.R. and Wood, D.M. (Coordinators) (1981) *Manual of Nearctic Diptera*, Vol. 1. Research Branch, Agriculture Canada, Monograph No. 27, Agriculture Canada, Ottawa, pp. 1–674.

Macedo, N., Mendonca Filho, A.F., Moreno, J.A. and Pinazza, A.H. (1984) Evaluation of the economic advantages of 10 years of biological control of *Diatraea* spp. through *Apanteles flavipes* Cameron, in the state of Alagoas (Brazil). *Entomology Newsletter* 16, 9–10.

McFarlane, J. (1984) *Nigeria Sorghum Entomology – 1983.* ICRISAT (International Crops Research Institute for the Semi-Arid Tropics)/SAFGRAD (Semi-Arid Food Grain Research and Development), Samaru, Zaria, Nigeria, 6 pp.

McFarlane, J. (1990) Damage assessment and yield losses in sorghum due to the stem borer *Busseola fusca* (Fuller) (Lepidoptera: Noctuidae) in northern Nigeria. *Tropical Pest Management* 36(2), 131–137.

Mackauer, M. (1976) Genetic problems on the production of biological control agents. *Annual Review of Entomology* 21, 369–385.

McNaughton, S.J., Tarrants, J.L., McNaughton, M.M. and Davis, R.H. (1985) Silica as a defense against herbivory and a growth promoter in African grasses. *Ecology* 66, 528–535.

Mahadevan, N.R. and Chelliah, S. (1986) Influence of intercropping legumes with sorghum on the infestation of the stem borer, *Chilo partellus* (Swinhoe) in Tamil Nadu, India. *Tropical Pest Management* 32, 162–163.

Mally C.W. (1920) The maize stalk borer *Busseola fusca* Fuller. *Bulletin of the Department of Agriculture of South Africa* 3, 111 pp.

Mallya, G.A. (1985) Maize entomology. In: *National Maize Research Programme 1984–1985.* Progress Report presented to National Maize Research Coordinating Committee by A.J. Moshi, 4 November 1985, pp. 41–43.

Mallya, G.A. (1986) Maize entomology. In: *Maize Improvement Project Progress Report 1985–1986*. UAC, Mbeya, Tanzania, pp. 27–31.

Maniania, N.K. (1991) Susceptibility of *Chilo partellus* (Swinhoe) (Lep., Pyralidae) eggs to entomopathogenic hyphomycetes. *Journal of Applied Entomology* 112(1), 53–58.

Maniania, N.K. (1993a) Effectiveness of the entomopathogenic fungus *Beauveria bassiana* (Bals.) Vuill. for control of the stem borer *Chilo partellus* (Swinhoe) in maize in Kenya. *Crop Protection* 12, 601–604.

Maniania, N.K. (1993b) Evaluation of three formulations of *Beauveria bassiana* (Bals.) Vuill. for control of the stem borer *Chilo partellus* (Swinhoe) (Lep., Pyralidae). *Journal of Applied Entomology* 115, 266–272.

Marandu, W.Y.F., Temu, A.E.M. and Kabango, D. (1987) Plant protection, In: *Maize Improvement Programme Annual Progress Report 1987*. UAC, Mbeya, Tanzania, p. 3.

Markham, R.H. (1985) *Natural enemies of* Acigona ignefusalis *and* Raghuva albipunctella *in Niger and Burkina Faso and Possibilities for Development of a Biological Control Strategy*. Report of a consultancy for ICRISAT Sahel Centre, Niamey, Niger, December 1985, 27 pp.

Marsh, P.M. (1993) Descriptions of new western hemisphere genera of the subfamily Doryctinae (Hymenoptera: Braconidae). *Contributions of the American Entomogical Institute* 28, 1–58.

Masi, L. (1917) Chalcididae of the Seychelle Islands. *Novitates Zoologicae* 24, 121–230.

Mason, W.R.M. (1978) A synopsis of the Nearctic Braconini with revisions of the Nearctic species of *Coeloides* and *Myosoma* (Hymenoptera: Braconidae). *Canadian Entomologist* 110, 721–768, figs 1–95.

Mason, W.R.M. (1981) The polyphyletic nature of *Apanteles* Foerster (Hymenoptera: Braconidae): a phylogeny and reclassification of Microgastrinae. *Memoirs of the Entomological Society of Canada* 115, 1–147, figs 1–110.

Mathez, F.C. (1972) *Chilo partellus* Swinh., *Chilo orichalcociliella* Strand (Lep., Crambidae) and *Sesamia calamistis* Hmps (Lep., Noctuidae) on maize in the Coast Province, Kenya. *Mitteilungen der Schweizerischen Entomologischen Gesellschaft* 45(4), 267–289.

Matthee, J.J. (1974) *Pests of Graminaceous Crops in South Africa*. Entomology Memoir, Department of Agricultural Technical Services, Pretoria, Republic of South Africa no. 40, 24 pp.

Matthews, M. (1991) Classification of the Heliothinae. *Natural Resources Institute Bulletin* 44, 1–198.

Matthews, M. and Jago, N.D. (1993) *Millet Pests of the Sahel. An Identification Guide*. Natural Resources Institute, Chatham, UK, 80 pp.

Mchowa, J.W. (1990) The effect of planting date on the incidence of the stalk borers, *Busseola fusca* (Fuller) Lepidoptera: Noctuidae and *Chilo partellus* (Swinhoe) Lepidoptera: Pyralidae in two varieties of sorghum. In: *Proceedings of the Sixth Regional Workshop on Sorghum and Millets for Southern Africa, 18–22 September 1989, Bulawayo, Zimbabwe: SADCC Conference*. ICRISAT, Patancheru, Andhra, Pradesh, pp 154–159.

Medler, J.T. (1980) *Insects of Nigeria – Check List and Bibliography*. Memoirs of the American Entomological Institute 30, Ann Arbor, Michigan, 919 pp.

Megenasa, T. (1982) Insect pests of sorghum in Ethiopia. In: *Proceedings of the Regional Workshop on Sorghum Improvement in Eastern Africa, 17–21 October 1982*. Sorghum Improvement Project, Nazareth and Debre Zeit, Ethiopia, pp. 54–64.

Melamed-Madjar, V., Meisner, J. and Ascher, K.R.S. (1989) Effects of neem on the corn borer, *Sesamia nonagrioides* Lefebvre (Lepidoptera: Noctuidae). *Zeitschrift für Pflanzenkrankheiten und Pflanzenschutz* 96, 521–525.

Menken, S.B.J. and Ulenberg, S.A. (1987) Biochemical characters in agricultural entomology. *Agricultural Zoology Reviews* 2, 305–360.

Mercer-Quarshie, H. (1970) Sorghum and millet improvement in Ghana. *African Soils/Sols Africains* 15, 175–184.

Mesnil, L.P. (1956) Trois nouveaux tachinaires d'Afrique. *Entomophaga* 1, 76–80.

Mesnil, L.P. (1968) Quelques espèces inédites de Tachinaires africains (Dipt. Tachinidae). *Stuttgarter Beiträge zur Naturkunde in Stuttgart* 187, 1–12.

Mihm, J.H., Smith, M.E. and Deutsch, J.A. (1988) Development of open-pollinated varieties, non-conventional hybrids and inbred lines of tropical maize with resistance to fall armyworm, *Spodoptera frugiperda* (Lepidoptera: Noctuidae) at CIMMYT. *Florida Entomologist* 71, 262–268.

Milner, J.D. (1965) Personal correspondence from Kampala, Uganda. Cited by Poinar, 1975.

Milner, J.E.D. (1967) Final report on a survey of the parasites of graminaceous stem-borers in East Africa. Commonwealth Institute of Biological Control (unpublished).

Minja, E.M. (1990) Management of *Chilo* spp. infesting cereals in eastern Africa. *Insect Science and its Application* 11(4–5), 489–499.

Mohyuddin, A.I. (1970) Notes on the distribution and biology of *Pediobius furvus* (Gah.) [Hym., Eulophidae] a parasite of graminaceous stem-borers. *Bulletin of Entomological Research* 59, 681–689.

Mohyuddin, A.I. (1971) Comparative biology and ecology of *Apanteles flavipes* (Cam.) and *A. sesamiae* as parasites of graminaceous borers. *Bulletin of Entomological Research* 61, 33–39.

Mohyuddin, A.I. (1972) Distribution, biology and ecology of *Dentichasmias busseolae* Heinr. (Hym., Ichneumonidae), a pupal parasite of graminaceous stem-borers (Lep., Pyralidae). *Bulletin of Entomological Research* 62, 161–168.

Mohyuddin, A.I. and Greathead, D.J. (1970) An annotated list of the parasites of graminaceous stem borers in East Africa, with a discussion of their potential in biological control. *Entomophaga* 15(3), 241–274.

Mohyuddin, A.I., Inayatullah, C. and King, E.G. (1981) Host selection and strain occurrence in *Apanteles flavipes* (Cameron) (Hymenoptera: Braconidae) and its bearing on biological control of graminaceous stem-borers (Lepidoptera: Pyralidae). *Bulletin of Entomological Research* 71, 575–581.

Momoi, S. (1968) Ichneumonidae (Hymenoptera) collected in paddy fields of the Orient, with descriptions of new species. Part 3. Subfamilies Ophioninae, Cremastinae and Ichneumoninae. *Mushi* 41, 215–220.

Morel, G., Vercambre, B., Bergoin, M. and Louis, C. (1981) Une polyédrose cytoplasmique chez le foreur ponctué *Chilo sacchariphagus*. *Agronomie Tropicale* 36, 378–381.

Morley, C. (1913) *The Fauna of British India including Ceylon and Burma; Hymenoptera 3; Ichneumonidae I. Ichneumones Deltoidei.* Taylor and Francis, London, 531 pp.

Morley, C. (1916) On some South African Ichneumonidae in the collection of the South African Museum. Part 1. *Annals of the South African Museum* 15, 353–400.

Mostafa, F.F. (1992) Morphological and mouth parts sensilla of the last larval instar of *Chilo agamemnon* Bles. (Lepidoptera: Crambidae). *Assiut Journal of Agricultural Sciences* 23(2), 141–150.

Moutia, L.A. (1934) The sugar-cane moth borers in Mauritius. *Bulletin of Entomological Research* 25, 33–46.

Moutia, L.A. and Courtois, C.M. (1952) Parasites of the moth-borers of sugar-cane in Mauritius. *Bulletin of Entomological Research* 43, 325–359.

Moutia, L.A. and Mamet, R. (1945) A review of twenty-five years of economic entomology in the Island of Mauritius. *Bulletin of Entomological Research* 36, 439–472.

Moutia, L.A. and Mamet, R. (1947) *An annotated List of Insects and Acarina of Economic Importance in Mauritius.* Bulletin of the Department of Agriculture of Mauritius, Science Series, no. 29, 43 pp.

Moyal, P. (1988) *Les Foreurs du maïs en zone des savanes de Côte d'Ivoire.* Collection Etudes et Thèses, Institut Français de Recherche Scientifique pour le Développement en Coopération, ORSTOM, Paris, 367 pp.

Moyal, P. (1995) Borer infestation and damage in relation to maize stand density and water stress in the Ivory Coast. *International Journal of Pest Management* 41, 114–121.

Moyal, P. and Tran, M. (1989) Etude morphologique des stades pré-imaginaux des lépidoptères foreurs du maïs en zone de savanes de Côte d'Ivoire. *Annales de la Société entomologique de France* 25(4), 461–472.

Moyal, P. and Tran, M. (1991a) Cob borer *Mussidia nigrivenella* (Lepidoptera; Pyralidae) of maize in Ivory Coast. I Morphological and biological data. *Insect Science and its Application* 12, 209–214.

Moyal, P. and Tran, M. (1991b) Cob borer *Mussidia nigrivenella* Ragonot (Lepidoptera; Pyralidae) of maize in Ivory Coast. II Ecological data. *Insect Science and its Application* 12, 215–223.

Moyal, P. and Tran, M. (1992) *Chilo aleniellus* (Lepidoptera: Pyralidae), a stem borer of maize in Côte d'Ivoire. *Bulletin of Entomological Research* 82, 67–72.

Mugoya, C.F. and Chinsembu, K.C. (1995) Potential of *Tephrosia vogelii* water extracts for controlling maize stalk borers and maize streak virus in Zambia. In: *Proceedings of the 10th Meeting and Scientific Conference of the African Association of Insect Scientists, 5–10 September 1993, Mombasa, Kenya.* ICIPE Science Press, Nairobi, Kenya.

Muyango, S. (1987) Les chenilles foreuses des tiges du maïs et du sorgho: écologie et lutte. In: *Seminaire sur les maladies et les ravageurs des principales cultures vivrières d'Afrique Centrale.* Centre Technique de Coopération Agricole et Rurale (CTA), Ede/Wageningen, pp. 117–126.

Nagarkatti, S. and Nair, K.R. (1973) The influence of wild and cultivated Gramineae and Cyperaceae on populations of sugarcane borers and their parasites in North India. *Entomophaga* 18, 419–430.

Nagarkatti, S. and Rao, V.P. (1975) Biology of and rearing technique for *Sturmiopsis parasitica* (Curr.) (Diptera, Tachinidae), a parasite of graminaceous borers in Africa. *Bulletin of Entomological Research* 65, 165–170.

Nayak, P. and Godse, D.B. (1986) Nuclear polyhedrosis of the rice skipper, *Parnara mathias* Fabr (Hesperiidae: Lepidoptera). *Current Science, India* 55, 743–744.

Ndoye, M. (1980) *Goniozus procerae* Risbec (Hymenoptères, Bethylidae), ectoparasite larvaire d'*Acigona ignefusalis* Hampson (Lépidoptères, Pyralidae, Crambinae). *Bulletin de l'IFAN sér. A* 42, 394–400.

Nesbitt, B.F., Beevor, P.S., Hall, D.R., Lester, R., Davis, J.C. and Seshu Reddy, K.V. (1979) Components of the sex pheromone of the female spotted stalk borer, *Chilo partellus* (Swinhoe) (Lepidoptera: Pyralidae): identification and preliminary field trials. *Journal of Chemical Ecology* 5, 153–163.

Ngi-Song, A.J., Overholt, W.A. and Ayertey, J.N. (1995) Suitability of African gramineous stemborers for the development of *Cotesia flavipes* and *Cotesia sesamiae* (Hymenoptera: Braconidae). *Environmental Entomology* 24(4), 978–984.

Ngi-Song, A.J., Overholt, W.A., Njagi, P.G.N., Dicke, M., Ayertey, J.N. and Lwande, W. (1996) Volatile infochemicals used in host and host habitat location by *Cotesia flavipes* Cameron and *Cotesia sesamiae* (Cameron) (Hymenoptera: Braconidae), larval parasitoids of stemborers on Graminae. *Journal of Chemical Ecology* 22, 307–323.

Nickel, J.L. (1964) *Biological Control of Rice Stem Borers: a Feasibility Study.* IRRI Technical Bulletin No. 2, Los Baños, 111 pp.

Nickel, J.L. (1967) The possible role of biotic factors in an integrated program for rice stem borer control. In: *The Major Insect Pests of the Rice Plant: Proceedings Symposium at IRRI, September 1964.* Johns Hopkins Press, Baltimore, pp. 443–453.

Nishida T. and Torii, T. (1970) *A Handbook of Field Methods for Research on Rice Stem-borers and Their Natural Enemies.* IBP Handbook No. 14, Blackwell Scientific Publications, Oxford, 132 pp.

Njokah, J.N. and Okhoba, M.M. (1985) Key research areas on irrigated rice in Kenya. In: Zan, K., Garrity, D.P. and Zimdahl, A. (eds) *Rice Improvement in Eastern, Central and Southern Africa. Proceedings of the International Rice Workshop at Lusaka, Zambia, 9–19 April 1984.* IRRI, Manila, pp. 91–96.

Nonveiller, G. (1984) Catalogue commenté et illustré des insectes du Cameroun d'intérêt agricole (appartitions, répartition, importance). *Institut pour la Protection des Plantes, Belgrade, Memoirs* 15, 1–210.

Noyes, J.S. (1982) Collecting and preserving chalcid wasps (Hymenoptera: Chalcidoidea). *Journal of Natural History* 16, 315–334.

Nunes, E., Sousa, D. and Sataric, I. (1985) Research on the constraints to maize production in Mozambique. In: *To Feed Ourselves. Proceedings of the First Eastern, Central and Southern Africa Regional Maize Workshop, Lusaka, Zambia, 10–17 March 1985,* ICRISAT, Patancheru.

Nwanze, K.F. (1981) Stem borers of cereals in Sahelian West Africa: distribution, relative importance and control./Les borers des tiges des céréales en Afrique de l'Ouest sahélienne: distribution, importance relative et controle. Biological control of pests: its potential in West Africa. In: *Proceedings of an International Conference, Dakar, Senegal, 9–13 February 1981.* USAID, pp. 115–123 (Eng. edn), 108–115 (Fr. edn).

Nwanze, K.F. (1988) Distribution and seasonal incidence of some major insect pests of sorghum in Burkina Faso. *Insect Science and its Application* 9, 313–322.

Nwanze, K.F. (1989) Insect pests of pearl millet in Sahelian West Africa 1. *Acigona ignefusalis* (Pyralidae, Lepidoptera): distribution, population dynamics and assessment of crop damage. *Tropical Pest Management* 35, 137–142.

Nwanze, K.F. (1991) Components for the management of two insect pests of pearl millet in Sahelian West Africa. *Insect Science and its Application* 12, 673–678.

Nwanze, K.F. and Harris, K.M. (1992) Insect pests of pearl millet in West Africa. *Review of Agricultural Entomology* 80, 1133–1155.

Nwanze, K.F. and Mueller, R.A.E. (1989) Management options for sorghum stem borers for farmers in the semi-arid tropics. In: *International Workshop on Sorghum Stem Borers, 17–20 November 1987.* ICRISAT, Patancheru, India.

Nyambo, B.T. and Mwangi, M. (1992) Management of maize and sorghum stalks and residues by small-scale farmers and its implications for IPM control strategies in Kenya. In: Gibson, R.W. and Sweetmore, A. (eds) *Proceedings of a Seminar on Crop Protection for Resource-poor Farmers, 4–8 November 1991.* pp. 55–63.

Nye, I.W.R. (1960) The insect pests of graminaceous crops in East Africa. *Colonial Research Studies* 31, 1–48.

Nzibande, L.M. (1993) Grain yield response to stem borer infestation. In: *Proceedings of the First National Sorghum and Pearl Millet Technology Transfer Workshop, Swaziland, 22–25 July 1993.* SADC/ICRISAT/SMIP, Bulawayo, Zimbabwe.

Ochieng, R.S., Onyango, F.O. and Bungu, M.D.O. (1985) Improvement of techniques for mass-culture of *Chilo partellus* (Swinhoe). *Insect Science and its Application* 6, 425–428.

Ochieng-Odero, J.P.R., Onyango, F.O., Kilori, J.T., Bungu, M.D.O. and Amboga, E.O. (1991) Insect rearing management as a prerequisite in the development of IPM for sustainable food production. *Insect Science and its Application* 12, 645–651.

Odindo, M.O. (1990) Potential of *Nosema* spp. (Microspora, Nosematidae) and viruses in the management of *Chilo* spp. (Lepidoptera, Pyralidae). *Insect Science and its Application* 11, 773–778.

Odindo M.O. (1994) Identification of insect pathogens common in the insectary. In: Ochieng-Odero, J.P.R. (ed.) *Techniques of Insect Rearing for the Development of Integrated Pest Management Strategies.* ICIPE Science Press, Nairobi, Kenya, pp. 247–271.

Odindo, M.O., Otieno, W.A., Oloo, G.W., Kilori, J. and Odhiambo, R.C. (1989) Prevalence of microorganisms in field-sampled borers on sorghum, maize, and cowpea in Western Kenya. *Insect Science and its Application* 10(2), 225–228.

Odindo, M.O., Amutalla, P.A., Opondo-Mbai, M. and Odero, T.A. (1993) Production of *Nosema marucae* for biological control of cereal stem borers. *Entomologia Experimentalis et Applicata* 67, 143–148.

Odjo, A.T. (1984) Contribution à l'étude de la biologie et de l'écologie de *Sesamia calamistis* Hampson (Lepidoptera: Noctuidae) en Côte d'Ivoire Centrale. Thèse de 3e cycle, Université de Paris (XI) Sud (Orsay), 109 pp.

Ogunwolu, E.O., Nwosu, K. and Ogunyebi, S.O. (1981) Stem borer damage in maize as affected by host plant density. *Journal of Agricultural Science (Cambridge)* 96, 695–697.

Ogwang, J.A. and Odindo, M.O. (1992) Studien zur ultrastruktur der transovarialen übertragung von *Nosema marucae* Odinho [*sic*] and Jura (Sporoz., Nosematidae) auf *Chilo partellus* Swinh. (Lep., Tortricidae) [*sic*]. *Anzeiger für Schädlingskunde Pflanzenschutz Umweltschutz* 65, 48–50.

Ogwaro K. (1983) Intensity levels of stem borers in maize and sorghum and the effect on yield under different cropping patterns. *Insect Science and its Application* 4(1–2), 33–47.

O'Hara, J.E. (1989) Systematics of the genus group taxa of the Siphonini. *Quaestiones Entomologicae* 25, 1–229.

Okech, S.H.O., Neukermans, L.M.N.R. and Chinsembu, K.C. (1994) Agroecological distribution of major stalk borers of maize in Zambia. *Insect Science and its Application* 15, 167–172.

Oloo, G.W. (1989) The role of local natural enemies in population dynamics of *Chilo partellus* (Swinh.) (Pyralidae) under subsistence farming systems in Kenya. *Insect Science and its Application* 10(2), 243–251.

Oloo, G.W. and Ogeda, K. (1990) The incidence of *Chilo partellus* (Swinh.) (Pyralidae) and the contribution of natural enemies to its mortality under intercropping system in Kenya. *Tropical Pest Management* 36, 244–248.

Omolo, E.O. (1983) Screening of local and exotic maize lines for stem borer resistance with special reference to *Chilo partellus. Insect Science and its Application* 4(1–2), 105–108.

Omolo, E.O. (1986) Stem borer incidence in sorghum intercropped with maize and cowpea tested in Kenya. In: *Sorghum and Millet Improvement in Eastern Africa: Proceedings of the 4th Regional Workshop, 22–26 July 1985, Soroti, Uganda.* SAFGRAD/ICRISAT, Nairobi, Kenya, pp. 367–378.

Omolo, E.O. and Seshu Reddy, K.V. (1985) Effect of different sorghum-based cropping systems on insect pests in Kenya. In: *Proceedings of the International Sorghum Entomology Workshop, 15–21 July 1984, Texas A&M University, College Station, Texas, USA.* ICRISAT, Patancheru, India, pp. 395–401.

Omwega, C.O., Kimani, S.W., Overholt, W.A. and Ogol, C.K.P.O. (1995) Evidence of the establishment of *Cotesia flavipes* (Hymenoptera: Braconidae) in continental Africa. *Bulletin of Entomological Research* 85, 525–530.

Omwega C.O., Overholt, W.A., Mbapila, J.C. and Kimani-Njogu, S.W. (1997) Establishment and dispersal of *Cotesia flavipes* Cameron (Hymenoptera: Braconidae), an exotic endoparasitoid of *Chilo partellus* Swinhoe (Lepidoptera: Pyralidae) in northern Tanzania. *African Entomology* (in press).

Onyango, F.O. and Ochieng-Odero, J.P.R. (1994) Continuous rearing of the maize stem borer *Busseola fusca* on an artificial diet. *Entomologia Experimentalis et Applicata* 73, 139–144.

Oshima, K. (1935) Infectivity of *Chilo simplex* Butler by *Nosema bombycis* and phagocytic action of the host haemolymphocyte against the parasite structure. *Zoological Magazine, Tokyo* 47, 607–626.

Otieno, W.A. (1986) Incidence of pathogens of crop borers in relation to their mortality in different intercropping systems. In: *ICIPE 13th Annual Report, 1985*, ICIPE, Nairobi, p. 17.

Overholt, W. (1991) Biological control of tsetse and crop pests. ICIPE Annual Report. 1991 (unpublished).

Overholt, W.A., Ngi-Song, A.J., Kimani, S.K., Mbapila, J., Lammers, P. and Kioko, E. (1994a) Ecological considerations of the introduction of *Cotesia flavipes* Cameron (Hymenoptera: Braconidae) for biological control of *Chilo partellus* (Swinhoe) (Lepidoptera: Pyralidae), in Africa. *Biocontrol News and Information* 15(2), 19–24.

Overholt, W.A., Ochieng, J.O., Lammers, P.M. and Ogedah, K. (1994b) Rearing and field release methods for *Cotesia flavipes* Cameron (Hymenoptera: Braconidae), a parasitoid of tropical gramineous stemborers. *Insect Science and its Application* 15(3), 253–259.

Overholt, W.A., Ogedah, K. and Lammers, P.M. (1994c) Distribution and sampling of *Chilo partellus* (Swinhoe) (Lepidoptera: Pyralidae) in maize and sorghum at the Kenya Coast. *Bulletin of Entomological Research* 84, 367–378.

Overholt, W.A., Ngi-Song, A.J., Omwega, C.O., Kimani-Njogu, S.W., Mbapila, J., Sallam, M.N. and Ofomata, V. (1997) A review of the introduction and establishment of *Cotesia flavipes* Cameron (Hymenoptera: Braconidae) in East Africa for biological control of cereal stem borers. *Insect Science and its Application* 17(1), 19–35.

Overholt, W.A. and Smith, J.W. Jr (1989) *Pediobius furvus* parasitization of overwintering generation southwestern corn borer pupae. *Southwestern Entomologist* 14, 35–40.

Pant, N.C., Gupta, P. and Nayar, J.K. (1960) Physiological studies on *Chilo zonellus* Swinhoe, a pest of maize crop. I. Growth on artificial diets. *Proceedings of the National Institute of Sciences of India (B)* 26, 379–383.

Pathak, R.S. and Olela, J.C. (1983) Genetics of host plant resistance in food crops with special reference to sorghum stem borers. *Insect Science and its Application* 4(1–2), 127–134.

Päts, P. (1992) Reproductive biology of the cereal stemborer *Chilo partellus*. PhD dissertation, Swedish University of Agricultural Sciences, Uppsala, Sweden, 97 pp.

Päts, P. (1996) Management of crop residues to reduce the aestivating population of stemborers in maize. *International Journal of Pest Management* 42, 151–156.

Peigler, R.S. (1996) Catalog of parasitoids of Saturniidae of the world. *Journal of Research on the Lepidoptera* 33(1994), 1–121, figs 1–7.

Phiri, G.S.N. (1995) Interaction of the spotted stem borer *Chilo partellus* (Swinhoe) with some alternative hosts and its larval parasitoid *Cotesia sesamiae* (Cameron) in Malawi. Unpublished PhD thesis, University of Reading, 347 pp.

Poinar, G.O., Jr (1975) *Entomogenous Nematodes. A Manual and Host List of Insect–Nematode Associations*. E.J. Brill, Leiden, 317 pp.

Polaszek, A. (1992) Cereal stem borers and their parasitoids in Africa. *Proceedings Experimental and Applied Entomology, NEV, Amsterdam* 3, 70–71.

Polaszek, A. and Dessart, P. (1996) Taxonomic problems in the *Aphanogmus hakonensis* species-complex (Hymenoptera: Ceraphronidae), common hyperparasitoids in biocontrol programmes against lepidopterous pests in the tropics. *Bulletin of Entomological Research* 86, 419–422.

Polaszek, A. and Kimani, S.W. (1990) *Telenomus* species (Hymenoptera: Scelionidae) attacking eggs of pyralid pests (Lepidoptera) in Africa: a review and guide to identification. *Bulletin of Entomological Research* 80, 57–71.

Polaszek, A. and Krombein, K.V. (1994) The genera of Bethylinae (Hymenoptera: Chrysidoidea). *Journal of Hymenoptera Research* 3, 91–105.

Polaszek, A. and LaSalle, J. (1995) The hyperparasitoids (Hymenoptera: Ceraphronidae, Encyrtidae, Eulophidae, Eurytomidae) of cereal stem borers in Africa (Lepidoptera: Noctuidae, Pyralidae) *African Entomology* 3, 131–146.

Polaszek, A. and Walker, A.K. (1991) The *Cotesia flavipes* species-complex: parasitoids of cereal stem borers in the tropics. *Redia* 74(3), 335–341.

Polaszek, A., Ubeku, J.A. and Bosque-Pérez, N.A. (1993) Taxonomy of the *Telenomus busseolae* species complex (Hymenoptera: Scelionidae) egg parasitoids of cereal stem borers (Lepidoptera: Noctuidae, Pyralidae). *Bulletin of Entomological Research* 83, 221–226.

Polaszek, A., Fitton, M.G., Bianchi, G. and Huddleston, T. (1994) The parasitoids of the African white rice borer, *Maliarpha separatella* Ragonot (Lepidoptera: Pyralidae). *Bulletin of Entomological Research* 84, 65–90.

Pollet, A. (1977) Les insectes ravageurs du riz en Côte d'lvoire II. La faune rencontrée sur riz irrigué en Côte d'Ivoire Centrale (Kotiessou). *Cahiers d'ORSTOM Série Biologie* 12(1), 3–23.

Pollet, A. (1978) Rice insects in Ivory Coast. IV. Determination of infestations in irrigated rice of Central Ivory Coast (in French). *Cahiers d'ORSTOM Série Biologie* 13, 87–99.

Pollet, A. (1981) *Maliarpha separatella* Ragonot (Pyralidae, Phycitinae) sur riz irrigué en Côte d'Ivoire Centrale (Kotiessou). Données biologiques et mécanismes régulateurs des populations naturelles. *Travaux et Documents de l'ORSTOM* 140, 230 pp.

Pollet, A., van Roon, N. and Mauritz, R. (1974) *Les Ravageurs du maïs en Côte d'Ivoire. II Inventaires qualitatifs et données quantitatives pour la Basse Côte.* ORSTOM, Adiopodoumé, 65 pp.

Pollet, A., van Roon, N. and Mauritz, R. (1978) Les ravageurs du maïs en Côte d'Ivoire. Données qualitatives et quantitatives pour la Basse Côte. *Cahiers d'ORSTOM, Série Biologie* 13(1), 71–85.

Poole, R.W. (1989) Noctuidae. *Lepidopterorum Catalogus (New Series)* 118 (3 parts). E.J. Brill, Leiden.

Potting, R.J.P. (1996) Hunting for hiding hosts. The behavioural ecology of the stemborer parasitoid *Cotesia flavipes*. PhD thesis, Wageningen Agricultural University, the Netherlands, pp. 1–125.

Potting, R.P.J., Overholt, W.A., Danso, F.O. and Takasu, K. (1997a) Foraging behaviour and life history of the stemborer parasitoid, *Cotesia flavipes. Journal of Insect Behaviour* 10, 13–29.

Potting, R.P.J., Vet, L.E.M. and Overholt, W.A. (1997b) Geographic variation in host selection behaviour and reproductive success in the stemborer parasitoid *Cotesia flavipes* (Hymenoptera: Braconidae). *Bulletin of Entomological Research* 87, 515–524.

Purseglove, J.W. (1972) *Tropical Crops – Monocotyledons.* Longman, London, 607 pp.

Quartey, S.Q. (1975) The biology of lepidopterous stem borers of rice in Eastern Ghana. MSc thesis, University of Ghana, Legon, 211 pp.

Quicke, D.L.J. (1987) The Old World genera of braconine wasps (Hymenoptera: Braconidae). *Journal of Natural History* 21, 43–157, figs 1–370.

Quicke, D.L.J. (1988) *Digonogastra*: the correct name for Nearctic *Iphiaulax* of authors (Hymenoptera: Braconidae). *Proceedings of the Entomological Society of Washington* 90, 196–200.

Quicke, D.L.J. and Wharton, R.A. (1989) *Myosoma nyanzaensis* sp. n. (Hymenoptera: Braconidae) parasitic on the stem-borer pest *Chilo partellus* (Swinhoe) (Lepidoptera: Pyralidae) in Kenya. *Bulletin of Entomological Research* 79, 1–5.

Rachie, K.O. and Majmudar, J.V. (1980) *Pearl Millet.* Pennsylvania University Press, Philadelphia, 307 pp.

Rachie, K.O. and Peters, L.V. (1977) *The Eleusines (A Review of World Literature).* ICRISAT, Hyderabad, 179 pp.

Ragonot, E.L. (1888) *Nouveaux genres et espèces de Phycitidae et Galleriidae.* Published by the Author, printed by Grandremy & Henon, Paris, 52 pp.

Rajabalee, M.A. and Governdasamy, M. (1988) Host specificity and efficacy of *Apanteles flavipes* (Cam.) and *A. sesamiae* (Cam.) (Hymenoptera: Braconidae) parasites of sugar cane moth borers in Mauritius. *Revue Agricole et Sucrière de l'Ile Maurice* 67, 78–80.

Rao, V.P. (1969) Biological control of stalk moth-borers in the old world – part I. *Indian Sugar* 18(11), 813–823.

Rao, V.P. (1970) Aims, objectives and future of biological control of rice stem borers. *Mushi* 44, 11–14.

Rao, V.P. and Nagaraja, H. (1969) *Sesamia* species as pests of sugar cane. In: Williams, J.R., Metcalf, J.R., Mungomery, R.W. and Mathes, R. (eds) *Pests of Sugar Cane.* Elsevier, Amsterdam.

Rao, V.P. and Nagarkatti, S. (1971) Can sugarcane borers in India be controlled by indigenous parasites? *Indian Sugar* 21(3), 3 pp.

Rao, Y.R.V.J., Rao, P.S.P., Varma, A. and Israel, P. (1971) Tests with an insect parasitic nematode DD-136 (Nematoda: Steinernematidae) against the rice stem borer, *Tryporyza incertulas* Walker. *Indian Journal of Entomology* 33, 215–217.

Rasnitsyn, A.P. (1980) (Origin and evolution of *Hymenoptera.*) *Trudy Paleontologicheskogo Instituta* 174, 1–190 (in Russian).

Rasnitsyn, A.P. (1988) An outline of the evolution of the hymenopterous insects (Order Vespida). *Oriental Insects* 22, 115–145.

Ratnadass, A., Doumbia, Y.O. and Hamadoun, A. (1992) *Neolimnus aegyticus* (Hemiptera, Cicadellidae) et *Sesamia penniseti* Tams et Bowden (Lepidoptera, Noctuidae), deux nouveaux ravageurs du sorgho de décrue dans la zone des lacs au nord du Mali. *Agronomie Tropicale* 46(4), 321–326.

Raulston, J.R. and Lingren, P.D. (1972) *Methods of Large Scale Rearing of the Tobacco Budworm.* USDA-ARS Production Research Report No. 145, Washington, DC, 10 pp.

Remillet, M. (1970). Donnés sur les nematodes entomophiles de la région ethiopienne. Unpublished report, ORSTOM, Abidjan, pp. 1–20.

Revington, J. (1986) This borer spreads rapidly through crops of maize and sorghum on the Highveld, but it can be controlled. *Farmer's Weekly*, 24 October.

Revington, J. (1987) Use of a synthetic sex pheromone in forecasting attacks on maize by maize stalkborer *Busseola fusca* (Fuller) (Lepidoptera: Noctuidae). MSc thesis, University of the Witwatersrand, Johannesburg, South Africa.

Revington, J., van Rensburg, J.B.J., Burghardt, G. and Knauf, W. (1984) Preliminary field trials with a pheromone based monitoring system for the maize stalkborer *Busseola fusca* (Fuller) (Lepidoptera: Noctuidae). *Journal of the Entomological Society of Southern Africa* 47, 107–113.

Riba, G., Marcandier, S., Richard, G. and Larget, I. (1983) Sensibilité de la pyrale du maïs (*Ostrinia nubilalis*) [Lep.: Pyralidae] aux hyphomycètes entomopathogènes. *Entomophaga* 28, 55–64.

Ripley, L.B. (1928) Top-dressing maize against stalk-borer. *Farming in South Africa* 10, 5 pp.

Ripley, L.B. and Hepburn, G.A. (1928) Top-dressing maize against stalk-borer. *Farming in South Africa* 6, 222.

Ripley, L.B. and Hepburn, G.A. (1929) Stalk-borer in maize: effect of top-dressing. *Farming in South Africa* 59, 4 pp.

Ripley, L.B. and Hepburn, G.A. (1934) Adhesives for cryolite suspensions. *Science Bulletin, Department of Agriculture, South Africa* 122, 12 pp.

Risbec, J. (1950a) Etat actuel des recherches entomologiques agricoles dans la région correspondant au secteur soudanais de recherches agronomiques. *Compte Rendus 1er Conference Internationale, Africanistes de l'Ouest* 1, 317–375.

Risbec, J. (1950b) I. La faune entomologique des cultures au Sénégal et au Soudan Français

& II. Contribution à l'étude des Proctotrupidae. *Travaux du Laboratoire d'Entomologie du Secteur Soudanais de Recherches Agronomiques*, Vols 1 and 2, Gouvernement Général de l'Afrique Occidentale Française, Paris, 639 pp.

Risbec, J. (1951) Les Chalcidoides d'A.O.F. *Mémoires de l'Institut Française d'Afrique Noire* 13, 5–409.

Risbec, J. (1956a) Hyménoptères parasites du Cameroun (2ᵉ contribution). *Bulletin de l'Institut Française de l'Afrique Noire, Dakar (A)* 18, 97–164.

Risbec, J. (1956b) Les parasites des insectes borers du riz au Cameroun. *Agronomie Tropicale* 11, 234–247.

Risbec, J. (1960) Les parasites des insectes d'importance en Afrique tropicale et à Madagascar. *Agronomie Tropicale* 15, 624–656.

Risch, S.J., Andow, D. and Altieri, M.A. (1983) Agroecosystem diversity and pest control: data, tentative conclusions, and new research directions. *Environmental Entomology* 12, 625–629.

Ritchie, A.R. (1932) Report of the government entomologist 1931. Appendix IV. In: *Tanganyika Territory, Annual Report of the Department of Agriculture, 1931*, pp. 83–86.

Rivnay, E. (1963) Present status of lepidopterous pests of maize and other graminaceous crops in Israel. *Plant Protection Bulletin FAO* 11(1), 3 pp.

Rivnay, E. (1967) A contribution to the biology of the maize borer *Chilo agamemnon* Bleszynski in Israel. *Israel Journal of Entomology* 2, 15–27.

Robertson, I.A.D. (1965) Notes on Diptera collected in Tanganyika. *Journal of the Entomological Society of Southern Africa* 28, 80–91.

Rouanet, G. (1984) Le maïs. *Le Technicien d'Agriculture Tropical* 2, 142 pp.

Rubtsov, I.A. (1976) Mermithids (Nematoda, Mermithidae) parasites of insects from Mongolia. *Nasekomye Mongolii* 4, 596–614.

Saad, A.S.A.K., Zeid, M. and El-Sebae, A.H. (1971) Chemical control of *Sesamia cretica* Led., *Chilo agamemnon* Bles. and *Ostrinia nubilalis* Hb. (Lepidoptera) and number of sprays required. *Zeitschrift für Angewandte Entomologie* 69, 91–98.

Sagnia, S.B. (1983) Possible integrated pest management tools for the effective control of cereal stem borers in the Gambia. *Insect Science and its Application* 4, 217–219.

Sampson, M.A. and Kumar, R. (1985) Life history, development and behaviour of *Eldana saccharina* Walker on sugarcane in southern Ghana. *Insect Science and its Application* 6, 135–143.

Sampson, M.A. and Kumar, R. (1986) Parasitism of *Descampsina sesamiae* (Mesnil) on *Sesamia* species in sugar-cane in southern Ghana. *Insect Science and its Application* 7, 543–546.

Sarhan, A.A. and Quicke, D.L.J. (1990) *Mesobraconoides psolopterus* (Hymenoptera: Braconidae), a larval parasitoid of the white rice borer, *Maliarpha separatella* (Lepidoptera: Pyralidae), in West Africa. *Bulletin of Entomological Research* 80, 217–222.

SASA (South African Sugar Association) (1994) *Eldana Borer – Control Measures*. Experiment Station Information Sheet, SASA, Mount Edgecombe, Natal, South Africa.

Saxena, K.N. (1985) Behavioural basis of plant resistance or susceptibility to insects. *Insect Science and its Application* 6, 303–313.

Saxena, K.N. (1990) Mechanisms of resistance/susceptibility of certain sorghum cultivars to the stem borer *Chilo partellus*: role of behaviour and development. *Entomologia Experimentalis et Applicata* 55, 91–99.

Saxena, K.N., Seshu Reddy, K.V., Nyambo, B.T. and Ngode, L. (1991) Integrated pest management in cereal and grain legume crops in Kenya. Presented at the XII International Plant Protection Congress, Rio de Janeiro, Brazil, 11–16 August 1991.

Schaefers, G.A. (1994) The changing role of pesticides in pest management. *Insect Science and its Application* 15, 583–598.

Scheibelreiter, G.K. (1980) Sugar-cane stem borers (Lepidoptera: Noctuidae and Pyralidae) in Ghana. *Zeitschrift für Angewandte Entomologie* 89, 87–99.

Scheltes, P. (1978) Ecological and physiological aspect of aestivation – diapause in the larvae of two Pyralid stalk borers of maize in Kenya. PhD thesis, Landbouwhogeschool, Wageningen.

Schulthess, F., Bosque-Pérez, N.A. and Gounou, S. (1991) Sampling lepidopterous pests on maize in West Africa. *Bulletin of Entomological Research* 81, 297–301.

Schulthess, F., Bosque-Pérez, N.A., Chabi-Olaye, A., Gounou, S., Ndemah, R. and Goergen, G. (1977) Exchange of natural enemies of lepidopteran cereal stemborers between African regions. *Insect Science and its Application* (in press).

Seshu Reddy, K.V. (1983a) Studies on stem borer complex of sorghum in Kenya. *Insect Science and its Application* 4(1–2), 3–10.

Seshu Reddy, K.V. (1983b) Sorghum stem borers in eastern Africa. *Insect Science and its Application* 4, 33–39.

Seshu Reddy, K.V. (1985a) Integrated approach to the control of sorghum stemborers. In: *Proceedings of the International Sorghum Entomology Workshop, 15–21 July 1984, College Station, Texas, USA.* International Crops Research Institute for the Semi-Arid Tropics, Patancheru, India, pp. 205–215.

Seshu Reddy, K.V. (1985b) Relative susceptibility and resistance of some sorghum lines to stem borers in western Kenya. *Insect Science and its Application* 6(3), 401–404.

Seshu Reddy, K.V. (1988) Assessment of on-farm yield losses in sorghum due to insect pests. *Insect Science and its Application* 9(6), 679–685.

Seshu Reddy, K.V. (1989) Sorghum stem borers in eastern Africa. In: *Proceedings of the International Workshop on Sorghum Stem Borers, 17–20 November 1987.* ICRISAT, Patancheru, India, pp. 33–40.

Seshu Reddy, K.V. (1990) Cultural control of *Chilo* spp. in graminaceous crops. *Insect Science and its Application* 11, 703–712.

Seshu Reddy, K.V. (1991) Insect pests of sorghum in Africa. *Insect Science and its Application* 12(5–6), 653–657.

Seshu Reddy, K.V. and Davis, J.C. (1978) A new medium for mass rearing of the sorghum stem borer *Chilo partellus* Swinhoe (Lepidoptera, Pyralidae) and its use in resistance screening. *Indian Journal of Plant Protection* 6, 48–55.

Seshu Reddy, K.V. and Masyanga, B.S.K. (1988) Effects of different proportions of sorghum/cowpea intercrop rows on crop borer incidence. In: *ICIPE 15th Annual Report 1987*, ICIPE, Nairobi, pp. 6–7.

Seshu Reddy, K.V. and Sum, K.O.S. (1991) Determination of economic injury level of the stem borer, *Chilo partellus* (Swinhoe) in maize, *Zea mays* L. *Insect Science and its Application* 12(1–3), 269–274.

Seshu Reddy, K.V. and Sum, K.O.S. (1992) Yield–infestation relationship and determination of economic injury level of the stem borer *Chilo partellus* (Swinhoe) in three varieties of maize, *Zea mays* L. *Maydica* 37, 371–376.

Seshu Reddy, K.V. and Unnithan, G.C. (1985) Alternative host plants of stem borers. In: *ICIPE 12th Annual Report, 1984*, ICIPE, Nairobi, p. 7.

Seshu Reddy, K.V. and Walker, P.T. (1990) A review of the yield losses in graminaceous crops caused by *Chilo* spp. *Insect Science and its Application* 11, 563–569.

Seshu Reddy, K.V., Sum, K.O.S. and Lubega, M.C. (1989) Empirical models for predicting yield loss in sorghum caused by *Chilo partellus* (Swinhoe). *Discovery and Innovation* 1, 90–94.

Sétamou, M. and Schulthess, F. (1995) The influence of egg parasitoids belonging to the *Telenomus busseolae* (Hymenoptera: Scelionidae) species complex on *Sesamia calamistis* (Lepidoptera: Noctuidae) populations in maize fields in southern Benin. *Biocontrol Science and Technology* 5, 69–81.

Sétamou, M., Schulthess, F., Bosque-Pérez, N.A. and Thomas-Odjo, A. (1993) Effect of plant nitrogen and silica on the bionomics of *Sesamia calamistis* Hampson (Lepidoptera: Noctuidae). *Bulletin of Entomological Research* 83, 405–411.

Sétamou, M., Schulthess, F., Bosque-Pérez, N.A. and Thomas-Odjo, A. (1995) The effect of stem and cob borers on maize subjected to different nitrogen and silica treatments, with special reference to *Sesamia calamistis* Hampson (Lepidoptera: Noctuidae). *Entomologia Experimentalis et Applicata* 77, 205–210.

Seyrig, A. (1935) Hymenoptera 2. Ichneumonidae: Cryptinae, Pimplinae, Tryphoninae et Ophioninae. *Mission scientifique de l'Omo* 3(18), 1–100 [= *Mémoires du Muséum national d'Histoire Naturelle* (*NS*) 4, 1–100 (1936)].

Shami, S. and Mohyuddin, A.I. (1987) Host selection by Indonesian strain of *Apanteles flavipes* (Cam.) and its suitability for various graminaceous borers in Pakistan. In: *Proceedings of the 23rd Annual Convention of the Pakistan Society of Sugar Technology, 25–27 July, Rawalpindi*, pp. 286–291.

Shanower, T.G., Schulthess, F. and Gounou, S. (1991) *Distribution and Abundance of Some Stem and Cob Borers in Benin.* Plant Health Management Division, Research Monograph No. 1, International Institute of Tropical Agriculture, Cotonou, Republic of Benin.

Shanower, T.G., Schulthess, F. and Bosque-Pérez, N.A. (1993a) Development and fecundity of *Sesamia calamistis* Hampson (Lepidoptera: Noctuidae) and *Eldana saccharina* Walker (Lepidoptera: Pyralidae). *Bulletin of Entomological Research* 83, 237–243.

Shanower, T.G., Schulthess, F. and Bosque-Pérez, N.A. (1993b) The effect of larval diet on the development of *Sesamia calamistis* Hampson (Lepidoptera: Noctuidae) and *Eldana saccharina* Walker (Lepidoptera: Pyralidae). *Insect Science and its Application* 14, 681–685.

Sharma, A.K., Saxena, J.D. and Subba Rao, B.R. (1966) A catalogue of the hymenopterous and dipterous parasites of *Chilo zonellus* (Swinhoe) (Crambidae: Lepidoptera). *Indian Journal of Entomology* 28, 510–542.

Sharma, H.C. and Davies, J.C. (1988) *Insect and Other Animal Pests of Millets.* International Crops Research Institute for the Semi-Arid Tropics, Patancheru, India, 86 pp., tables and plates.

Shorey, H.H. and Hale, R.L. (1965) Mass rearing of the larvae of nine noctuid species on a simple artificial medium. *Journal of Economic Entomology* 58, 522–524.

Sibanda, S. (1985) The use of sorghum and millets for feeding livestock. In: *Proceedings of the Second Regional Workshop on Sorghum and Millets for Southern Africa, 23–27 September 1985, Gaborone, Botswana.* ICRISAT, pp. 228–247.

Siddiq, S.A. (1972) Graminaceous stem borers in the northern province of Sudan 1. Ecological studies. *Zeitschrift für Angewandte Entomologie* 71, 376–381.

Sigwalt, B. and Pointel, J.G. (1980) Statut spécifique et séparation des *Apanteles* du sous-groupe *flavipes* [Hym.: Braconidae] utilisées en lutte biologique. *Annales de la Societé Entomologique de France* (*NS*) 16, 109–128.

Sikorowski, P.P. (1984) Pathogens and microbial contaminants: their occurrence and control. In: Sikorowski, P.P. and Lindig, D.H. (eds) *Bollweevil Mass Rearing Technology.* Mississippi University Press, pp. 115–168.

Silvie, P. (1991) Effet du parasitisme naturel observé au Tchad chez deux Lépidoptères phyllophages du cottonier: *Syllepte derogata* (Crambidae) et *Cosmophila flava* (Noctuidae). *Entomophaga* 36, 431–441.

Singh, P. (1983) A general purpose laboratory diet mixture for rearing insects. *Insect Science and its Application* 4, 357–362.

Singh, P. (1985) Multiple species rearing diets. In: Singh, P. and Moore, R.F. (eds) *Handbook of Insect Rearing,* Vol. I. Elsevier Science Publishers, Amsterdam, the Netherlands, pp. 19–44.

Singh, P. and Ashby, M.D. (1985) Insect rearing management. In: Singh, P. and Moore, R.F. (eds) *Handbook of Insect Rearing,* Vol. 1. Elsevier Science Publishers, Amsterdam, the Netherlands, pp. 19–44.

Singh, P. and Leppla, N.C. (eds) (1984) *Advances and Challenges in Insect Rearing.* USDA Technical Bulletin, Washington, DC.

Singh, P. and Moore, R.F. (eds) (1985) *Handbook of Insect Rearing,* Vols I and II. Elsevier Science Publishers, Amsterdam, the Netherlands.

Singh, G. and Sandhu, G.S. (1979) Exploitation of the ovipositional behaviour of *Chilo partellus* (Swinhoe) for pest management. *Journal of Research of Punjab Agricultural University* 16, 476–478.

Sithanantham, S., Kamau, G.M., Oendo, A., Ajala, O.S., Saxena, K.N., Mwangi, K.M. and Muli, B. (1995) Adaptive testing of environment-friendly options for maize stalk borer management in Coastal Kenya. Paper presented at the Second Crop Science Conference for Eastern and Southern Africa, 19–24 February 1995, Blantyre, Malawi.

Sithole, S.Z. (1987a) Maize insect pests in Zimbabwe. In: *Toward Insect Resistant Maize for the Third World. Proceedings of the International Symposium on Methodologies for Developing Host Plant Resistance to Maize Insects.* Centro Internacional de Majoramiento de Maiz y Trigo, Mexico City, pp. 286–288.

Sithole, S.Z. (1987b) The effect of date of planting on shootfly and stem borer infestations on sorghum. In: *Proceedings of the Third Regional Workshop on Sorghum and Millets for Southern Africa, 6–10 October 1986, Lusaka, Zambia.* ICRISAT, Patancheru, pp. 174–183.

Sithole, S.Z. (1988) The effect of protecting sorghum at different growth stages on stem borer infestation and yield. In: *Proceedings of the Fourth Regional Workshop on Sorghum and Millets for Southern Africa, 21–24 September 1987, Bulawayo, Zimbabwe.* ICRISAT, Patancheru, pp. 201–208.

Sithole, S.Z. (1989b) Sorghum stem borers in southern Africa. In: *International Workshop on Sorghum Stem Borers, 17–20 November 1987.* International Crops Research Institute for the Semi-Arid Tropics, Patancheru, India, pp. 41–47.

Sithole, S.Z. (1989c) The effect of date of planting on shootfly and stem borer infestations on sorghum. In: *Proceedings of the Third Regional Workshop on Sorghum and Millets for Southern Africa, Lusaka, Zambia, 6–10 October 1986.* ICRISAT, Patancheru, pp. 174–183.

Sithole, S.Z. (1990) Status and control of the stem borer, *Chilo partellus* Swinhoe (Lepidoptera: Pyralidae) in southern Africa. *Insect Science and its Application* 11, 481–488.

Sithole, S.Z. (1994) Distribution and economic importance of sorghum stem borers in Zimbabwe. PhD thesis, University of Zimbabwe, Harare, Zimbabwe.

Sithole, S.Z. and Makombe, G. (1989) Economic analysis for the application of dipterex 2.5 percent granules to control stemborers attacking sorghum in the communal areas of Zimbabwe. In: *Proceedings of the Fifth Regional Workshop on Sorghum and Millets in Southern Africa, 21–23 September 1988, Maseru, Lesotho.* ICRISAT, Patancheru, pp. 204–211.

Skidmore, P. (1985) The biology of the Muscidae of the world. *Series Entomologica* 29, 1–550.

Skoroszewski, R.W. and van Hamburg, H. (1987) The release of *Apanteles flavipes* (Cameron) (Hymenoptera: Braconidae) against stalk-borers of maize and grain-sorghum in South Africa. *Journal of the Entomological Society of Southern Africa* 50, 249–255.

Skovgard, H and Päts, P. (1996) Effects of intercropping on maize stemborers and their natural enemies. *Bulletin of Entomological Research* 86, 599–607.

Smith, J.W., Rodriguez-del-Bosque L.A. and Agnew, C.W. (1990) Biology of *Mallochia pyralidis* (Hymenoptera: Ichneumonidae), an ectoparasite of *Eoreuma loftini* (Lepidoptera: Pyralidae) from Mexico. *Annals of the Entomological Society of America* 83, 961–966.

Smith J.W., Jr, Barau, A.D., Goldman, A. and Mareck, J.H. (1993a) The role of technology in agricultural intensification: the evolution of production systems in the northern Guinea savanna of Nigeria. *Economic Development and Cultural Change* 42, 537–554.

Smith, J.W., Wiedenmann, R.N. and Overholt, W.A. (1993b) *Parasites of Lepidopteran Stemborers of Tropical Gramineous Plants.* ICIPE Science Press, Nairobi, 89 pp.

Smithers, C.N. (1960a) Moisture and pupation in *Busseola fusca* (Hmps.) (Lep.: Noctuidae). *Journal of the Entomological Society of Southern Africa* 23, 225–227.

Smithers, C.N. (1960b) Some recent observations on *Busseola fusca* (Fuller) (Lep., Noctuidae) in Southern Rhodesia. *Bulletin of Entomological Research* 50, 809–819.

Spaull, M.A. (1990) Field tests to control the pyralid, *Eldana saccharina*, with an entomogenous nematode, *Heterorhabditis* sp. *Proceedings of the South African Sugar Technologists' Association* 64, 103–106.

Spaull, M.A. (1991) *Heterorhabditis* and *Steinernema* species (Nematoda: Rhabditida) for the control of a sugarcane stalk borer in South Africa. *Phytophylactica* 23, 213–215.

Spaull, M.A. (1992) On the use of methylcellulose polymer to increase the effectiveness of a *Heterorhabditis* species against the sugarcane stalk borer, *Eldana saccharina*. *Fundamental and Applied Nematology* 15, 457–461.

Spencer, D.S.C. and Sivakumar, M.V.K. (1987) Pearl millet in African agriculture. In: *Proceedings of the International Pearl Millet Workshop, ICRISAT, 7–11 April 1986*. ICRISAT, Patancheru, pp. 19–31.

Starks, K.J. (1969) Some cereal crop insects in East Africa. East African Agriculture and Forestry Research Organization, Serere Research Station, Uganda (unpublished).

Subasinghe, S.M. and Amarasena, J. (1988) Temporal distribution and chemical control of the maize stem-borer, *Chilo partellus* Swinhoe (Lep.: Pyralidae). *Tropical Agriculturist* 144, 123–133.

Subba Rao, B.R. (1970) A redescription of *Exoristobia* Ashmead 1904 and description of two new species (Hymenoptera: Encyrtidae). *Proceedings of the Royal Entomological Society of London (B)* 39, 112–113.

Swaine, G. (1957) The maize and sorghum stalkborer, *Busseola fusca* (Fuller), in peasant agriculture in Tanganyika territory. *Bulletin of Entomological Research* 48, 711–722.

Szépligeti, G.V. (1908) 8. Hymenoptera; 3. Braconidae und Ichneumonidae. In: Sjöstedt, Y. (ed.) *Wissenschaftliche Ergebnisse der Schwedisches Zoologische Expedition nach dem Kilimandjaro, dem Meru und den umgebenden Massaisteppen Deutsch-Ostafrikas 1905–1906 unter Leitung von Prof. Dr. Yngve Sjöstedt*, Vol. 2, pp. 25–95.

Takasu, K. and Overholt, W.A. (1996) Oviposition behaviour of *Cotesia flavipes* and *Goniozus indicus*, larval parasitoids of the lepidopteran stemborer *Chilo partellus*. *Japanese Journal of Applied Entomology and Zoology* 40, 308–310.

Tams, W.H.T. (1932) New species of African Heterocera. *Entomologist* 65, 1241–1249.

Tams, W.H.T. and Bowden, J. (1953) A revision of the African species of *Sesamia* Guenée and related genera (Lepidoptera – Agrotidae). *Bulletin of Entomological Research* 43, 645–678.

Taneja, S.L. and Nwanze, K.F. (1989) Assessment of yield loss of sorghum and pearl millet due to stem borer damage. In: *ICRISAT International Workshop on Sorghum Stem Borers, 17–20 November 1987*. ICRISAT, Patancheru, India, pp. 95–104.

Taneja, S.L. and Nwanze, K.F. (1990) Mass rearing of *Chilo* spp. on artificial diets and its use in resistance screening. *Insect Science and its Application* 11, 605–616.

Tateishi, I. and Murata, T. (1955) On the infestation period of parasitic fungus (*Isaria farinosa* (Dicks.) Fr.) to *Chilo simplex* Butler. *Proceedings of the Association for Plant Protection, Kyushu* 1, 75–76.

Tavakilian, G. (1977) Le genre *Chilo* en Côte d'Ivoire. *Cahiers d'ORSTOM Série Biologie* 12, 47–54.

Taylor, L.R. (1961) Aggregation, variance and the mean. *Nature* 189, 732–735.

Thyagarajan, G. (1988) Pesticide safety, storage and disposal. In: Prinsley, R.T. and Terry, P.G. (eds) *Crop Production for Small Scale Farms in East and Central Africa – A Review*. Commonwealth Science Council, London, pp. 89–96.

Tobias, V.I. (1986) Agathidinae. In: Medvedev, G.S. (ed.) Opredelitel nasekomych Evropeiskoi tchasti SSSR 3, Perepontchatokrylye 4. *Opr. Faune SSSR* 145, 1–501, figs 1–263. (Translation 1995, Lebanon, USA, pp. 480–507).

Toguebaye, B.S. and Bouix, G. (1983) *Nosema manierae* sp. n., Microspridie parasite de

Chilo zacconius Bleszynski 1970 (Lepidoptera: Pyralidae), hôte naturel, et *Heliothis armigera* (Huebner 1808) (Lepidoptera: Noctuidae), hôte expérimental cycle évolutif et ultrastructure. *Zeitschrift für Parasitenkunde* 69, 191–205.

Torii, T. (1975) Feasibility of using the entomophilic nematode, DD-136, as a biotic insecticide against stem borers. In: *Approaches to Biological Control.* Japanese Committee for the International Biological Programme, University of Tokyo Press, Tokyo, Japan, pp. 87–88.

Torto, B., Addae-Mensah, I. and Moreka, L. (1992). Antifeedant activity of *Piper guineense* Scum & Thonn. amides against larvae of the sorghum stem borer *Chilo partellus* (Swinhoe). *Insect Science and its Application* 13, 705–708.

Townes, H. (1969) The genera of Ichneumonidae, part 1. *Memoirs of the American Entomological Institute* 11, 300 pp.

Townes, H. (1970a) The genera of Ichneumonidae, part 2. *Memoirs of the American Entomological Institute* 12, 537 pp.

Townes, H. (1970b) The genera of Ichneumonidae, part 3. *Memoirs of the American Entomological Institute* 13, 307 pp.

Townes, H. (1971) The genera of Ichneumonidae, part 4. *Memoirs of the American Entomological Institute* 17, 372 pp.

Townes, H. and Chiu, S.C. (1970) The Indo-Australian species of *Xanthopimpla* (Ichneumonidae). *Memoirs of the American Entomological Institute* 14, 372 pp.

Townes, H. and Townes, M. (1973) A catalogue and reclassification of the Ethiopian Ichneumonidae. *Memoirs of the American Entomological Institute* 19, 416 pp.

Townsend, C.H.T. (1941) *Curranomyia* nom. nov. for *Rhodesina* Curr. (preocc.). *Transactions of the Rhodesian Science Association* 38, 115–116.

Tran, M. (1981) Reconnaissance des principaux foreurs des tiges du riz, du maïs et de la canne à sucre en Côte d'Ivoire. *Initiations Documentations Techniques, ORSTOM* 48

Tran, V.L. (1977a) Les ravageurs du riz en Côte d'Ivoire. Systématique des espèces et de leurs parasites. Biologie de *Chilo diffusilineus* (J. de Joannis) et de *Chilo zacconius* (Bleszynski) (Lepidoptera, Pyralidae). Thèse de Doctorat de 3e cycle, Université de Paris VI, 120 pp.

Tran, V.L. (1977b) Morphologie des pièces génitales et nervation alaire des principales pyrales foreurs du riz en Côte d'Ivoire. Descriptions de quelques Hyménoptères parasites. *Cahiers ORSTOM, série Biologique* 12, 29–45.

Ukwungwu, M.N. (1985a) Effect of nitrogen and carbofuran on gall midge (GM) and white stem borer (SB) infestation in Nigeria. *International Rice Research Newsletter* 10, 19–20.

Ukwungwu, M.N. (1985b) Effect of insecticides and variety on stem borer (SB) incidence. *International Rice Research Newsletter* 10, 20.

Ukwungwu, M. (1987) Seasonal changes in the stem borer (SB) *Maliarpha separatella* populations. *International Rice Research Newsletter* 12(3), 34–35.

Ullyett, G.C. (1935) Note on *Apanteles sesamiae*, Cam., a parasite of the maize stalk-borer (*Busseola fusca* Fuller) in South Africa. *Bulletin of Entomological Research* 26, 253–262.

Unnithan, G.C. (1987) Development and reproductive biology of the maize stem borer *Busseola fusca* Fuller (Lepidoptera, Noctuidae). *Journal of Applied Entomology* 104, 172–179.

Unnithan, G.C. and Saxena, K.N. (1990) Factors determining pheromone trapping of the stem-borer *Chilo partellus* (Swinhoe) (Lep., Pyralidae). *Journal of Applied Entomology* 110, 183–193.

Unnithan, G.C. and Seshu Reddy, K.V. (1989) Incidence, diapause and carryover of the cereal stem borers in Rusinga Island, Kenya. *Tropical Pest Management* 35(4), 414–419.

Usua, E.J. (1968) Temperature and relative humidity effects on the development of immature stages of the maize stem borers *Busseola fusca* and *Sesamia calamistis. Journal of Economic Entomology* 61, 1091–1093.

Usua, E.J. (1969) Description of the larvae of *Sesamia penniseti* Bowden and *S. botanephaga* Tams & Bowden (Lepidoptera: Noctuidae). *Bulletin of the Entomological Society of Nigeria* 2, 72–76.

Usua, E.J. (1973) The induction of diapause in the maize stem borer *Busseola fusca. Entomologia Experimentalis et Applicata* 16, 322–328.

Usua, E.J. (1987) Description of the larvae and pupae of some important lepidopterous stem borers of cereals. *Occasional Publications, Entomological Society of Nigeria* 29, 1–29.

Uvah, I.I. and Ajayi, O. (1991) Chemical control of sorghum stem borers in the Nigerian savanna. Presented at the SAFGRAD Inter-Networks Conference, 7–14 March 1991, Niamey, Niger, 14 pp.

van Achterberg, C. (1979) A revision of the Amicrocentrinae, a new subfamily (Hymenoptera, Braconidae), with the description of the final larval instar *Amicrocentrum curvinervis* by J.R.T. Short. *Tijdschrift voor Entomologie* 122, 1–28.

van Achterberg, C. (1988) Revision of the subfamily Blacinae Foerster (Hymenoptera, Braconidae). *Zoologische Verhandelingen* 249, 1–324.

van Achterberg, C. (1990) Revision of the Western Palaearctic Phanerotomini (Hymenoptera: Braconidae). *Zoologische Verhandelingen* 255, 1–106.

van Achterberg, C. (1993a) Illustrated key to the subfamilies of Braconidae (Hymenoptera: Ichneumonoidea). *Zoologische Verhandelingen* 283, 1–189.

van Achterberg, C. (1993b) Revision of the genus *Tropobracon* Cameron (Hymenoptera: Braconidae). *Zoologische Mededelingen* 67, 49–62.

van Achterberg C. and Polaszek, A. (1996) The parasites of cereal stem borers (Lepidoptera: Cossidae, Crambidae, Noctuidae, Pyralidae) in Africa, belonging to the family Braconidae (Hymenoptera: Ichneumonoidea). *Zoologische Verhandelingen* 304, 1–123.

van den Berg, H. (1993) *Natural Control of* Helicoverpa armigera *in Smallholder Crops in East Africa.* Wageningen Agricultural University, the Netherlands.

van den Berg, J. (1998) Use of a moth production index to determine the impact of various sorghum varieties in the management of *Chilo partellus* (Lepidoptera: Pyralidae) in Southern Africa. *Insect Science and its Application* (in press).

van den Berg, J. and van der Westhuizen, M.C. (1995) Development of a chemical control strategy for *Chilo partellus* (Lepidoptera: Pyralidae) in grain sorghum. *South Africa Journal of Plant and Soil* 12, 105–107.

van den Berg, J. and van der Westhuizen, M.C. (1998) The potential value of stem borer resistance in the management of *Busseola fusca* (Lepidoptera: Noctuidae) and *Chilo partellus* (Lepidoptera: Pyralidae) populations in sorghum. *Insect Science and its Application* (in press).

van den Berg, J. and van Rensburg, J.B.J. (1991a) Unavoidable losses in insecticidal control of *Chilo partellus* (Swinhoe) (Lepidoptera: Pyralidae) in maize and grain sorghum. *South African Journal of Plant and Soil* 8, 12–16.

van den Berg, J. and van Rensburg, J.B.J. (1991b). The effect of plant density on the injuriousness of *Busseola fusca* (Fuller) (Lepidoptera: Noctuidae) in grain sorghum. *South African Journal of Plant and Soil* 8, 85–87.

van den Berg, J. and van Rensburg, J.B.J. (1991c) Infestation and injury levels of stem borers in relation to yield potential of grain sorghum. *South African Journal of Plant and Soil* 8, 127–131.

van den Berg, J. and van Rensburg, J.B.J. (1992) Chemical control of *Chilo partellus* (Lepidoptera: Pyralidae) larvae behind leaf sheaths of grain sorghum. *Applied Plant Science* 6, 28–30.

van den Berg, J. and van Rensburg, J.B.J. (1993) Importance of persistence and synergistic effects in the chemical control of *Chilo partellus* (Lepidoptera: Pyralidae) of grain sorghum. *Applied Plant Science* 7, 5–7.

van den Berg, J. and van Rensburg, J.B.J. (1996) Comparison of various directional insecticide sprays against *Busseola fusca* (Lepidoptera: Noctuidae) and *Chilo partellus* (Lepidoptera: Pyralidae) in sorghum and maize. *South Africa Journal of Plant and Soil* 13, 51–54.

van den Berg, J., van Rensburg, J.B.J. and Pringle, K.L. (1991) Comparative injuriousness of *Busseola fusca* (Lepidoptera: Noctuidae) and *Chilo partellus* (Lepidoptera: Pyralidae) on grain sorghum. *Bulletin of Entomological Research* 82, 137–142.

van den Berg, J., van Rensburg, J.B.J. and van der Westhuizen, M.C. (1993) Status of resistance to *Chilo partellus* (Lepidoptera: Pyralidae) in South African grain sorghum hybrids. *South African Journal of Plant and Soil* 10, 174–177.

van den Berg, J., van Rensburg, G.D.J. and van der Westhuizen, M.C. (1994) Host plant resistance and chemical control of *Chilo partellus* (Swinhoe) and *Busseola fusca* (Fuller) in an integrated pest management system on grain sorghum. *Crop Protection* 13, 307–310.

van den Berg, J., van Rensburg, G.D.J. and van der Westhuizen, M.C. (1997) Economic threshold levels for *Chilo partellus* (Lepidoptera: Pyralidae) control on resistant and susceptible sorghum plants. *Bulletin of Entomological Research* 87, 89–93.

Vanderzant, E.S. and Reiser, R. (1956) Aseptic rearing of the pink bollworm in synthetic media. *Journal of Economic Entomology* 49, 7–10.

van Emden, F.I. (1960) Keys to the Ethiopian Tachinidae. III. Macquartiinae. *Proceedings of the Zoological Society of London* 134, 313–487.

van Hamburg, H. (1976) The bionomie en ekonomiese belangrikheid vandie graansorghum-stamruspe, *Chilo partellus* (Swinhoe) (Lepidoptera: Pyralidae). DSc thesis, Pretoria University, Pretoria, South Africa.

van Hamburg, H. (1979) The grain-sorghum stalk-borer, *Chilo partellus* (Swinhoe) (Lepidoptera: Pyralidae): seasonal changes in adult populations in grain sorghum in the Transvaal. *Journal of the Entomological Society of Southern Africa* 42, 1–9.

van Hamburg, H. (1980) The grain sorghum stalk-borer, *Chilo partellus* (Swinhoe) (Lepidoptera: Pyralidae); survival and location of larvae at different infestation levels in plants of different ages. *Journal of the Entomological Society of Southern Africa* 43, 71–76.

van Hamburg, H. (1987) A biological control approach to pest management on grain crops with special reference to the control of stalkborers. *Technical Communication, Department of Agriculture and Water Supply, Republic of South Africa* 212, 52–55.

van Lenteren, J.C. and Steinberg, S. (1991) A preliminary list of criteria for quality control of beneficial arthropods used commercially in greenhouse crops. In: Biggler, F. (ed.) *Proceedings of the Global IOBC Working Group on Quality Control of Mass Reared Arthropods, 25–28 March.* IOBC, Wageningen, the Netherlands, pp. 195–199.

van Miltenburg, L. (1968) *Plagen van gierst en cowpea in Niger. Het effect van een mengteelt van cowpea op de boorder* Acigona ignefusalis *Hmps. in gierst.* Praktijkverslag, Department of Entomology WAU, Wageningen, 82 pp.

van Rensburg, G.D.J. (1990) Comparison of chemical control approaches against the maize stalk borer *Busseola fusca* (Lepidoptera: Noctuidae), in maize in South Africa. *Phytophylactica* 22, 87–91.

van Rensburg, G.D.J. and Bate, R. (1987) Preliminary studies on the relative abundance and distribution of the stalk borers *Busseola fusca* and *Chilo partellus. Technical Communication, Department of Agriculture and Water Supply, Republic of South Africa* 212, 49–52.

van Rensburg, G.D.J. and Malan, E.M. (1982) Control of sorghum pests and phytotoxic effects of carbofuran on five hybrids of grain sorghum. *Phytophylactica* 14, 159–163.

van Rensburg, J.B.J. (1988) Efficacy of cloethocarb for control of *Busseola fusca, Cicadulina* spp. and nematodes in maize. *Applied Plant Science* 2, 63–67.

van Rensburg, J.B.J. (1992) Evaluation of pheromone trapping systems in relation to light trap captures of the maize stalk borer, *Busseola fusca* (Fuller) (Lepidoptera: Noctuidae). *South African Journal of Plant and Soil* 9, 144–149.

van Rensburg, J.B.J. (1997) Seasonal moth flight activity of the maize stem borer, *Busseola fusca* (Fuller) (Lepidoptera: Noctuidae) in small farming areas of South Africa. *Applied Plant Science* 11(2) (in press).

van Rensburg, J.B.J. and Malan, C. (1992) Resistance of maize genotypes to the maize stalk borer, *Busseola fusca* (Fuller) (Lepidoptera: Noctuidae). *Technical Communication, Department of Agricultural Development of South Africa* 232, 98–99.

van Rensburg, J.B.J. and Pringle, K.L. (1989) A sequential sampling technique for surveys of eggs laid by the maize stalk borer, *Busseola fusca* (Fuller) (Lepidoptera: Noctuidae). *Journal of the Entomological Society of Southern Africa* 52, 223–228.

van Rensburg J.B.J. and van den Berg, J. (1992a) Infestation patterns of stalk borers *Busseola fusca* (Fuller) (Lep.: Noctuidae) and *Chilo partellus* (Swinhoe) (Lep.: Pyralidae). *Journal of the Entomological Society of Southern Africa* 55, 197–212.

van Rensburg J.B.J. and van den Berg, J. (1992b) Stem borers in grain sorghum: I. Injury patterns with time after crop emergence. *South African Journal of Plants and Soil* 9, 73–80.

van Rensburg, J.B.J. and van den Berg, J. (1992c) Timing of insecticide application for stem borer control in maize and grain sorghum. *Applied Plant Science* 6, 24–27.

van Rensburg, J.B.J. and van den Berg, J. (1992d) Impact of insecticide application on numbers of *Busseola fusca* (Fuller) and *Chilo partellus* (Swinhoe) at early growth stages of maize and grain sorghum. *Applied Plant Science* 6, 70–72.

van Rensburg, J.B.J. and van Rensburg, G.D.J. (1993) Laboratory production of *Busseola fusca* (Fuller) (Lepidoptera: Noctuidae) and techniques for the detection of resistance in maize plants. *African Entomology* 1(1), 25–28.

van Rensburg, J.B.J. and Walters, M.C. (1978) The efficacy of systemic insecticides applied to the soil for the control of *Cicadulina mbila* (Naudé) (Hem.: Cicadellidae), the vector of maize streak disease, and the maize stalk borer, *Busseola fusca* (Fuller) (Lep.: Noctuidae). *Phytophylactica* 10, 49–52.

van Rensburg, J.B.J., Walters, M.C. and Stemmet, G.P. (1978) A preliminary study on the application of carbofuran granules to the soil for the control of grain sorghum pests. *Phytophylactica* 10, 28–30.

van Rensburg, J.B.J., Walters, M.C. and Giliomee, J.H. (1985) Geographical variation in the seasonal moth flight activity of the maize stalk borer, *Busseola fusca* (Fuller), in South Africa. *South African Journal of Plant and Soil* 2, 124–126.

van Rensburg, J.B.J., Walters, M.C. and Giliomee, J.H. (1987) Ecology of the maize stalk borer, *Busseola fusca* (Fuller) (Lepidoptera: Noctuidae). *Bulletin of Entomological Research* 77, 255–269.

van Rensburg, J.B.J., Walters, M.C. and Giliomee, J.H. (1988a) Mortality in natural populations of the maize stalk borer, *Busseola fusca* (Fuller) (Lepidoptera: Noctuidae) in South Africa. *Phytophylactica* 20, 17–19.

van Rensburg, J.B.J., Walters, M.C. and Giliomee, J.H. (1988b) The comparative abundance and in-season distribution of larval infestations of *Busseola fusca* (Fuller) (Lepidoptera: Noctuidae) and *Chilo partellus* (Swinhoe) (Lepidoptera: Pyralidae) on maize. *Journal of the Entomological Society of Southern Africa* 51, 138–140.

van Rensburg, J.B.J., Walters, M.C. and Giliomee, J.H. (1988c) Response of maize to levels and times of infestation by *Busseola fusca* (Fuller) (Lepidoptera: Noctuidae). *Journal of the Entomological Society of Southern Africa* 51, 283–291.

van Rensburg, J.B.J., Walters, M.C. and Giliomee, J.H. (1988d) Plant population and cultivar effects on yield losses caused by the maize stalk borer, *Bacilli fusca* (Lepidoptera: Noctuidae). *South African Journal of Plant and Soil* 5, 215–218.

van Rensburg, J.B.J., Walters, M.C. and Giliomee, J.H. (1989) Selective oviposition by the maize stalk borer, *Busseola fusca* (Fuller). *Journal of the Entomological Society of Southern Africa* 52, 105–108.

van Rensburg, N.J. and van Hamburg, H. (1975) Grain sorghum pests: an integrated control approach. In: *Proceedings 1st Congress, Entomological Society of Southern Africa.* Entomological Society of Southern Africa, Pretoria, pp. 151–162.

Vayssière, P. and Mimeur, J. (1925) Les pyrales du cotonnier (*Sylepta derogata* F. et *Glyphodes indica* Saund.) en Afrique occidentale française. *Agronomie coloniale* 12(90), 255–268.

Vercambre, B. (1977a) Premières recherches en vue d'une lutte biologique contre les insectes nuisibles au riz irrigué au Sénégal. *Déliberations du séminaire ADRAO sur la protection des plantes en riziculture,* 21–29 May 1973, Monrovia (Liberia), pp. 121–128.

Vercambre, B. (1977b) Initial research for biological control of pests on irrigated rice in Senegal. In: *WARDA Seminar Proceedings No. 4 Plant Protection for the Rice Crop, Monrovia, Liberia, March 1977,* pp. 117–123.

Vercambre, B. (1977c) 1. Recherches effectuées sur les foreurs riz irrigué en Casamance (synthèse des résultats obtenus). In: *Réunion sur les foreurs des graminées. Bouaké, Côte d'Ivoire, 8–11 novembre, Annexe XI,* 10 pp.

Vercambre, B. (1982) *Diopsis thoracica* West (Dipt. Diopsidae), important ravageur du riz en Afrique de l'Ouest: données bio-écologiques et application à la lutte intégrée. *Agronomie Tropicale* 37, 89–98.

Vercambre, B., Bordat, D. and Djiba, S. (1990) Statut et distinction des deux principales espèces de *Chilo* spp. (Lepidoptera, Pyralidae) sur riz en Casamance (Sénégal). *Agronomie Tropicale* 45(2), 131–138.

Verma, A.N. and Singh, S.P. (1989) Cultural control of sorghum stem borers. In: *International Workshop on Sorghum Stem Borers, Patancheru, India, 17–20 November 1987.* ICRISAT, Patancheru, pp. 81–87.

Vermeulen, J.B., Krause, M., Nel, A., Hollings, N. and Greyling, J. (1992) *A Guide to the Use of Pesticides and Fungicides in the Republic of South Africa.* Department of Agriculture, Government Printer, Pretoria.

Viggiani, G. (1972) Richerche sugli Hymenoptera Chalcidoidea, XXXV. Nuovi Tricogrammatidi d'Israele (*Paraittys latipennis,* n. gen., n. sp. e *Oligosita gerlingi,* n. sp.), contreperti su varie specie. *Bolletino del Laboratorio Entomologia Agraria Filippo Silvestri* 30, 152–157.

Vinson, J. (1942) Biological control of *Diatraea mauriciella* Wlk., in Mauritius. I. Investigations in Ceylon in 1939. *Bulletin of Entomological Research* 33, 39–65.

Waage, J.K. and Schulthess, F. (1989) Relevant research activities in support of sustainable pest management. In: Yaninek, H.S. and Herren, H.R. (eds) *Biological Control: A Sustainable Solution to Crop Pest Problems in Africa. Proceedings of the Inaugural Conference and Workshop, 5–9 December 1989.* IITA, Ibadan, Nigeria.

Wahl, D.B. (1993) Cladistics of the genera of Mesochorinae (Hymenoptera: Ichneumonidae). *Systematic Entomology* 18, 371–387.

Wahl, R.O. (1926) The maize-stalk borer (*Busseola fusca,* Fuller). *Farming in South Africa,* November.

Waiyaki, J.N. (1974) The ecology of *Eldana saccharina* Walker, and associated loss in cane yield at Arusha-Chini, Moshi, Tanzania. *Proceedings, International Society of Sugarcane Technologists* 15(1), 457–462.

Walker, A.K. (1994) Species of Microgastrinae (Hymenoptera: Braconidae) parasitizing lepidopterous cereal stem borers in Africa. *Bulletin of Entomological Research* 84, 421–434.

Walker, F. (1865) List of the specimens of Lepidopterous insects in the collection of the British Museum, supplement. *Catalogue of Ledidoptera Heterocera, 7th Series* 31 (Suppl.), 1–706.

Walker, P.T. (1957) The progress of stalk borer control in East Africa. Miscellaneous Report 191. T/C/3/1/383 Colonial Pesticides Research, Arusha, Tanganyika. In: *Proceedings of the IVth International Congress of Crop Protection, Hamburg, September 1957,* Vol. 1, pp. 761–763.

Walker, P.T. (1960) A survey of the use of maize stalk borer control methods in East Africa. *East African Agricultural Journal* 25, 165–168.

Walker, P.T. (1961) The distribution of granules by a commercial hand-operated distributor and their retention by maize plants. *Journal of Agricultural Engineering and Research* 6, 112–118.

Walker, P.T. (1972) The effect of physical properties of granules on their insecticidal performance. *Pesticides Science* 3, 805–809.

Walker, P.T. (1981) The relation between infestation by lepidopterous stem borers and yield of maize: methods and results. *EPPO Bulletin* 11, 101–106.

Walker, P.T. and Hodson, M.J. (1976) Developments in maize stem-borer control in East Africa, including the use of insecticide granules. *Annals of Applied Biology* 84, 111–114.

Walters, H.S. and Kfir, R. (1993) Development of *Nosema partelli* sp. n. (Protozoa: Microsporida: Nosematidae) in the stem borer, *Chilo partellus* (Swinhoe) (Lepidoptera: Pyralidae). *African Entomology* 1, 57–62.

Walters, M.C. (1974) Mechanisms of resistance to insects and application to maize stalk borer. *Technical Communication, Department of Agricultural and Technical Services, South Africa* 132, 83–87.

Walters, M.C. (1975) Evolution in tillage techniques and impact on entomological research, with special reference to the maize stalk borer, *Busseola fusca* (Fuller). In: *Proceedings of the First Congress of the Entomological Society of Southern Africa*. Entomological Society of Southern Africa, Pretoria, pp. 235–244.

Walters, M.C. and Drinkwater, T.W. (1975) Preliminary studies on the application of systemic insecticides to the soil for the control of the maize stalk borer, *Busseola fusca* (Fuller) (Lep.: Noctuidae). *Phytophylactica* 7, 121–123.

Warui, C.M. and Kuria, J.N. (1983) Population incidence and the control of maize stalk borers, *Chilo partellus* (Swinhoe), *C. orichalcociliellus* Strand and *Sesamia calamistis* Hmps. in Coast Province, Kenya. *Insect Science and its Application* 4(1–2), 11–18.

Waterston, J. (1915) Notes on African Chalcidoidea – III. *Bulletin of Entomological Research* 6, 231–247.

Watmough, R.H. and Kfir, R. (1995) Predation on pupae of *Helicoverpa armigera* Hbn. (Lep., Noctuidae) and its relation to stem borer numbers in summer grain crops. *Journal of Applied Entomology* 119, 679–688.

Way, M.J. and Kfir, R. (1997) The stem borer *Chilo partellus* (Swinhoe) in sugarcane in southern Africa. *African Entomology* 5, 170–172.

Wen, C.S. and Sun, C.X. (1989) Notes on Microsporidia (Protozoa: Microspora) from the silkworm *Bombyx mori* and the sugarcane stem borer *Chilo infuscatellus* Snellen in China. *Scientia Agricultura Sinica* 22, 15–19.

Wen, J.Z. and Sun, C.X. (1988) Two new species of *Nosema* (Microspora: Nosematidae). *Acta Zootaxonomica Sinica* 13, 105–111.

Wharton, R.A., Smith, J.W. Jr, Quicke, D.L.J. and Browning, H.W. (1989) Two new species of *Digonogastra* Viereck (Hymenoptera: Braconidae) parasitic on Neotropical pyralid borers (Lepidoptera) in maize, sorghum and sugarcane. *Bulletin of Entomological Research* 79, 401–410, figs 1–19.

Whitcomb, W.H. (1981) The use of predators in insect control. In: Pimentel, D. (ed.) *CRC Handbook of Pest Management in Agriculture*, Vol. II, pp. 105–123.

Wiedenmann, R.N. and Smith, J.W. Jr. (1993) Functional response of the parasite *Cotesia flavipes* (Hymenoptera: Braconidae) at low densities of the host *Diatraea saccharalis* (Lepidoptera: Pyralidae). *Environmental Entomology* 22, 848–858.

Williams, J.R. (1953) The larvae and pupae of some important Lepidoptera. *Bulletin of Entomological Research* 43, 691–701.

Williams, J.R. (1958) A list of Hymenoptera (excluding Formicidae) recorded from Mauritius. *Mauritius Institute Bulletin* 5(4), 108–128.

Williams, J.R. (1969) Cane pests. In: *Mauritius Sugar Industry Research Institute Annual Report 1969*, pp. 77–80.

Williams, J.R. (1978) *An Annotated Check List of the Invertebrates (Insects, Mites, Nematodes) of Sugar Cane in Mauritius.* Mauritius Sugar Industry Research Institute, Occasional Paper No. 30, 22 pp.

Williams, J.R. (1980) *Entomological Parasite–Host Records from Mauritius*, revised edn. Mauritius Sugar Industry Research Institute, Occasional Paper No. 32, 39 pp.

Williams, J.R. (1983) The sugarcane stem borer *(Chilo sacchariphagus)* in Mauritius. *Revue Agricole et Sucrière de l'Ile Maurice* 62(1), 5–23.

Williams, J.R. and Ma net, J.R. (1954) Insects and other invertebrates which feed upon sugarcane in Ma uritius and Réunion and their natural enemies. *Bulletin of the Department of Agriculture of Mauritius* 89, 1–19.

Woodley, N.E. (1994) A new species of *Lydella* (Diptera: Tachinidae) from Mexico with a discussion of the definition of the genus. *Bulletin of Entomological Research* 84, 131–136.

Wright, R.J. (1984) Evaluation of crop rotation for control of Colorado potato beetles (Coleoptera: Chrysomelidae) in commercial potato fields on Long Island. *Journal of Economic Entomology* 77, 1254–1259.

Yang, H., Guo, S.D., Li, J.X., Chen, X.J. and Fan, Y.L. (1991) Transgenic rice plants produced by direct uptake of the delta-endotoxin protein gene from *Bacillus thuringiensis* into rice protoplasts. In: *Rice Genetics II: Proceedings of the Second International Rice Genetics Symposium, 14–18 May 1990.* pp. 605–611.

Yen, D.F. and Ooi, W.H. (1967) An entomogenous fungus of rice stem borer, *Chilo suppressalis* Walker, in Taiwan. *Plant Protection Bulletin, Taiwan* 9, 15–20.

Yitaferu, K. and Gebre-Amlak, A. (1994) Phenology and survival of the maize stalk borer, *Busseola fusca* in eastern Ethiopia. *Insect Science and its Application* 15, 177–184.

Youdeowei, A. (1989) Major arthropod pests of food and industrial crops of Africa and their economic importance. In: Yaninek, J.S. and Herren, H.R. (eds) *Biological Control: a Sustainable Solution to Crop Pest Problems in Africa.* IITA, Ibadan, Nigeria, pp. 31–50.

Youdeowei, A and Service, M.W. (1983) *Pest and Vector Management in the Tropics.* Longman, London.

Youm, O. (1990) Evaluation of natural enemies associated with the millet stem borer, *Haimbachia ignefusalis* (Hampson) (Lepidoptera: Pyralidae) in Niger. PhD dissertation, Texas A&M University, College Station, USA, 145 pp.

Youm, O. and Beevor, P.S. (1995) Field evaluation of pheromone-baited traps for *Coniesta ignefusalis* (Lepidoptera: Pyralidae) in Niger. *Journal of Economic Entomology* 88, 65–69.

Youm, O. and Gilstrap, F.E. (1994) Habitat site selection, crop damage and oviposition preference by *Coniesta* (= *Haimbachia*) *ignefusalis* (Lepidoptera: Pyralidae). *International Journal of Pest Management* 40, 231–236.

Youm, O., Gilstrap, F.E. and Teetes, G.L. (1990) Pesticides in traditional farming systems in West Africa. *Journal of Agricultural Economics* 7, 171–181.

Youm, O., Beevor, P.S. and Hall, D.R. (1993a) Trap design studies with the pheromone of *Coniesta ignefusalis* (Hampson) (Lepidoptera: Pyralidae) in sub-Saharan Africa. *IOBC WPRS Bulletin* 16, 58–63.

Youm, O., Mamalo, A.K. and Nwanze, K.F. (1993b) Bio-ecology and integrated management of the millet stem borer (Lepidoptera: Pyralidae): a review and highlights of current research at the ICRISAT Sahelian Center. In: Youm, O. and Kumar, K.A. (eds) *Proceedings of the Regional Pearl Millet Workshop, 19–21 September 1990.* ICRISAT Sahelian Center, Niamey, Niger, pp. 55–63.

Zelazny, B., Chiarappa, L. and Kenmore, P. (1985) Integrated pest control in developing countries. *FAO Plant Protection Bulletin* 33, 147–158.

Zettel, H. (1990) Zwei neue *Phanerotoma*-Arten aus Südafrica (Hymenoptera, Braconidae: Cheloninae). *Linzer Biologische Beitrage* 22, 335–340.

Zwart, K.W.R. (1973) New Ichneumonidae, parasitic upon the rice borer *Rupela albinella* (Cr.) (Lep., Pyralidae) in Surinam, with a key to the species of *Strabotes* (Hym., Ichneumonidae). *Entomologische Berichten, Amsterdam* 33, 231–240.

Glossary

abscissa (pl. **abscissae**): a point on a wing vein that is marked by the intersection of other veins.

aedeagus: the intromittent organ of male insects, located below the scaphium and enclosed in a sheath; typically it is a sclerotized tube; often incorrectly referred to as the penis, which is a term restricted to the muscular intromittent organ of animals other than arthropods.

anellus: part of the aedeagus; in Lepidoptera, the area between the two arms of the vinculum is closed by a membrane that supports the aedeagus. This part of the membrane is the anellus and may bear sclerotizations with specific value (Fig. 3).

anteroventral: toward the front and lower surface.

antrum: a sclerotized, ring-shaped structure delimiting the first part of the copulatory duct, the ductus burae; a part of the female genitalia, it is one of the sclerotized structures round the sinus vaginalis and the ostium bursae, which together form the sterigma (Fig. 5).

anus: the opening at the end of the digestive tract, through which body wastes are passed.

apodous: lacking feet, as in many larvae.

apomorphic: a term used in phylogenetics indicating that a body structure is considered to be derived from a more primitive state into a more advanced condition.

apophyses: any rod-like, tubular or otherwise elongate processes of the body wall of an insect, either internal or external.

apophyses anteriores: in female genitalia of Lepidoptera, these are rod-like sclerotizations on the eighth abdominal segment (Fig. 5).

apophyses posteriores: in female genitalia, the rod-like sclerotizations at the base of the papillae anales are the apophyses posteriores (Fig. 5).

areolet: in Hymenoptera, the small cell in the centre of the forewing, the first radial-sector cell (A in Fig. 558).

arista: on the antenna of Diptera, a specialized bristle near the base of the apical antennal segment, usually situated dorsally. The arista varies from minutely pubescent to plumose or pectinate (Fig. 623).

arrhenotokous: in Hymenoptera, males are haploid while females are diploid. In those species displaying arrhenotokous parthenogenesis, which are the most common, unfertilized females produce males only. See also **thelytokous**, **deuterotokous**.

asetose: without obvious setae.

axillae: two small, subtriangular sclerites at the lateral basal angles of the mesoscutum (q.v.) and adjoining the scutellum (q.v.) in many Hymenoptera.

biordinal: in the proleg of caterpillars (Lepidoptera), having the crochets (curved spines, usually arranged in circles) of two different, alternating lengths.

bipectinate: feather-like, often used to describe form of antennae (Figs 41–43).
brachypterous: with very abbreviated wings.
bursa copulatrix: part of the female genitalia, this is the copulatory pouch. Two major parts, the ductus bursae and corpus bursae, together may be referred to as the bursa copulatrix (Fig. 5).
carina (pl. **carinae**): a raised or elevated ridge or keel.
cells: any fully or partially enclosed area of insect wing bounded by veins.
chaetosemata: in Lepidoptera, a small circular cluster of setae, often on a raised prominence and usually immediately behind the ocelli, although variations do occur.
chaetotaxy: the study of the arrangement and nomenclature of the setae or bristles on the exoskeleton of insects.
clava: in Hymenoptera, the club or enlarged apical flagellomeres of the antenna (Fig. 93).
clypeus: the part of the head situated antero-ventrally but above the mouth and below the frons, often separated from the latter by the fronto-clypeal suture (cl in Fig. 554).
conjunctivum: in Lepidoptera, the hearing organs consist of a thin transparent membrane (tympanum) delimited by another opaque membrane, the conjunctivum (Fig. 80).
coriaceous: with a leather-like surface or sculpture.
cornutus (pl. **cornuti**): the male aedeagus consists of an invaginated membrane, the vesica. The vesica may carry horn-like or spiny sclerotizations, called cornuti. These often have specific value (Fig. 3).
corona: crown-like process; in the genitalia of male Lepidoptera, a row of spines on the valve apex.
corpus bursae: a membranous pouch often carrying sclerotizations, called signa; it is the terminal part of the female copulatory duct, preceded by the ductus bursae (Fig. 5).
costa: as a general entomological term, it refers to any raised ridge with a rounded crest, e.g. the thickened leading edge of a wing. In Lepidoptera, it also refers to the dorsal part of the two lateral valves of the male genitalia.
countertympanal hood: a flap at the anterior of the abdomen associated with the tympanum, or ear, of noctuids.
cremaster: apex of the last abdominal segment; in lepidopteran pupae, it is the terminal spine or hook with which they suspend themselves.
crochets: curved spines or hooks on the prolegs of lepidopteran larvae and on the cremaster (q.v.) of pupae (Fig. 92).
cubitus: the fifth longitudinal vein of the insect wing, immediately posterior to the media, extending from the base and usually branching before reaching the outer margin.
deuterotokous: in Hymenoptera, males are haploid while females are diploid. In those species displaying deuterotokous parthenogenesis, unfertilized females produce females almost exclusively, but also produce the occasional male. See also **arrhenotokous**, **thelytokous**.
diapause: the state of arrested development and reduced metabolic rate (suspended animation) which many insects go through during their development, particularly as larvae.
digitus (pl. **digiti**): small, finger-like process diverging from the paramere in Hymenoptera (Fig. 103).
discal cell: in Lepidoptera, the usually enlarged cell extending from the basal part of the wing toward the centre.
dorsellum: a medial expansion of the metanotum.
ductus ejaculatorius: the terminal part of the invaginated membrane which forms the male aedeagus and through which the seminal fluid is discharged.
ductus seminalis: the tube connecting the bursa copulatrix with the oviduct, usually originating just beneath the ring-shaped structure, the antrum (Fig. 3).
ectoparasite/ectoparasitoid: an external parasite. See also **parasitoid**.
endoparasite/endoparasitoid: an internal parasite. See also **parasitoid**.
epicranium: in larval Lepidoptera, the upper part of the head capsule, above the frons.

episternum: the anterior and larger lateral thoracic sclerite between the sternum and the notum.

epistome: in Diptera, the part of the face just above the mouthparts, the oral margin (Figs 623, 624).

eruciform: of caterpillar-like form or appearance.

filiform: thread-like, often used to describe antennae.

flagellomeres: in the antennae of Hymenoptera, the divisions or segments composing the flagellum.

flagellum: in antennae, the part beyond the pedicel (q.v.).

fovea (pl. **foveae**): a pit or depression with well-marked sides.

frenate: possession a frenulum.

frenulum: a spine or group of bristles arising at the base of the hindwings in Lepidoptera and which unites the hind and fore wings in flight. In the male, a single bristle arises from the base of the anterior edge of the hindwing and engages in a tongue-like structure, the retinaculum, which arises from the subcostal vein of the fore wing; in the female, the frenulum consists of several bristles and engages in a tuft of scales on the forewing directed anteriorly from the cubital vein, where it forms the posterior part of the discal cell.

frons: an unpaired sclerite on the head lying between the epicranial sutures and bearing the median ocellus (q.v.); the upper, anterior part of the head capsule between the epicranium (q.v.) and the clypeus (q.v.).

funicle: that part of the antenna lying between the club (clava) at the apex and the ring joints (annelli) following the pedicel (Fig. 93).

gaster: in Hymenoptera (Apocrita), the rounded part of the abdomen (= metasoma, the last seven or eight segments) posterior to the mesosoma (thorax + propodeum).

gena (pl. **genae**): in Diptera, the cheek, the part of the head on each side, below and behind the compound eyes, between the frontal and occipital sulci.

gnathos: part of the genitalia of male Lepidoptera, a pair of structures which are found beneath (and supporting) the tuba analis and attached to the tegumen.

halteres: in Diptera, the hindwings are modified into a characteristic pair of knobbed balancing organs, the halteres, which are used to maintain stability in flight.

hamuli: in Hymenoptera, minute hooks on the leading edge of the hindwings, used to unite them, while in flight, to the forewings.

harpe: in the male genitalia of Lepidoptera, one of a pair of clasping organs derived as appendages of the ninth abdominal sternite, articulating with the caudal margin of the vinculum (q.v.), also called valves (q.v.) (Fig. 3).

holometabolous: with a life cycle involving a complete transformation from egg, through larva, pupa and adult, with each stage separate.

hyperparasite/hyperparasitoid (= **superparasitoid**): a parasitoid of another parasitoid.

hypoclypeal depression: in Hymenoptera, a distinct curved depression at the intersection of the clypeus and labrum; the bottom of the hyopoclypeal depression consists of the concave labrum and a depressed part of the clypeus.

hypognathous (= **orthognathous**): the head is arranged vertically with the mouthparts directed downwards, ventrally (ventrad).

hypopleuron: in Diptera, the lower part of the mesepimeron (q.v.), a sclerite on the thorax located just above the hind coxae (Fig. 629).

hypopygium: the last visible sternum of the abdomen; the lower plate of the anal opening.

idiobiont: the larva develops inside the host, but the activity of the host is permanently impaired by injection of venom etc. to produce paralysis.

jugal lobe: the backward-pointing lobe at the base of the hindwing in some Hymenoptera (Fig. 101).

juxta: part of the male genitalia of Lepidoptera, it is a sclerotized plate, forming a ventral extension of the anellus and sited below the aedeagus, which it supports (Fig. 3).

katepisternal setae: in Diptera, setae found on the katepisternum, the lower part of the episternum (q.v.) on the thorax (Fig. 602).

katepisternum: the lower part of the episternum (Fig. 602).

koinobiont: the larva develops inside the host, but without impairing its mobility and the host remains active.

lamella vaginalis: in the female genitalia of Lepidoptera, thin, plate- or leaf-like lobes at the apex, often minutely spined.

laminae volsellares: in Hymenoptera, a median pair of genital appendages within the aedeagus (Fig. 103).

meron: an area at the base of the coxae; the posterior part of the meso- and metathoracic coxae in many insects.

mesal: at or near the midline of the body.

mesepimeron: the epimeron (the area of a thoracic pleuron posterior to the pleural suture) of the mesothorax (Fig. 95).

mesopleuron: in Hymenoptera, the lateral sclerite of the mesothorax, below the wing (Fig. 95).

mesoscutellum: also called the scutellum, it lies posterior to the mesoscutum on the dorsal surface of the thorax; together these two parts form the mesonotum.

mesoscutum: the anterior portion of the mesonotum, divided from the posterior scutellum (Fig. 95).

mesoseries: referring to the arrangement of crochets (q.v.) on the prolegs of caterpillars, arranged in a row near the midline of the proleg.

mesosoma: in the Apocrita (Hymenoptera), in order to maintain the distinction between the true thorax and the apparent thorax (i.e. including the propodeum), the thorax + propodeum together are referred to as the mesosoma (Fig. 94).

metanotum: upper surface of the third thoracic ring, i.e. behind the scutellum in Hymenoptera (Fig. 94).

metasoma: in Hymenoptera (Apocrita), the rounded part of the abdomen (the last seven or eight segments) posterior to the mesosoma (thorax + propodeum); also called the gaster.

metathorax: the third or posterior ring or segment of the thorax, bearing the hindlegs and second pair of wings.

microchaetae: small setae or bristles (opposite of macrochaetae).

moniliform: beaded like a necklace, often used to describe antennae.

mystax: in Diptera, a patch of hairs or bristles above the mouth, above the vibrissae.

notauli: in Hymenoptera, on the dorsal surface of the pronotum there are often a pair of posteriorly convergent impressed lines, dividing the mesoscutum into two side-lobes and a midlobe (Figs 125, 131).

ocelli (sing. **ocellus**): simple eyes of insects and other arthropods.

orbicular: round and flat.

orthognathous (= **hypognathous**): the head is arranged vertically, with the mouthparts directed downwards, ventrally (ventrad).

ostium bursae: the genital aperture or opening of the bursa copulatrix in Lepidoptera; it is situated in the intersegmental membrane of the seventh and eighth segments of the abdomen.

ovipositor: the tubular egg-laying apparatus, forming the external genitalia of the female; often concealed but may project beyond the end of the body.

papilla: a small, nipple-like mound.

papillae anales: in the female, the caudal part of the abdomen ends in a pair of papillae anales; they delimit the anus and the ovipositor (Fig. 5).

parafacial: in Diptera, the part of the face between the facial ridges (the ridges which border the antennal grooves) and the eyes.
parameres: two lateral processes or lobes on the ninth sternite, close to the base of the aedeagus.
paraphyletic: a term used in phylogenetics indicating that a group of taxa is considered to have evolved from more than one ancestral taxon.
parapsidal lines: in Hymenoptera, on the mesoscutum (lateral to the notauli) may be found a second pair of impressed marks, the parapsidal lines.
parasite/parasitoid: an organism that lives on or at the expense of another; parasitoids are specialized parasites which feed on or in another living animal for some time, while consuming its tissues, before eventually killing the host organism.
paraspiracular carina: in Hymenoptera, a carina which is more or less in the shape of an inverted Y just medial to the spiracle on the propodeum.
parastigma: see **pterostigma**.
pecten: a comb-like structure, often of stout spines or setae, on various parts of the body.
pedicel: in Hymenoptera, the second segment of the geniculate antenna, between the scape and the funicle; also used for the stem of the abdomen, between the thorax and the gaster (ants) (Fig. 93).
penellipse: on the proleg of caterpillars, the shape formed when a part less than half of a uniserial circle of crochets is absent.
penis: more correctly called the aedeagus; the male genital or intromittent organ.
petiole: in Hymenoptera, the first segment of the slender pedicel between the mesosoma and metasoma.
pinaculum (pl. **pinacula**): in caterpillars, a swollen, seta-bearing papilla forming a flattened plate.
plesiomorphic: a term used in phylogenetics indicating that a body structure is considered to reflect a more primitive or ancestral state.
plica: a fold or wrinkle.
plical lobe: the basal fold of the wing, forming a lobe on the inner, hind edge demarcated by a crease or fold.
porrected: projected forwards horizontally from the head.
prepectus: in Hymenoptera, a triangular or subrectangular sclerite along the anteroventral margin of the mesepisternum, below the first thoracic spiracle, set off by a furrow or suture (Fig. 125).
prognathous: the head is horizontal with the jaws directed forward.
propleuron: the lateral portion, or pleuron, of the prothorax (Fig. 96).
propodeum: in Hymenoptera (Apocrita), the first abdominal segment has become fused to the thorax and forms the posterior portion of the mesosoma (Fig. 94).
prostigma (= **stigma**): in the forewing of Hymenoptera, a small, dark area along the costal margin.
prothoracic shield: also called the cervical shield, a chitinous plate on the prothorax of caterpillars immediately behind the head.
pterostigma: an opaque, thickened spot on the costal margin of the wing, near the tip.
reniform: kidney-shaped.
retinaculum: in Lepidoptera, the loop on the hind edge of the forewing arising from the subcostal vein into which the bristles of the frenulum are fitted to lock the wings together. In Hymenoptera, chitinous, movable scales serving to move the sting or to prevent its being extended too far.
sacculus: in Lepidoptera, the two lateral valves of the male aedeagus may have different forms or processes, such as spines, hooks or bristles. The dorsal part is called the costa, the ventral part, near the base, the sacculus (Fig. 3).

saccus: in the male genitalia of Lepidoptera where the two arms of the vinculum (a U- or V-shaped sclerite derived from the ninth abdominal sternite) fuse, the median part at the base of the vinculum is called the saccus (Fig. 4).
sagitate: in the shape of an arrowhead.
scape (= **scapus**): the long basal segment of an antenna, usually applied to the three basal joints in Hymenoptera (Fig. 93).
scaphium: part of the tuba analis is sometimes sclerified: the dorsal part is called the scaphium, the ventral part the subscaphium.
scapus: see **scape**.
sclerotized: when the body wall or cuticle becomes hardened.
scrobe: groove into which an appendage folds away, e.g. the antennae frequently fold into scrobes in many insect groups; in Hymenoptera they are usually circular depressions on the frons, in which the scapes of the antennae revolve.
scutellum (or **mesoscutellum**): it lies posterior to the mesoscutum on the dorsal surface of the thorax; together these two parts form the mesonotum (Fig. 95).
setae (sing. **seta**): often referred to as hairs, bristles or macrotrichia, these are hollow, slender, hair-like structures produced as an extension of the dermal layer.
setose: with a covering of setae or hairs.
setulae: small, stiff bristles or setae.
signum (pl. **signa**): in female Lepidoptera, the first part of the copulatory duct, the ductus bursae, ends in the corpus bursae, a membranous pouch often carrying sclerotizations called signa.
sinus vaginalis: in female Lepidoptera, the area around the genital aperture or ostium bursae is the sinus vaginalis.
sterigma: in the genitalia of female Lepidoptera, in the area around the sinus vaginalis and the ostium bursae, occur variously sclerotized structures of great taxonomic importance. For the whole complex of these, the name sterigma is used as a general term.
sternum: a sclerite on the ventral side of a body segment; in specific and general usage, the underside of the insect thorax, between the coxal cavities.
stigma (pl. **stigmata**): a pigmented wing spot.
subcostal: the usually unbranched, longitudinal vein between the costa and the radius.
subscaphium: in the aedeagus of Lepidoptera, part of the tuba analis (q.v.) may sometimes be sclerotized: the dorsal part is called scaphium, the ventral part subscaphium.
subscutellum (**postscutellum**): in Diptera, the ventral, often swollen lobe below the hind margin of the scutellum.
suranal plate: in Lepidoptera, a tergite lying above the anus of a caterpillar.
tegula: a small scale-like sclerite overlying the extreme base of the costa of the front wing.
tegumen: in the aedeagus of male Lepidoptera, the two lateral valves articulate on a hood-like or ring-shaped structure, the dorsal part of which is the tegumen.
thelytokous: in Hymenoptera, males are haploid while females are diploid. In those species displaying thelytokous parthenogenesis, unfertilized females produce only females. See also **arrhenotokous**, **deuterotokous**.
triordinal: on the prolegs of caterpillars, when the crochets (q.v.) are of three different lengths.
tuba analis: in the male genitalia of Lepidoptera, lying underneath the uncus, is the membranous tuba analis, which terminates in the anus.
tympanum: an opaque, drumskin-line membrane. In Lepidoptera, the paired tympanal organs are usually found ventrally at the base of the abdomen or on each side of the metathorax. It is usually associated with a flap at the anterior of the abdomen, the countertympanal hood.
uncus: in the male genitalia of Lepidoptera, Diptera and other groups, the uncus is a downward-pointing curved hook, forming the most posterior, middorsal, unpaired

structure of the aedeagus. The uncus originates on and articulates with the median part of the tegumen (q.v.) (Figs 3, 4).

uniordinal: on the prolegs of caterpillars, when the crochets (q.v.) are of a uniform length.

valves: characteristic for the male genitalia of Lepidoptera are the two lateral valves. The valves may have different forms or processes, such as spines, hooks or bristles (Fig. 3).

valvula: in Hymenoptera, the valve is the expanded, plate-like galea of the maxilla; the vagina in Diptera.

venation: the pattern of veins in the wings of Lepidoptera, Hymenoptera and other insects, often of great taxonomic value.

vermiform: worm-like, may be applied to larvae.

vesica: the terminal part of the aedeagus consists of an invaginated membrane, the vesica (Fig. 3).

vibrissae: in Diptera, curved bristles situated between the mystax (q.v.) and the antennae (Fig. 625).

vinculum: in the aedeagus of male Lepidoptera, the two lateral valves articulate on a hood-like or ring-shaped structure, the ventral part of which is the vinculum; it is derived from the ninth abdominal sternite (Fig. 4).

Figures

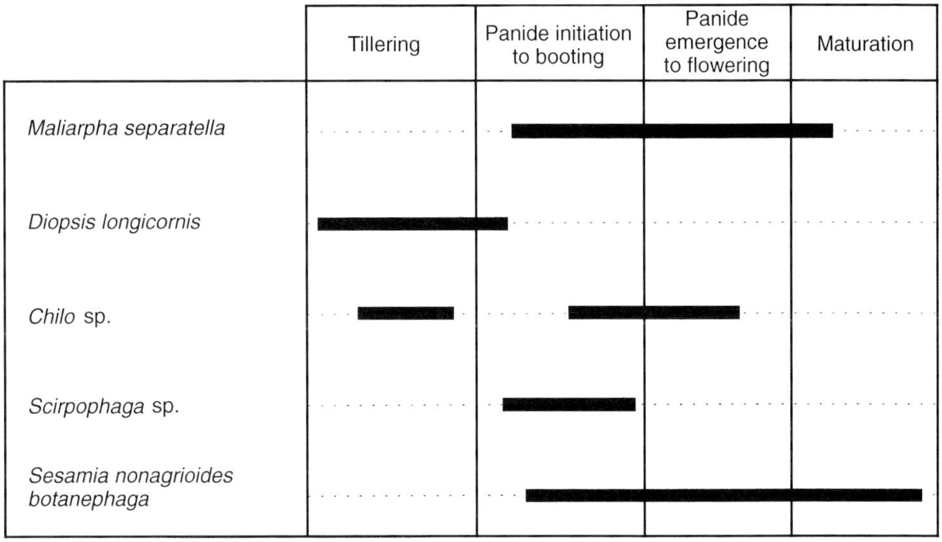

Fig. 1. Growth stages of irrigated rice attacked by five stem borer species in the Ivory Coast (from Pollet, 1977).

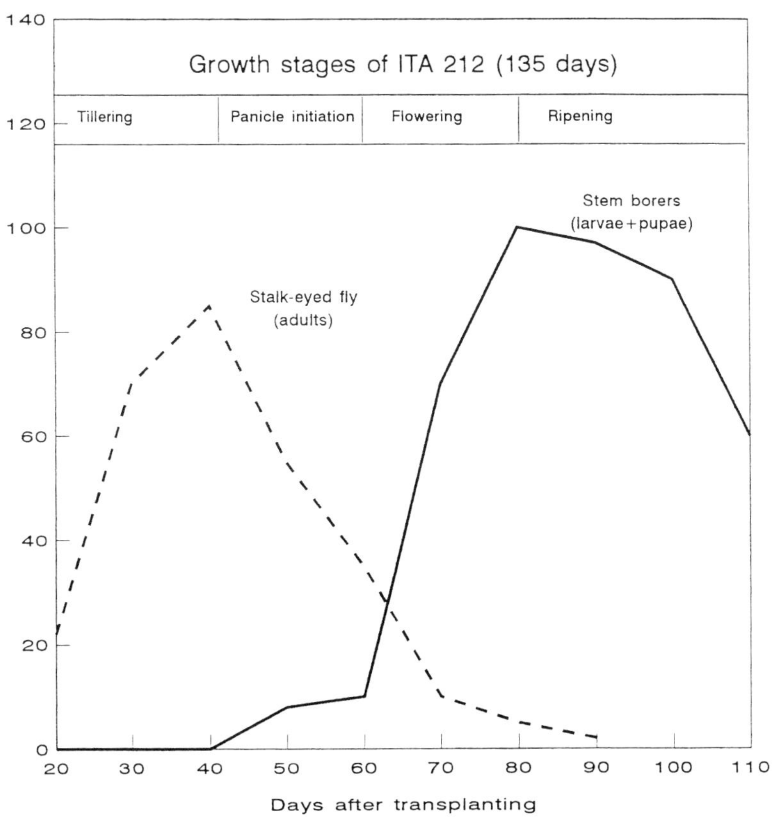

Fig. 2. Relationship between irrigated rice-crop phenology and field populations of stalk-eyed fly (*Diopsis* spp.) adults and lepidopterous stem-borer larvae and pupae at IITA, Abadan, Nigeria (from Alam, 1988).

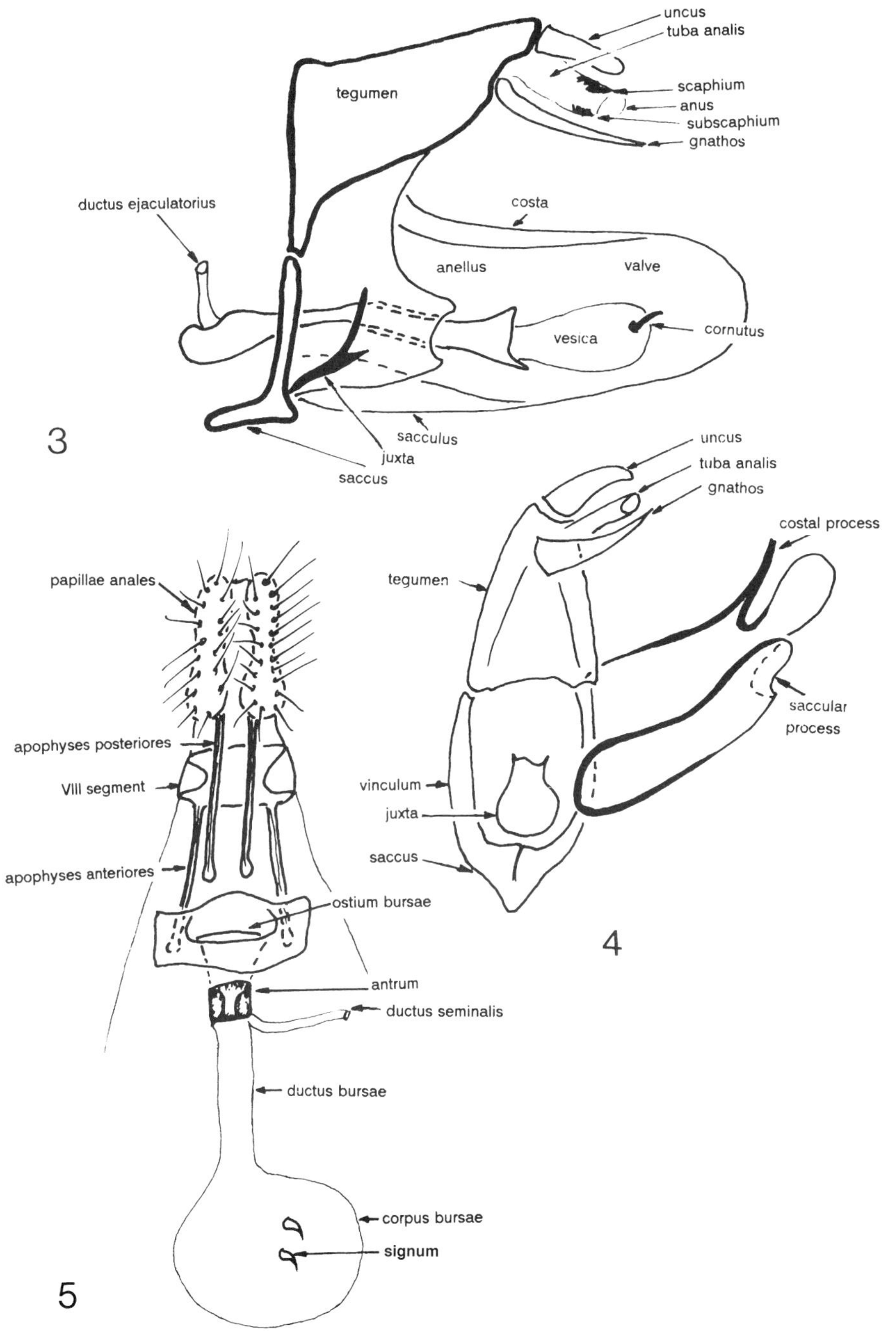

Figs 3–5. Genitalia of moths. (3) Male genitalia (lateral view). (4) Male genitalia. (5) Female genitalia.

Figs 6–9. (6) Frenulum (male). (7) Frenulum (female). (8) Tympanal organ Noctuidae, in metathorax. (9) Tympanal organ Pyraloidea, on first sternite of abdomen.

Figs 10–11. Venation of hind wing. (10) Noctuidae. (11) Pyraloidea. (After Lewvanich, 1981.)

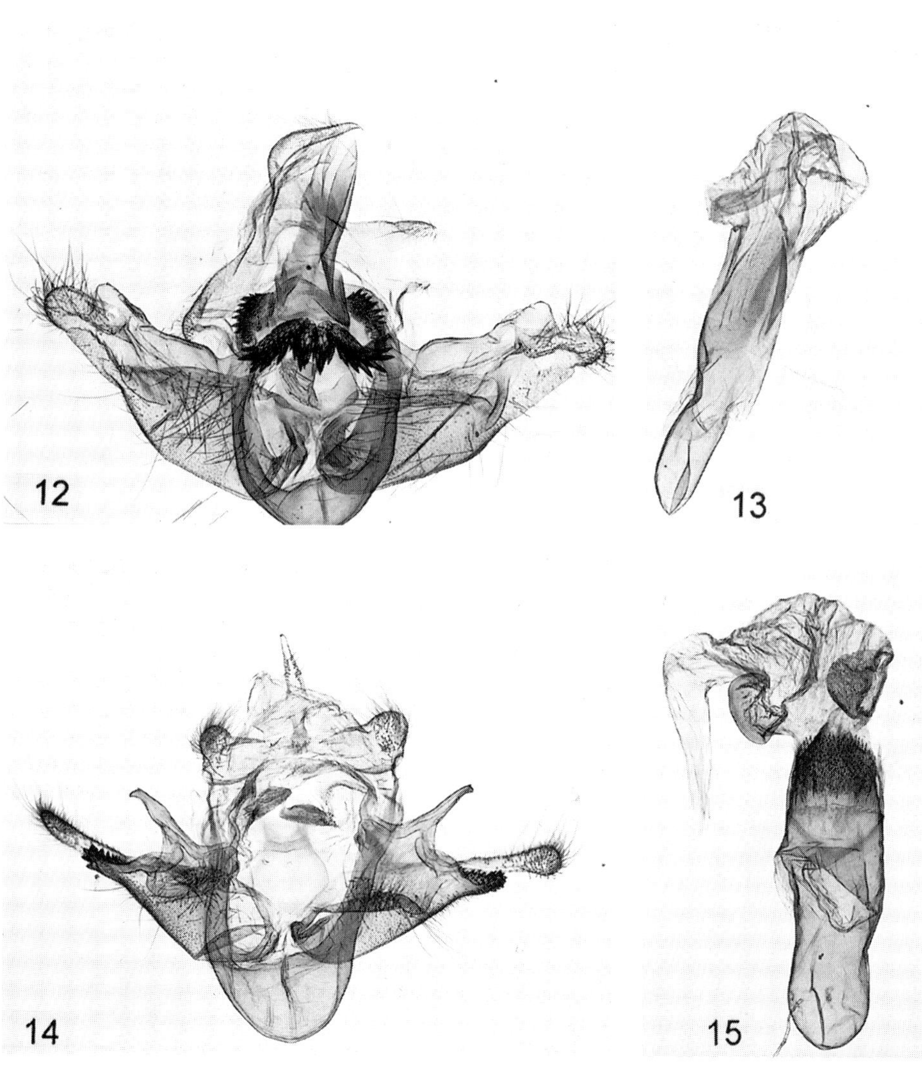

Figs 12–15. Male genitalia. (12) *Sesamia cretica.* (13) Aedeagus, same species. (14) *Sesamia calamistis.* (15) Aedeagus, same species.

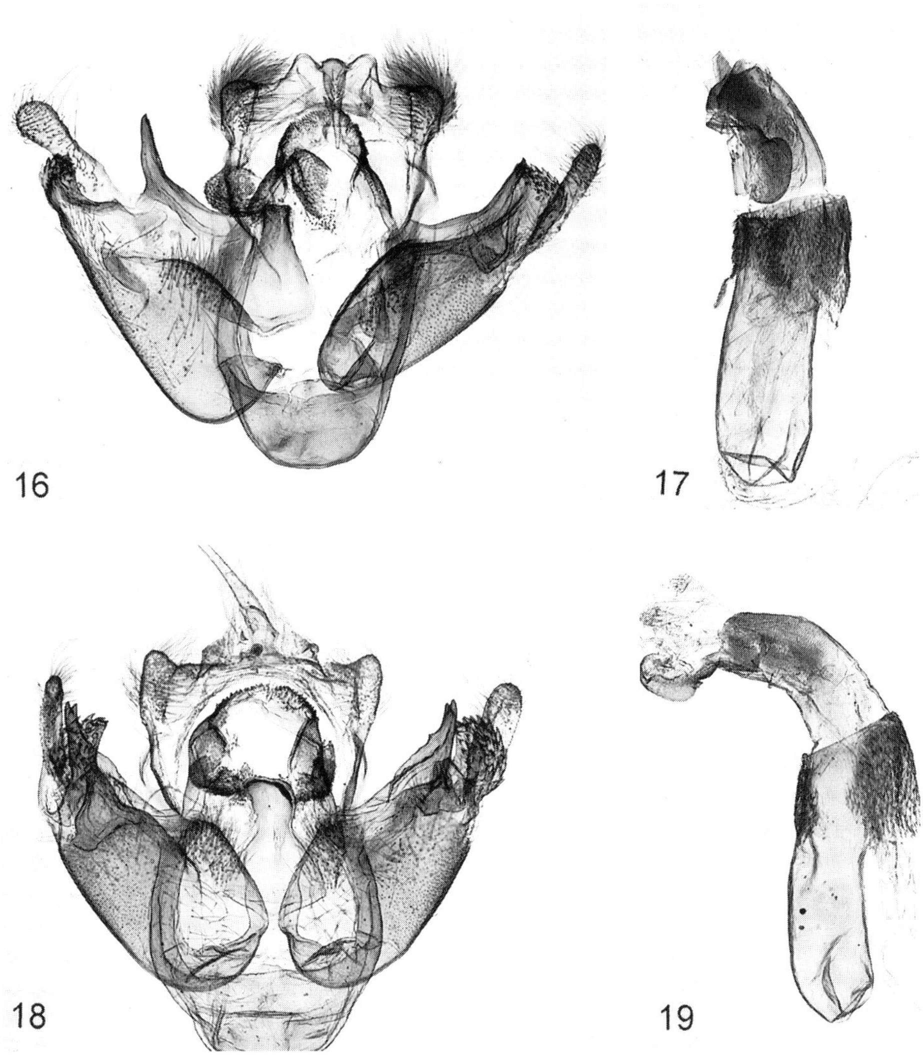

Figs 16–19. Male genitalia. (16) *Sesamia penniseti*. (17) Aedeagus, same species. (18) *Sesamia poephaga*. (19) Aedeagus, same species.

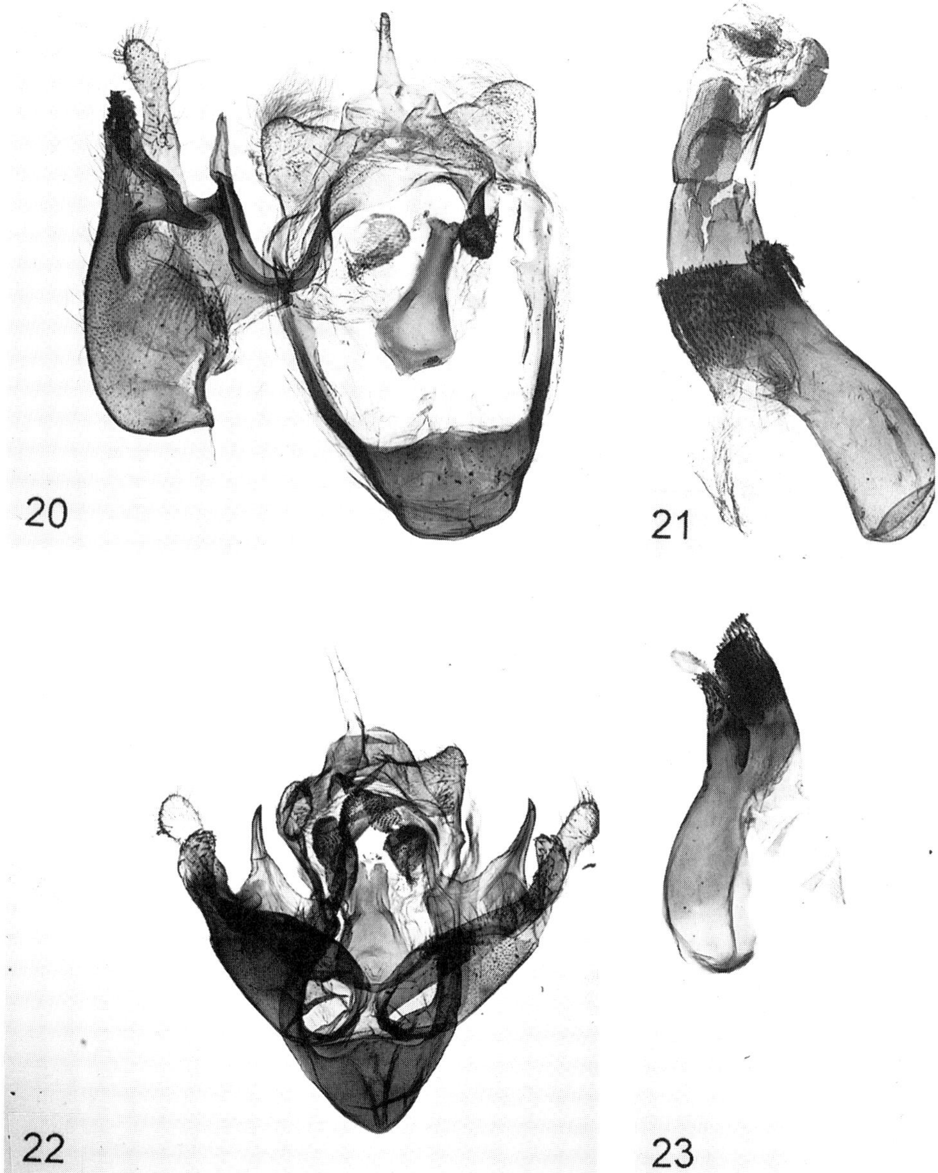
20 21 22 23

Figs 20–23. Male genitalia. (20) *Sesamia poebora*. (21) Aedeagus, same species. (22) *Sesamia nonagrioides botanephaga*. (23) Aedeagus, same species.

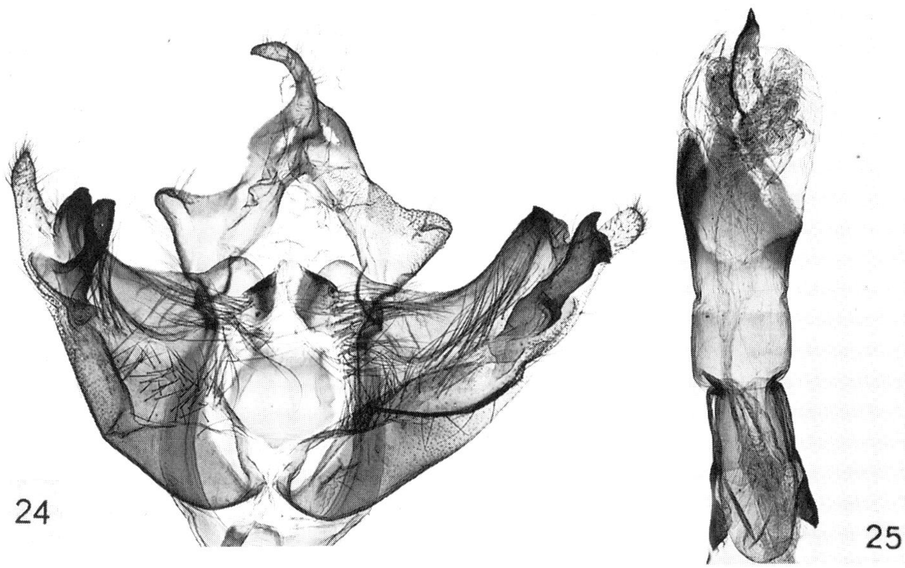

Figs 24, 25. Male genitalia. (24) *Busseola fusca*. (25) Aedeagus, same species.

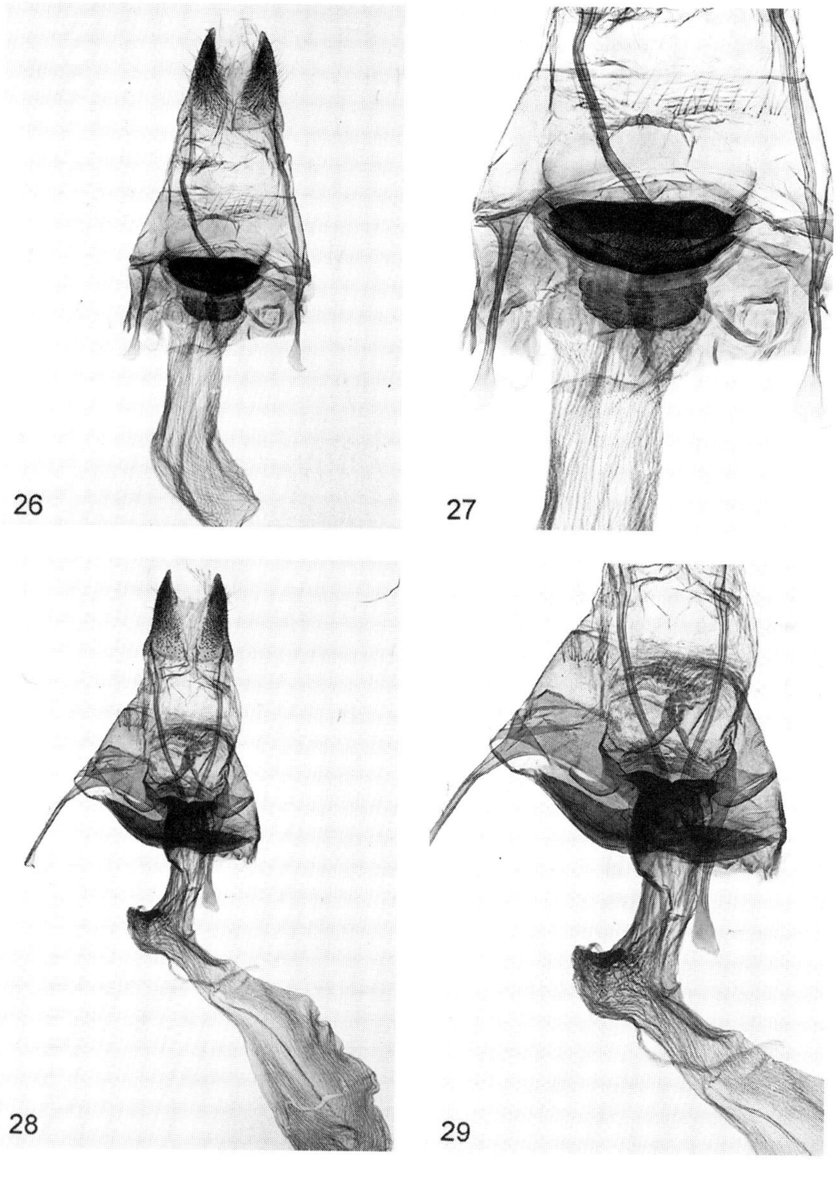

Figs 26–29. Female genitalia. (26, 27) *Sesamia cretica*. (28, 29) *Sesamia calamistis*.

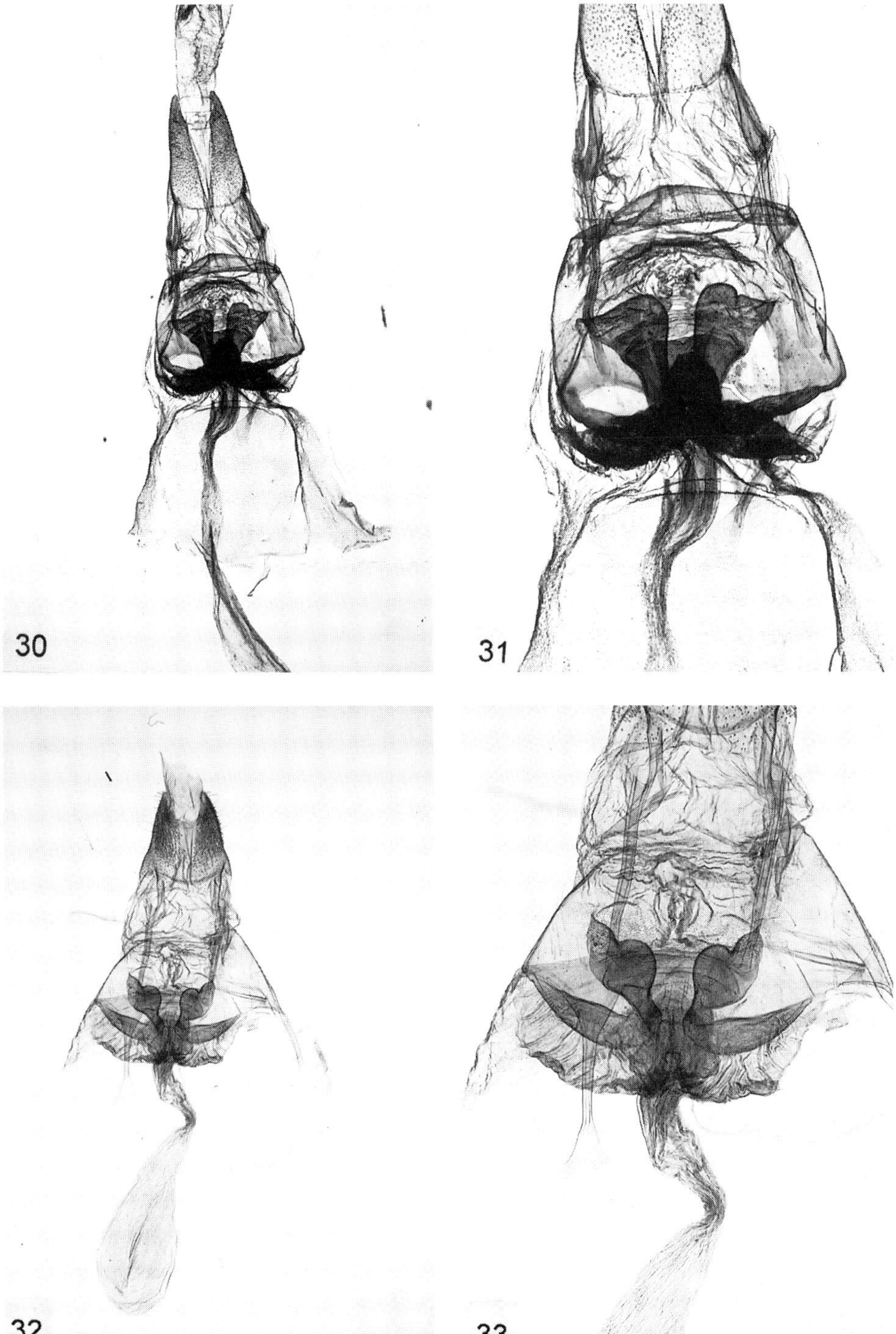

Figs 30–33. Female genitalia. (30, 31) *Sesamia penniseti*. (32, 33) *Sesamia poephaga*.

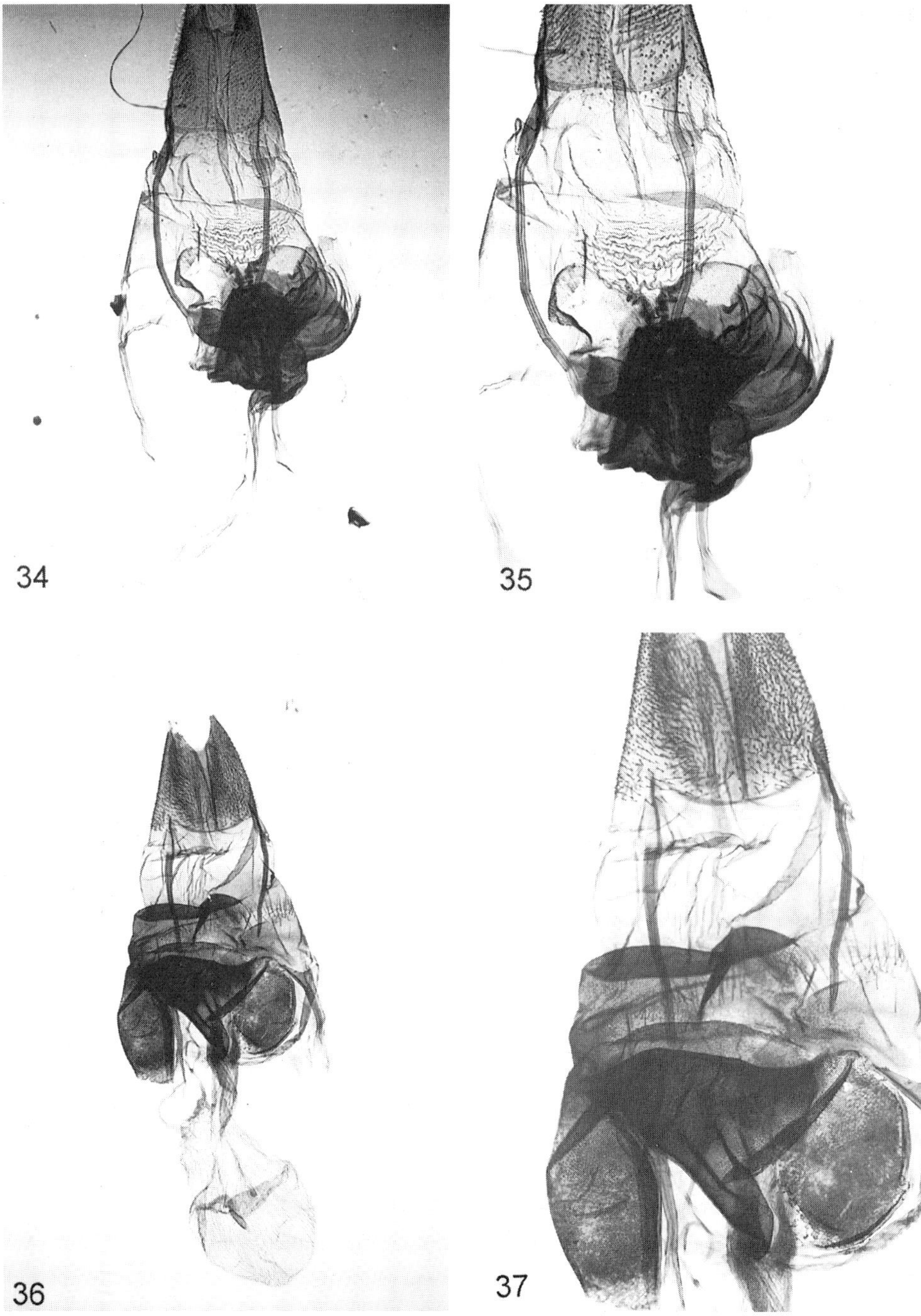

Figs 34–37. Female genitalia. (34, 35) *Sesamia nonagrioides botanephaga*. (36, 37) *Busseola fusca*.

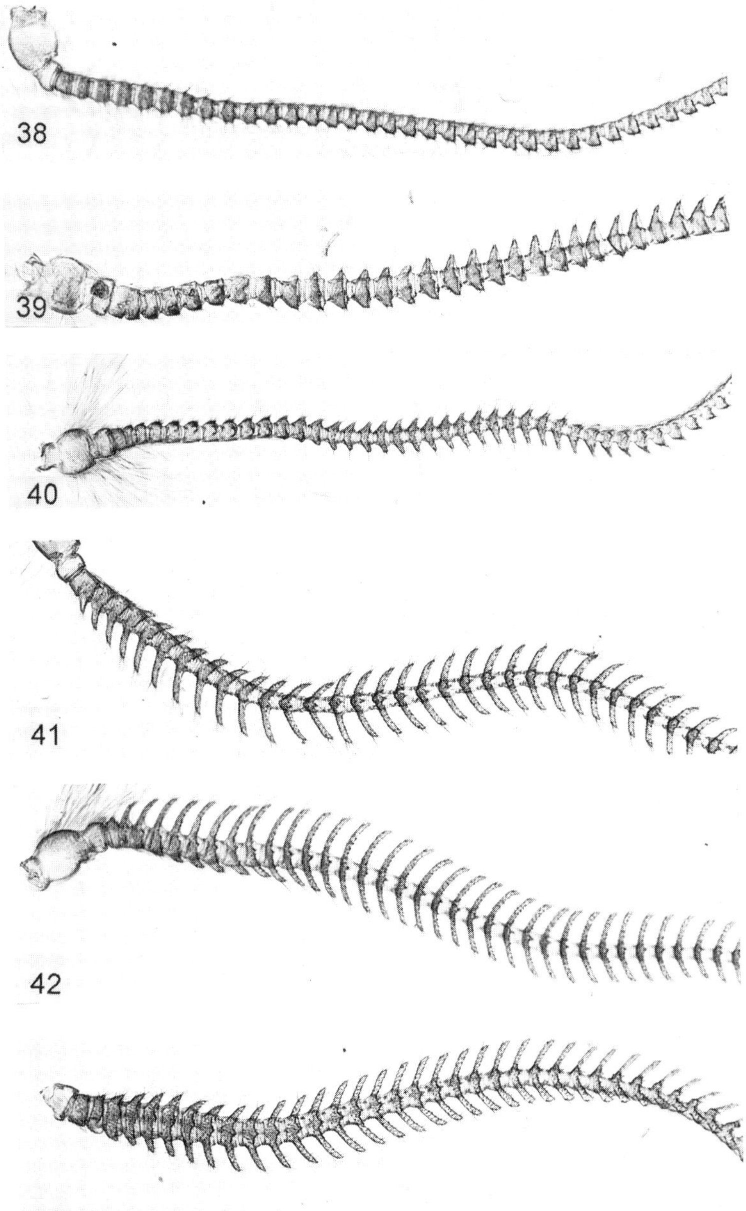

Figs 38–43. Male antennae. (38) *Sesamia cretica.* (39) *Sesamia poephaga.* (40) *Sesamia penniseti.* (41) *Sesamia nonagrioides botanephaga.* (42) *Sesamia calamistis.* (43) *Busseola fusca.*

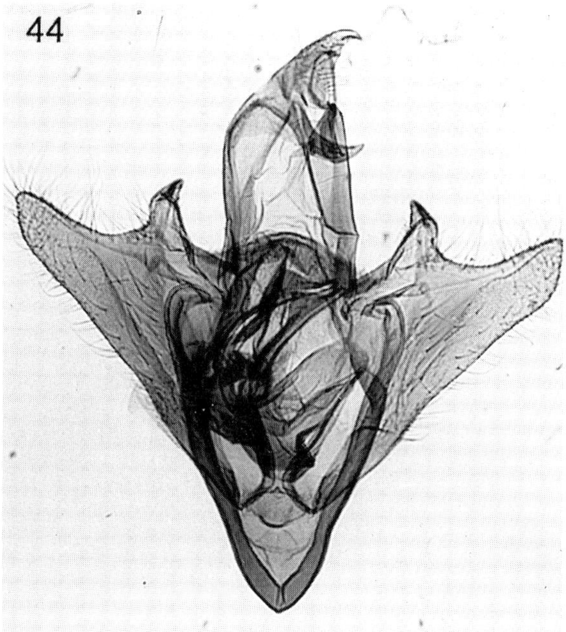

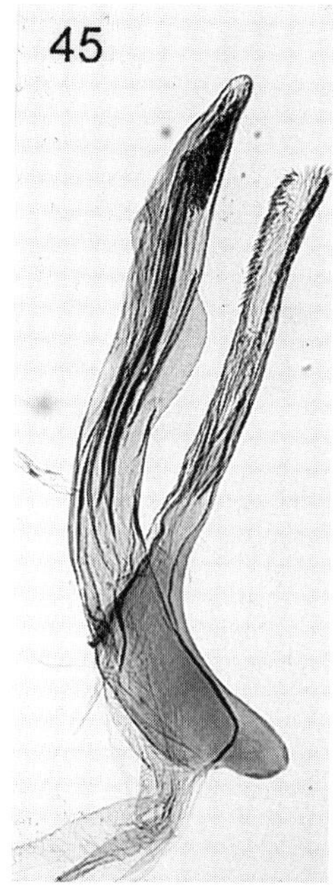

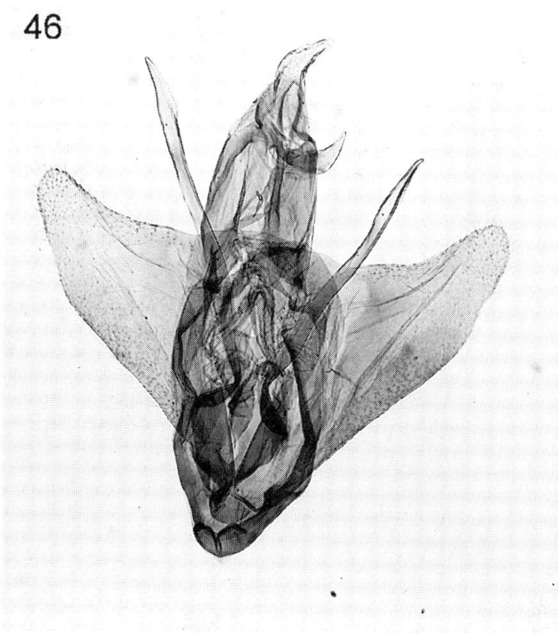

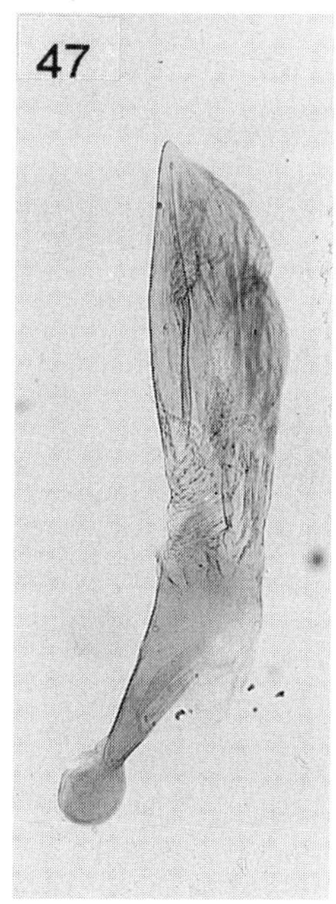

Figs 44, 45. Male genitalia *Chilo partellus*.
Figs 46, 47. Male genitalia *Chilo zacconius*.

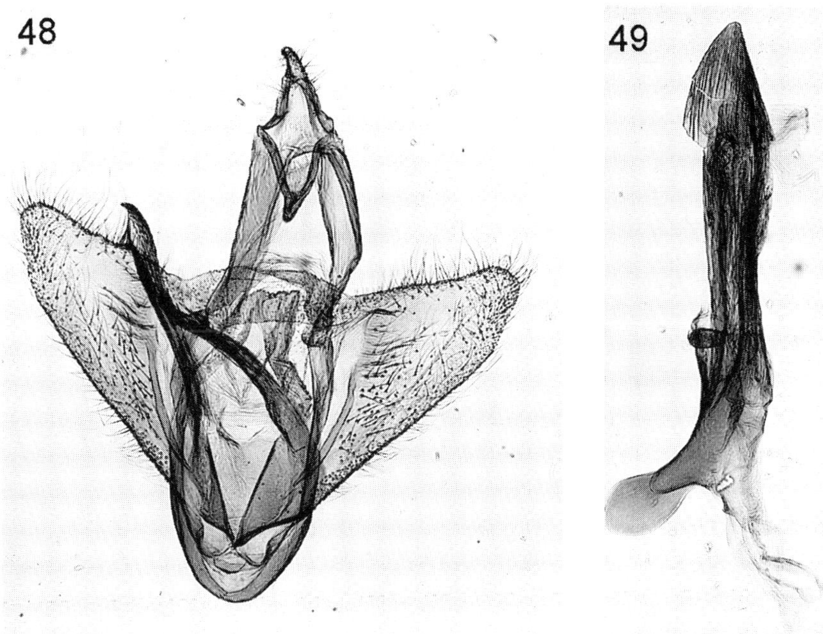

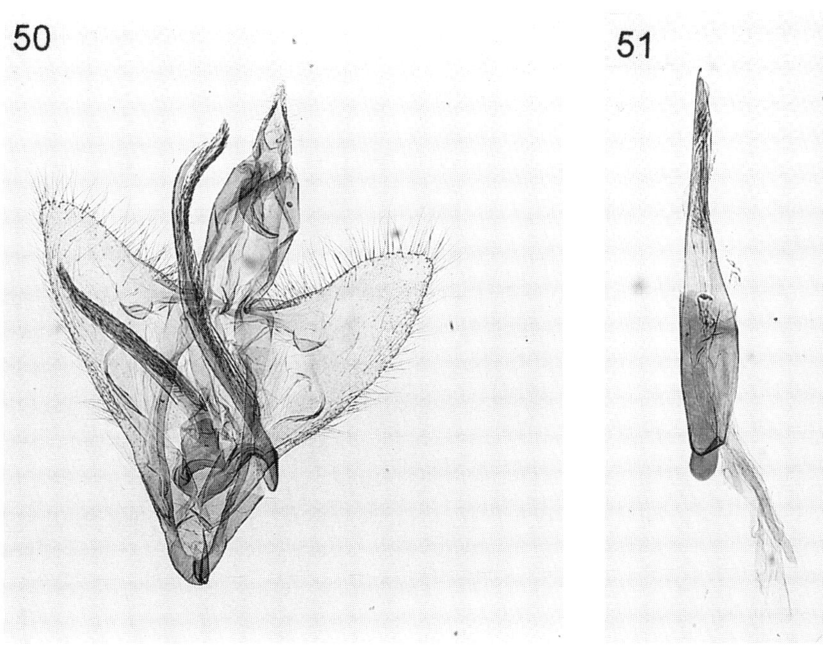

Figs 48, 49. Male genitalia *Chilo diffusilineus*.
Figs 50, 51. Male genitalia *Chilo aleniellus*.

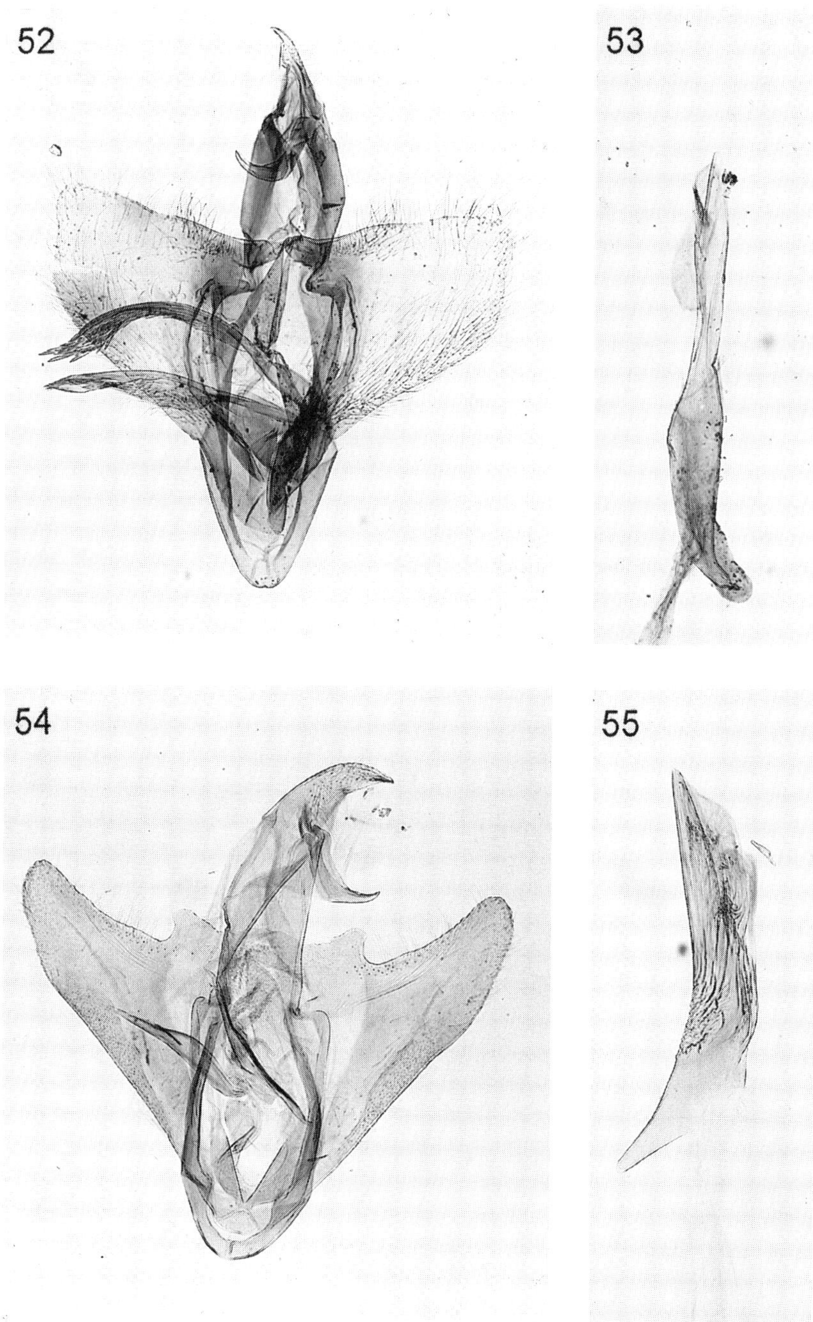

Figs 52, 53. Male genitalia *Chilo orichalcociliellus*.
Figs 54, 55. Male genitalia *Chilo agamemnon*.

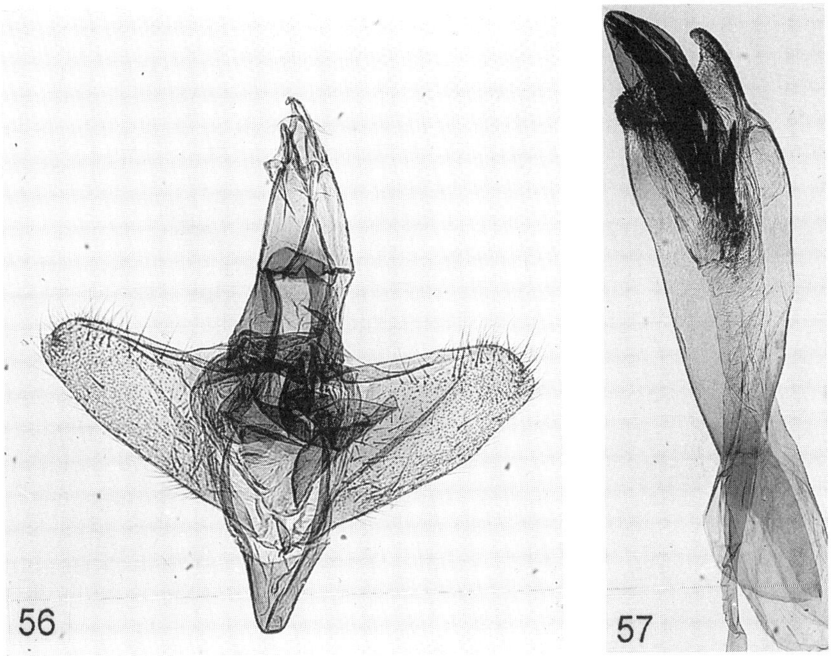

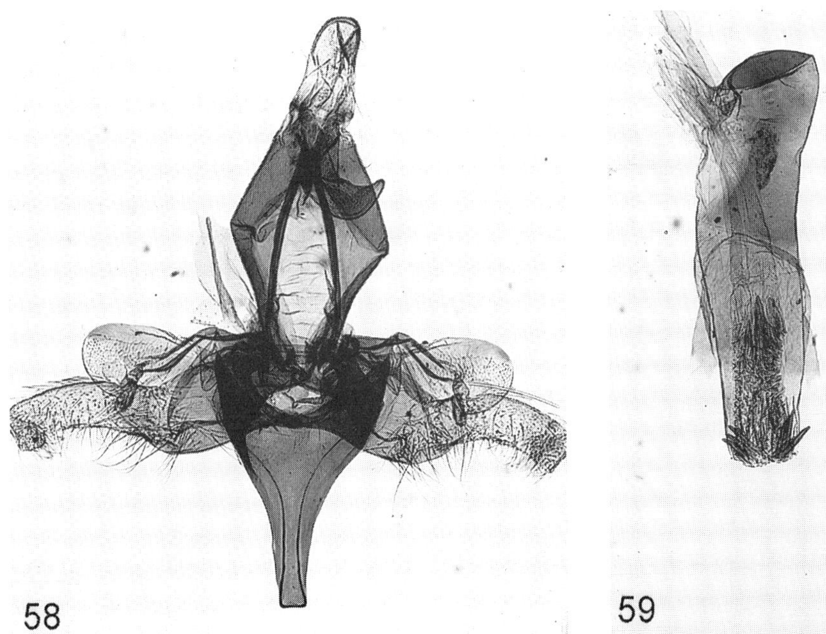

Figs 56, 57. Male genitalia *Chilo sacchariphagus indicus.*
Figs 58, 59. Male genitalia *Coniesta ignefusalis.*

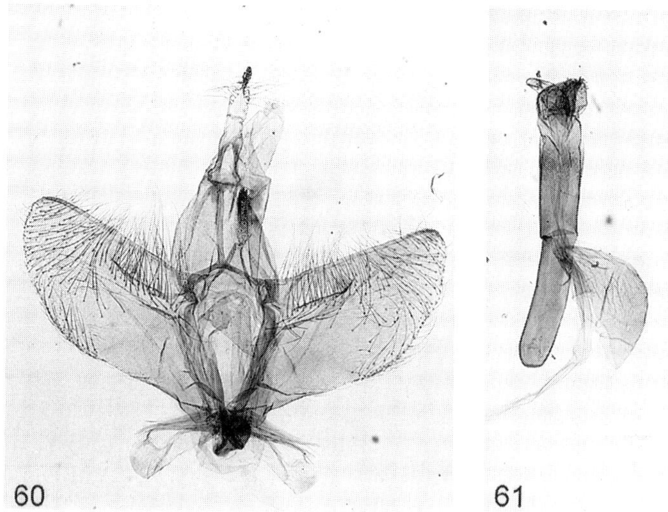

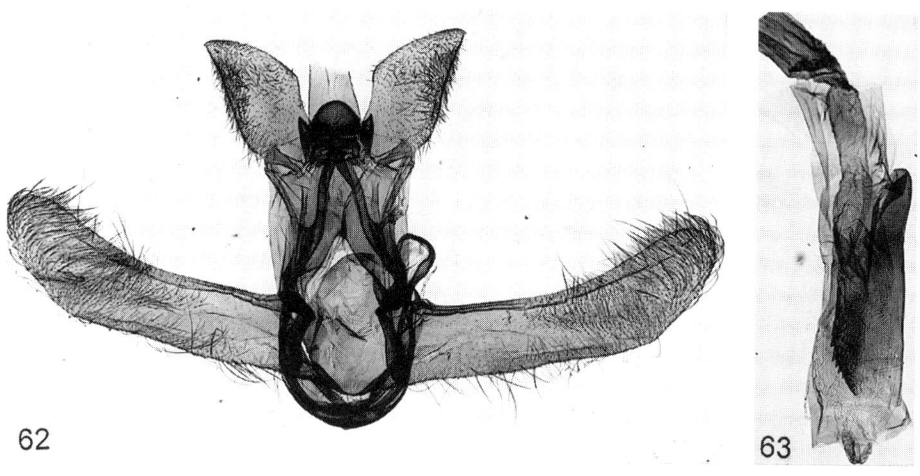

Figs 60, 61. Male genitalia *Adelpherupa flavescens.*
Figs 62, 63. Male genitalia *Maliarpha separatella.*

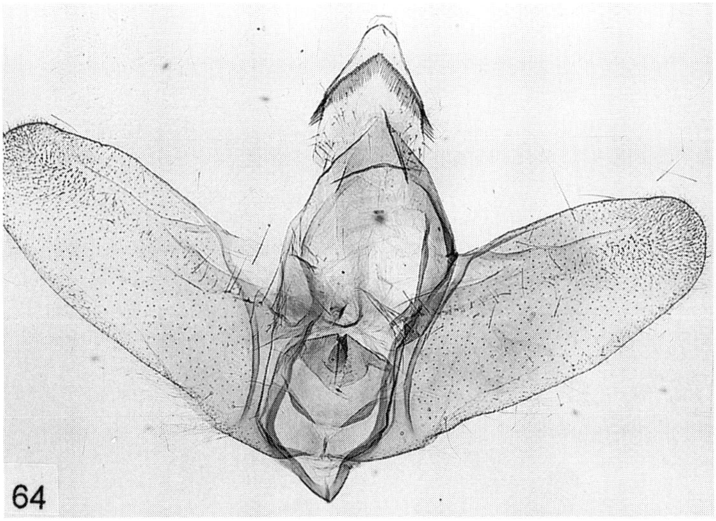
64

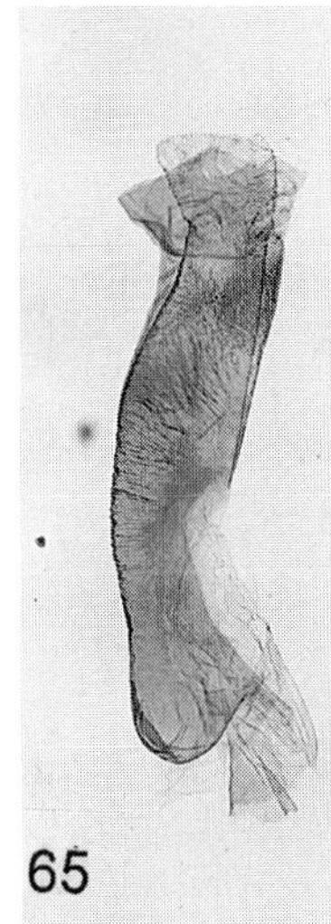
65

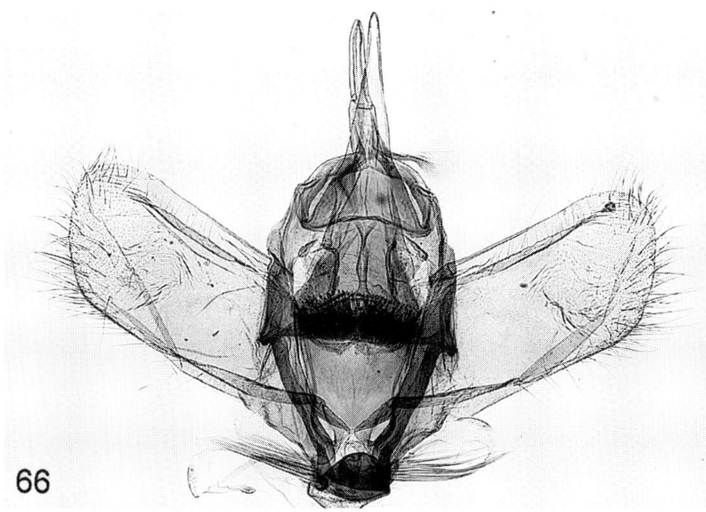
66

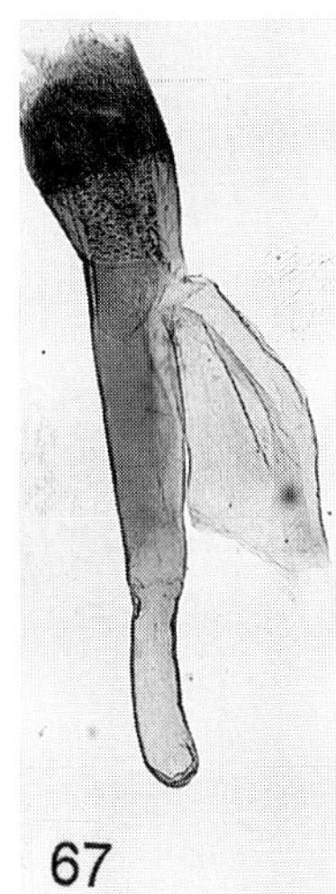
67

Figs 64, 65. Male genitalia *Eldana saccharina*.
Figs 66, 67. Male genitalia *Scirpophaga marginepunctella*.

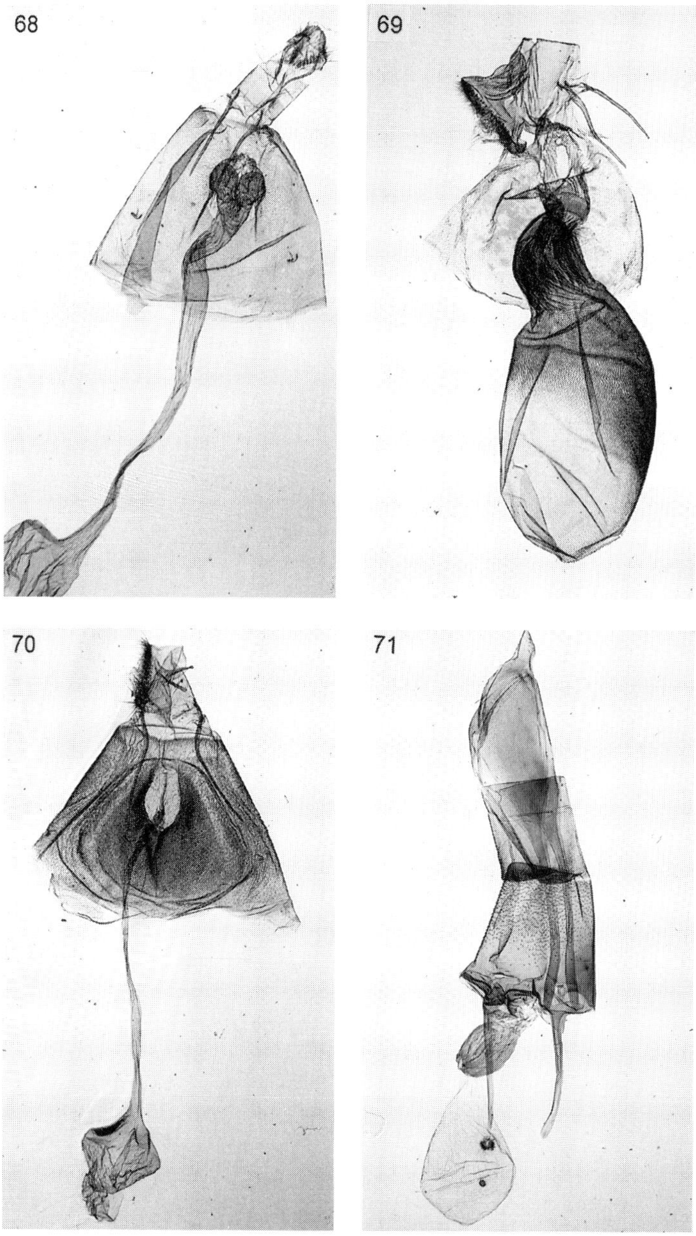

Figs 68–71. Female genitalia. (68) *Chilo partellus*. (69) *Chilo sacchariphagus indicus*. (70) *Chilo aleniellus*. (71) *Coniesta ignefusalis*.

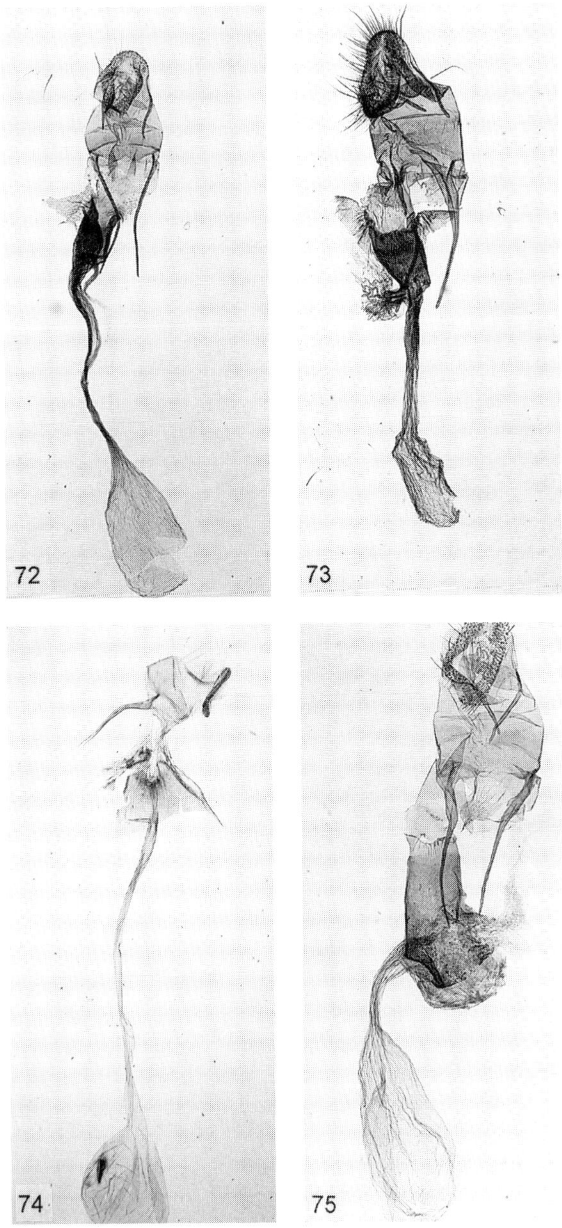

Figs 72–75. Female genitalia. (72) *Chilo diffusilineus*. (73) *Chilo agamemnon*. (74) *Chilo orichalcociliellus*. (75) *Chilo zacconius*.

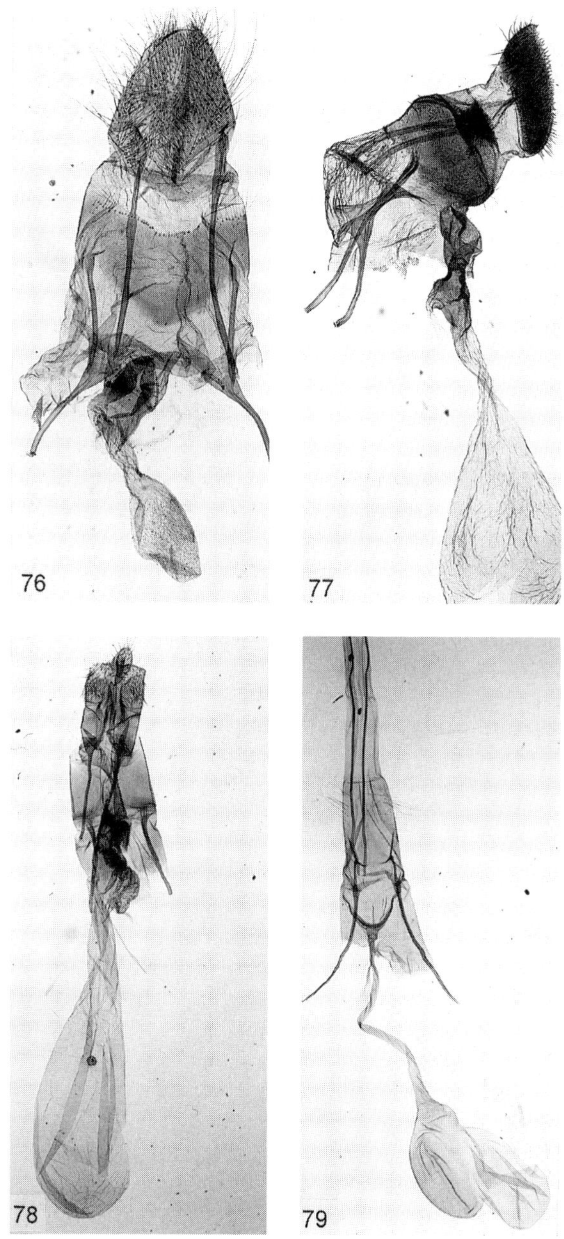

Figs 76–79. Female genitalia. (76) *Scirpophaga marginepunctella*. (77) *Adelpherupa albescens*. (78) *Maliarpha separatella*. (79) *Eldana saccharina*.

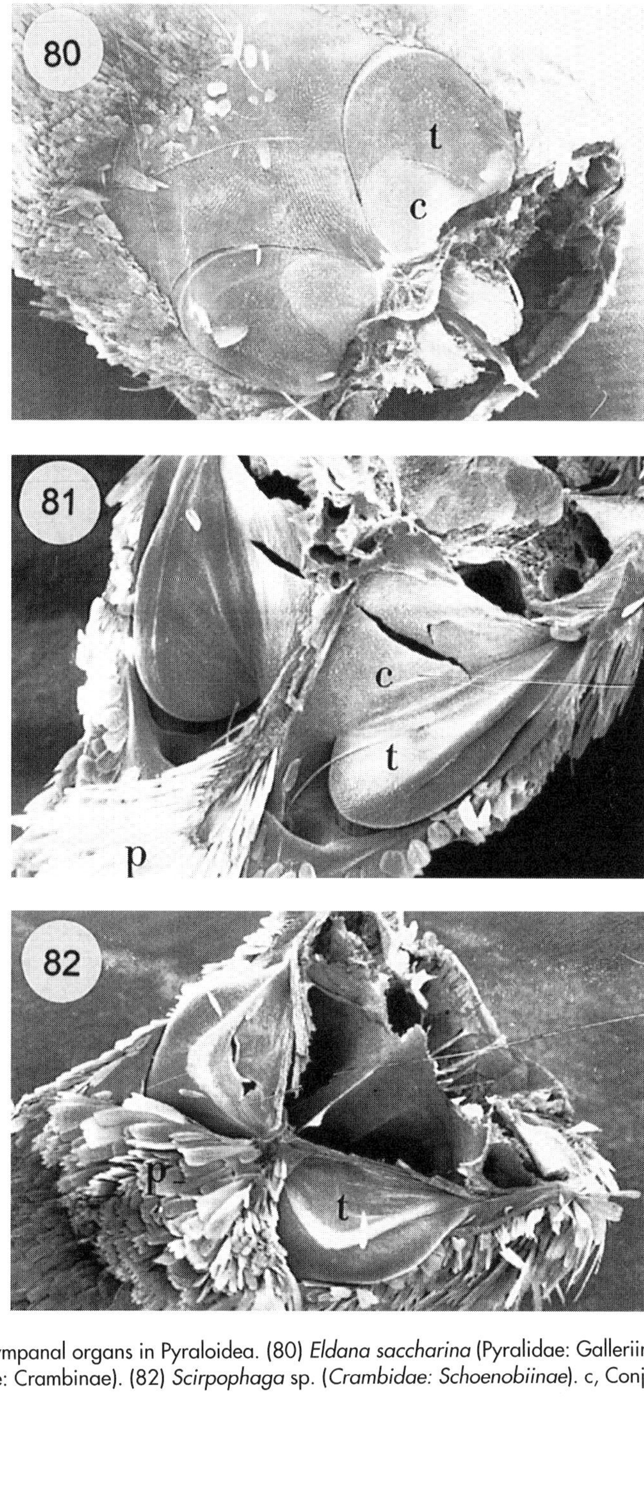

Figs 80–82. Tympanal organs in Pyraloidea. (80) *Eldana saccharina* (Pyralidae: Galleriinae). (81) *Chilo* sp. (Crambidae: Crambinae). (82) *Scirpophaga* sp. (*Crambidae: Schoenobiinae*). c, Conjonctivum; t, tympanum.

83

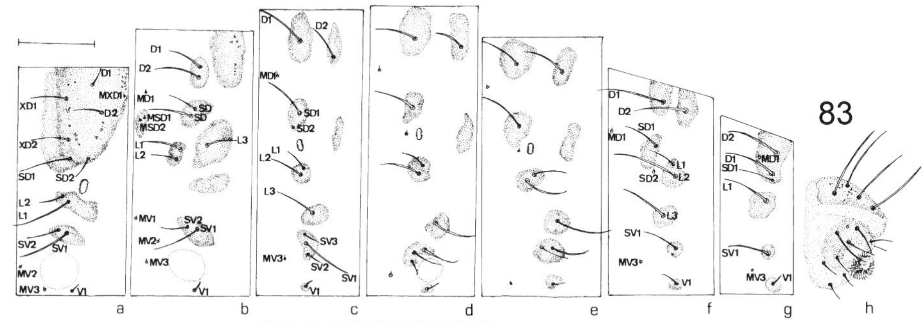

84

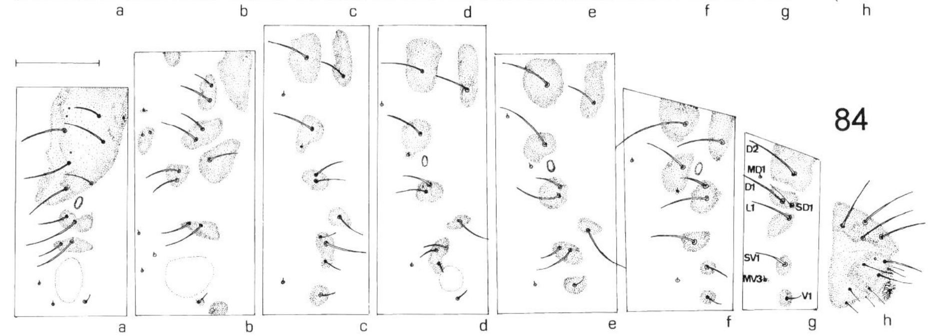

Figs 83, 84. Setal maps. (83) *Chilo partellus*. (84) *Chilo orichalcociliellus*. a, Prothorax; b, mesothorax (≈ metathorax); c, abdominal segment 1 (≈ 2); d, abdominal segment 3 (≈ 4, 5, 6); e, abdominal segment 7; f, abdominal segment 8; g, abdominal segment 9; h, abdominal segment 10.

85

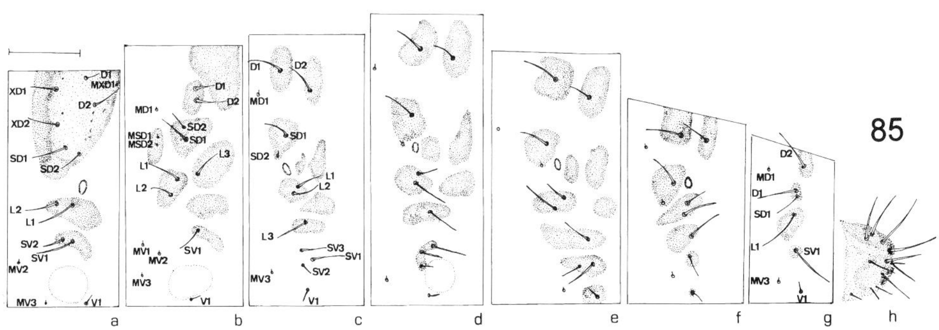

86

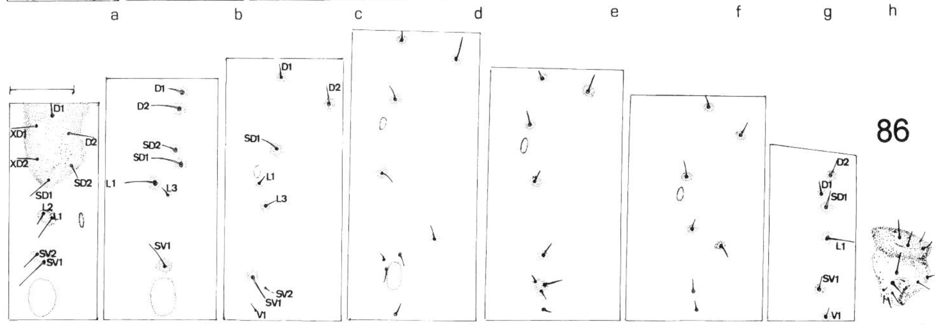

Figs 85, 86. Setal maps. (85) *Coniesta ignefusalis*. (86) *Scirpophaga* sp. a, Prothorax; b, mesothorax (≈ metathorax); c, abdominal segment 1 (≈ 2); d, abdominal segment 3 (≈ 4, 5, 6); e, abdominal segment 7; f, abdominal segment 8; g, abdominal segment 9; h, abdominal segment 10.

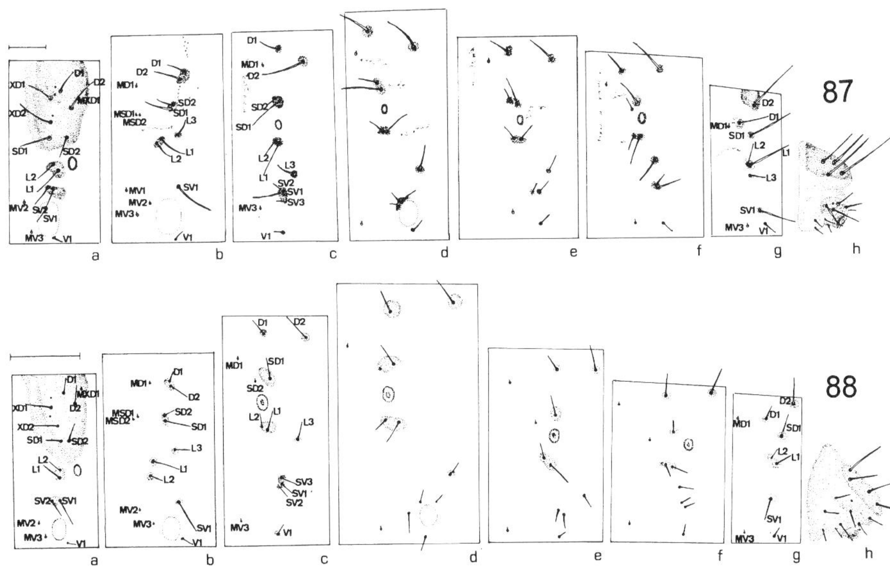

Figs 87, 88. Setal maps. (87) *Eldana saccharina.* (88) *Maliarpha separatella.* a, Prothorax; b, mesothorax (≈ metathorax); c, abdominal segment 1(≈ 2); d, abdominal segment 3(≈ 4, 5, 6); e, abdominal segment 7; f, abdominal segment 8; g, abdominal segment 9; h, abdominal segment 10.

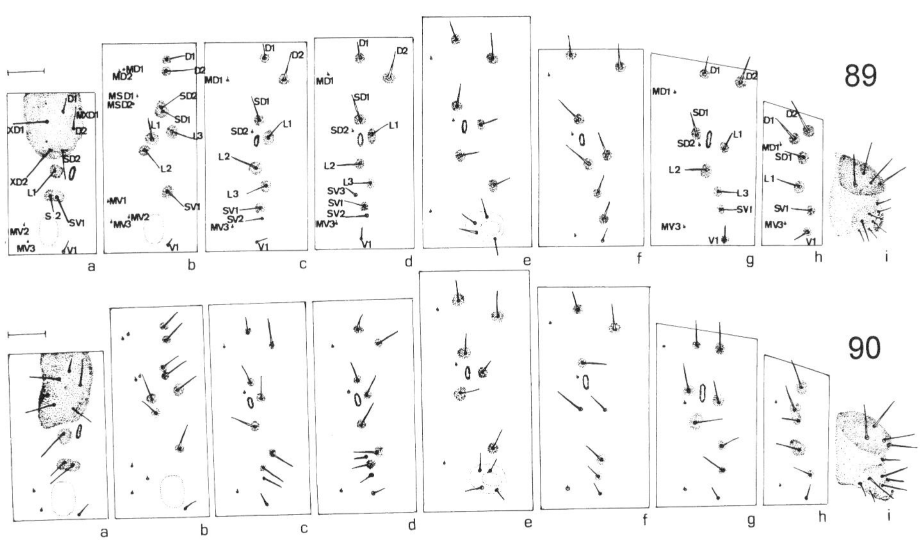

Figs 89, 90. Setal maps. (89) *Busseola fusca.* (90) *Sesamia calamistis.* a, Prothorax; b, mesothorax (≈ metathorax without MD2); c, abdominal segment 1; d, abdominal segment 2; e, abdominal segment 3(≈ 4, 5, 6); f, abdominal segment 7; g, abdominal segment 8; h, abdominal segment 9; i, abdominal segment 10.

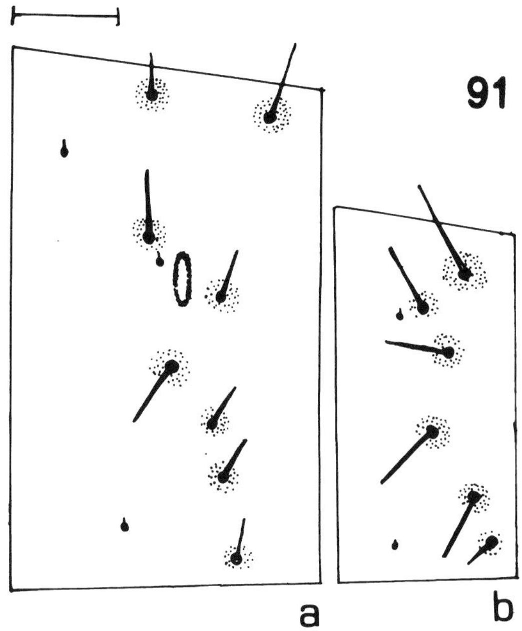

Fig. 91. Setal map, *Sesamia cretica.* a, Abdominal segment 8; b, abdominal segment 9.

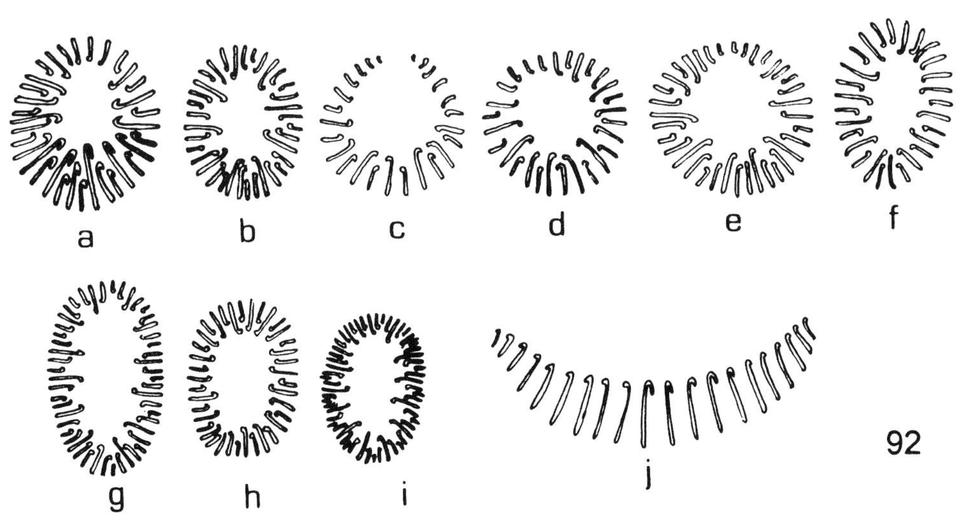

Fig. 92. Arrangement of abdominal crochets. a, *Chilo partellus*; b, *Chilo orichalcociliellus*; c, *Chilo zacconius*; d, *Chilo diffusilineus*; e, *Coniesta ignefusalis*; f, *Scirpophaga* sp.; g, *Eldana saccharina*; h, i, *Maliarpha separatella*; j, *Sesamia calamistis*.

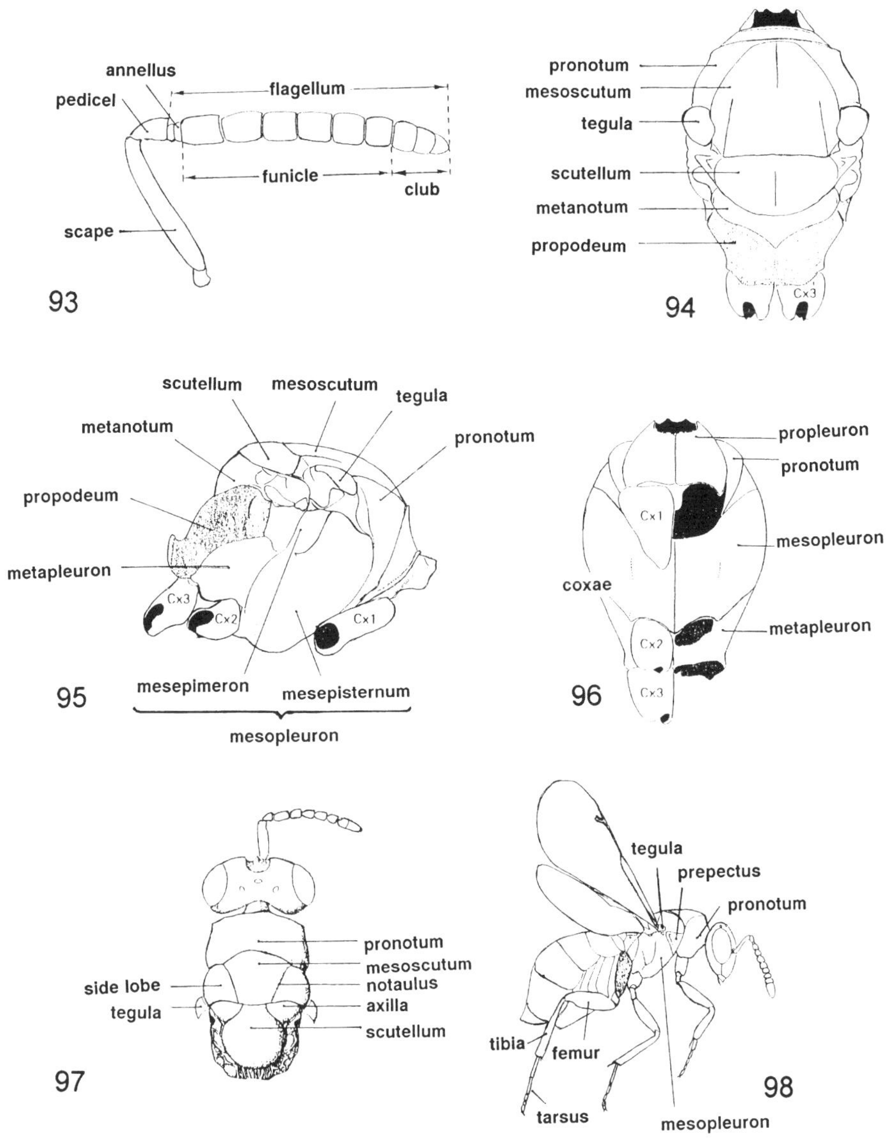

Figs 93–98. Morphological terminology in Hymenoptera: (93) Antenna (Pteromalidae). (94) Mesosoma (Vespidae), dorsal view. (95) Lateral view. (96) Ventral view. (97) Mesosoma (Chalcidoidea). (98) Lateral habitus (Chalcidoidea).

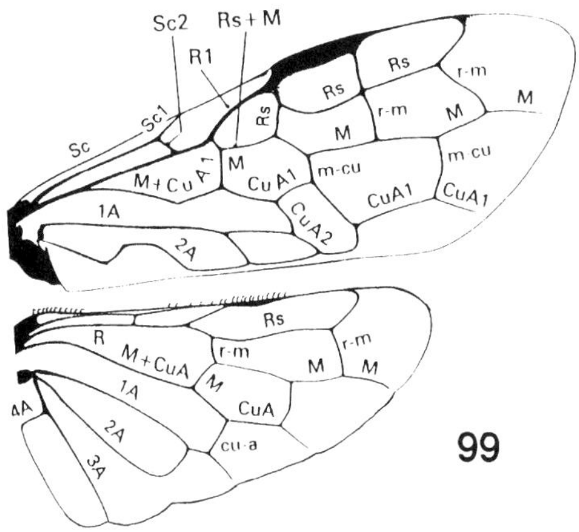

99

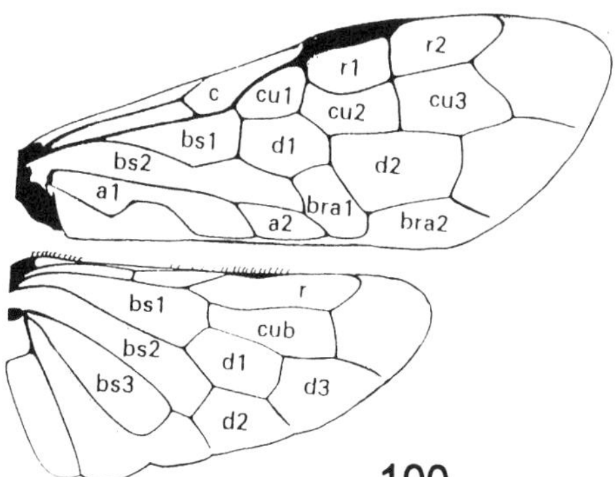

100

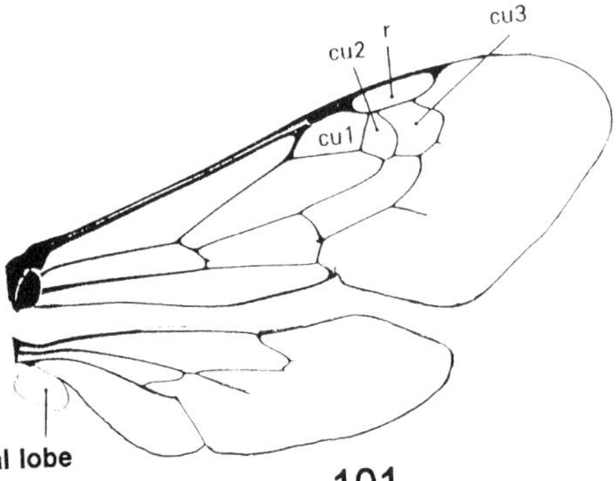

101

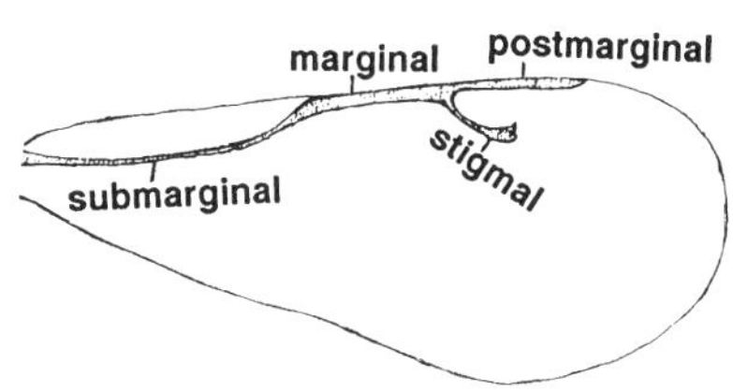

102

Figs 99–102. Wing terminology in Hymenoptera: (99) Wing venation, Symphyta. (100) Wing cells (Symphyta). (101) Aculeata, showing jugal lobe. (102) Wing venation (Chalcidoidea).

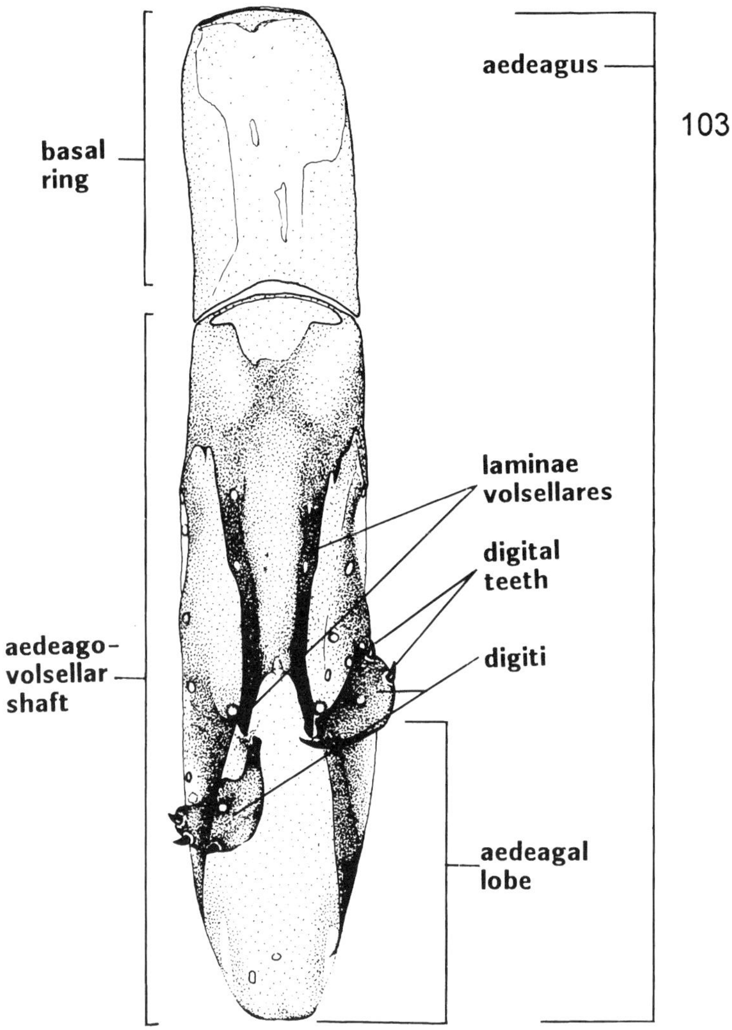

Fig. 103. Male genitalia of *Telenomus* sp.

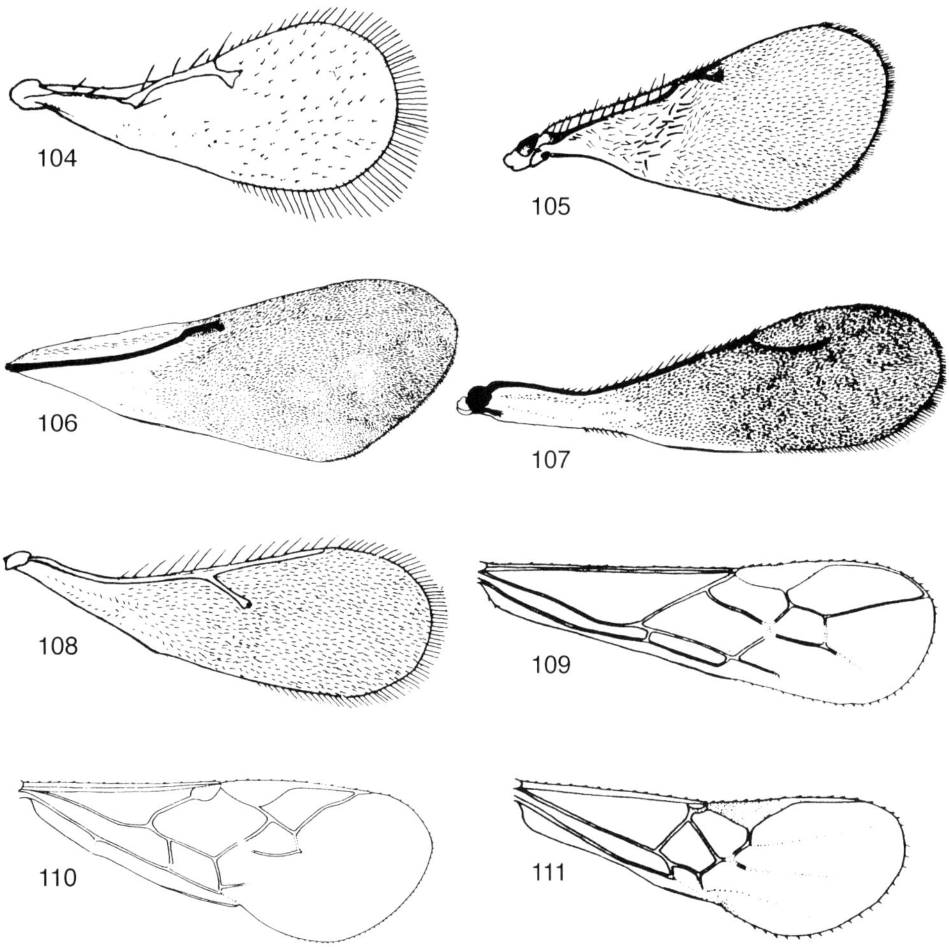

Figs 104–111. Fore wings (104) *Lathromeris* sp. (Trichogrammatidae). (105) *Exoristobia* sp. (Encyrtidae). (106) *Psilochalcis* sp. (Chalcididae). (107) *Aphanogmus* sp. (Ceraphronidae). (108) *Telenomus* sp. (Scelionidae). (109) *Rhaconotus* sp. (Braconidae). (110) *Chelonus* sp. (Braconidae). (111) *Cotesia* sp. (Brasconidae).

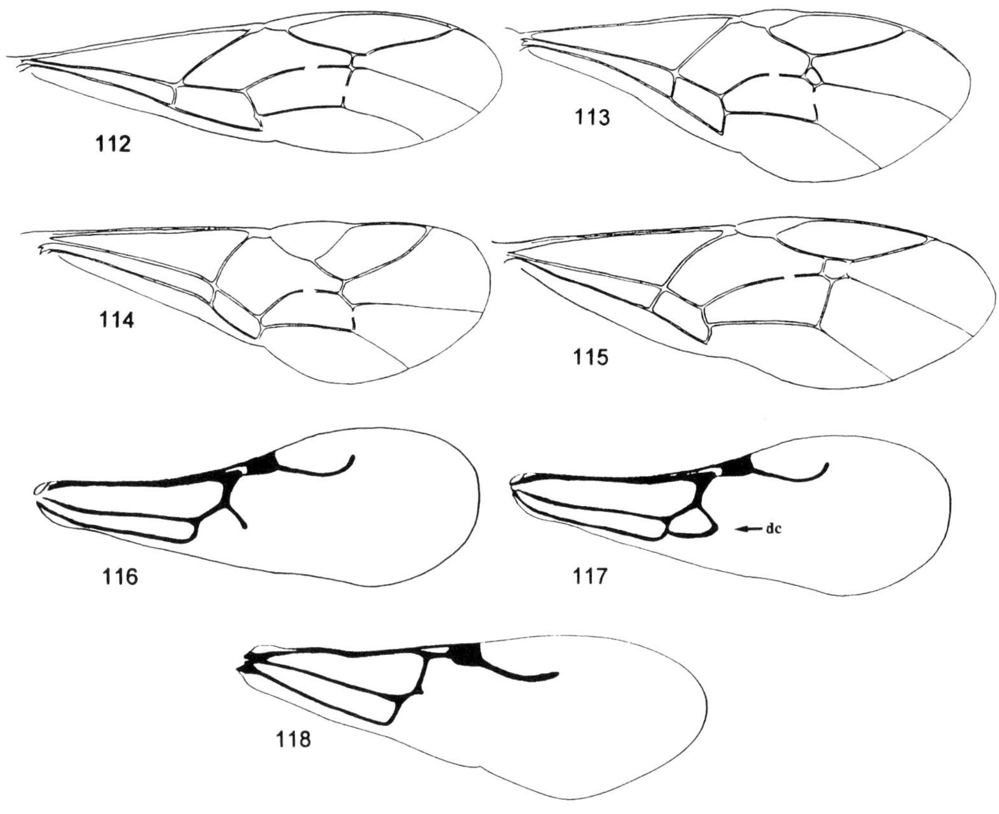

Figs 112–118. Fore wings (112) *Amauromorpha* sp. (Ichneumonidae). (113) *Venturia* sp. (Ichneumonidae). (114) *Pristomerus* sp. (Ichneumonidae). (115) *Vadonia* sp. (Ichneumonidae). (116, 117) *Goniozus* spp. (Bethylidae), dc = discoidal cell. (118) *Odontepyris* sp. (Bethylidae).

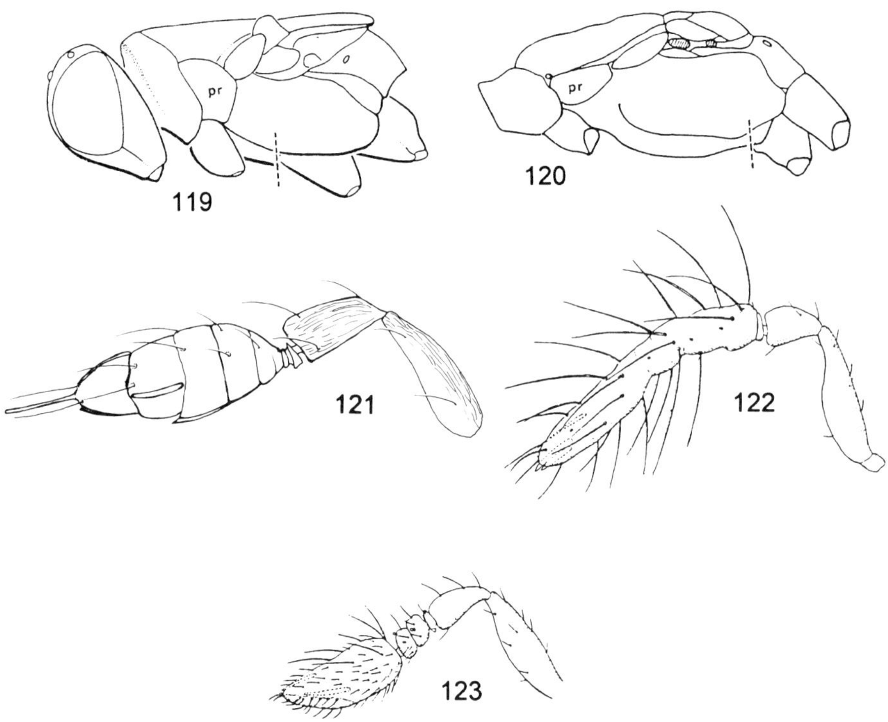

Figs. 119–123. (119) Encyrtidae mesosoma, lateral view (pr = prepectus). (120) Eupelmidae (♀) mesosoma, lateral view (pr = prepectus). (121) *Lathoromeris* sp. (Trichogrammatidae) antenna. (122) *Trichogramma* sp. (male) (Trichogrammatidae) antenna. (123) *Trichogramma* sp. (female) (Trichogrammatidae) antenna.

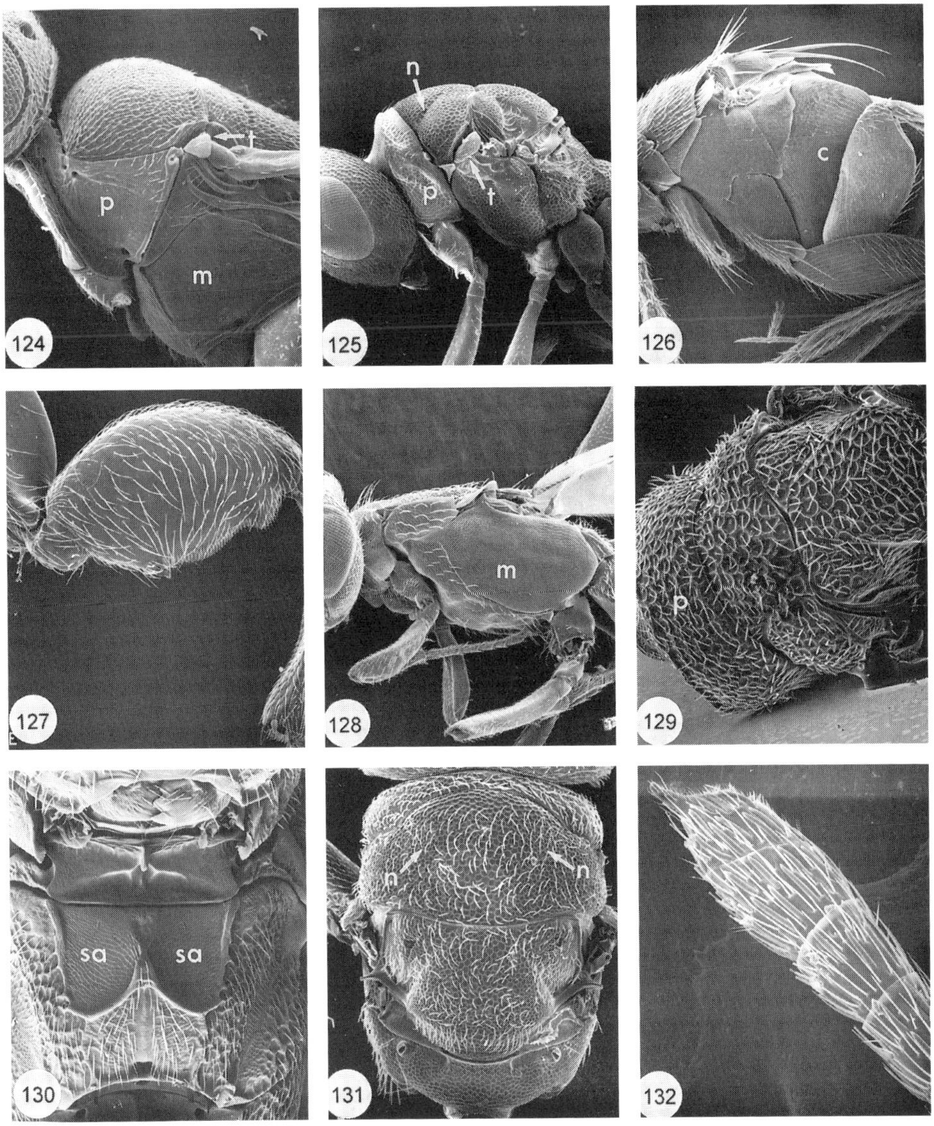

Figs 124–132. (124) *Aphanogmus* sp. mesosoma, lateral view (m = mesopleuron, p = pronotum, t = tegula). (125) Chalcidoid mesosoma, lateral view (p = pronotum, t = tegula). (126) *Elasmus* sp. (Elasmidae) mesosoma, lateral view (c = hind coxa). (127) *Psilochalcis soudanensis* (Chalcididae) hind femur. (128) Eupelmidae (♀) mesosoma, lateral view (m = mesopleuron). (129) *Eurytoma oryzivora* (Eurytomidae) mesosoma, dorsal view (p = pronotum). (130) *E. oryzivora* mesosoma, ventral view (sa = subpleural area). (131) *Norbanus* sp. (Pteromalidae) mesosoma, dorsal view (n = notauli). (132) *Norbanus* sp. (♀) (Pteromalidae) antennal club.

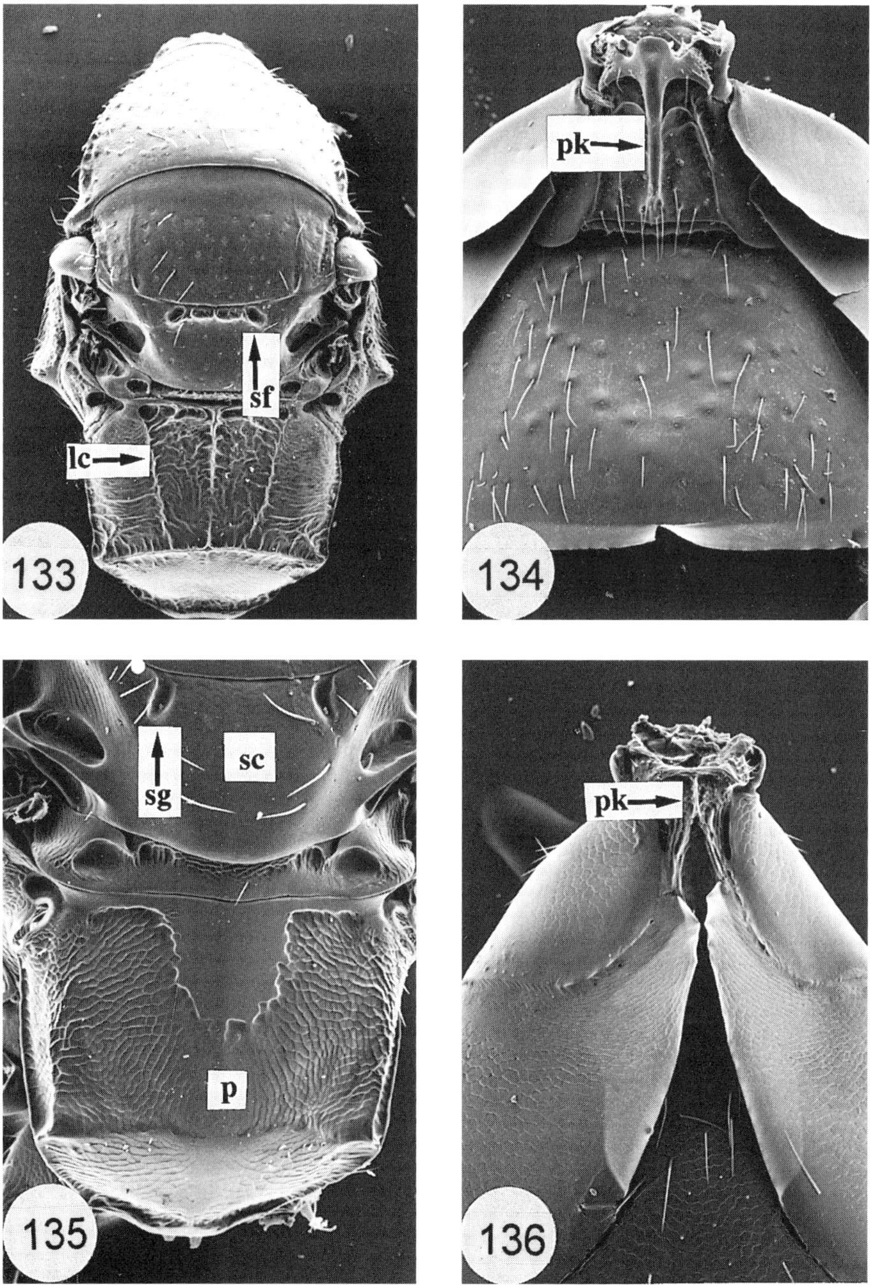

Figs 133–136. Bethylidae. (133) *Odontepyris* sp. mesosoma (thorax + propodeum). (134) Ventral metasoma (gaster). (135) *Goniozus* sp., scutellum and propodeum. (136) *Goniozus* sp. ventral metasoma. (sc = Scutellum; lc = lateral carina; p = propodeum; pk = petiolar keel; sf = scutellar foveae; sg = scutellar grooves.)

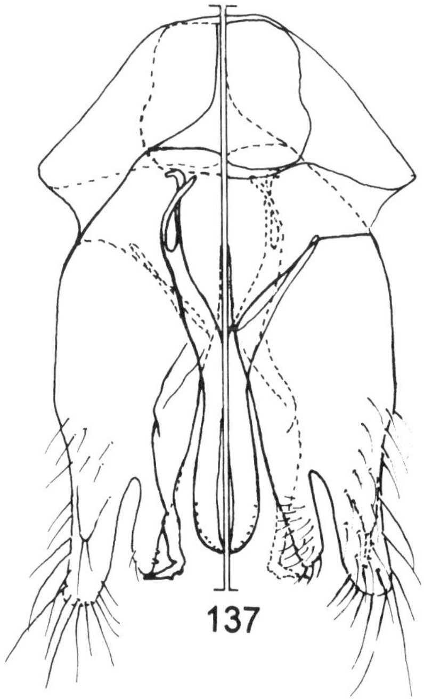

137

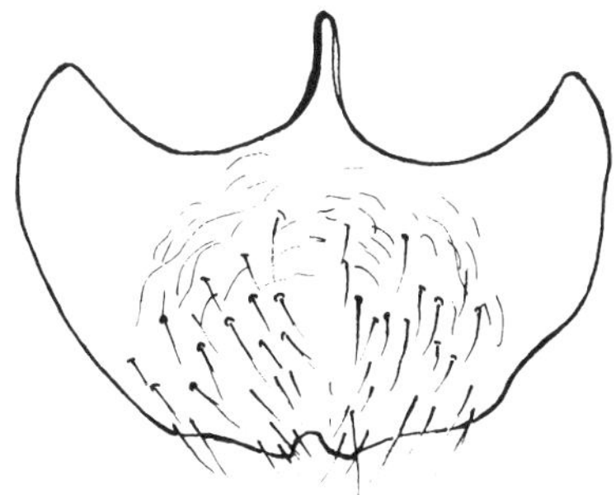

138

Fig. 137, 138. *Goniozus indicus*. (137) Male genitalia. (138) Male subgenital plate.

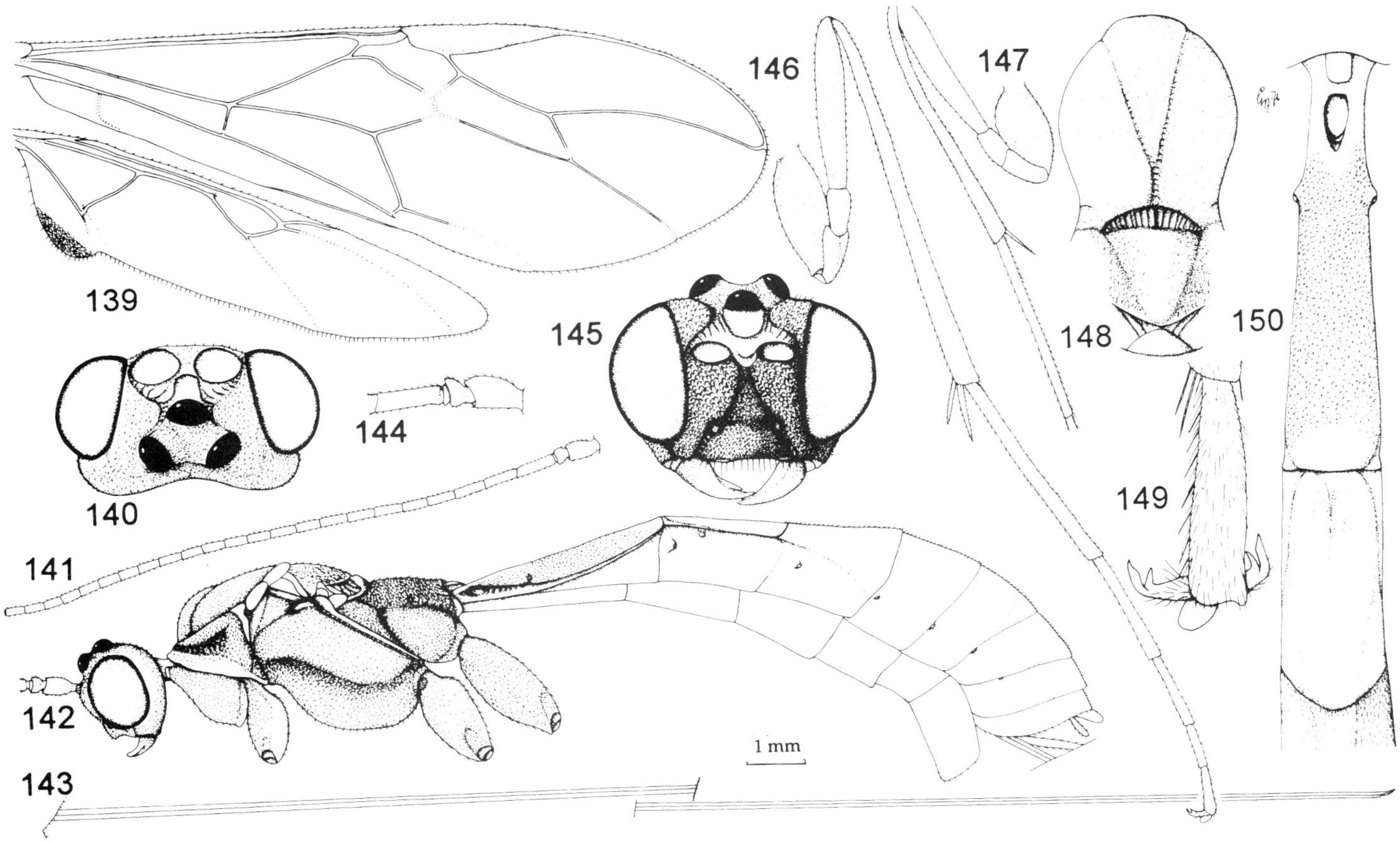

Figs 139–150. *Amicrocentrum curvinervis* ♀, holotype. (139) Wings. (140) Head, dorsal aspect. (141) Antenna (142) Habitus, lateral aspect. (143) Ovipositor. (144) Base of antenna, lateral aspect. (145) Head, frontal aspect. (146) Hind leg. (147) Middle leg. (148) Thorax, dorsal aspect. (149) Hind telotarsus, with full sight on outer hind claw. (150) First and second metasomal tergites, dorsal aspect. 139, 141–143, 146, 147: 1.0 × scale line; 140, 144, 145, 148, 150: 2.0 ×; 149: 5.0 ×.

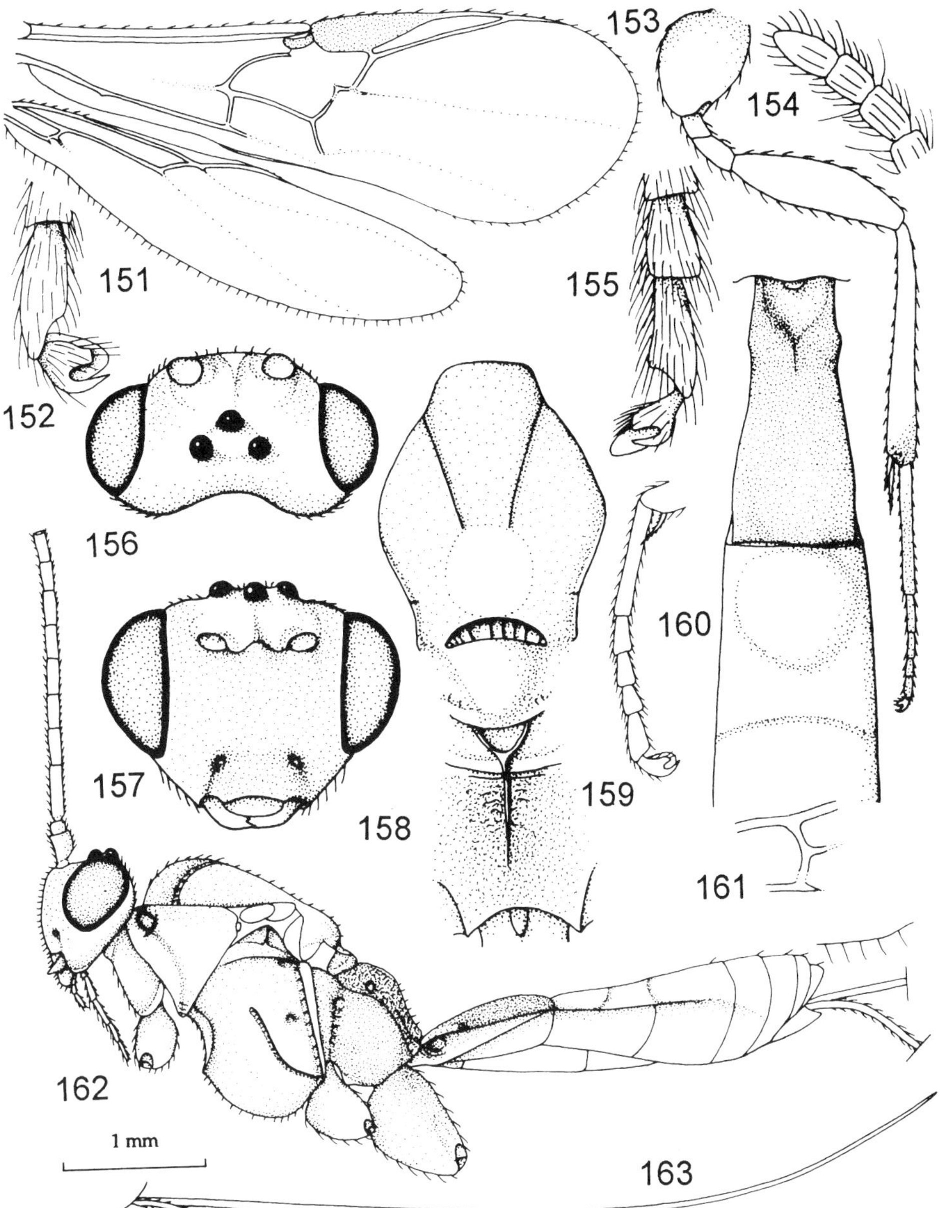

Figs 151, 152, 156–163. *Bassus costalis*, ♀, holotype of *Camptothlipsis costalis*. **Figs 153–155**, Bassus sp. near *B. costalis*, ♀, Kenya, Voi River Valley. (151) Wings. (152) Inner fore claw. (153) Hindleg. (154) Apex of antenna. (155) Outer hind claw. (156) Head, dorsal aspect. (157) Head, frontal aspect. (158) Mesosoma, dorsal aspect. (159) Fore tarsus. (160) First and second metasomal tergites, dorsal aspect. (161) Detail of vein 2-CU of hind wing. (162) Habitus, lateral aspect. (163) Ovipositor. 151, 153, 162, 163: 1.0 × scale line; 152, 154, 155, 161: 5.0 ×; 156–160: 2.0 ×.

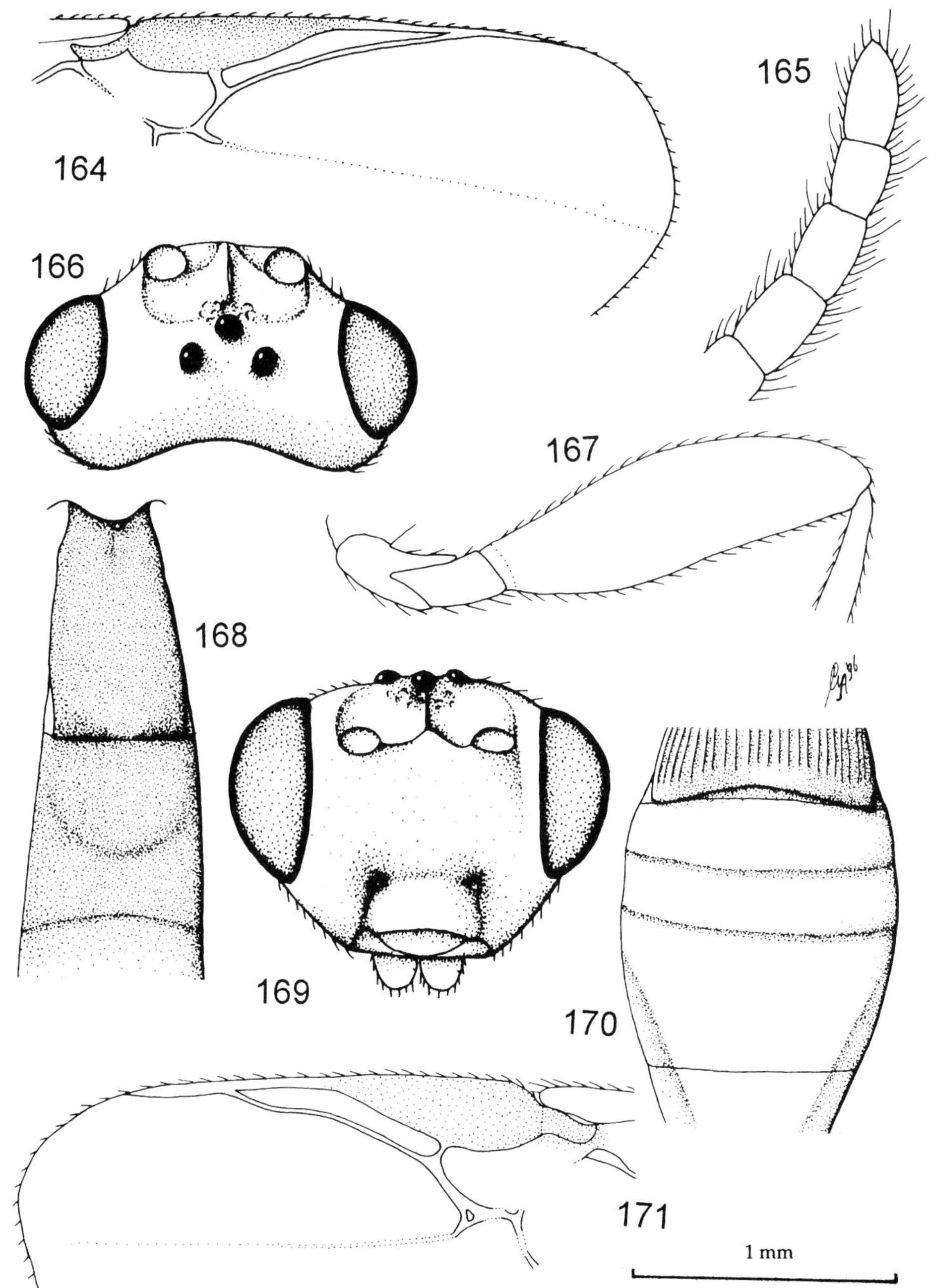

Figs 164–169. *Bassus sublevis*, ♀, lectotype. **Figs 170, 171**, *Bassus* sp. A. (164, 171) Detail of marginal cell of fore wing. (165) Apex of antenna. (166) Head, dorsal aspect. (167) Hind femur. (168) First and second metasomal tergites, dorsal aspect. (169) Head, frontal aspect. (170) Second and third metasomal tergites, dorsal aspect. 164, 168: 1.0 × scale line; 165: 3.4 ×; 166, 169: 1.5 ×; 167: 2.3 ×; 170: 1.4 ×; 171: 1.2 ×.

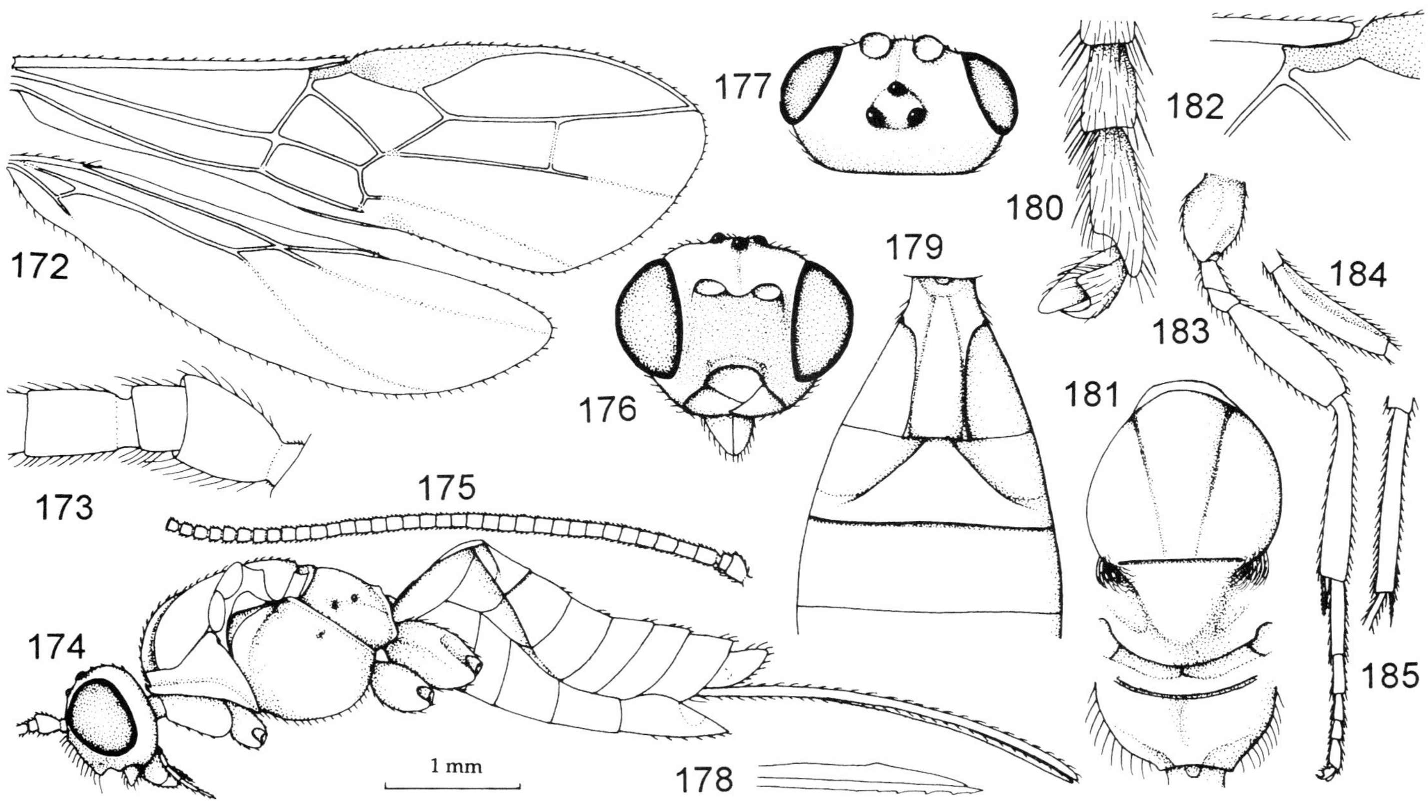

Figs 172–185. *Amyosoma flavistigma*, ♀, holotype. (172) Wings. (173) Base of antenna. (174) Habitus, lateral aspect. (175) Antenna. (176) Head, frontal aspect. (177) Head, dorsal aspect. (178) Apex of ovipositor. (179) First to third metasomal tergites, dorsal aspect. (180) Outer hind claw. (181) Mesosoma, dorsal aspect. (182) Detail of vein 1-SR of fore wing. (183) Hindleg. (184) Hind femur, dorsal aspect. (185) Hind tibia, dorsal aspect. 172, 174, 175, 183–185: 1.0 × scale line; 173: 5.0 ×; 176, 177, 179, 181: 1.6 ×; 178, 180: 5.4 ×; 182: 2.0×.

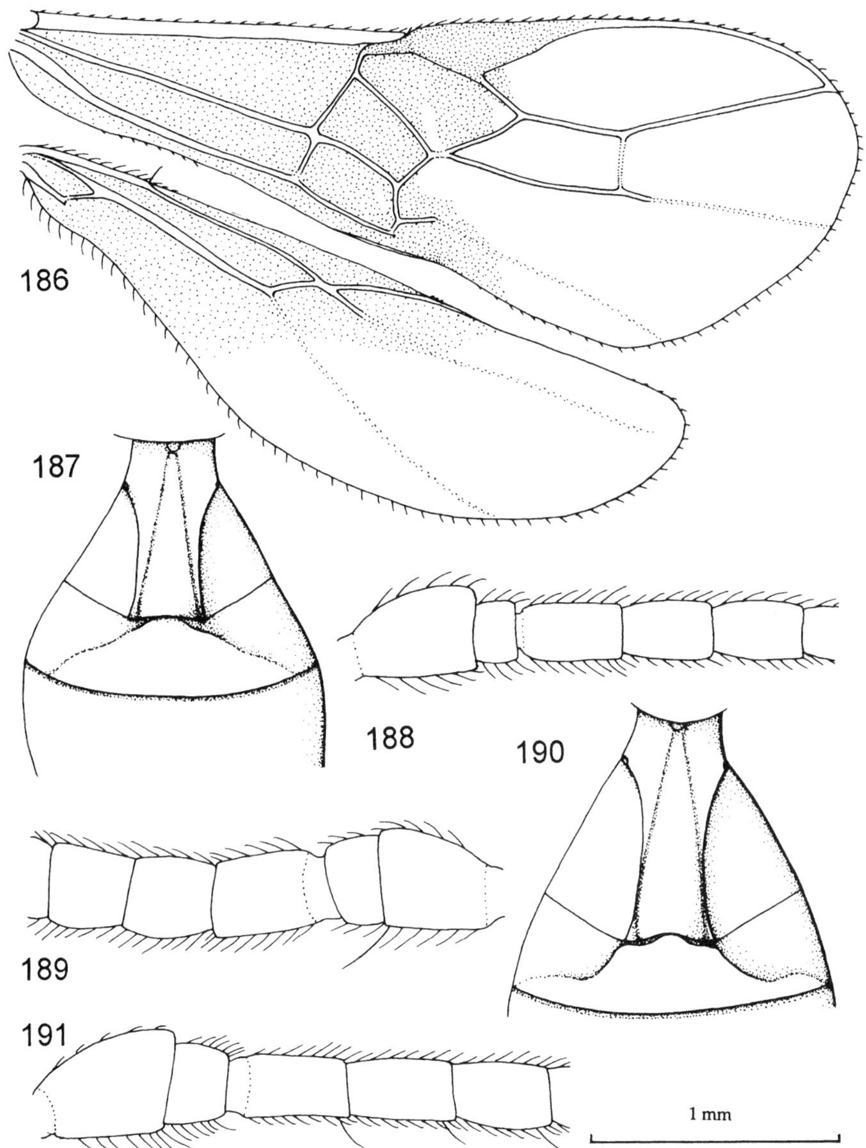

Figs 186–188. *Amyosoma nyanzaense*, ♀, Kenya, Kisumu. **Figs 189, 190.** *Amyosoma leuzerae*, ♀; **Fig. 191**. *Amyosoma chinense*, ♀. (186) Wings. (187, 190) First and second metasomal tergites, dorsal aspect. (188, 189, 191) Base of antenna. 186: 1.0 × scale line; 187: 1.5 ×; 188, 189, 191: 3.0 ×; 190: 1.3 ×.

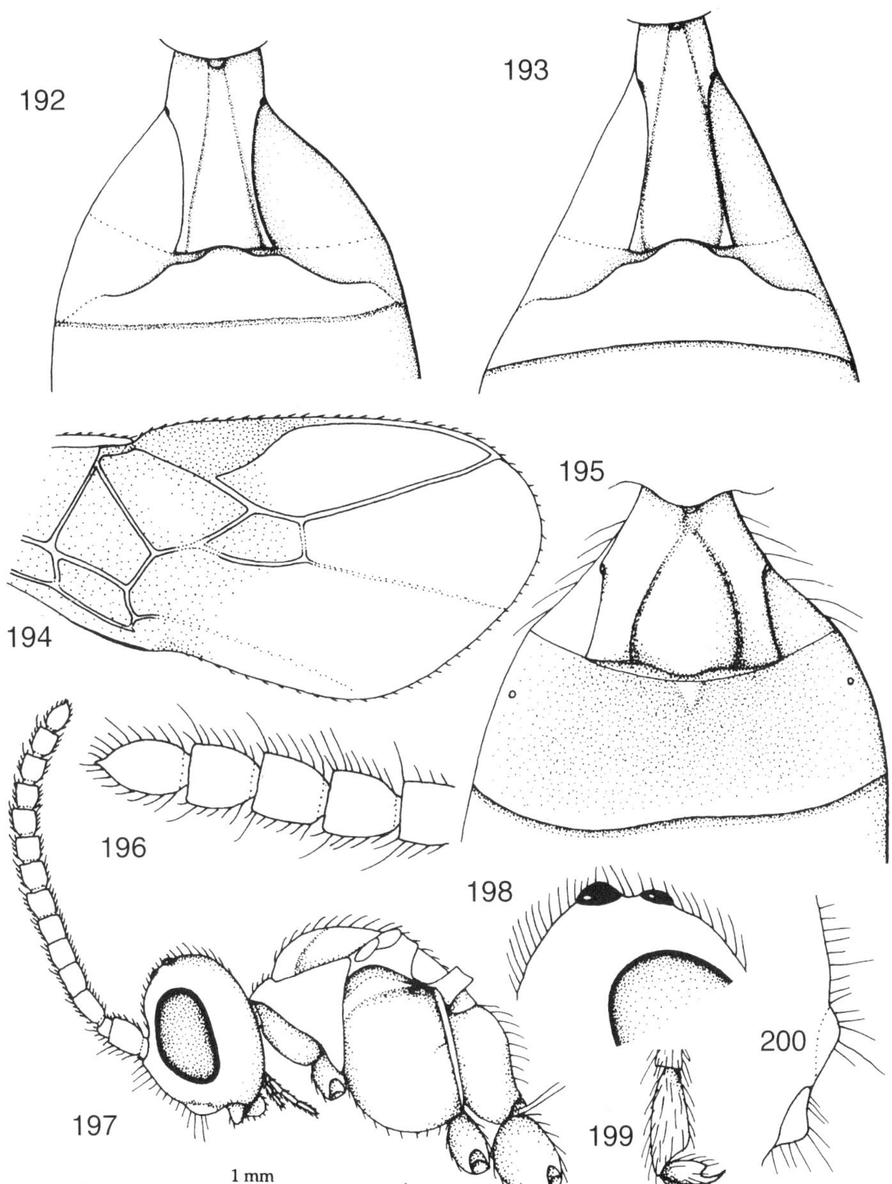

Figs 192, 193. *Amyosoma chinense*, ♀. **Figs 194–200**, *Bracon* (*Habrobracon*) *hebetor*, ♀. (192, 193, 195) First and second metasomal tergites, dorsal aspect. (194) Apical half of fore wing. (197) Head and mesosoma, lateral aspect. (196) Apex of antenna. (198) Vertex, lateral aspect. (199) Outer hind claw. (200) Clypeus, lateral aspect. 192–194, 197: 1.0 × scale line; 195: 1.3 ×; 196, 198–200: 2.3 ×.

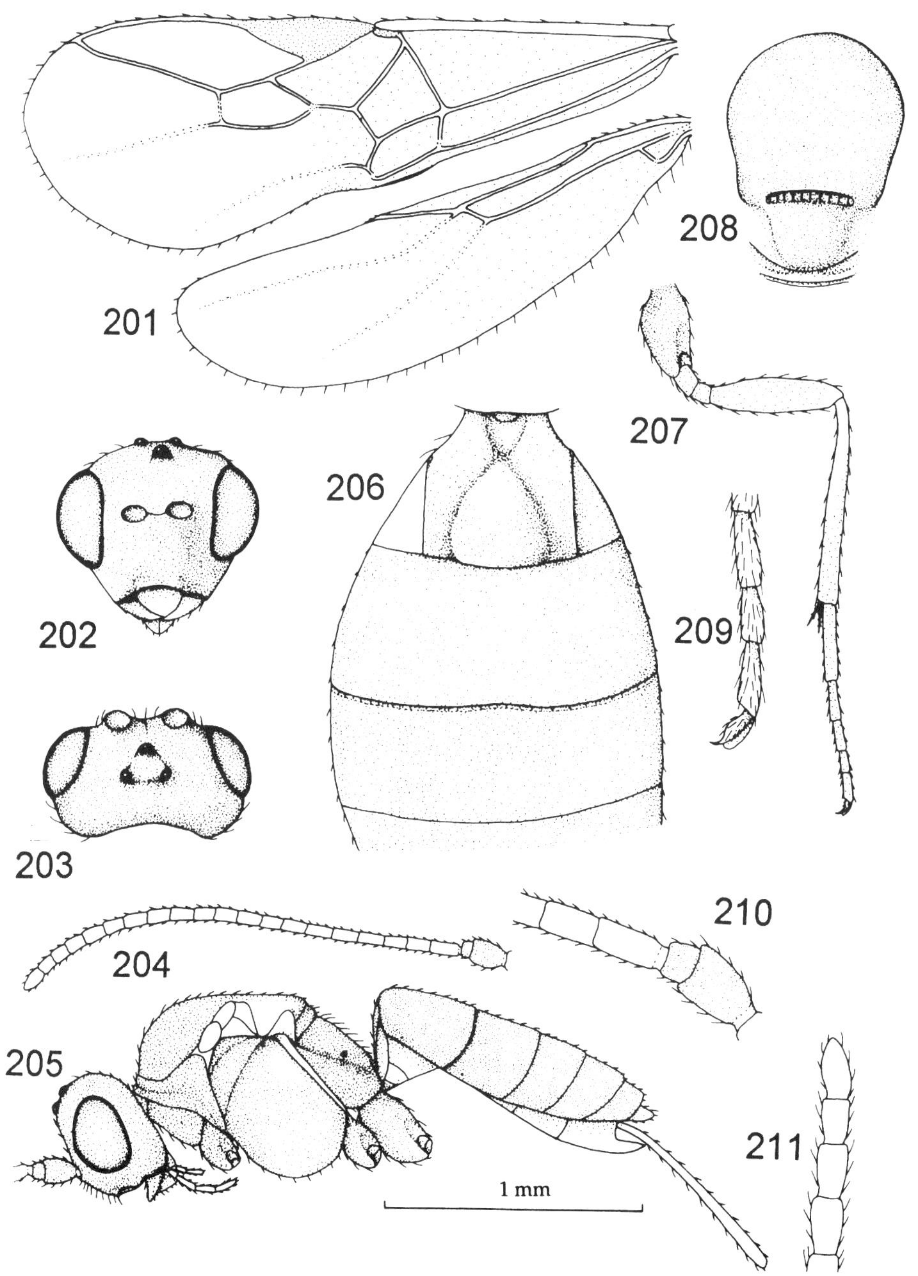

Figs 201–211. *Bracon* (*Habrobracon*) *gelechiae*, ♀. (201) Wings. (202) Head, frontal aspect. (203) Head, dorsal aspect. (204) Antenna. (205) Habitus, lateral aspect. (206) First to third metasomal tergites, dorsal aspect. (207) Hind leg. (208) Thorax, dorsal aspect. (209) Outer hind claw. (210) Base of antenna;. (211) Apex of antenna. 201, 204, 205, 207: 1.0 × scale line; 202, 203, 208: 1.2 ×; 206: 2.0 ×; 209–211: 5 ×.

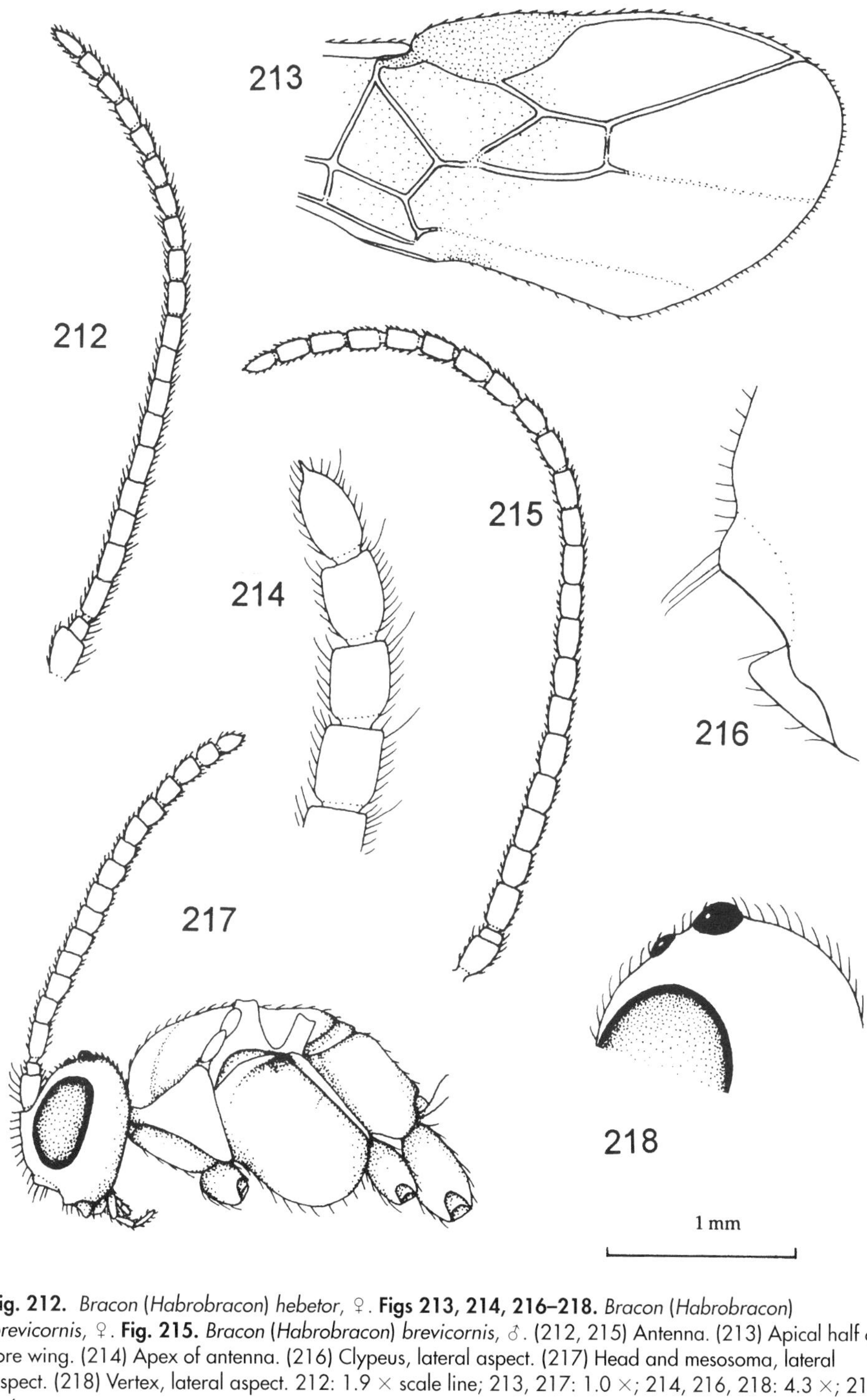

Fig. 212. *Bracon* (*Habrobracon*) *hebetor*, ♀. **Figs 213, 214, 216–218.** *Bracon* (*Habrobracon*) *brevicornis*, ♀. **Fig. 215.** *Bracon* (*Habrobracon*) *brevicornis*, ♂. (212, 215) Antenna. (213) Apical half of fore wing. (214) Apex of antenna. (216) Clypeus, lateral aspect. (217) Head and mesosoma, lateral aspect. (218) Vertex, lateral aspect. 212: 1.9 × scale line; 213, 217: 1.0 ×; 214, 216, 218: 4.3 ×; 215: 1.6 ×.

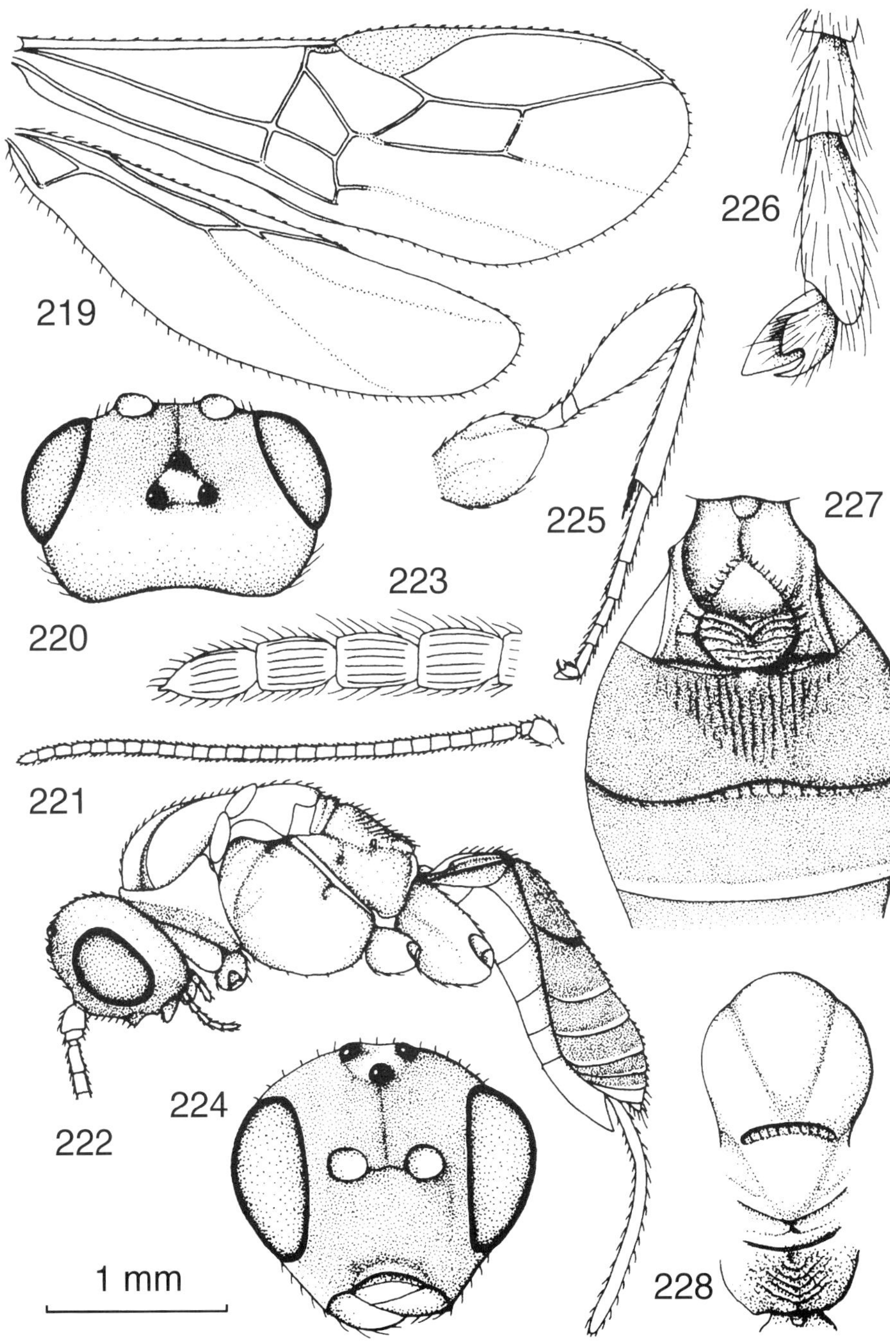

Figs 219–228. *Bracon* (*Bracon*) *sesamiae*, ♀, lectotype. (219) Wings. (220) Head, dorsal aspect. (221) Antenna (222) Habitus, lateral aspect. (223) Apex of antenna. (224) Head, frontal aspect. (225) Hindleg. (226) Inner hind claw. (227) First to third metasomal tergites, dorsal aspect. (228) Mesosoma, dorsal aspect. 219, 221, 222, 225: 1.0 × scale line; 220, 224: 1.8 ×; 223, 226: 4.2 ×; 227: 1.5 ×; 228: 1.3 ×.

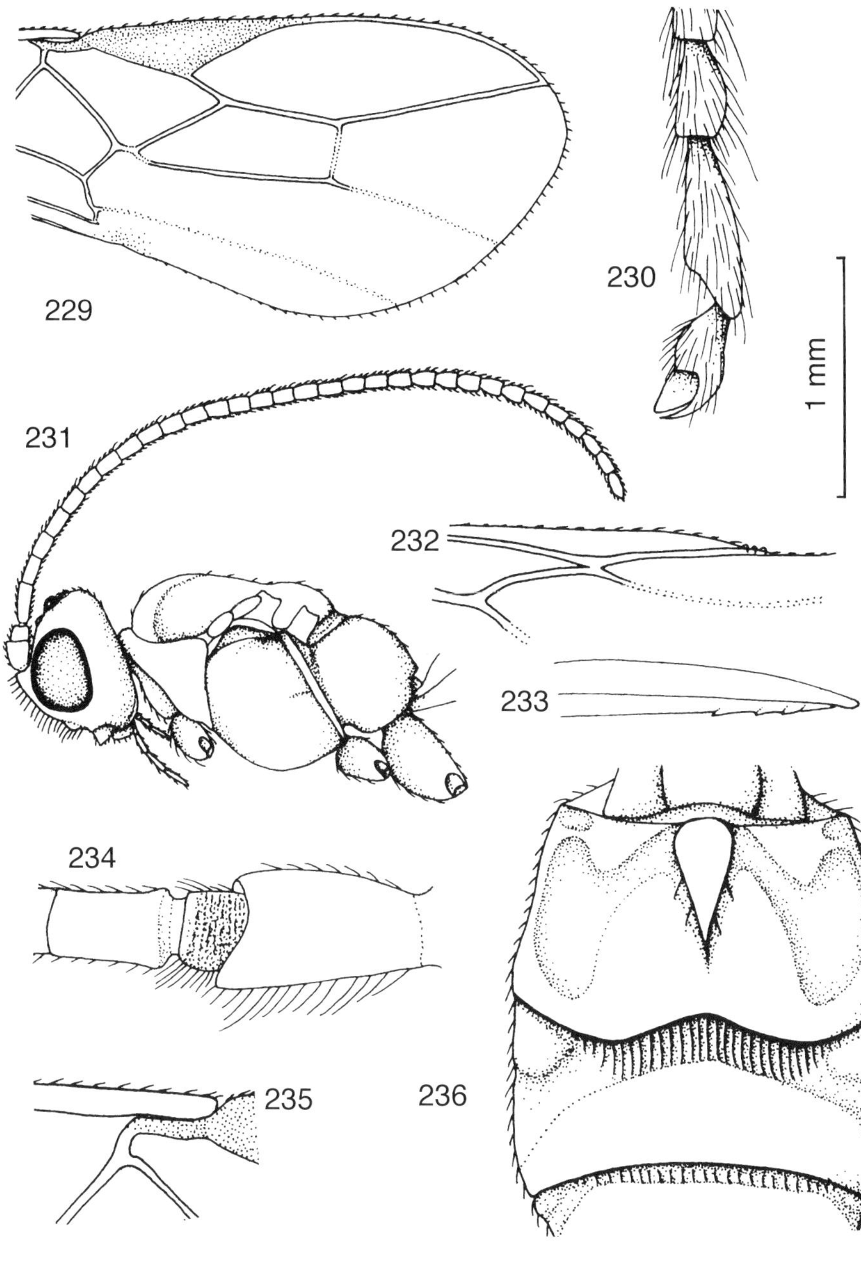

Figs 229–231. *Bracon* (*Bracon*) *testaceorufatus*, ♀. **Figs 232–236.** *Digonogastra kimballi*, ♀. (229) Apical half of fore wing. (230) Outer hind claw. (231) Head and mesosoma, lateral aspect. (232) Detail of vein 1r-m of hind wing. (233) Apex of ovipositor. (234) Scapus, lateral aspect. (235) Detail of vein 1-SR of fore wing. (236) Second and third metasomal tergites, dorsal aspect. 229: 1.2 × scale line; 230, 233: 3.3 ×; 231, 232, 235: 1.0 ×; 234: 2.2 ×; 236: 1.1 ×.

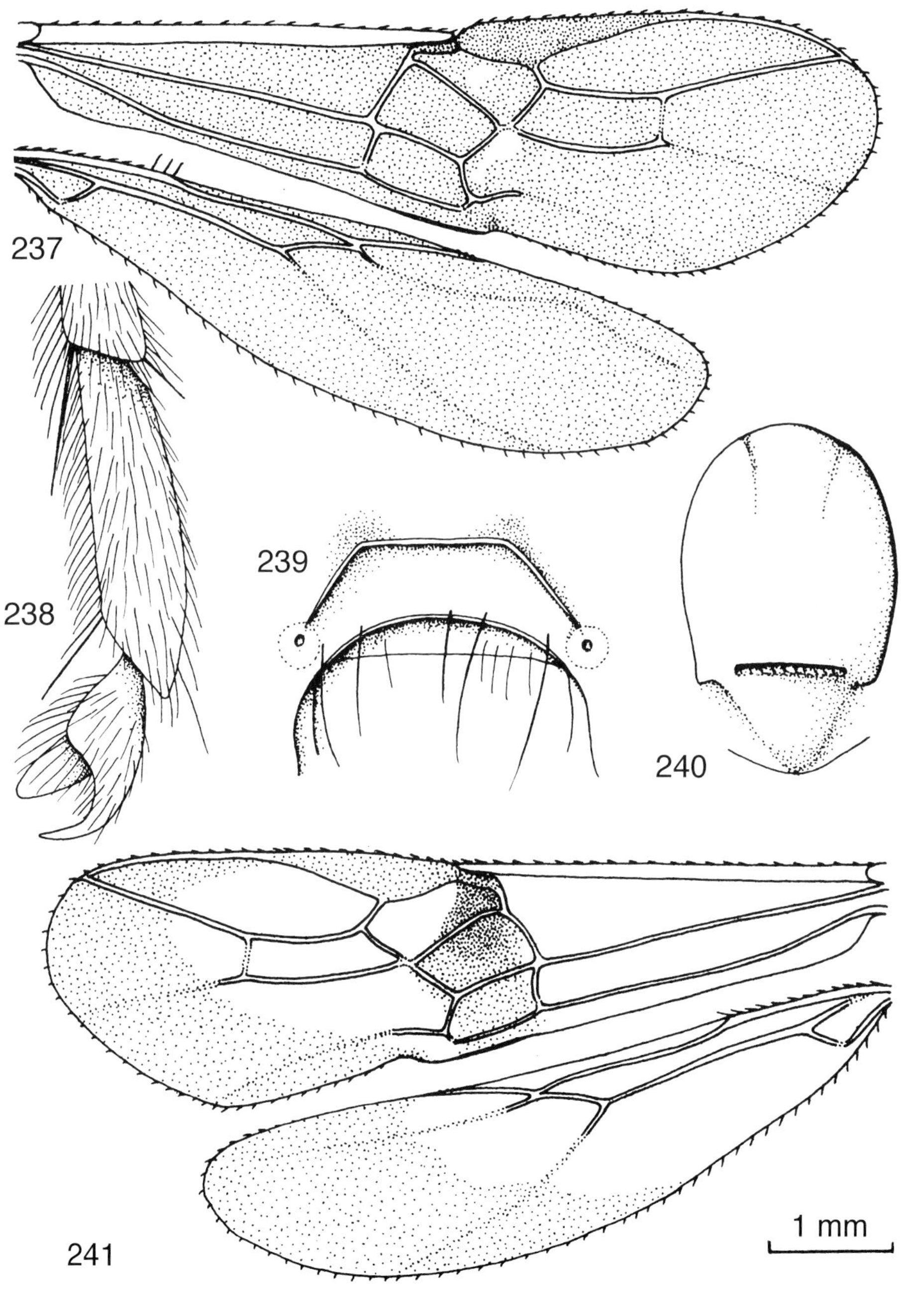

Figs 237, 238. *Digonogastra kimballi*, ♀. **Figs 239–241.** *Iphiaulax dubiotusus*, ♀. (237, 241) Wings. (238) Outer hind claw. (239) Clypeus, frontal aspect. (240) Mesosnotum, dorsal aspect. 237, 241: 1.0 × scale line; 238, 239: 6.7 ×; 240: 1.4 ×.

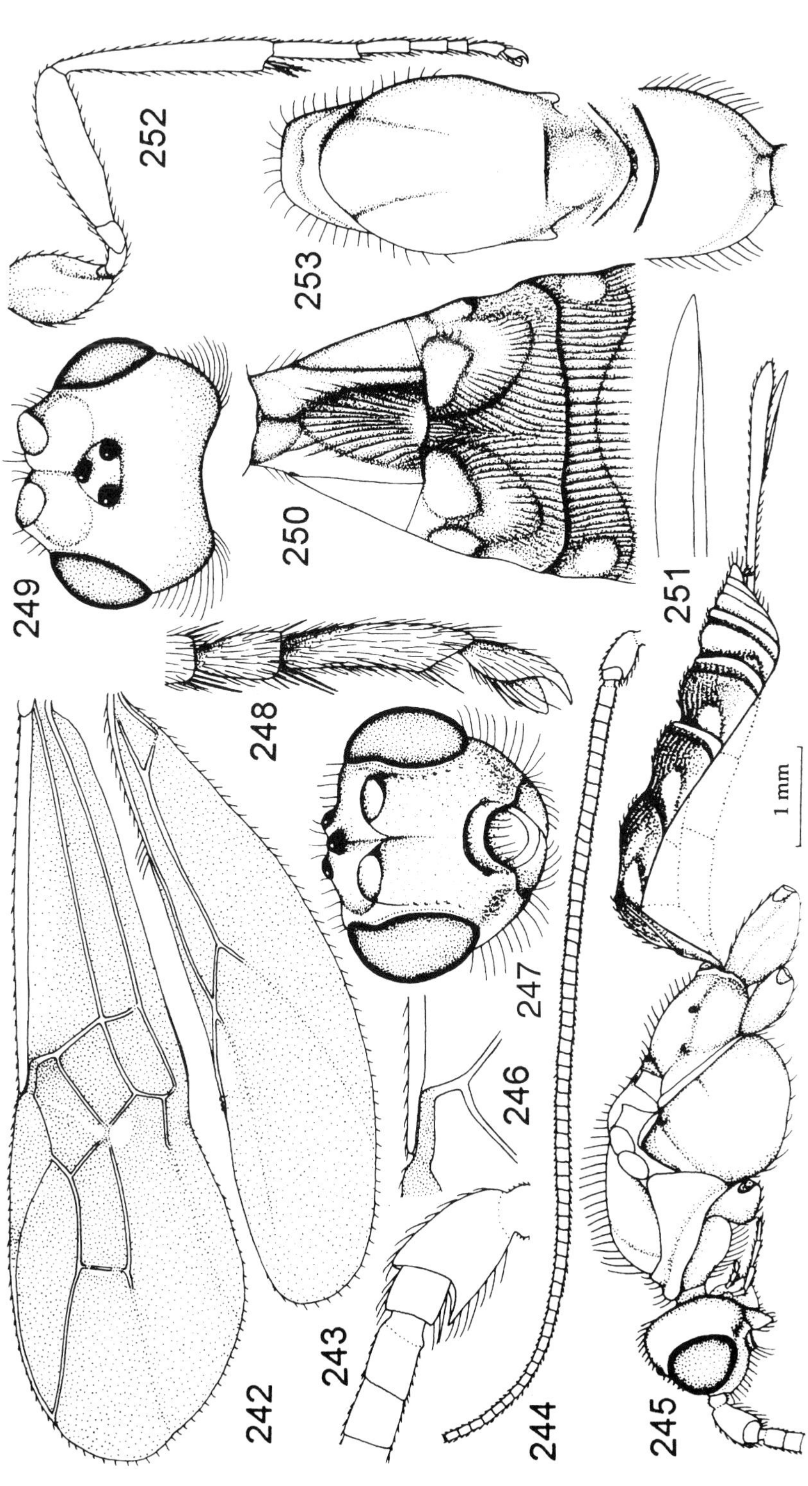

Figs 242–253. *Iphiaulax pilisoma*, ♀, holotype. (242) Wings. (243) Scapus, lateral aspect. (244) Antenna. (245) Habitus, lateral aspect. (246) Detail of vein 1-SR of fore wing. (247) Head, frontal aspect. (248) Outer hind claw. (249) Head, dorsal aspect. (250) First and second metasomal tergites, dorsal aspect. (251) Apex of ovipositor. (252) Hindleg. (253) Mesosoma, dorsal aspect. 242, 244, 245, 252: 1.0 × scale line; 243: 3.2 ×; 246: 2.0 ×; 247, 249: 2.2 ×; 248: 5.0 ×; 250, 253: 1.8 ×; 251: 5.9 ×.

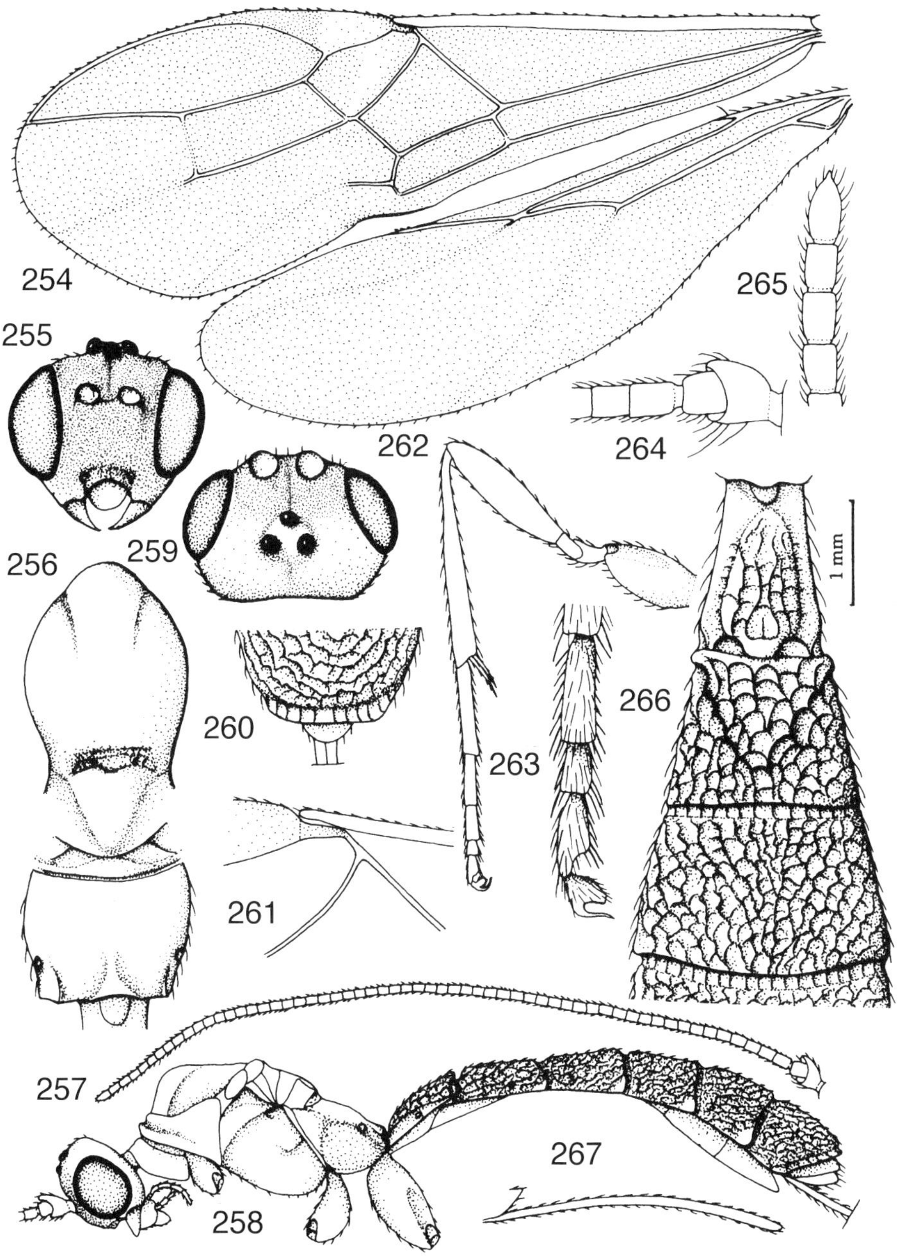

Figs 254–267. *Mesobraconoides psolopterus*, ♀, holotype. (254) Wings. (255) Head, frontal aspect. (256) Mesosoma, dorsal aspect. (257) Antenna. (258) Habitus, lateral aspect. (259) Head, dorsal aspect. (260) Apex of sixth metasomal tergite, dorsal aspect. (261) Detail of vein 1-SR of fore wing. (262) Hindleg; (263) Outer hind claw. (264) Scapus, outer lateral aspect. (265) Apex of antenna. (266) First to third metasomal tergites, dorsal aspect. (267) Ovipositor sheath. 254, 257, 258, 262, 267: 1.0 × scale line; 255, 256, 259–261, 264, 266: 2.0 ×; 263: 4.3 ×; 265: 5.0 ×.

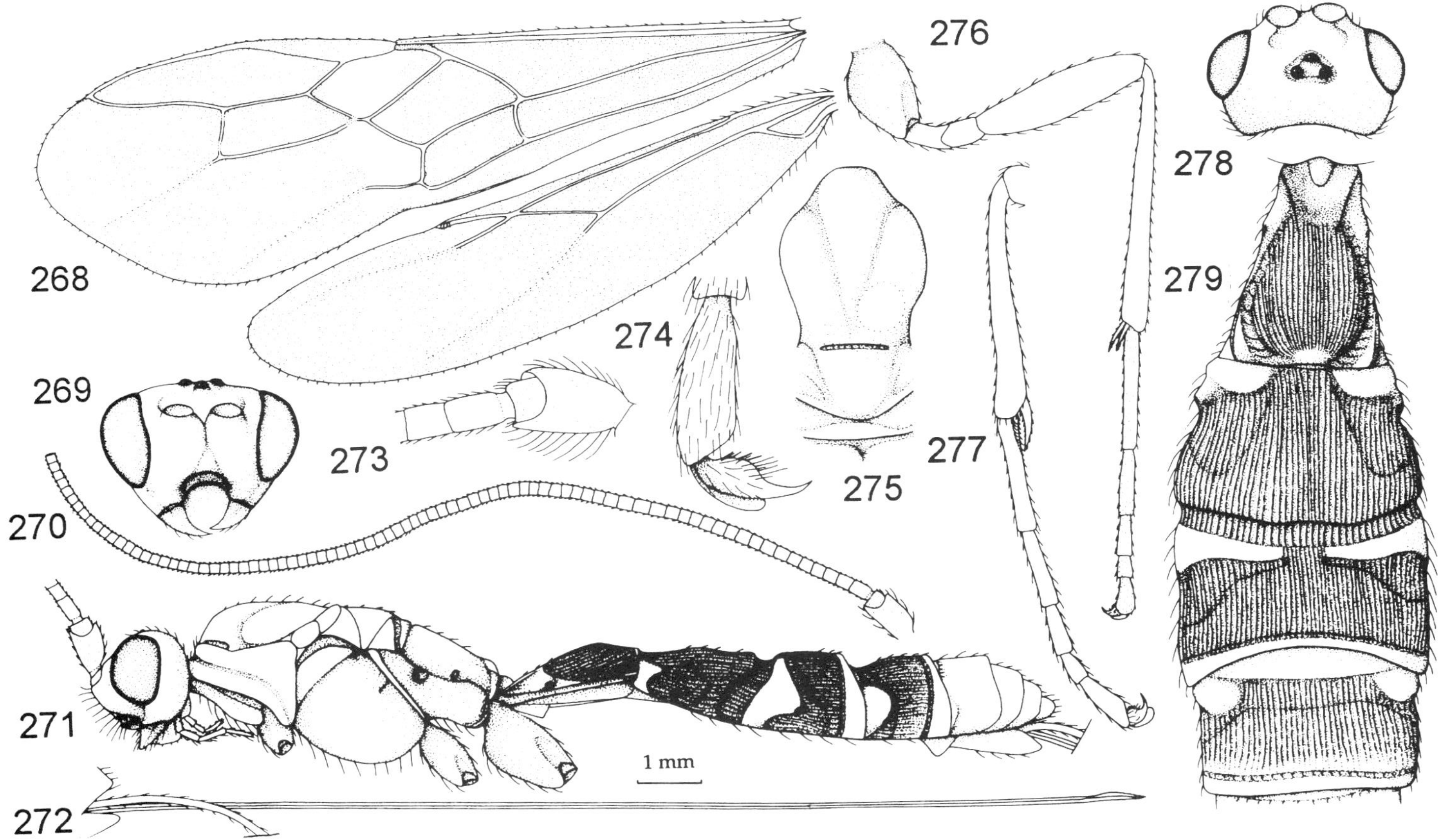

Figs 268–279. *Stenobracon* (*Euvipio*) *rufus*, ♀, holotype. (268) Wings. (269) Head, frontal aspect. (270) Antenna. (271) Habitus, lateral aspect. (272) Ovipositor. (273) Scapus, outer lateral aspect. (274) Inner middle claw. (275) Thorax, dorsal aspect. (276) Hindleg. (277) Fore tibia and tarsus. (278) Head, dorsal aspect. (279) First to fourth metasomal tergites, dorsal aspect. 268, 270–272, 276: 1.0 × scale line; 269, 275, 278: 1.4 ×; 273: 2.5 ×; 274: 5.0 ×; 277: 2.0 ×; 279: 1.3 ×.

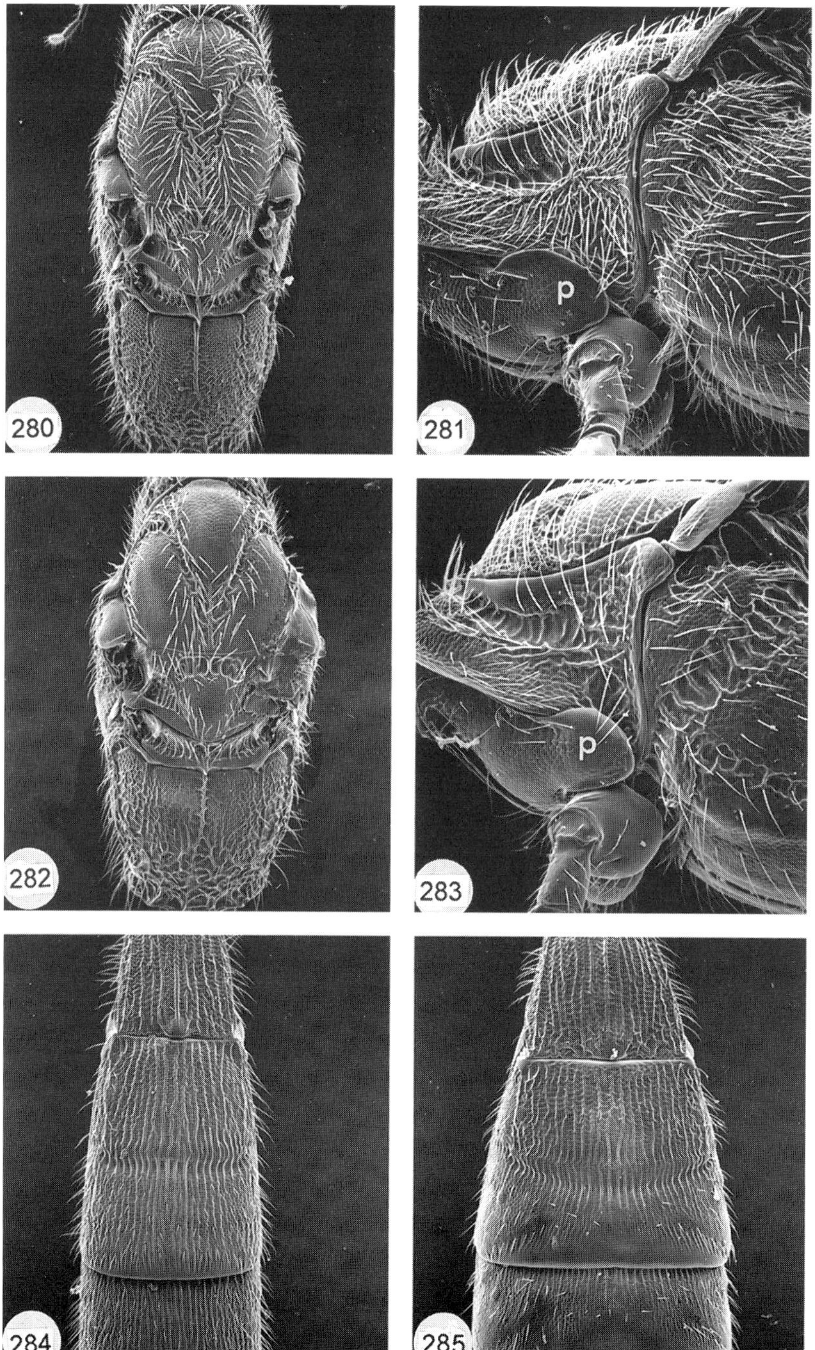

Figs 280, 281, 284. *Rhaconotus scirpophagae.* **Figs 282, 283, 285.** *Rhaconotus carinatus.* (280, 282) Mesosoma, dorsal aspect. (281, 283) Anterior half of mesosoma, lateral aspect. (284, 285) First to third metasomal tergites, dorsal aspect. 281, 283: p = posterior flange of propleuron.

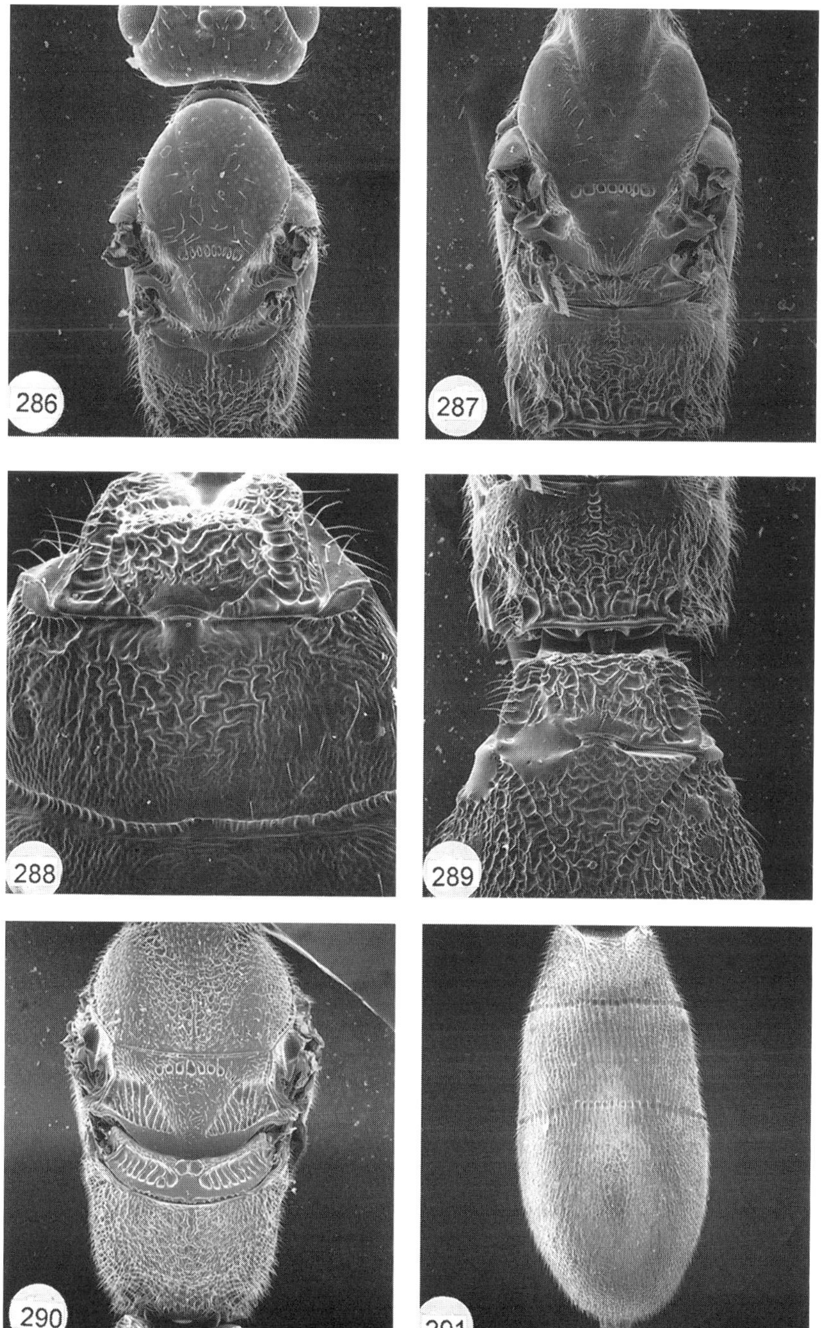

Figs 286, 288. *Bracon* (*Bracon*) *testaceorufatus*. **Figs 287, 289.** *Tropobracon antennatus*. **Figs 290, 291.** *Phanerotoma saussurei*. (286, 287, 290) Mesosoma, dorsal aspect. (288, 289) First and second metasomal tergites, dorsal aspect. (291) Metasoma, dorsal aspect.

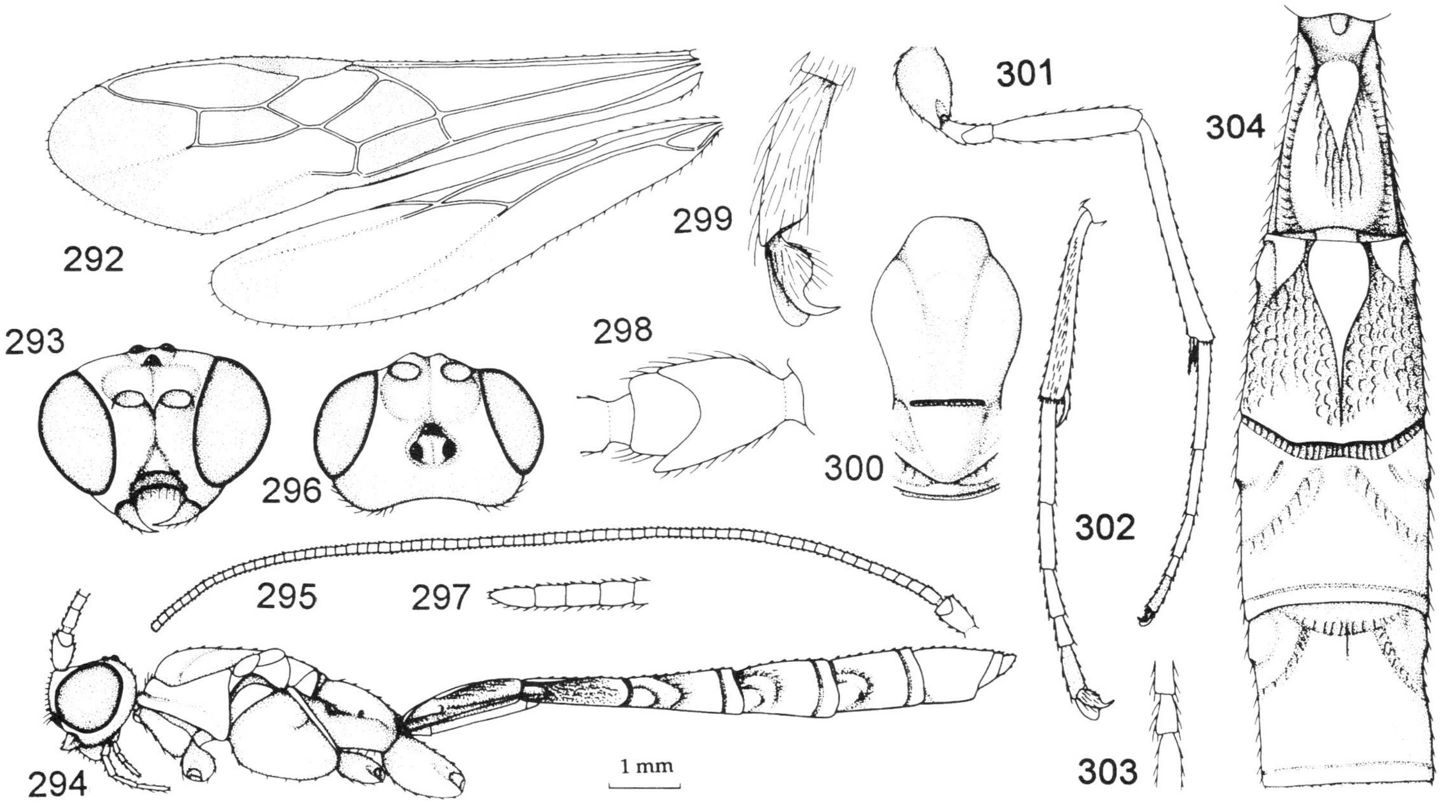

Figs 292–304. *Stenobracon* (*Stenobracon*) *oculatus*, ♀. (292) Wings. (293) Head, frontal aspect. (294) Habitus, lateral aspect. (295) Antenna. (296) Head, dorsal aspect. (297) Apex of antenna. (298) Scapus, outer lateral aspect. (299) Outer hind claw. (300) Thorax, dorsal aspect. (301) Hindleg. (302) Fore tibia and tarsus. (303) Fourth hind tarsal segment, dorsal aspect. (304) First to fourth metasomal tergites, dorsal aspect. 292, 294, 295, 301: 1.0 × scale line; 293, 296: 2.0 ×; 297–299: 5.0 ×; 300, 302–304: 2.0 ×.

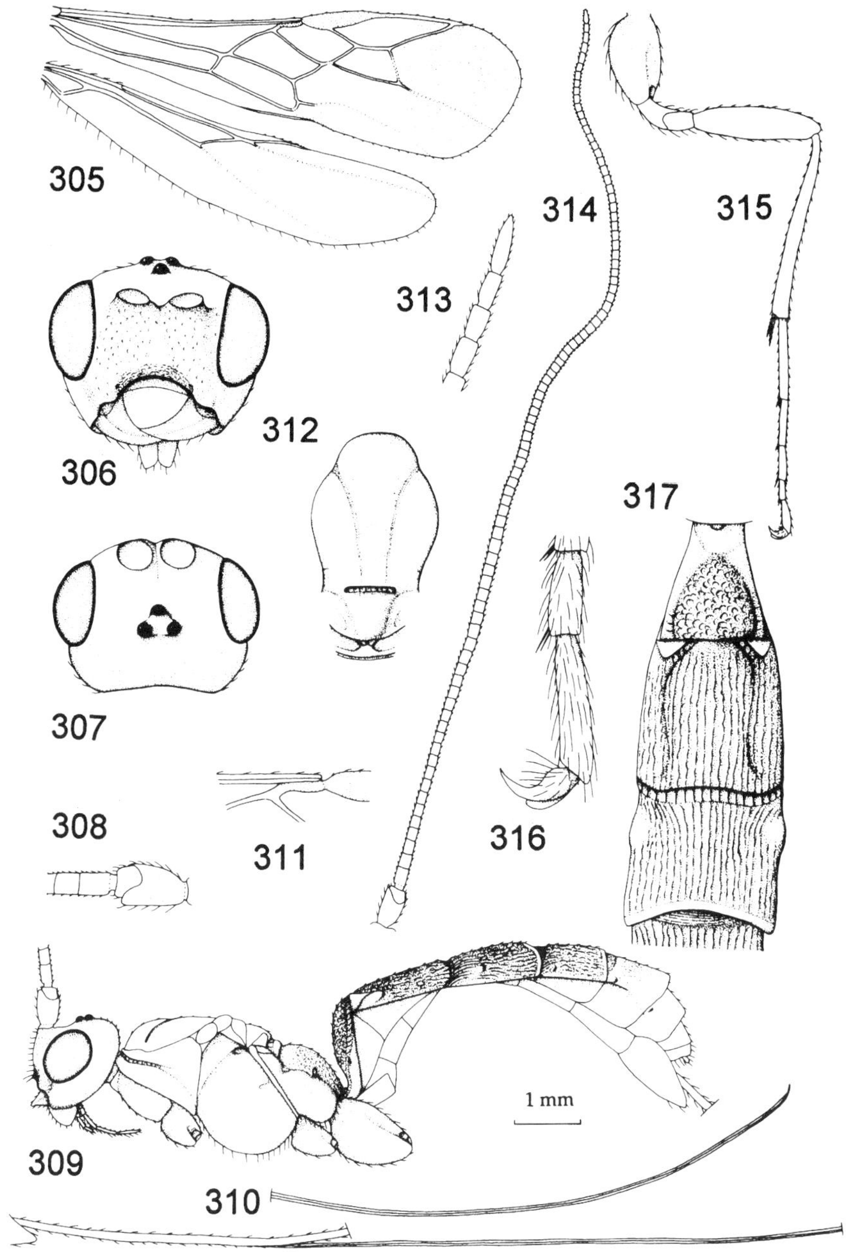

Figs 305–317. *Odesia longicornis*, ♀, holotype. (305) Wings. (306) Head, frontal aspect. (307) Head, dorsal aspect. (308) Scapus, outer lateral aspect. (309) Habitus, lateral aspect. (310) Ovipositor. (311) Detail of vein 1-SR of fore wing. (312) Thorax, dorsal aspect. (313) Apex of antenna. (314) Antenna. (315) Hindleg. (316) Outer hind claw. (317) First to third metasomal tergites, dorsal aspect. 305, 309, 310, 314, 315: 1.0 × scale line; 306–308, 311: 2.0 ×; 312, 317: 1.5 ×; 313, 316: 5.0 ×.

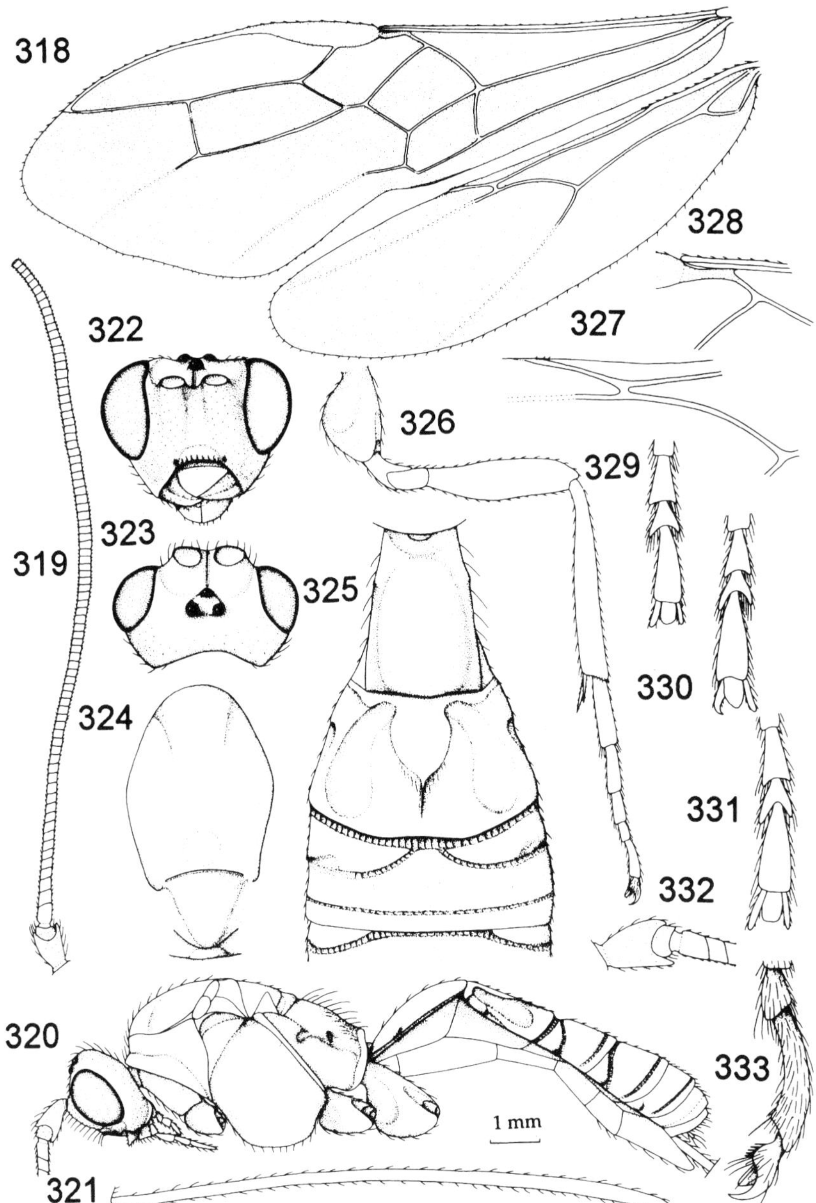

Figs 318–333. *Bathyaulax cyanogaster*, ♀. (318) Wings. (319) Antenna. (320) Habitus, lateral aspect. (321) Ovipositor sheath. (322) Head, frontal aspect. (323) Head, dorsal aspect. (324) Thorax, dorsal aspect. (325) First to third metasomal tergites, dorsal aspect. (326) Hindleg. (327) Detail of vein 1r-m of hind wing. (328) Detail of vein 1-SR of fore wing. (329) Apical half of fore tarsus, dorsal aspect. (330) Dorsal aspect of midtarsus. (331) Hind tarsus. (332) Scapus, outer lateral aspect. (333) Outer hind claw. 318–321, 326: 1.0 × scale line; 322–325: 1.5 ×; 327–332: 2.0 ×; 333: 3.1 ×.

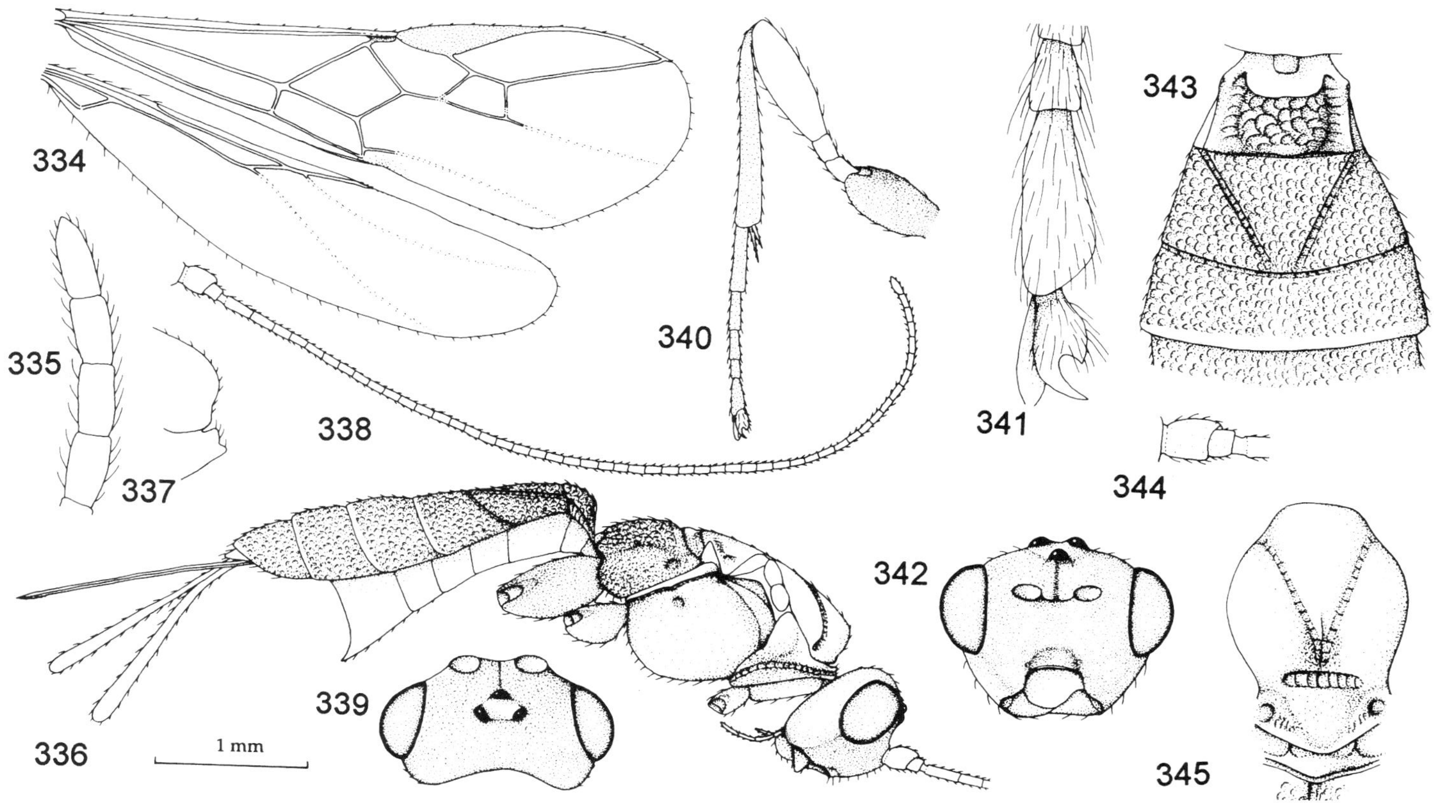

Figs 334–345. *Tropobracon luteus.* (334) Wings. (335) Apex of antenna. (336) Habitus, lateral aspect. (337) Mesoscutum and pronotum anteriorly, lateral aspect. (338) Antenna. (339) Head, dorsal aspect. (340) Habitus, lateral aspect. (341) Outer hind claw. (342) Head, frontal aspect. (343) First to third metasomal tergites, dorsal aspect. (344) Scapus, outer lateral aspect. (345) Thorax, dorsal aspect. 346, 347, 349, 350: 1.0 × scale line; 335, 341: 5.0 ×; 334, 337: 2.0 ×; 339, 342, 343, 345: 1.8 ×.

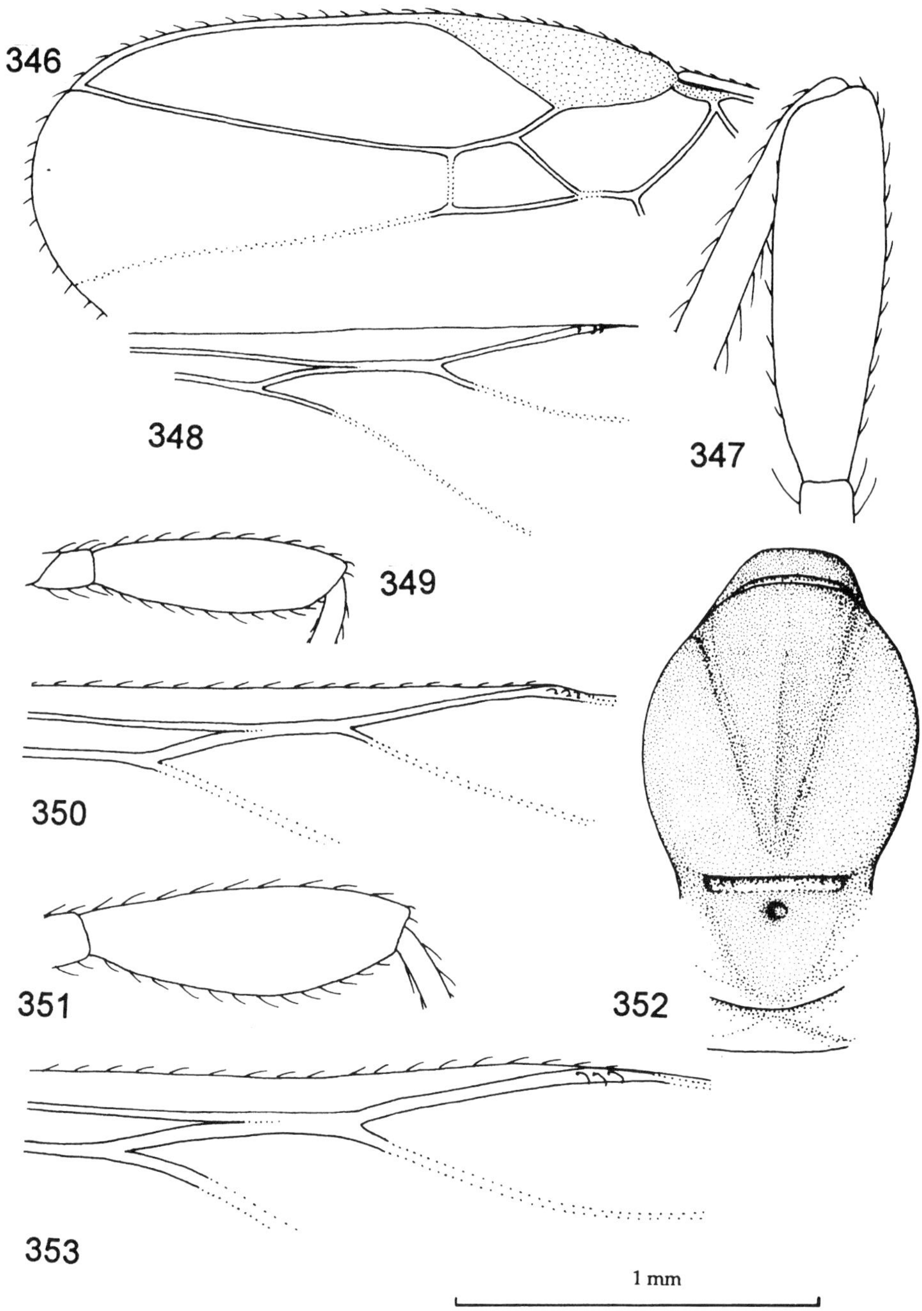

Figs 346, 347, 349, 350. *Tropobracon persimilis*, ♀. **Figs 348, 351–353.** *Tropobracon antennatus*, ♀. (346) Anterior part of fore wing. (347, 349, 351) Hind femur. (348, 350, 353) Detail of veins 1r-m and 1-SC+R of hind wing. (352) Mesoscutum, dorsal aspect. 346, 349, 351, 352: 1.0 × scale line; 347, 350, 353: 1.5 ×; 348: 1.2 ×.

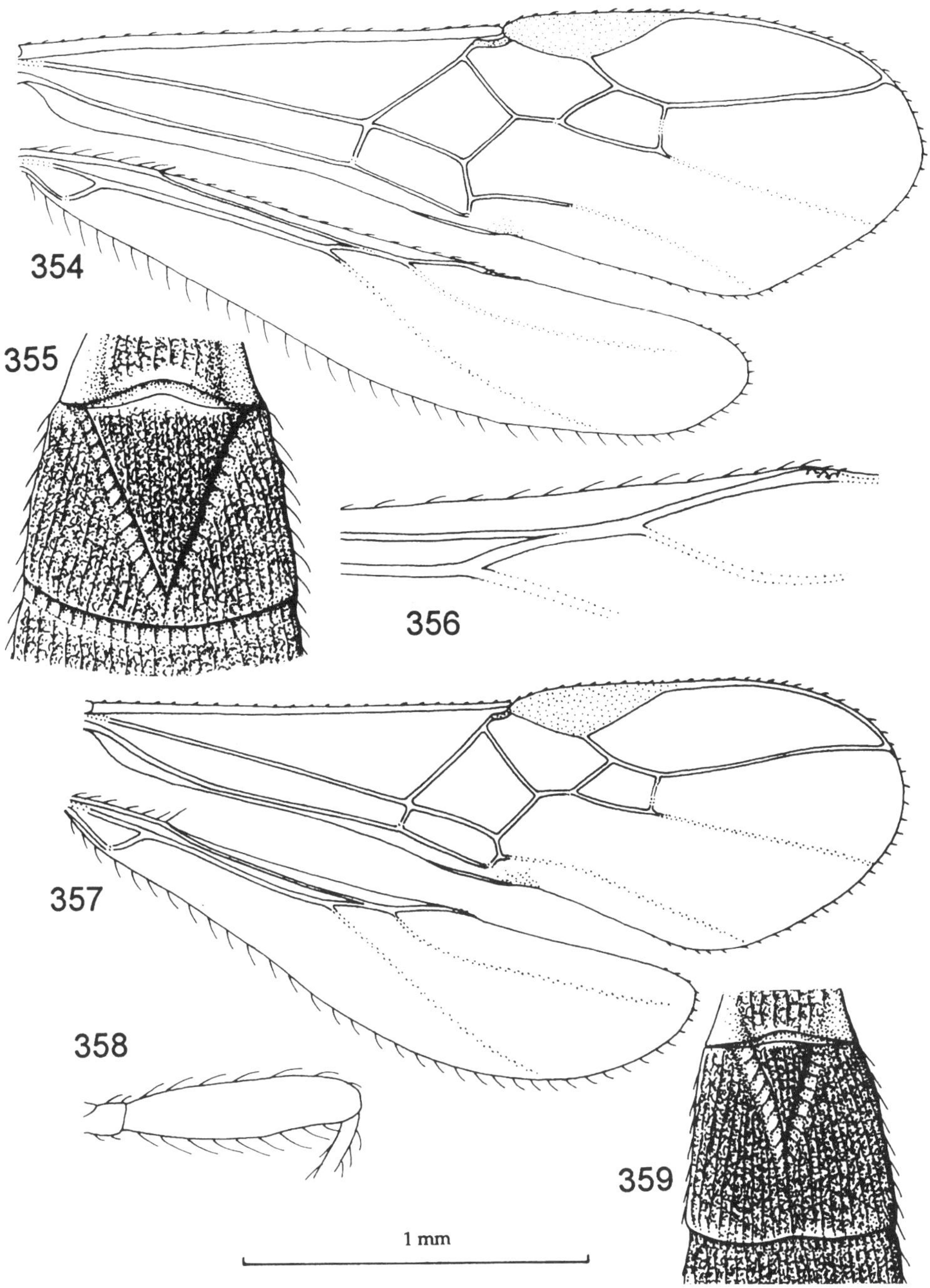

Figs 354, 355. *Tropobracon antennatus*, ♀. **Figs 356–359**. *Tropobracon comorensis* ♂.
(354, 357) Wings. (355, 359) Second metasomal tergite, dorsal aspect. (356) Detail of veins 1r-m and 1-SC+R of hind wing. (358) Hind femur. 354, 357: 0.7 × scale line; 355, 358, 359: 1.0 ×; 356: 1.5 ×.

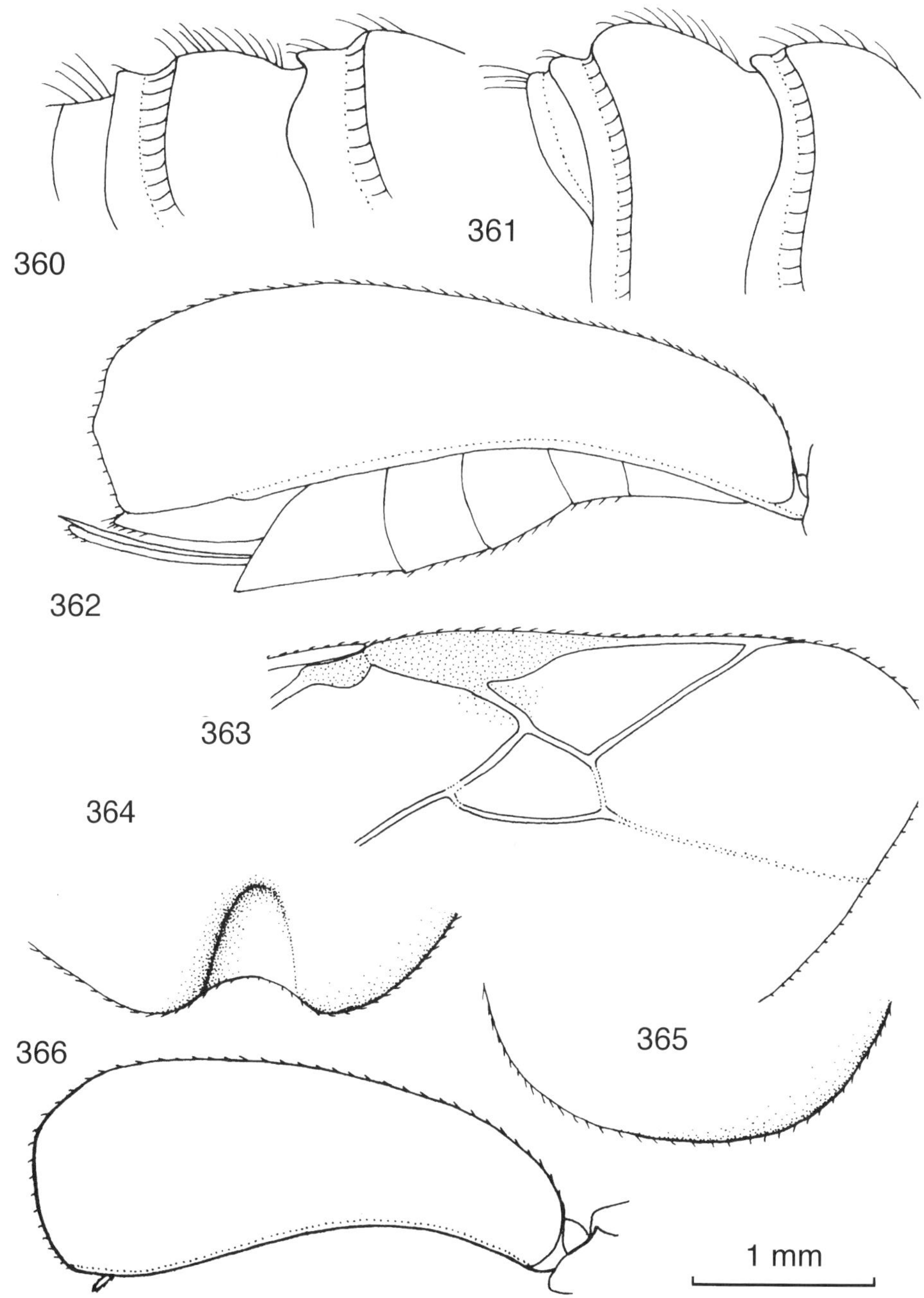

Fig. 360. *Iphiaulax pilisoma*, ♀; **Fig. 361.** *Iphiaulax dubiotusus*. **Figs 362–365.** *Chelonus maudae*. **Fig. 366.** *Chelonus curvimaculatus*. (360, 361) Detail of fifth metasomal tergite, lateral aspect. (362, 366) Metasoma, lateral aspect. (363) Apical part of fore wing. (364, 365) apex of metasoma, posterodorsal aspect. 360, 361: 2.0 × scale line; 362: 1.0 ×; 363: 1.4 ×; 364, 365: 2.0 ×; 366: 1.6 ×.

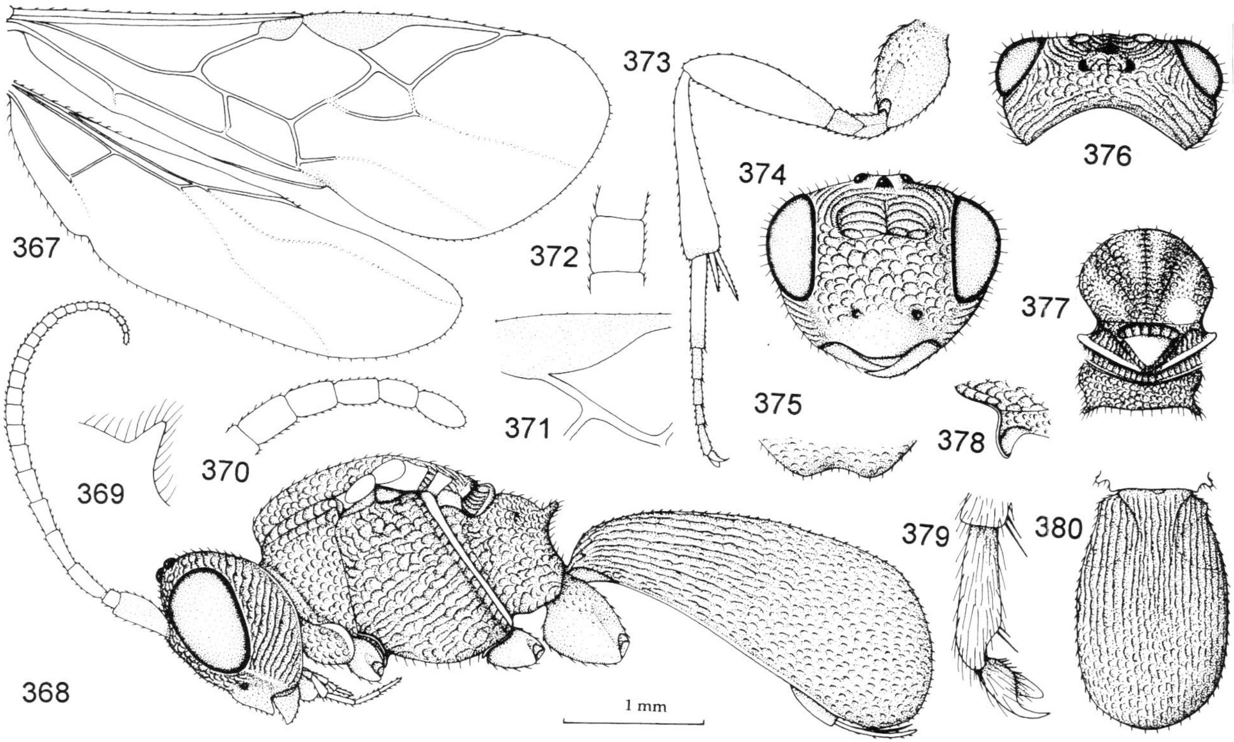

Figs 367–380. *Chelonus oculator*, ♀. (367) Wings. (368) Habitus, lateral aspect. (369) Detail of propodeal tubercle, lateral aspect. (370) Apex of antenna. (371) Detail of vein r of fore wing. (372) Detail of sixteenth antennal segment. (373) Hindleg. (374) Head, frontal aspect. (375) Apex of metasoma, posterodorsal aspect. (376) Head, dorsal aspect. (377) Mesosoma, dorsal aspect. (378) Detail of lateral carina of mesoscutum, posterior part. (379) Outer hind claw. (380) Metasoma, dorsal aspect. 367, 368, 373–376: 1.0 × scale line; 369, 371, 378: 2.0 ×; 370, 372, 379: 5.0 ×; 377, 380: 0.6 ×.

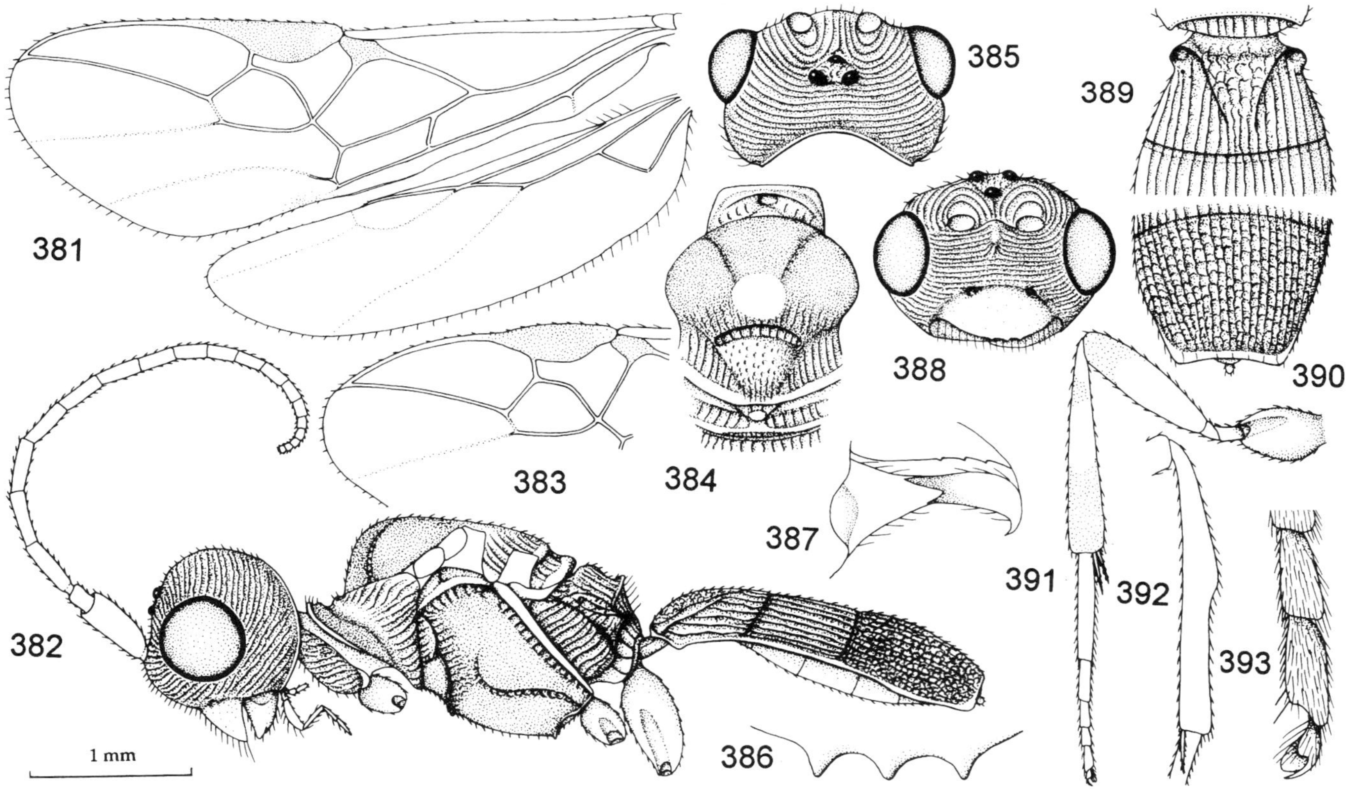

Figs 381–393. *Phanerotoma rufescens*, ♀. (381) Wings. (382) Habitus, lateral aspect. (383) Detail of apical part of fore wing. (384) Thorax, dorsal aspect. (385) Head, dorsal aspect. (386) Ventral margin of clypeus. (387) Detail of mandible, lateroventral aspect. (388) Head, frontal aspect. (389) First metasomal tergite, dorsal aspect. (390) Third metasomal tergite, dorsal aspect. (391) Hindleg. (392) Middle tibia, posterior aspect. (393) Outer hind claw. 381–385, 388–391: 1.0 × scale line; 387, 392: 2.0 ×; 386, 393: 5.0 ×.

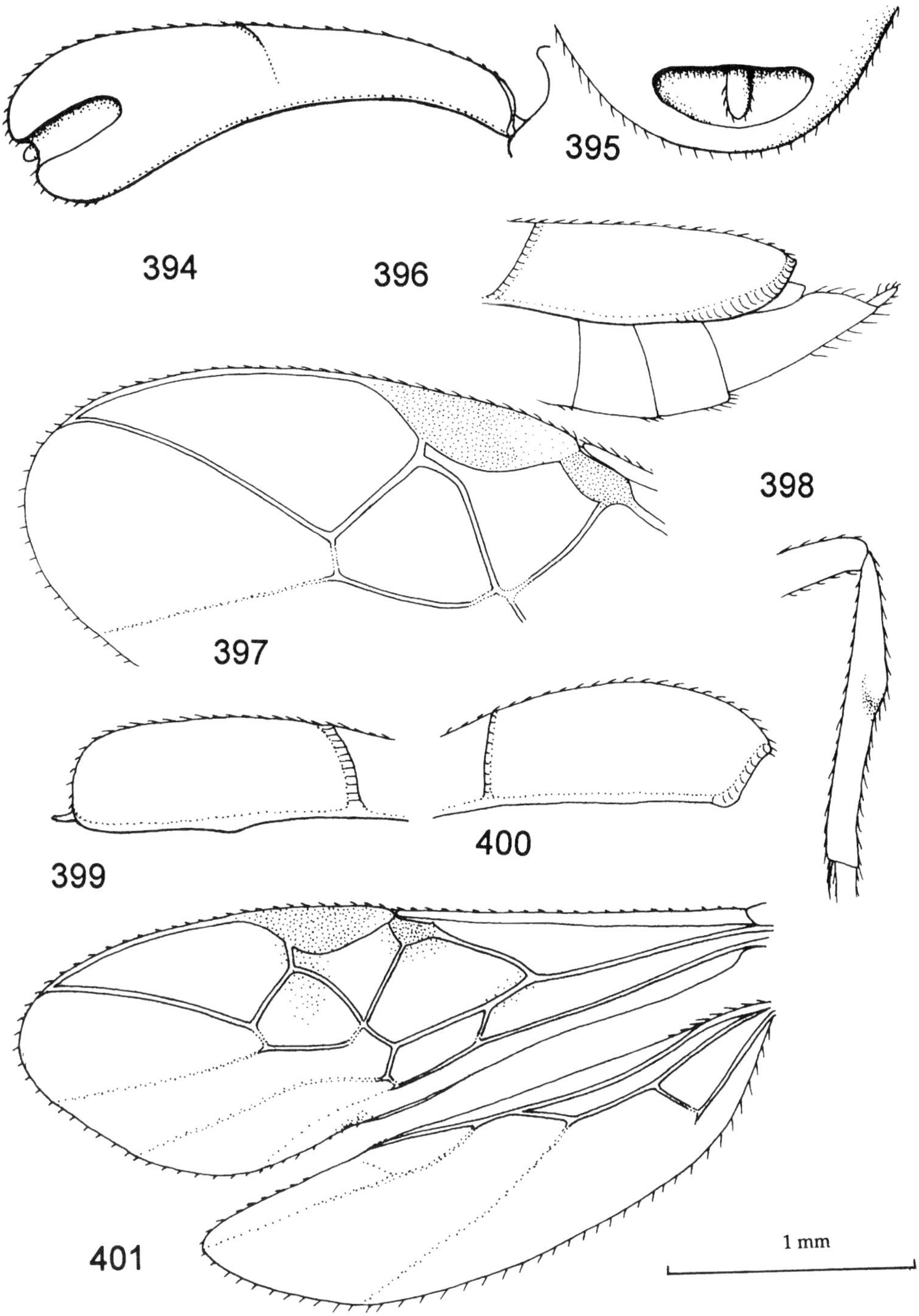

Figs 394, 395. *Chelonus curvimaculatus.* **Figs 396, 397.** *Phanerotoma hendecasisella.*
Figs 398–401. *Phanerotoma leucobasis.* (394) Metasoma, lateral aspect. (395) Apex of metasoma, posteroventral aspect. (396, 399, 400) Third metasomal tergite, lateral aspect. (397) Apical part of fore wing. (398) Middle tibia, frontal aspect. (401) Wings. 394: 1.1 × scale line; 395–400: 1.4 ×; 401: 1.0 ×.

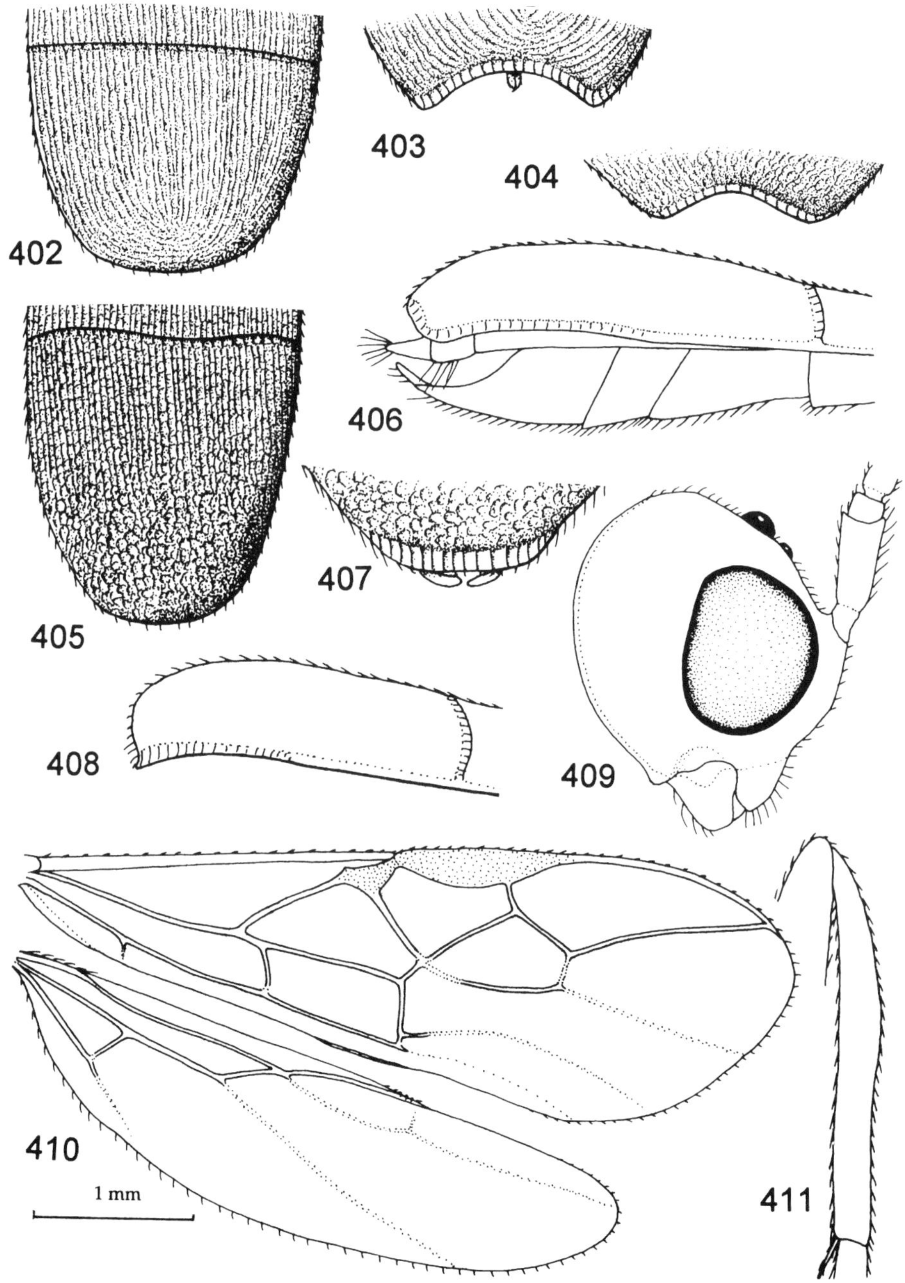

Figs 402–404. *Phanerotoma leucobasis*, ♀ (404 = ♂). **Figs 405–411.** *Phanerotoma saussurei*, ♀. (402, 405) Third metasomal tergite, dorsal aspect. (403, 404, 407) Apex of metasoma, posterodorsal aspect. (406) Third tergite, lateral aspect. (409) Head, lateral aspect. (410) Wings. (411) Middle tibia, frontal aspect. 402–403: 2.0 × scale line; 405, 408: 1.5 ×; 406: 1.8 ×; 407: 2.2 ×; 409: 1.9 ×; 410: 1.0 ×; 411: 2.3 ×.

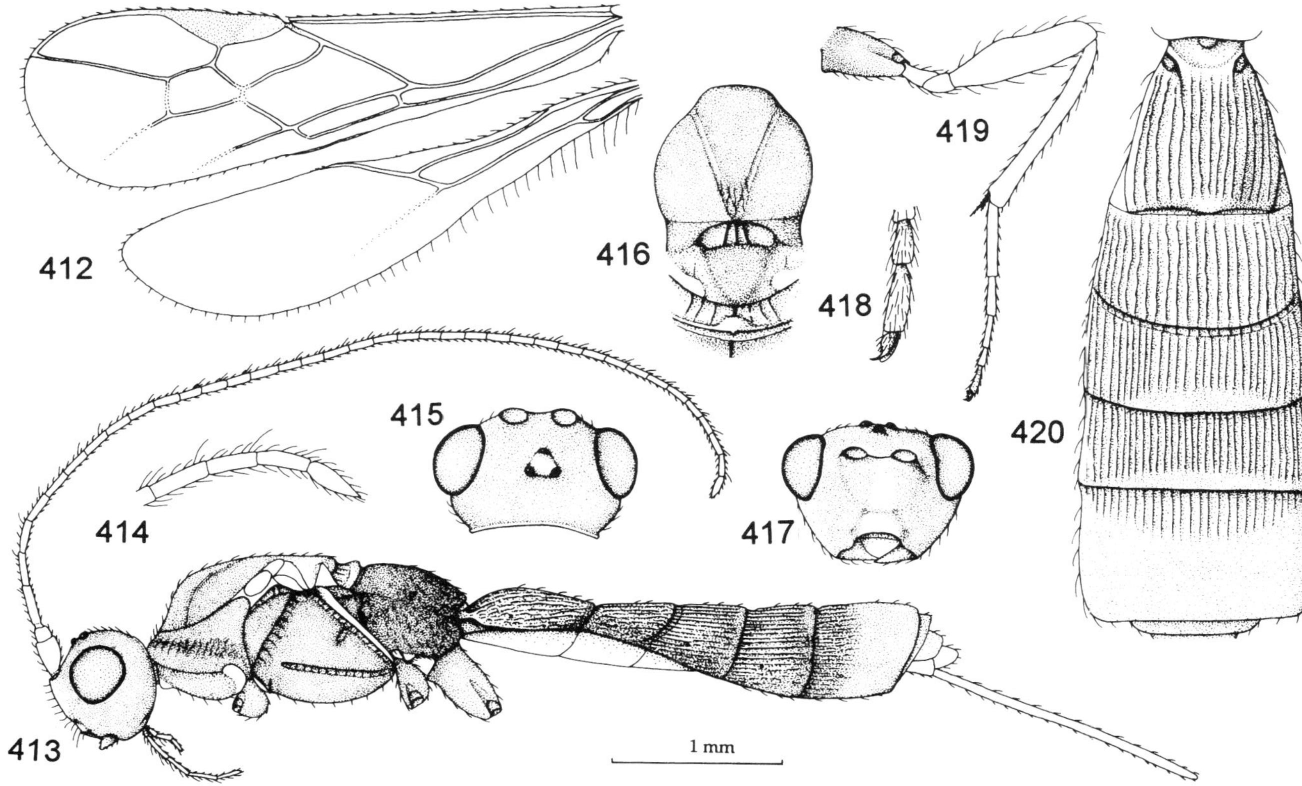

Figs 412–420. *Rhaconotus ollivieri*, ♀. (412) Wings. (413) Habitus, lateral aspect. (414) Apex of antenna. (415) Head, dorsal aspect. (416) Thorax, dorsal aspect. (417) Head, frontal aspect. (418) Outer hind claw. (419) Hindleg. (420) First to fifth metasomal tergites, dorsal aspect. 412, 413, 419: 1.0 × scale line; 414, 418, 2.5 ×; 415–417, 420: 1.3 ×.

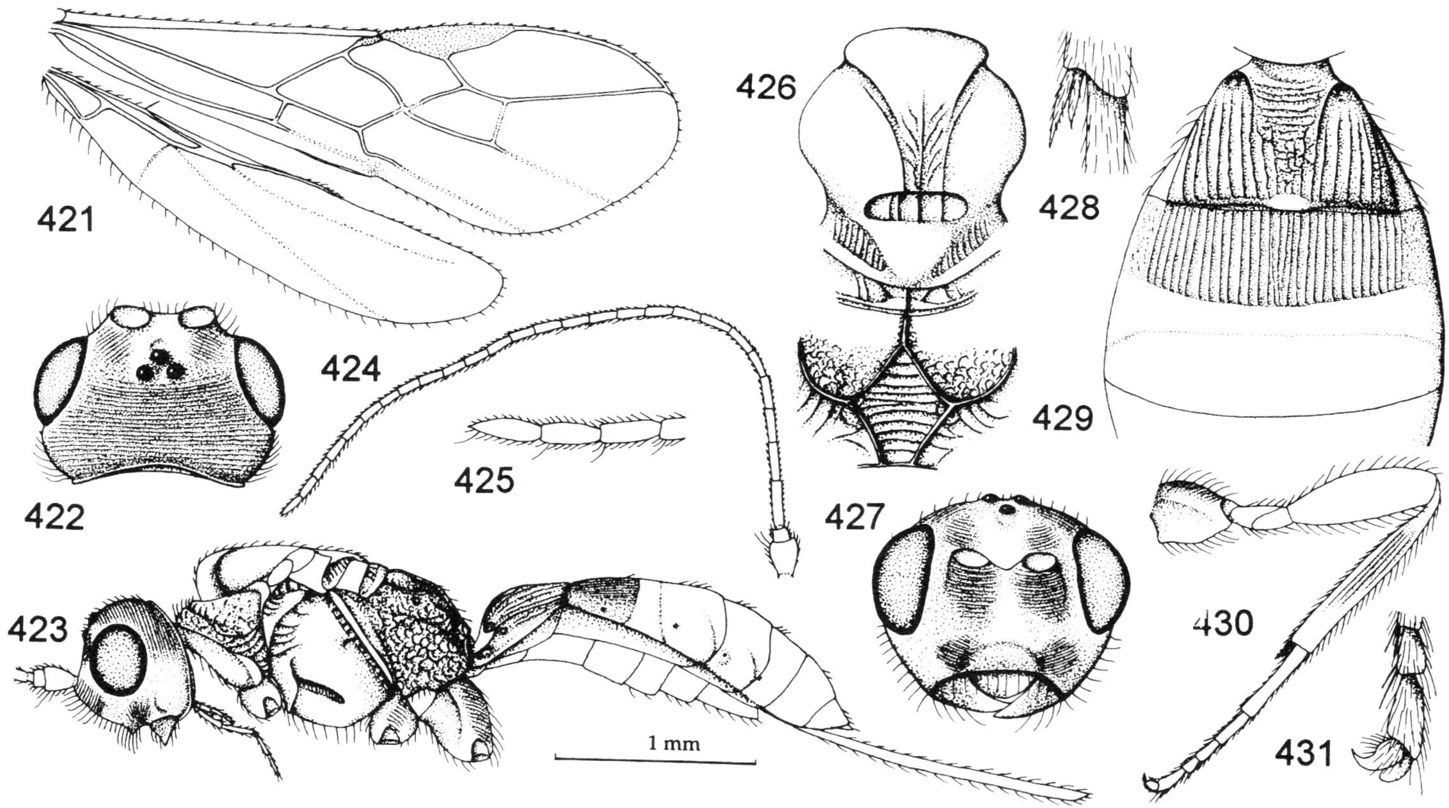

Figs 421–431. *Parallorhogas pyralophagus*, ♀. (421) Wings. (422) Head, dorsal aspect. (423) Habitus, lateral aspect. (424) Antenna. (425) Apex of antenna. (426) Mesosoma, dorsal aspect. (427) Head, frontal aspect. (428) Hind tibial spurs. (429) First to third metasomal tergites, dorsal aspect. (430) Hindleg. (431) Outer hind claw. 421, 423, 424, 430: 1.0 × scale line; 422, 426, 427, 429: 1.6 ×; 425, 428, 431: 2.5 ×.

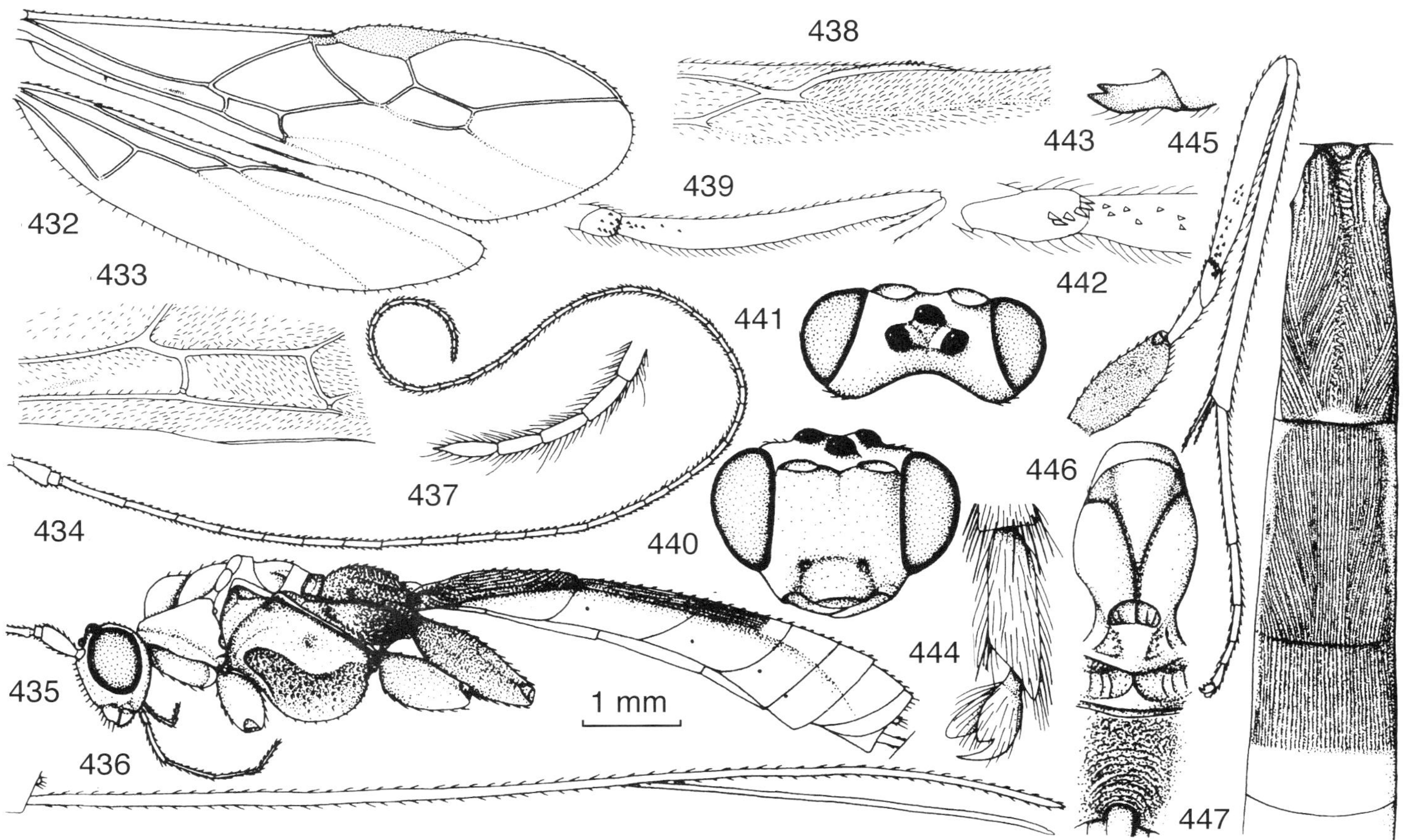

Figs 432–447. *Macrocentrus sesamivorus*, ♀. (432) Wings. (433) Detail of first subdiscal and apical half of subbasal cells of fore wing. (434) Antenna. (435) Habitus, lateral aspect. (436) Ovipositor. (437) Apex of antenna. (438) Detail of vein 2-SC+R of hind wing. (439) Fore femur, lateral aspect. (440) Head, frontal aspect. (441) Head, dorsal aspect. (442) Hind trochantellus. (443) Mandible, dorsal aspect. (444) Inner hind claw. (445) Hindleg. (446) Mesosoma, dorsal aspect. (447) First to third metasomal tergites, dorsal aspect. 432, 434–436, 445: 1.0 × scale line; 433, 438: 2.4 ×; 437: 5.0 ×; 439, 447: 1.8 ×; 440, 441: 1.6 ×; 442: 2.6 ×; 443: 7.3 ×; 444: 5.8 ×; 446: 1.2 ×.

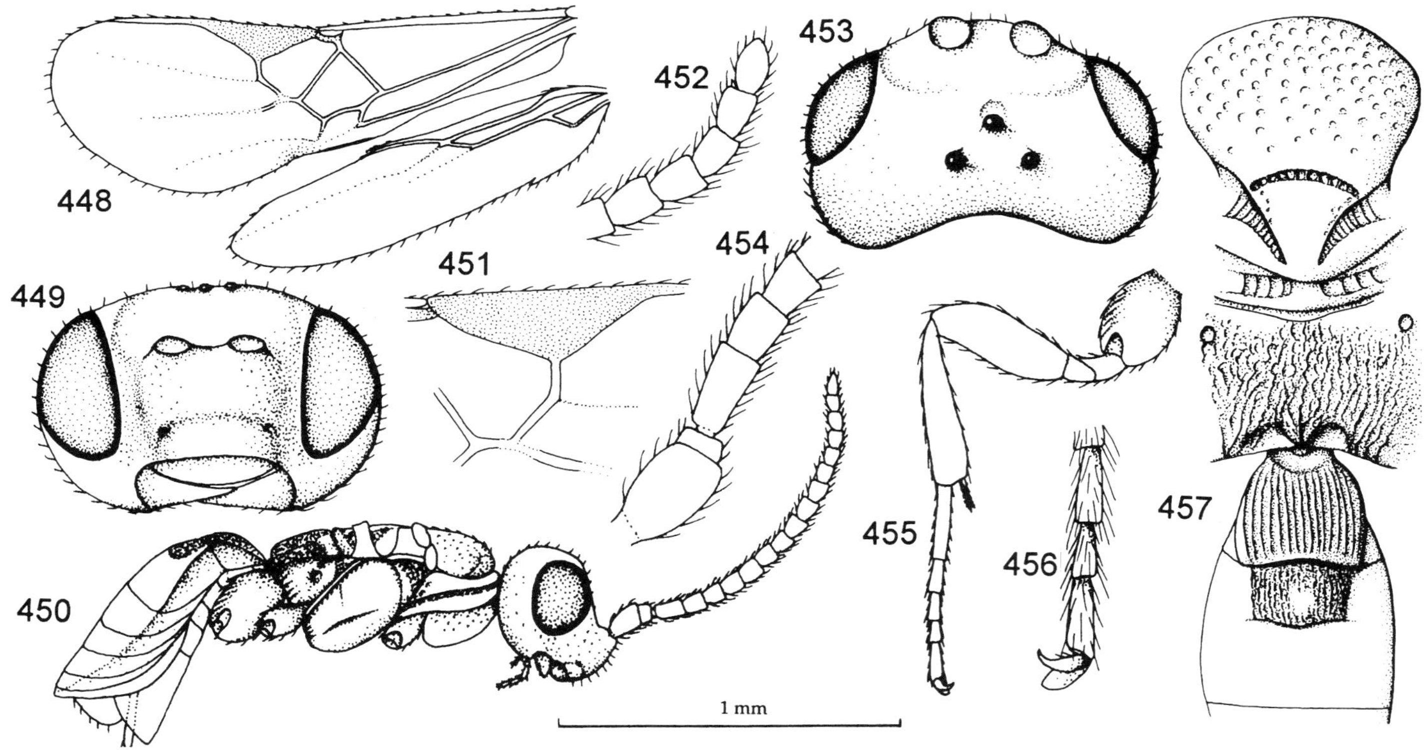

Figs 448–457. *Cotesia flavipes*, ♀. (448) Wings. (449) Head, frontal aspect. (450) Habitus, lateral aspect. (451) Detail of veins r and 2-SR of fore wing. (452) Apex of antenna. (453) Base of antenna. (454) Head, dorsal aspect. (455) Hindleg. (456) Outer hind claw. (457) Mesosoma and first to third metasomal tergites, dorsal aspect. 448, 450, 455: 1.0 × scale line; 449, 454: 1.8 ×; 451: 1.9 ×; 452, 453, 456: 2.5 ×; 457: 1.5 ×.

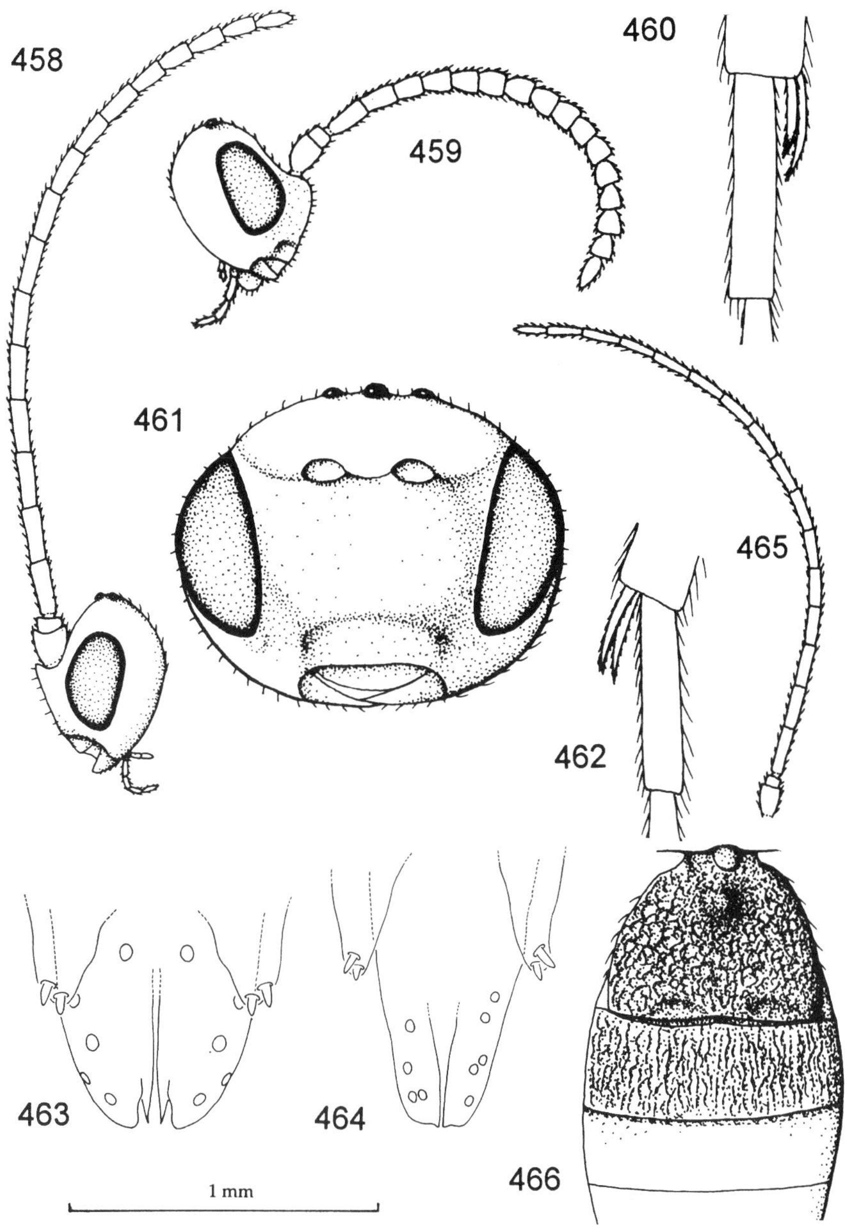

Fig. 458. *Cotesia ruficrus*, ♀. **Figs 459–461, 463.** *Cotesia sesamiae*. **Figs 462, 464, 465.** *Cotesia flavipes*. **Fig. 466.** *Dolichogenidea oryzae*. (458, 459) Head and antenna, lateral aspect. (460, 462) Hind tibial spurs. (461) Head, frontal aspect. (463, 464) Apex of aedeagus. (465) Antenna. (466) First to third metasomal tergites, dorsal aspect. 458: 1.0 × scale line; 459, 466: 1.2 ×; 460–462: 2.3 ×; 465: 1.1 ×; 463, 464.

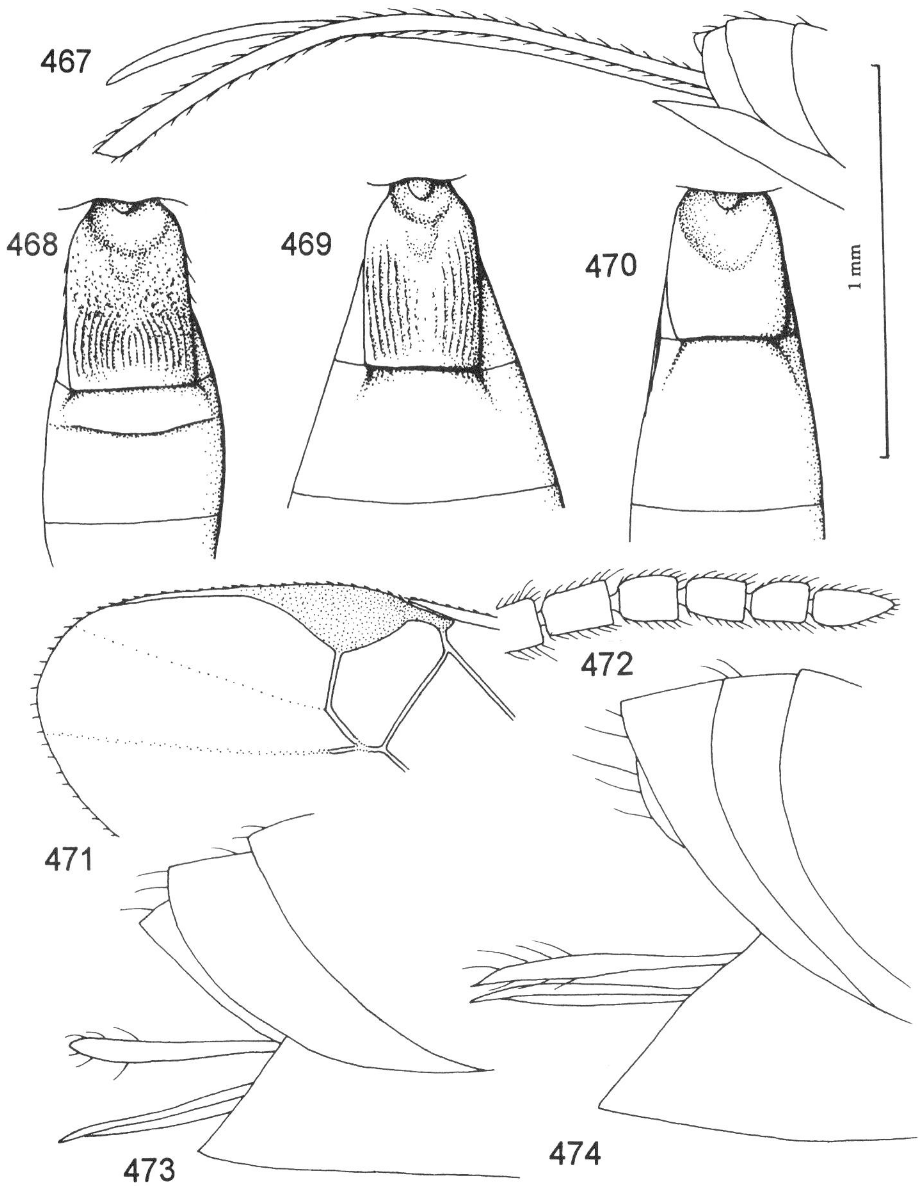

Figs 467, 468. *Dolichogenidea polaszeki.* **Fig. 469.** *Dolichogenidea fuscivora.*
Figs 470, 473. *Dolichogenidea cameroonensis.* **Figs 471, 472, 474.** *Dolichogenidea aethiopica.*
(467, 473, 474) Apex of metasoma, lateral aspect. (468–470) First to third metasomal tergites, dorsal aspect. (471) Detail of apical part of fore wing. (472) Apex of antenna. 467–469: 1.0 × scale line; 470: 1.6 ×; 471: 0.8 ×; 472–474: 1.9 ×.

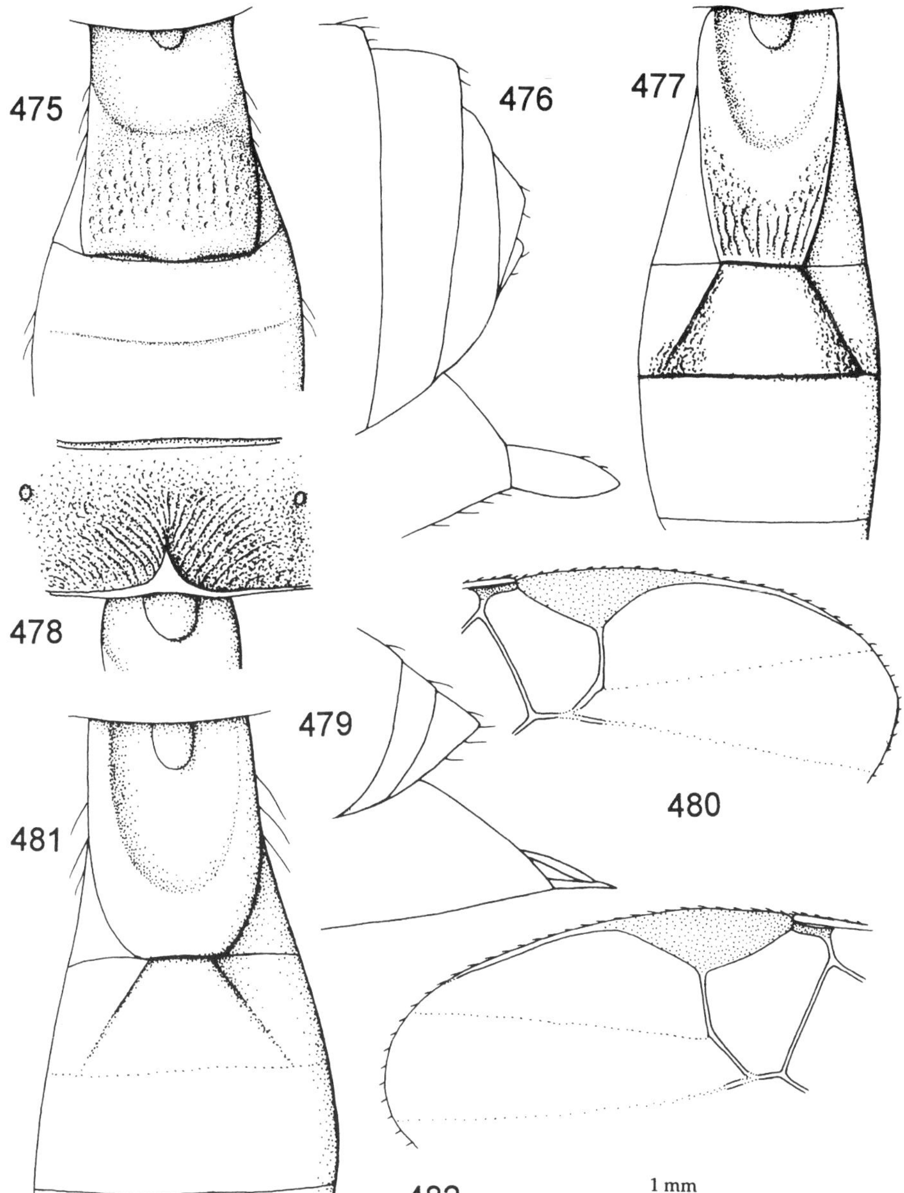

Fig. 475. *Dolichogenidea aethiopica*, ♀; **Figs 476–478, 480.** *Glyptapanteles maculitarsis*, ♀; **Figs 479, 481, 482.** *Glyptapanteles africanus*, ♀. (475, 477, 481) First to third metasomal tergites, dorsal aspect. (476, 479) Apex of metasoma. (478) Propodeum, dorsal aspect. (480, 482) Apical part of fore wing. 475, 476, 479: 2.5 × scale line; 477, 478: 2.1 ×; 480: 1.0 ×; 481: 3.0 ×; 482: 1.5 ×.

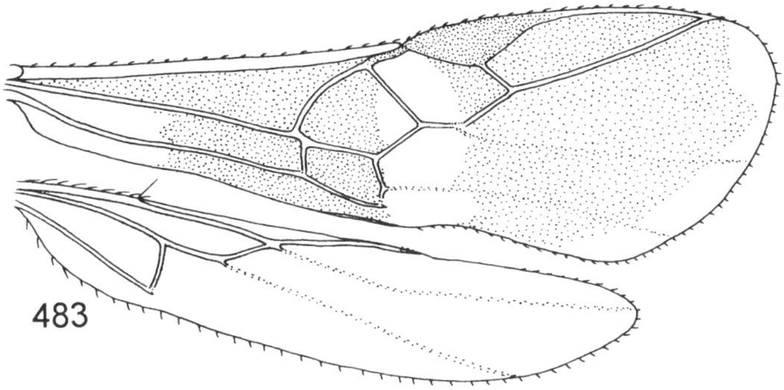

483

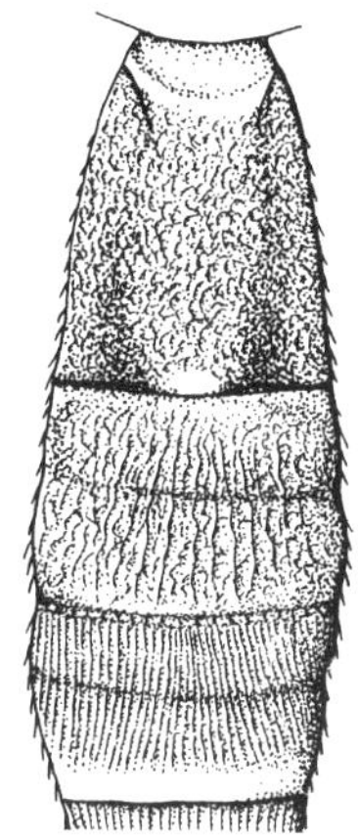

484

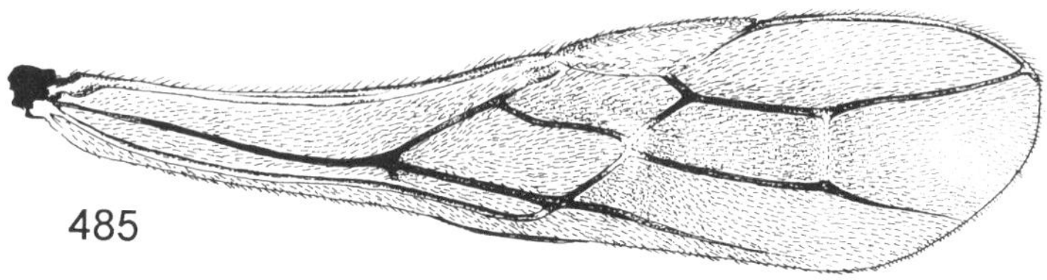

485

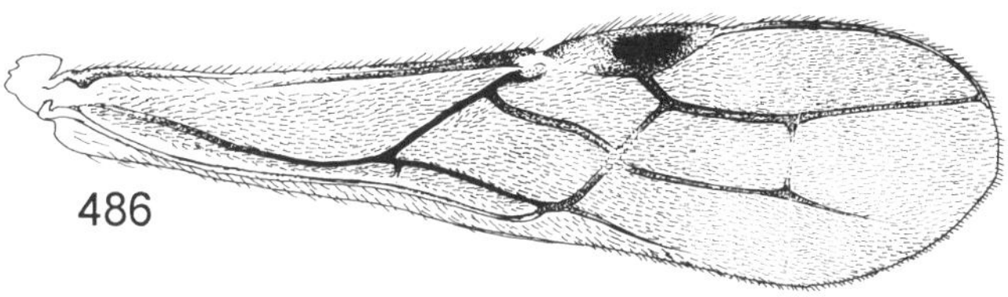

486

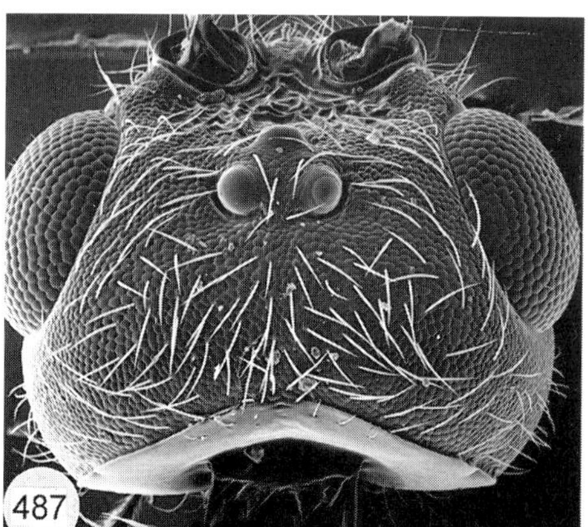

487

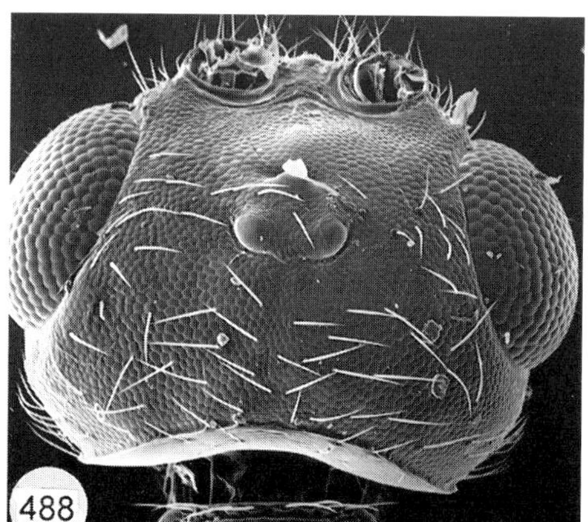

488

Figs 483, 484. *Orgilus bifasciatus*, ♀. **Figs 485, 487.** *Rhaconotus scirphophagae*.
Figs 486, 488. *Rhaconotus carinatus*. (483) Wings. (484) First to third metasomal tergites, dorsal aspect. (485, 486) Fore wing. (487, 488) Head, dorsal aspect. 483: 1.1 × scale line; 484: 1.0 ×.

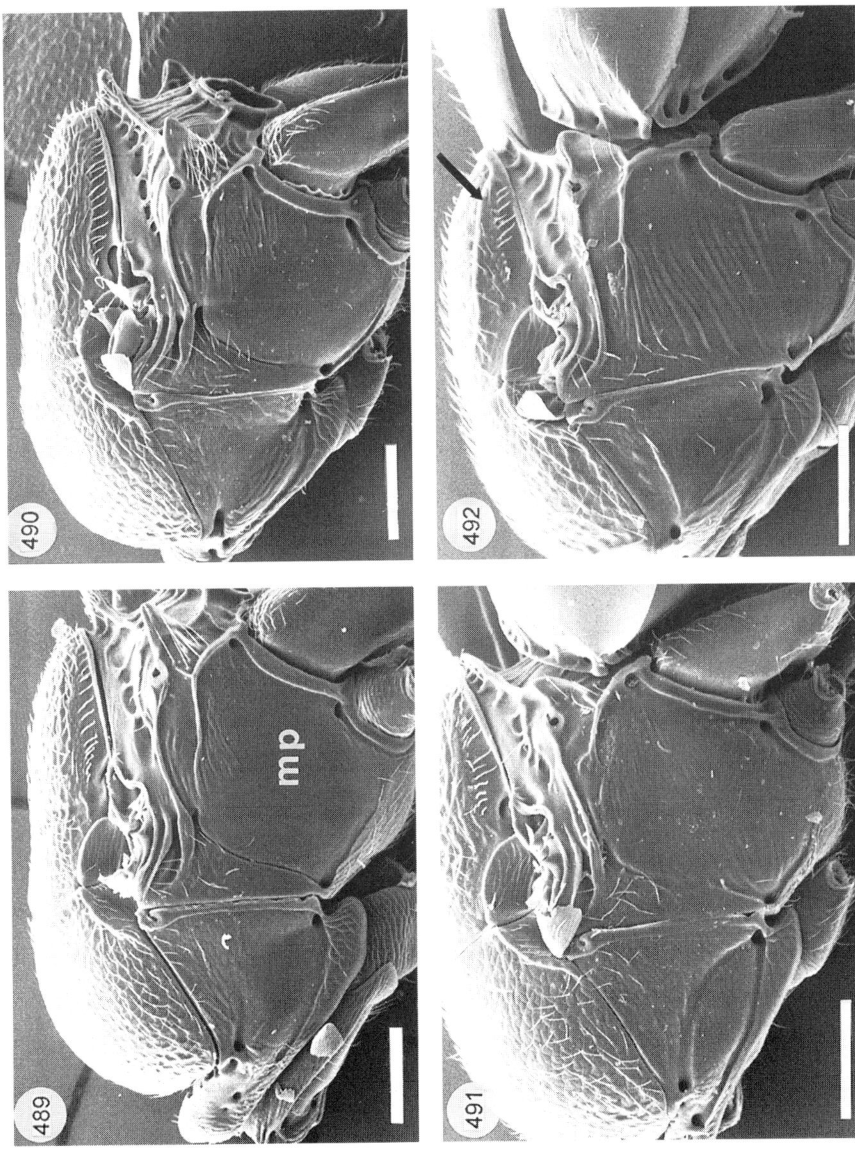

Fig 489. *Aphanogmus fijiensis*, ♀. **Fig. 490.** *Aphanogmus reticulatus*, ♀. **Fig. 491.** *Aphanogmus trasides*, ♀. **Fig. 492.** *Aphanogmus hakonensis*, ♀. Lateral mesosoma. Scale bar = 0.1 mm.

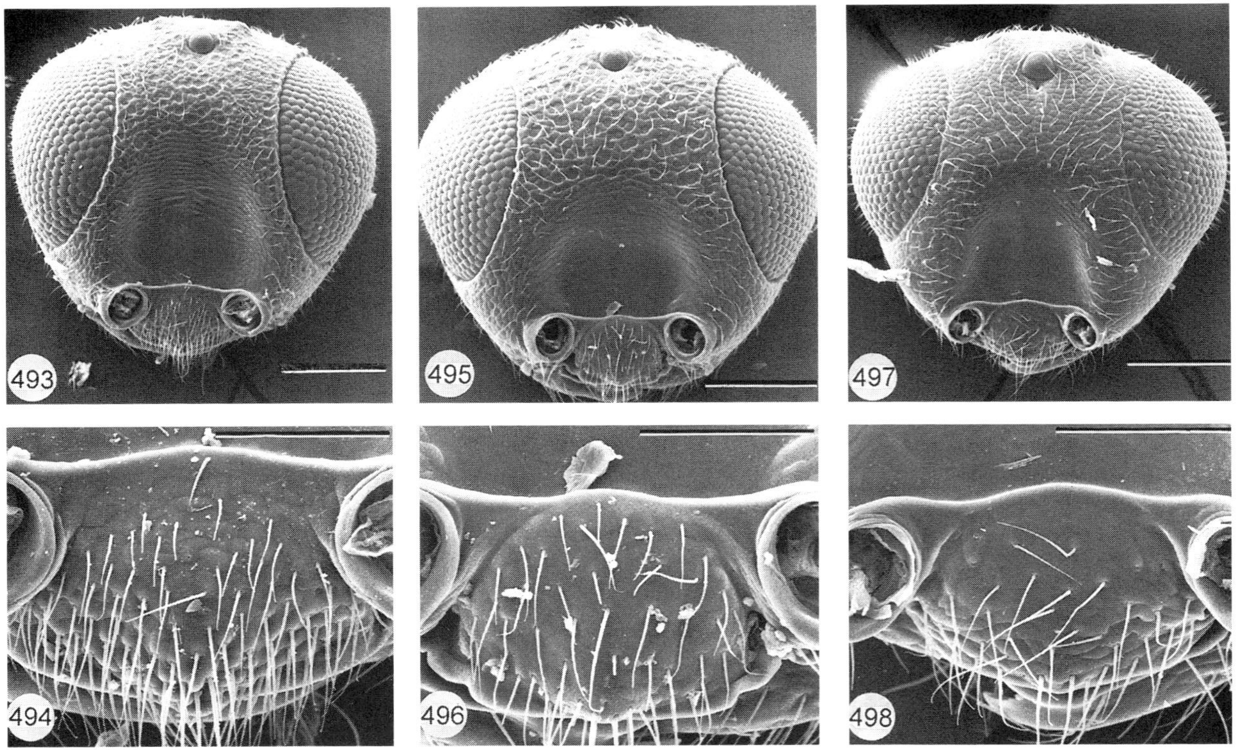

Figs 493, 494. *Aphanogmus fijiensis*, ♀. **Figs 495, 496.** *Aphanogmus reticulatus*, ♀. **Figs 497, 498.** *Aphanogmus trasides*, ♀. (493, 495, 497) Head. (494, 496, 498) Clypeus. Scale bars Figs 493, 495, 497 = 0.1 mm; Figs 494, 496, 498= 0.05 mm.

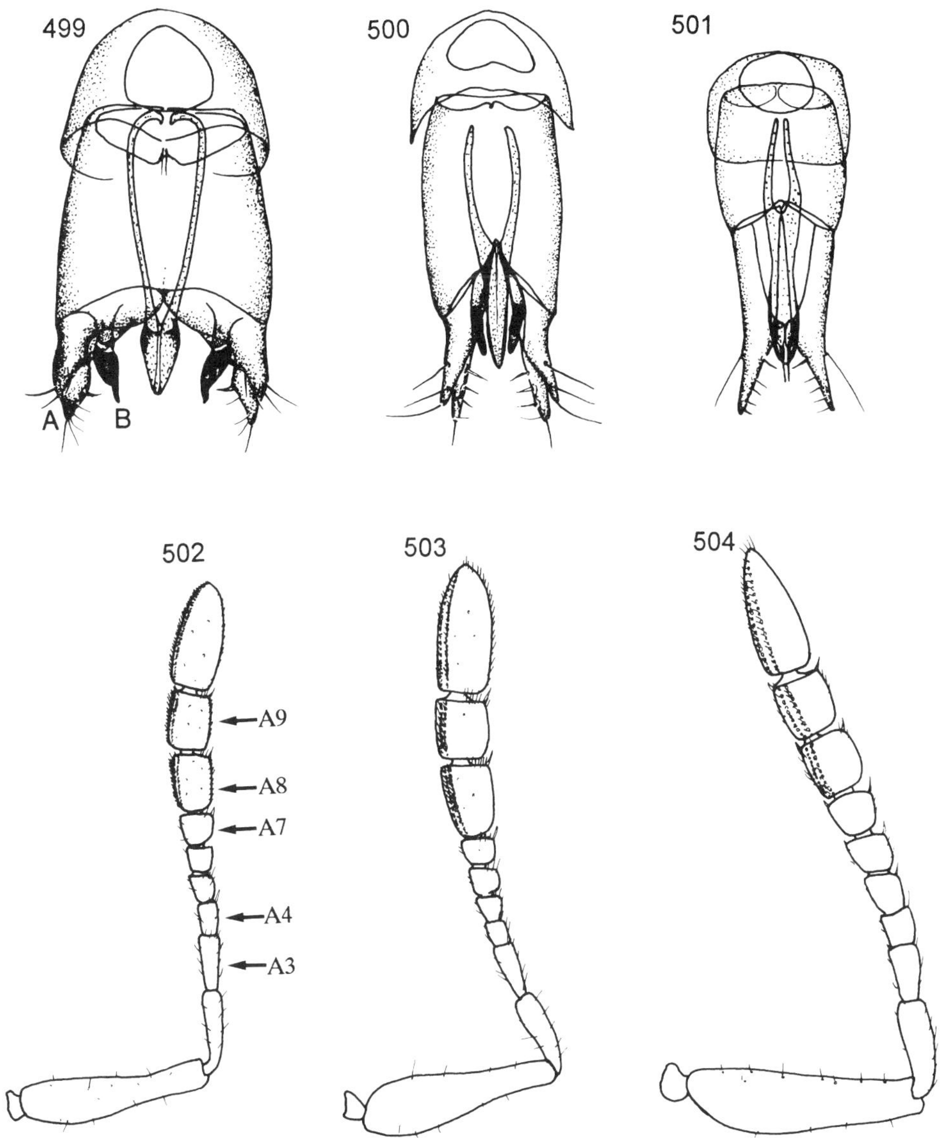

Figs 499, 502. *Aphanogmus fijiensis.* **Figs 500, 503.** *Aphanogmus reticulatus.* **Figs 501, 504.** *Aphanogmus trasides.* (499–501) Male genitalia. (502–504) Antenna ♀.

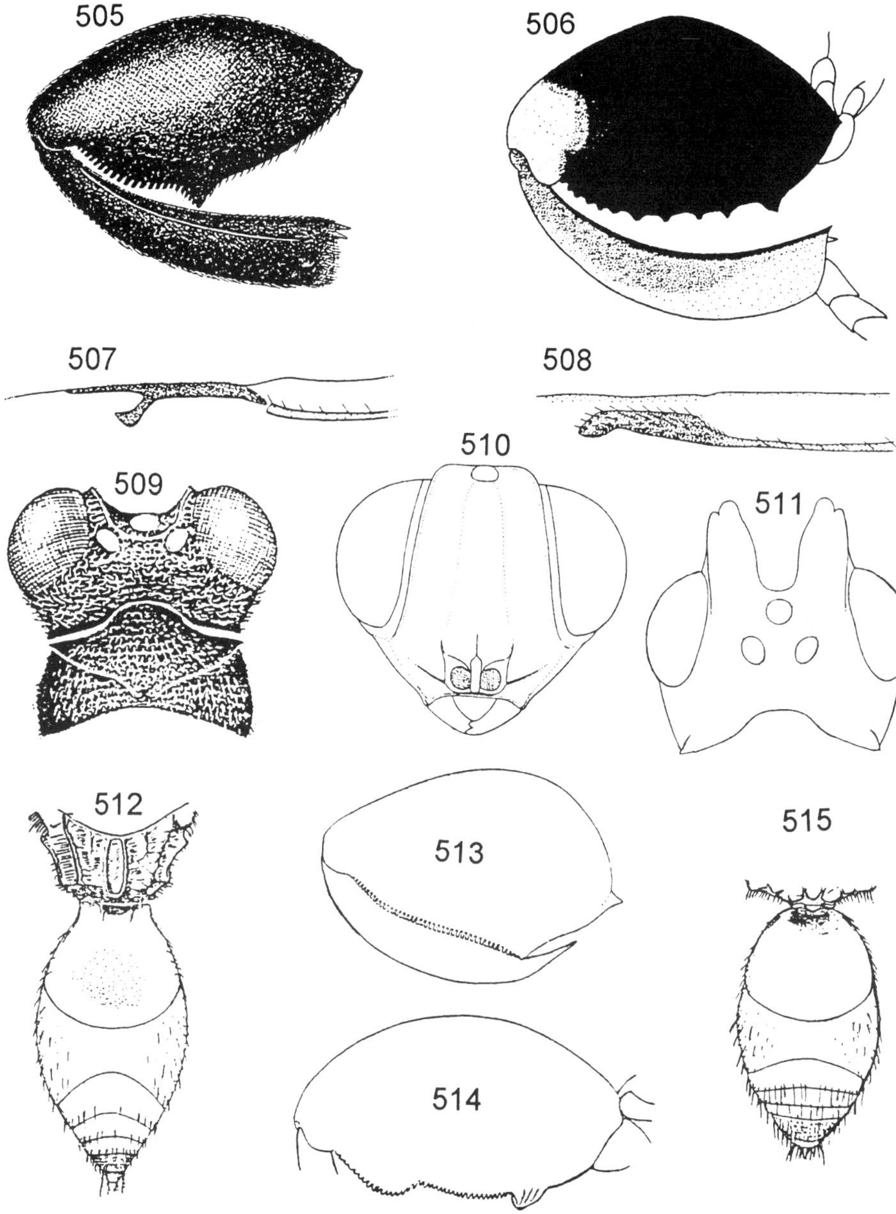

Figs 505–515. Chalcididae. (505, 506) Hind femora. (507) *Brachymeria* sp., fore wing venation. (508) *Psilochalcis* sp., fore wing venation. (509–510) *Hockeria* sp., head. (511) *Dirhinus* sp., head. (512) *Proconura* sp., metasoma. (513) Chalcididae, hind femora. (514) *Psilochalcis soudanensis*, hind femora. (515) *Psilochalcis soudanensis* metasoma.

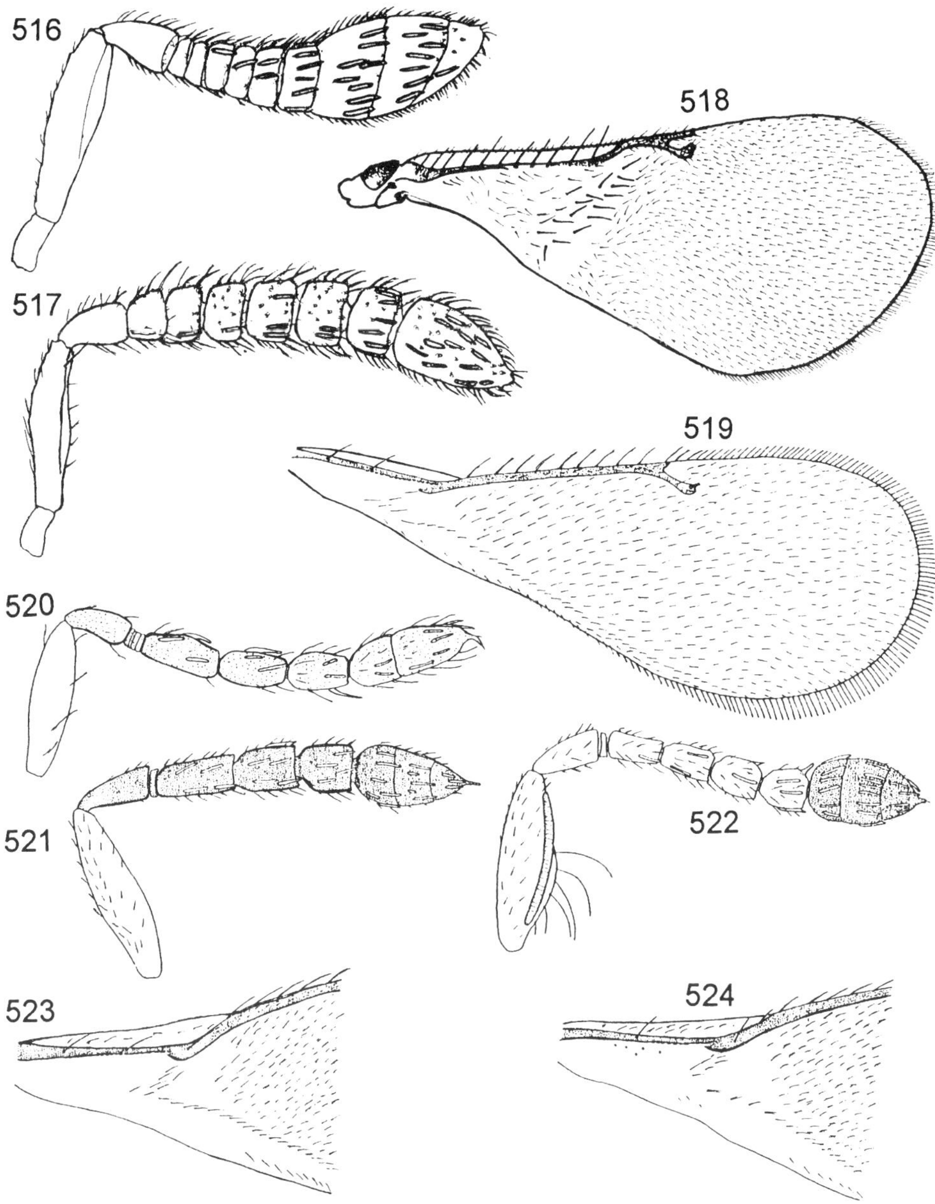

Figs 516–518. *Exoristobia dipterae*. (516) Antenna, female. (517) Antenna, male. (518) Fore wing.
Figs 519, 520. *Aprostocetus theioneurus*. (519) Fore wing. (520) Antenna, female.
Figs 521, 522, 524. *Tetrastichus atriclavus*. (521) Antenna, female. (522) Antenna, male. (524) Fore wing, base. **Fig. 523.** *Tetrastichus howardi*, fore wing, base.

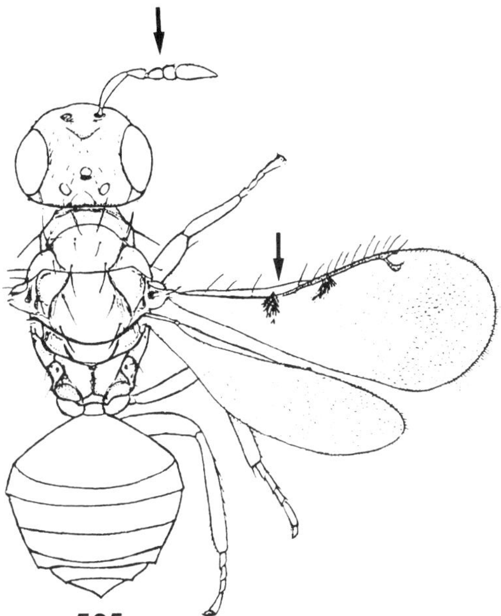

525

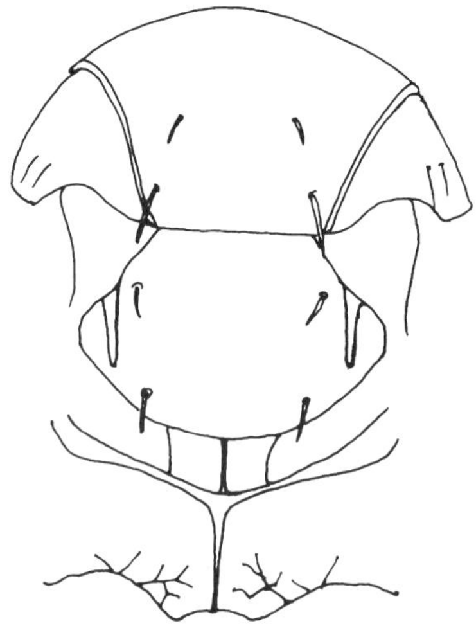

526

Fig. 525. *Trichospilus* sp. **Fig. 526.** *Tetrastichomyia* sp., mesosoma.

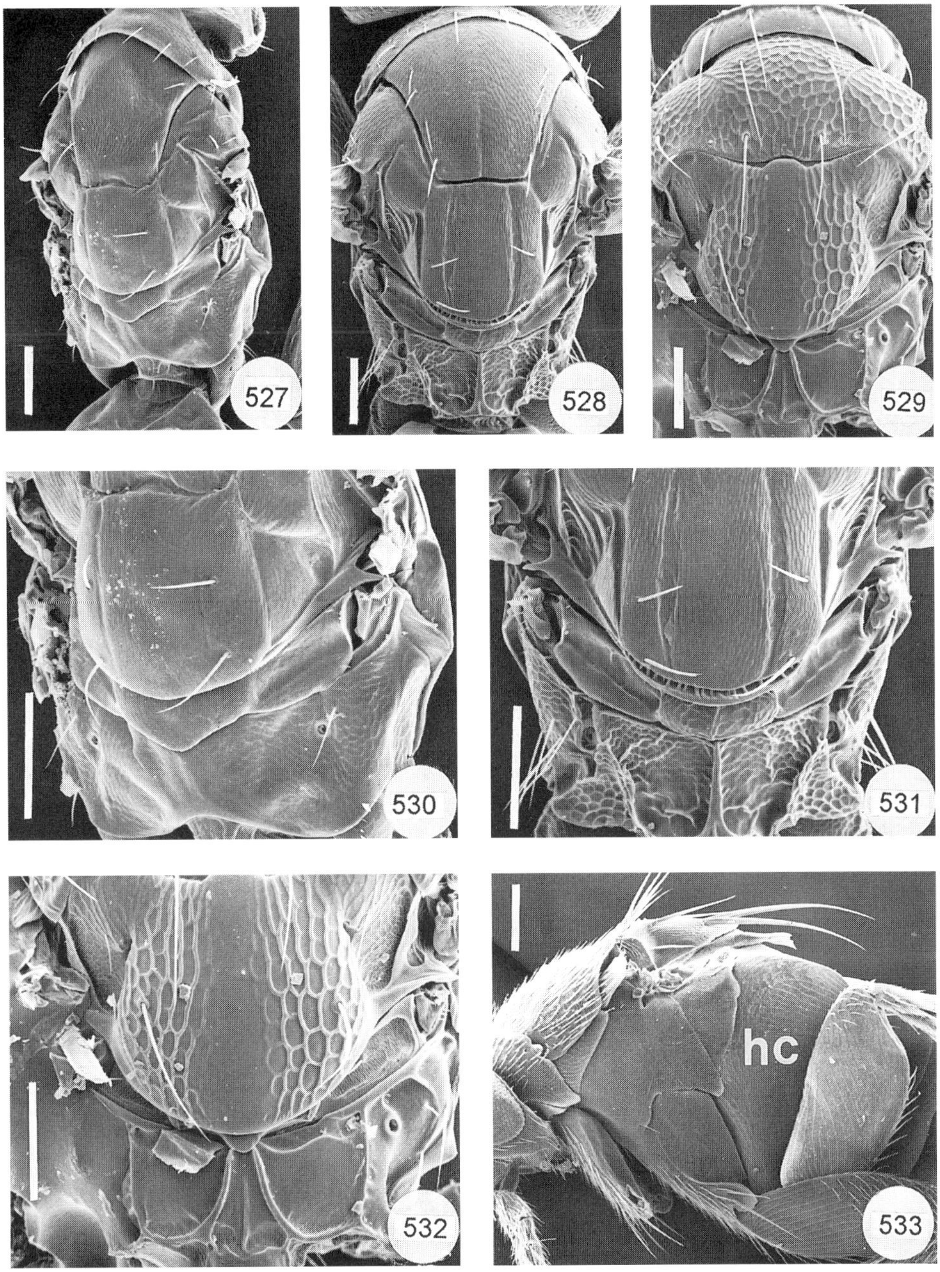

Figs 527–533. (527) *Aprostocetus theioneurus*, dorsal mesosoma. (528) *Tetrastichus atriclavus*, dorsal metasoma. (529) *Pediobius homoeus*, dorsal mesosoma. (530) *Aprostocetus theioneurus*, scutellum and propodeum. (531) *Tetrastichus atriclavus*, scutellum and propodeum. (532) *Pediobius homoeus*, scutellum and propodeum. (533) *Elasmus* sp., lateral mesosoma. Scale bars = 0.1 mm.

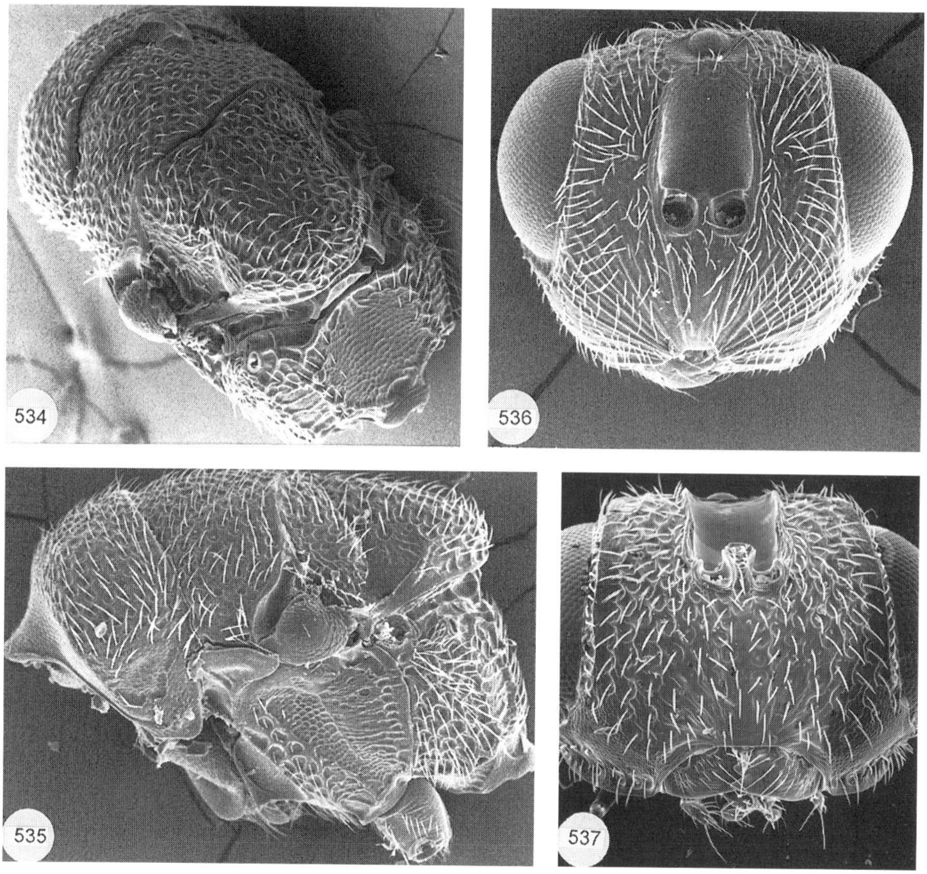

Figs 534–537. *Eurytoma.* (534) *Eurytoma braconidis*, female dorsal mesosoma. (535) *Eurytoma braconidis*, female mesosoma. (536) *Eurytoma braconidis*, female head. (537) *Eurytoma oryzivora* (female), head.

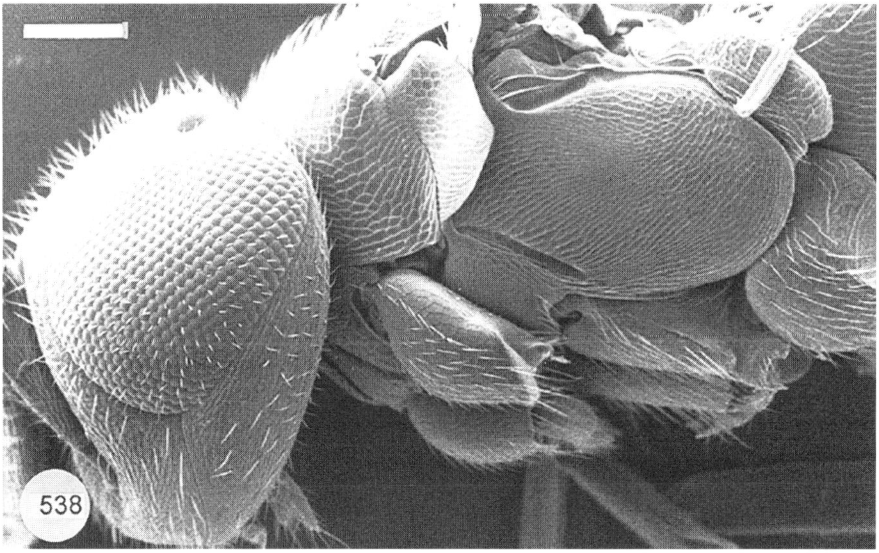

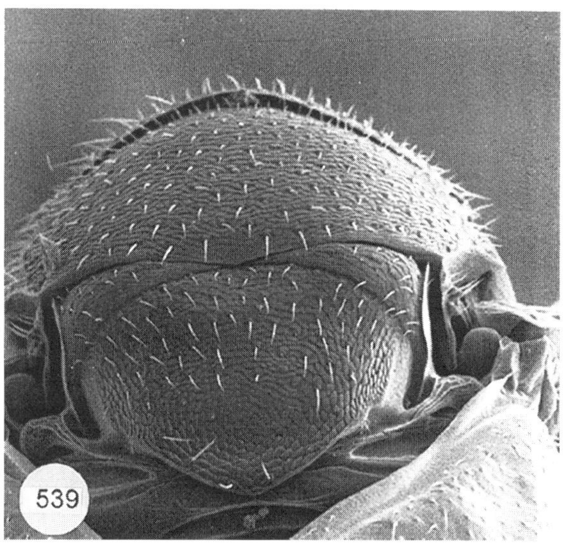

Figs 538, 539. *Exoristobia dipterae*, mesosoma, lateral and dorsal aspects.

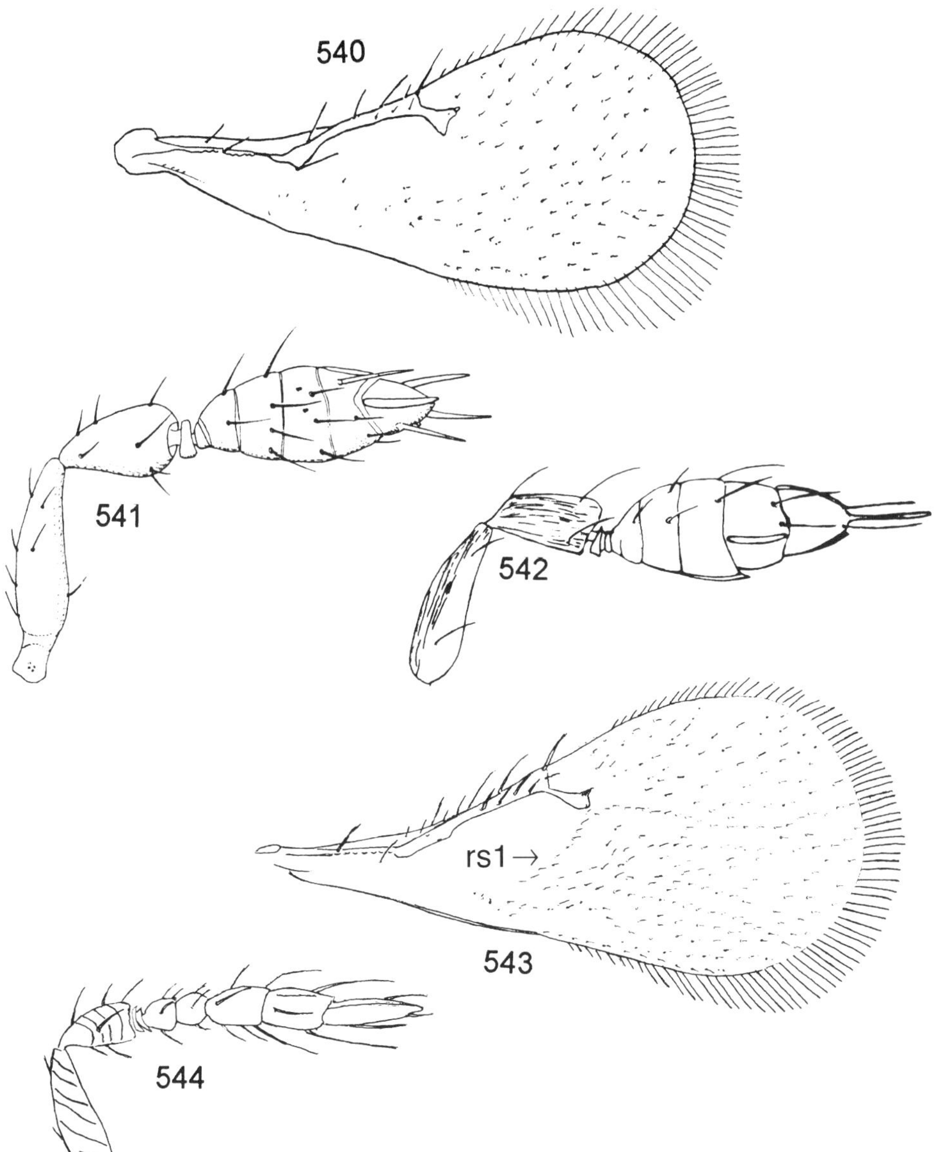

Figs 540–544. Trichogrammatidae. (540) *Lathromeris ovicida*, fore wing. (541) Male antenna. (542) Female antenna. (543) *Paracentrobia* sp., fore wing. (544) Female antenna.

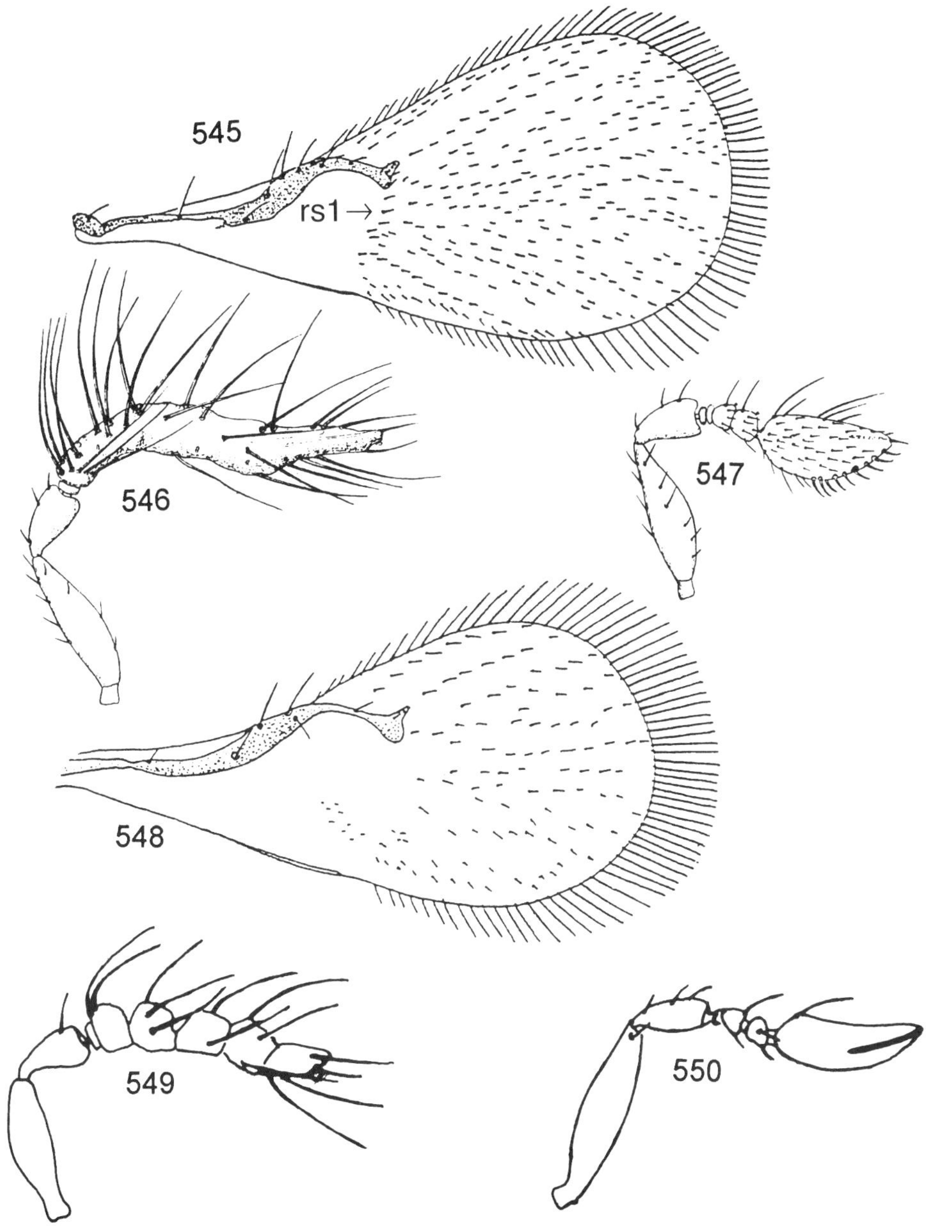

Figs 545–550. Trichogrammatidae. (545) *Trichogramma* sp., fore wing. (546) Male antenna. (547) Female antenna. (548) *Trichogrammatoidea* sp., fore wing. (549) Male antenna. (550) Female antenna.

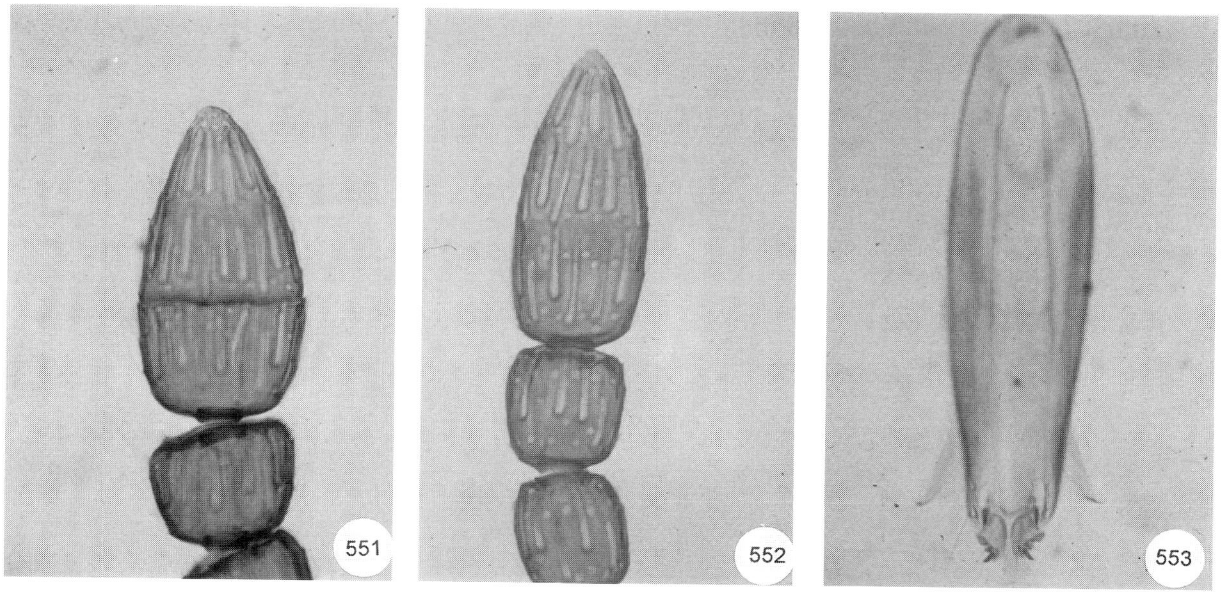

Figs 551–553. *Eurytoma*. (551) *Eurytoma braconidis*, female, apex of antenna. (552) *Eurytoma* sp., female, apex of antenna. (553) *Eurytoma braconidis*, male genitalia.

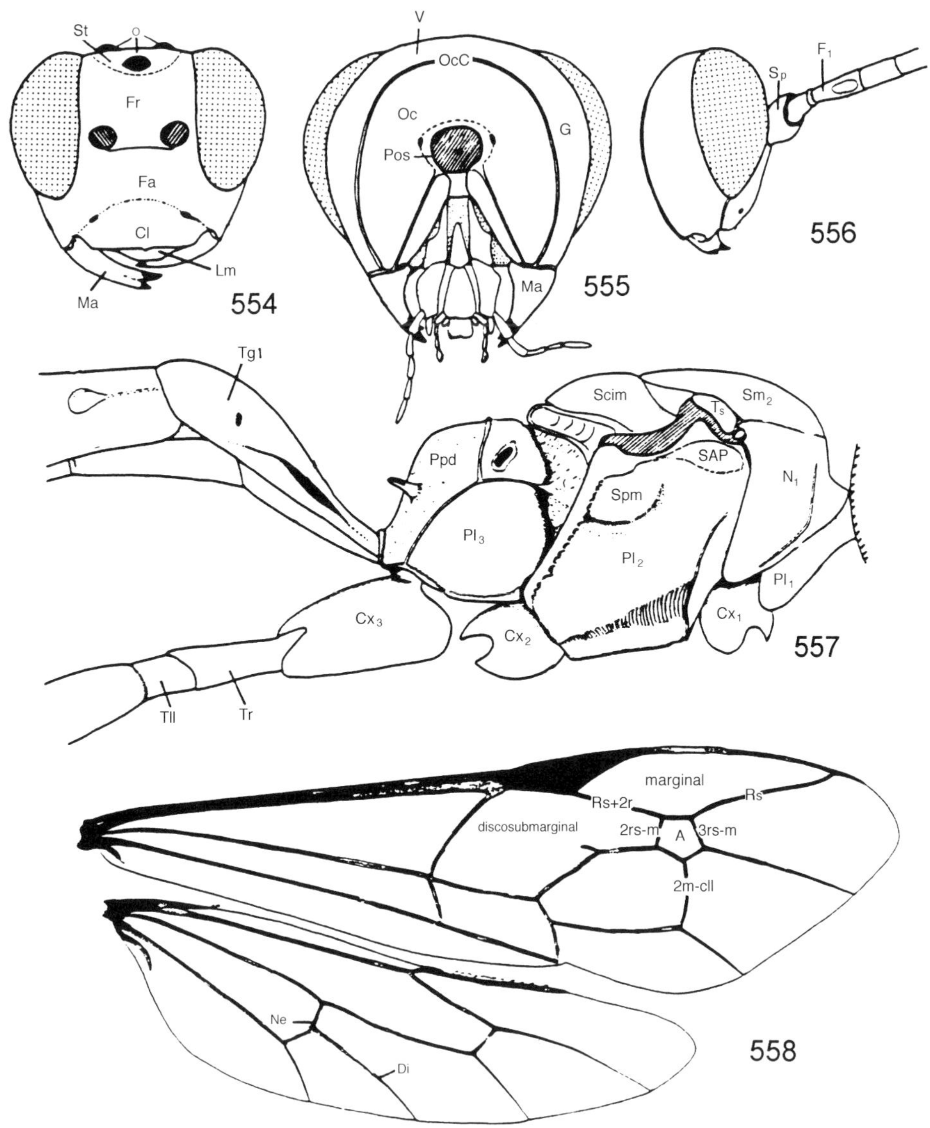

Figs 554–558. Ichneumonidae, morphological terminology. (554) Head, anterior view. (555) Head, posterior view. (556) Head and base of antenna, lateral view. (557) Mesosoma and anterior part of metasoma, lateral view. (558) Wings. 2m-cu, second recurrent vein; 2rs-m, first intercubital vein; 3rs-m, second intercubital vein; A, areolet; Cl, clypeus; Cx_{1-3}, fore, mid- and hind coxae; Di, discoidella; F_1, 1st flagellar segment; Fa, face; Fr, frons; G, gena; Lm, labrum; Ma, mandible; N_1, pronotum; Ne, nervellus; O, ocellus; Oc, occiput; OcC, occipital carina; Pl_1, propleuron; Pl_2, mesopleuron; Pl_3, metapleuron; Pos, postocciput; Ppd, propodeum; Rs, Rs+2r, radial veins; SAP, subalar prominence; Sclm, scutellum; Sm_2, mesoscutum; Sp, scape; Spm, speculum; St, stemmaticum; Tg, tegula; Tg_1, tergite 1; Tg_2, tergite 2; Tll, trochantellus; Tr, trochanter; V, vertex.

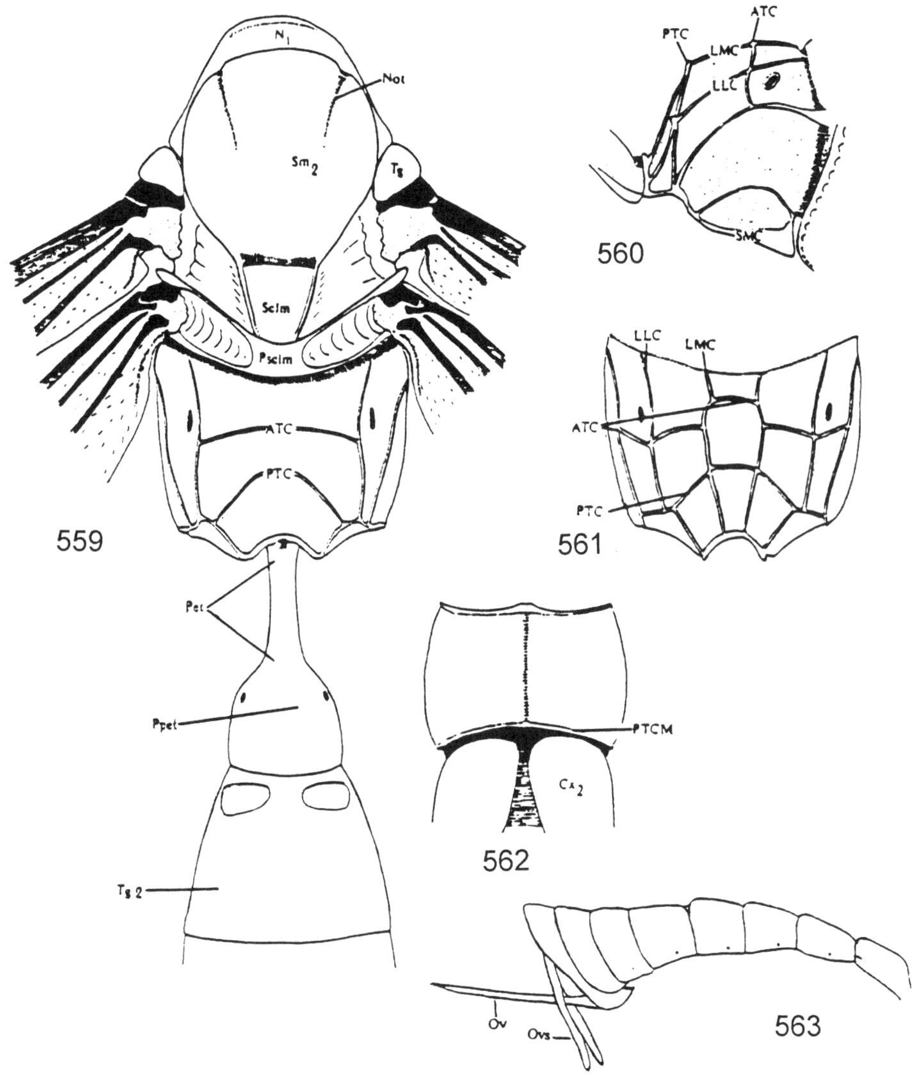

Figs 559–563. Ichneumonidae, morphological terminology. (559) Mesosoma and anterior part of metasoma, dorsal view. (560) Propodeum, lateral view; (561) Propodeum, dorsal view. (562) Mesosternal region, ventral view. (563) Metasoma/gaster, lateral view. ATC, anterior transverse carina; Cx_{1-3}, fore, mid- and hind coxae; LLC, lateral longitudinal carina; LMC, lateromedian longitudinal carina; Ov, ovipositor; Ovs, ovipositor sheath; N_1, pronotum; Not, notaulus; Pet, petiole; Ppet, postpetiole; Psclm, postscutellum or dorsellum; PTC, posterior transverse carina; PTCM, posterior transverse carina of the mesosternum; Sm_2, mesoscutum; SMC, submetapleural carina; Tg, tegula.

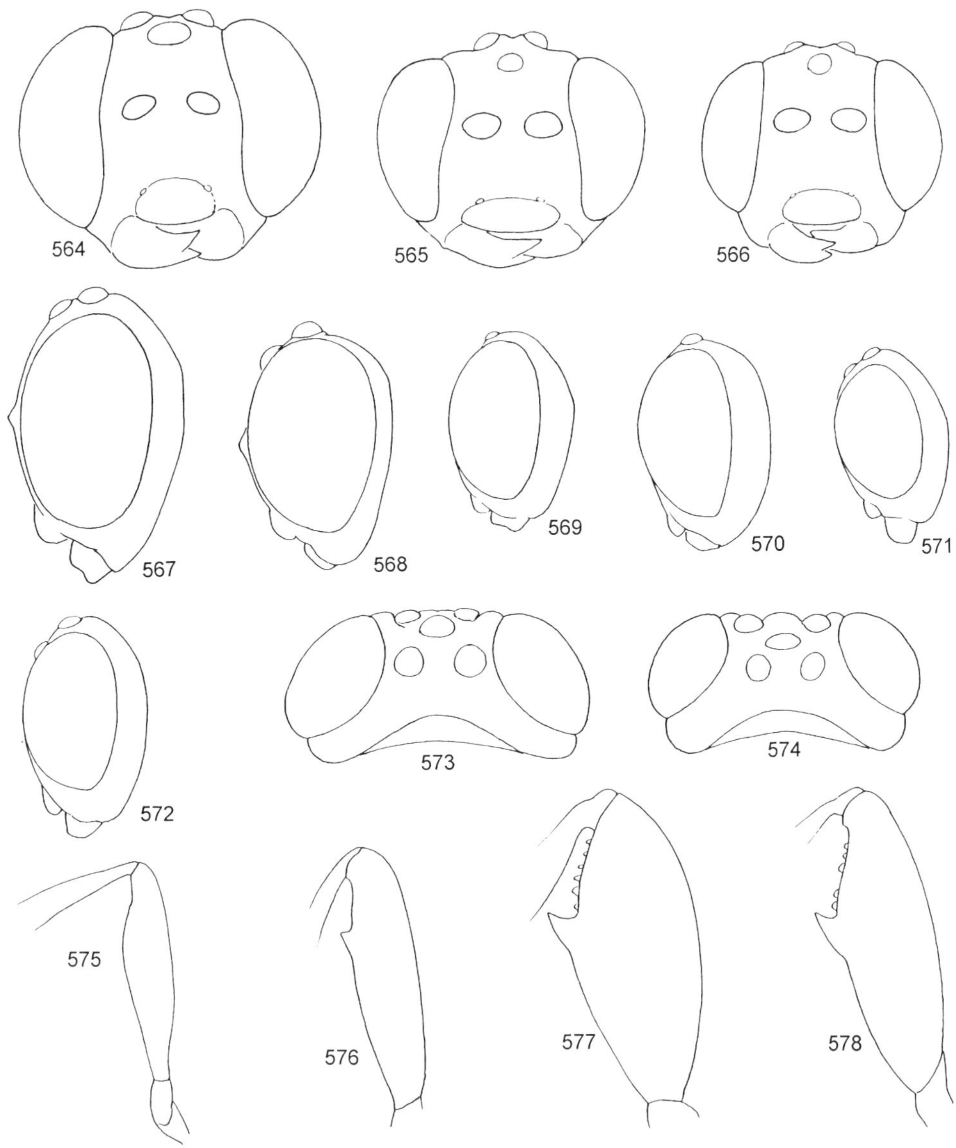

Figs 564–578. *Pristomerus* spp. (564). *P. caris* (male) head, anterior. (565) *P. bullis* (male) head, anterior. (566) *P. africator* (male) head, anterior. (567) *P. caris* (female) head, lateral. (568) *P. caris* (male) head, lateral (569) *P. bullis* (female) head, lateral. (570) *P. bullis* (male) head, lateral. (571) *P. africator* (female) head, lateral. (572) *P. africator* (male) head, lateral. (573) *P. caris* (male) head, dorsal. (574) *P. bullis* (male) head, dorsal. (575). *P. caris* (female) hind femur. (576) *P. bullis* (female) hind femur. (577) *P. bullis* (male) hind femur. (578) *P. africator* (male) hind femur.

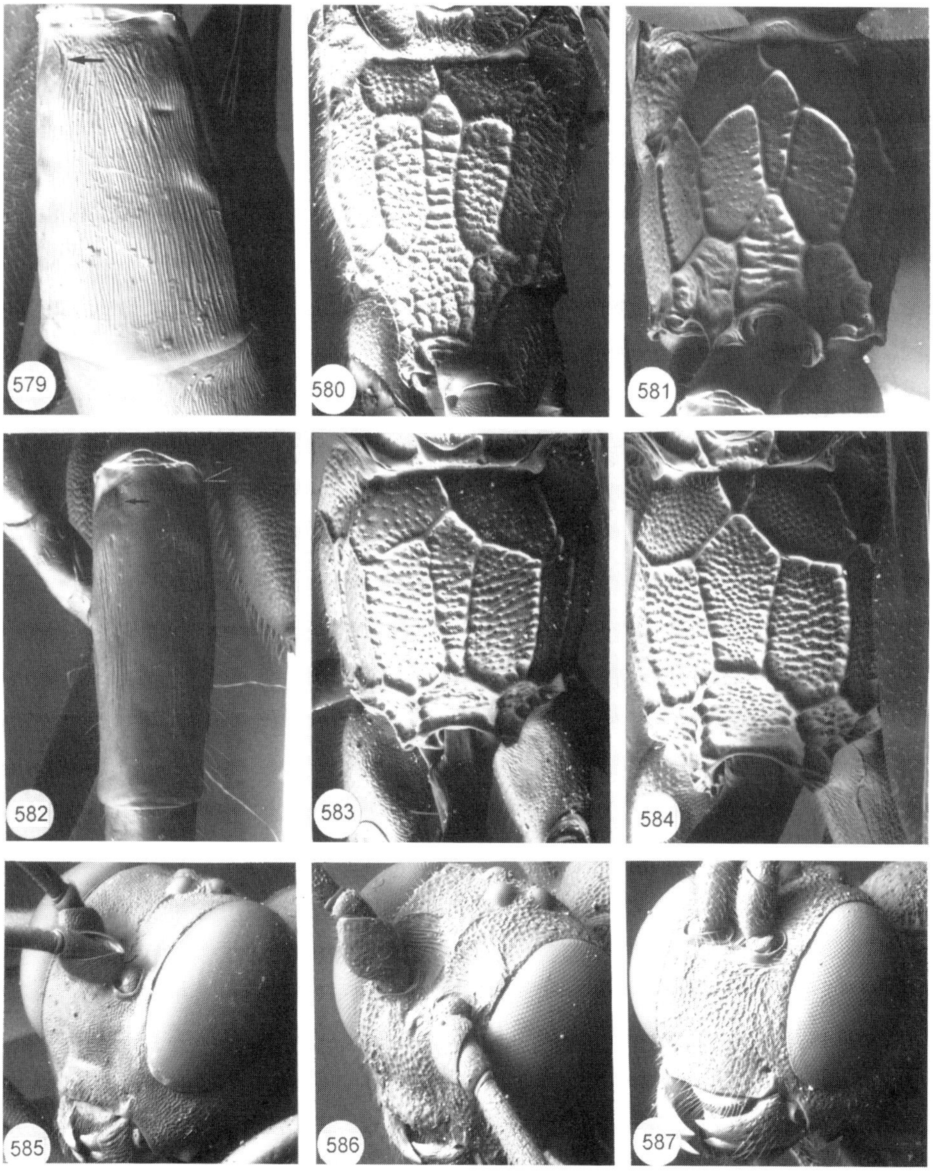

Figs 579–587. (579) *Pristomerus africator*, tergite 2. (580) *Venturia jordanae*, propodeum. (581). *Pristomerus bullis*, propodeum. (582). *Pristomerus caris*, tergite 2. (583) *Pristomerus caris*, propodeum. (584) *Pristomerus africator*, propodeum. (585). *Amauromorpha* sp., head. (586). *Gambroides ?nimbipennis*, head. (587) *Venturia jordanae*, head.

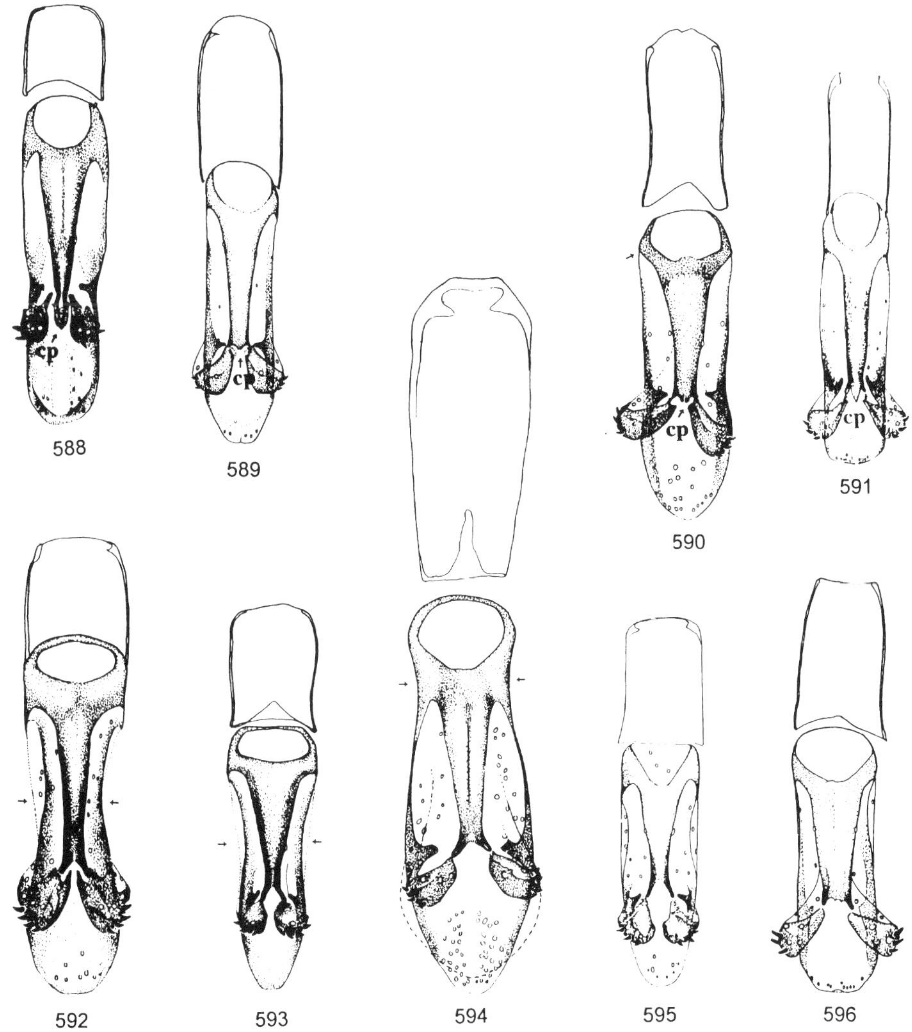

Figs. 588–596. *Telenomus* male genitalia. (588) *T. nemesis*. (589) *T. bini*. (590) *T. creusa*. (591) T. busseolae. (592) *T. thestor*. (593) *T. soudanensis*. (594) *T. nephele/T. versicolor* (= dotted line). (595) *T. isis*. (596) *T. applanatus*.

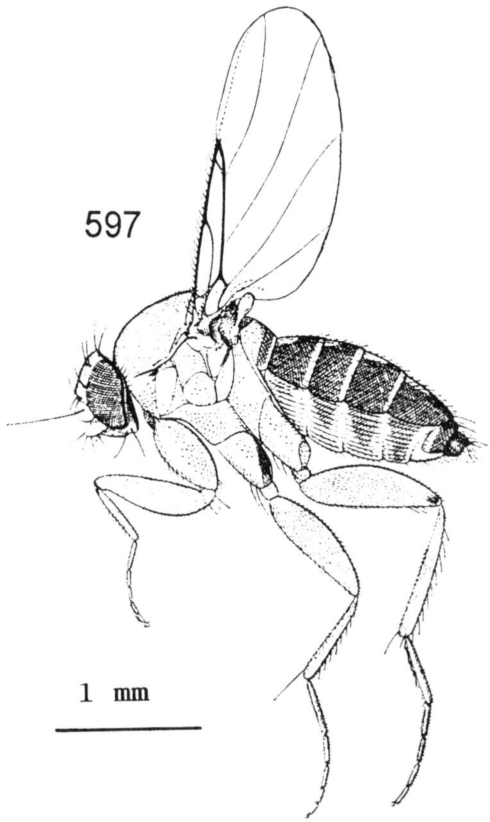

Fig. 597. Phoridae: *Megaselia scalaris* (BMNH copyright).

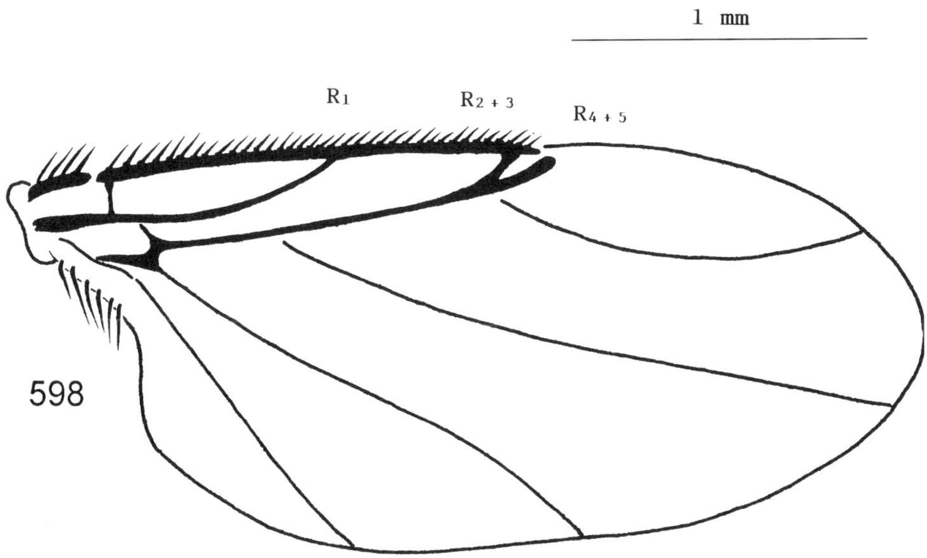

598

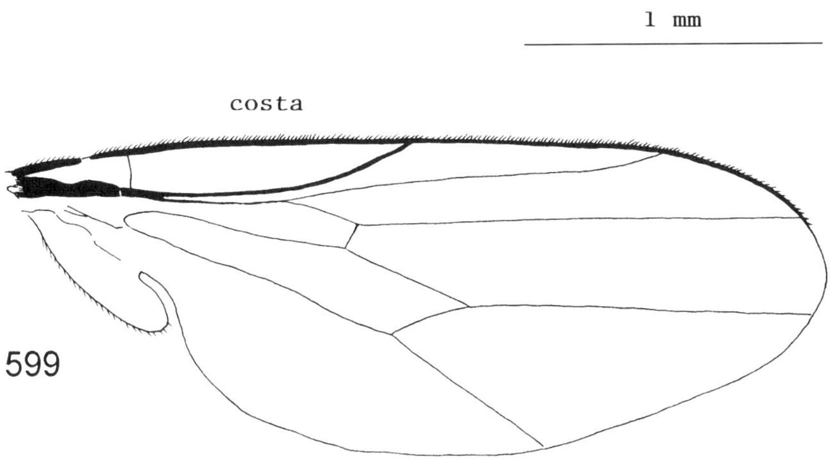

599

Figs 598, 599. Wings. (598) *Megaselia scalaris*. (599) *Chloropidae*.

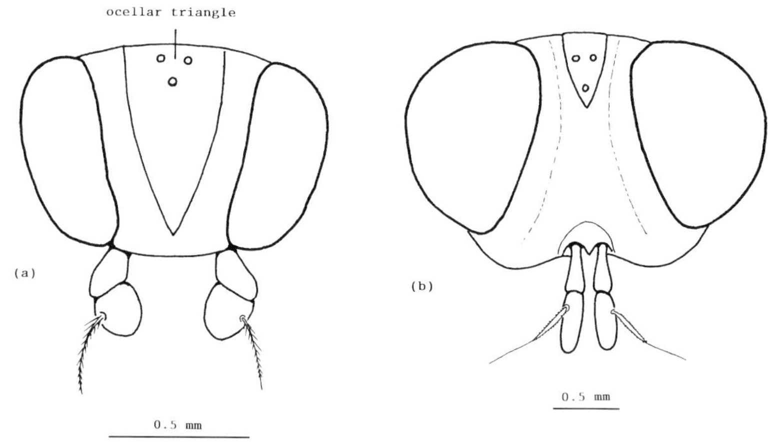

600

Fig. 600. Head of a chloropid (a) and a tachinid (b), showing large and small ocellar triangles, respectively.

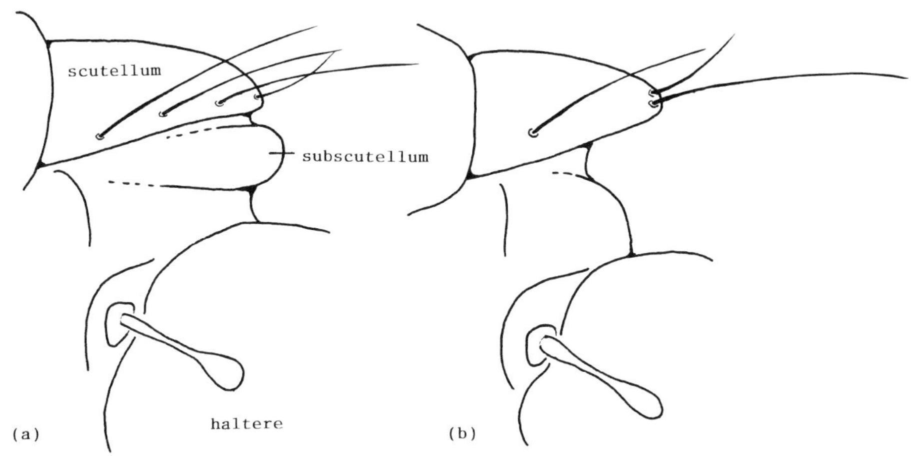

601

Fig. 601. Lateral view of tachinid (a) and sarcophagid (b), showing convex subscutellum in the former and halteres in both.

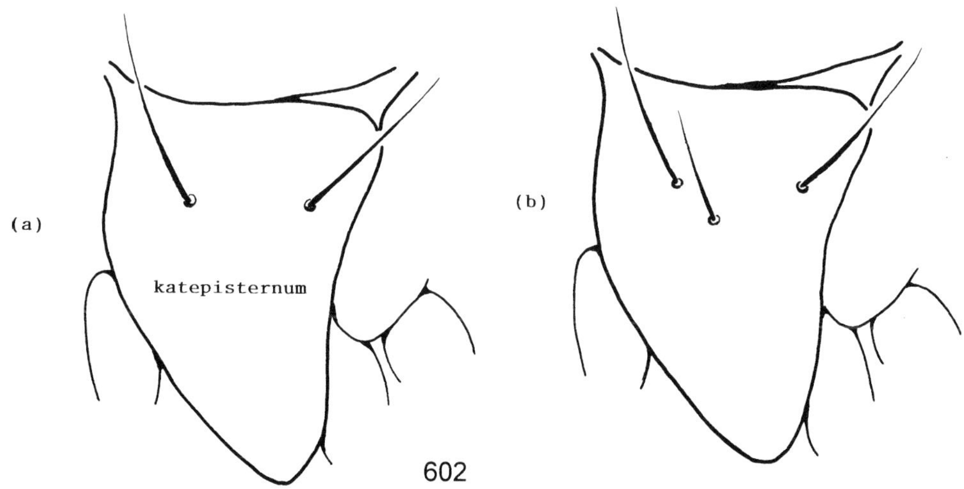

Fig. 602. Katepisternum with two (a) and three (b) (sternopleural) setae.

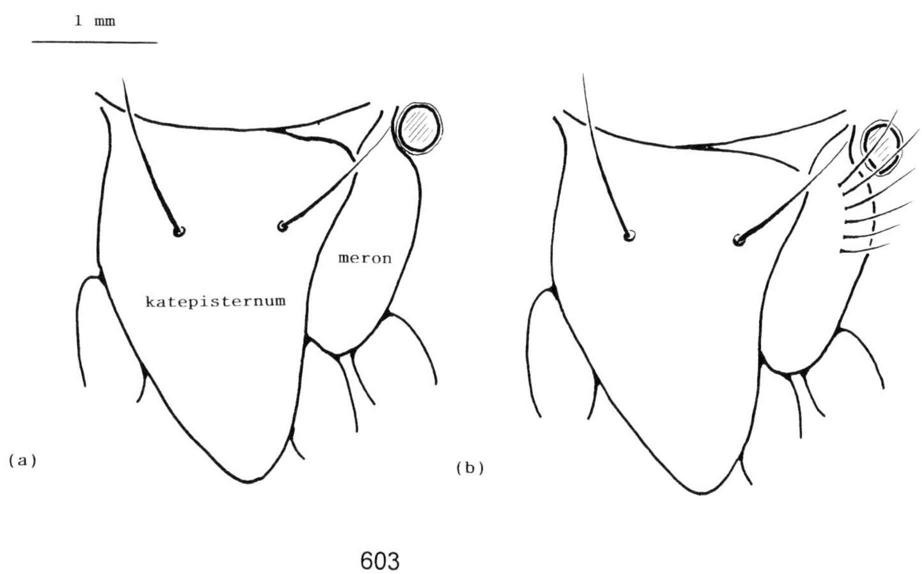

Fig. 603. Lateral views of meron without (a) and with (b) a distinct row of bristles.

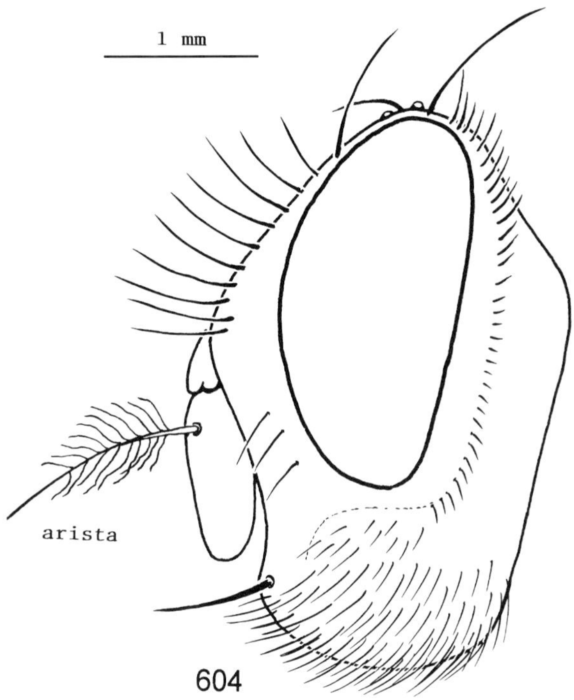

Fig. 604 Sarcophagid head.

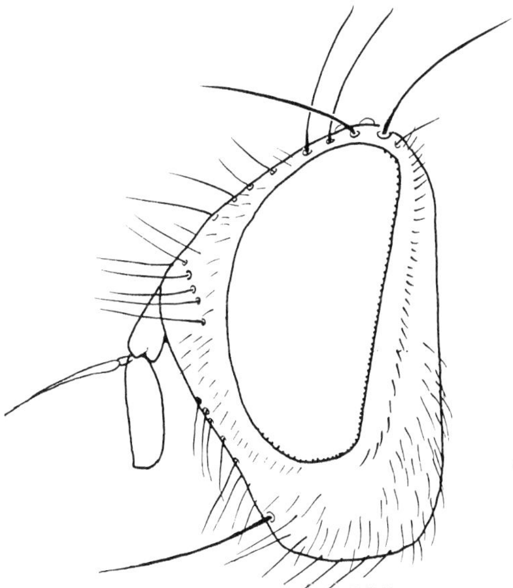

605

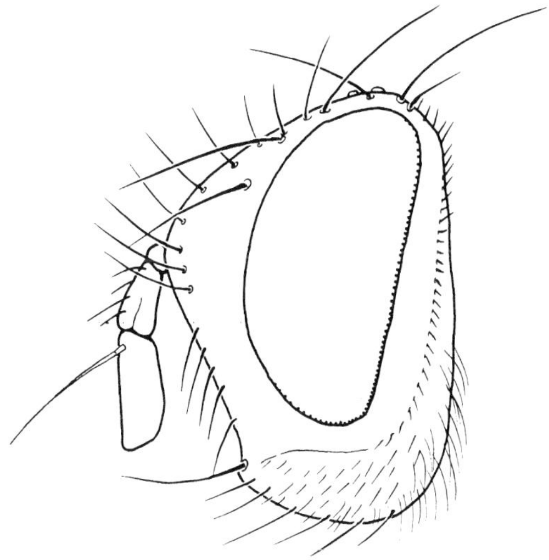

606

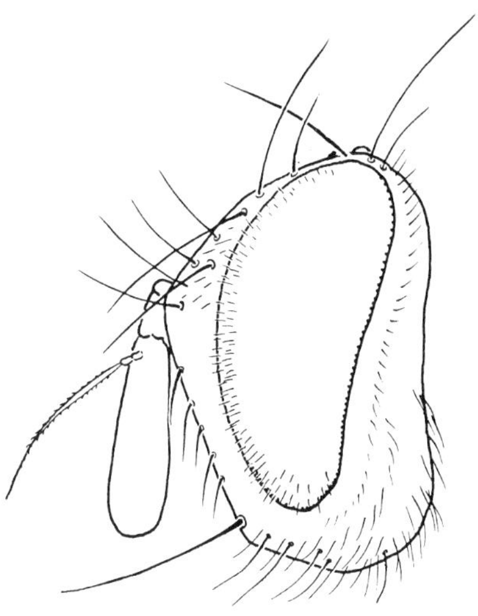

607

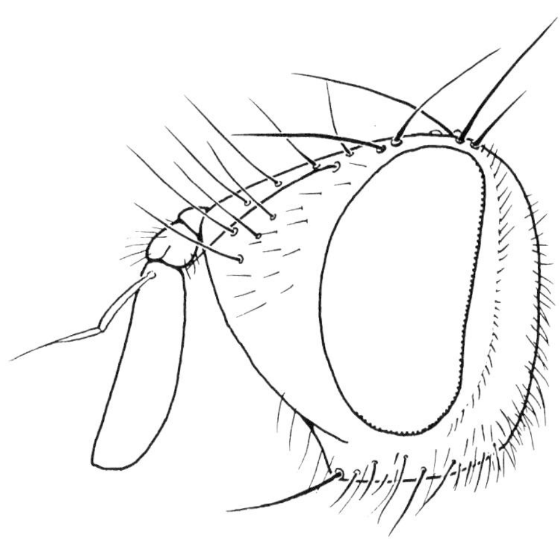

608

Figs 605–608. Heads. (605) *Sturmiopsis parasitica*. (606) *Descampsina sesamiae*. (607) *Schembria eldana*. (608) *Lydella sesamiae*.

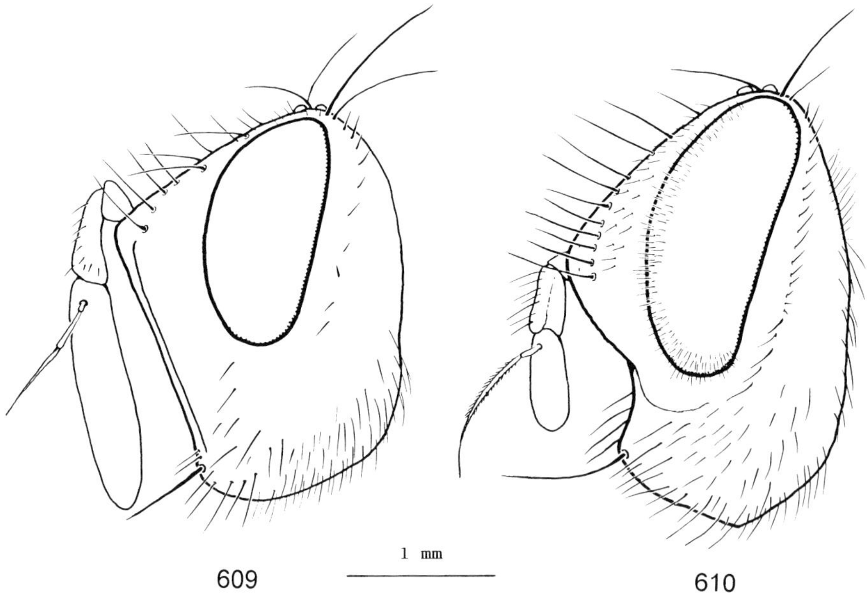

Figs 609, 610. Heads. (609) *Chromatophania picta*. (610) *Nemoraea discoidalis*.

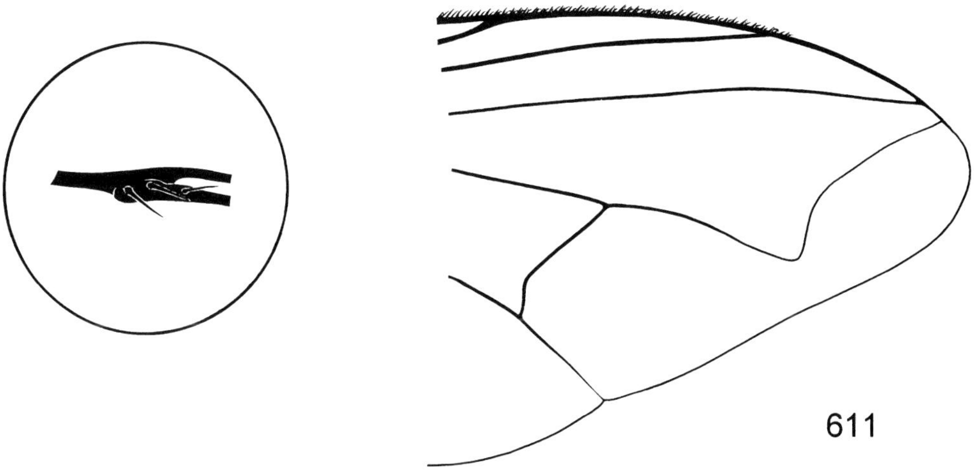

611

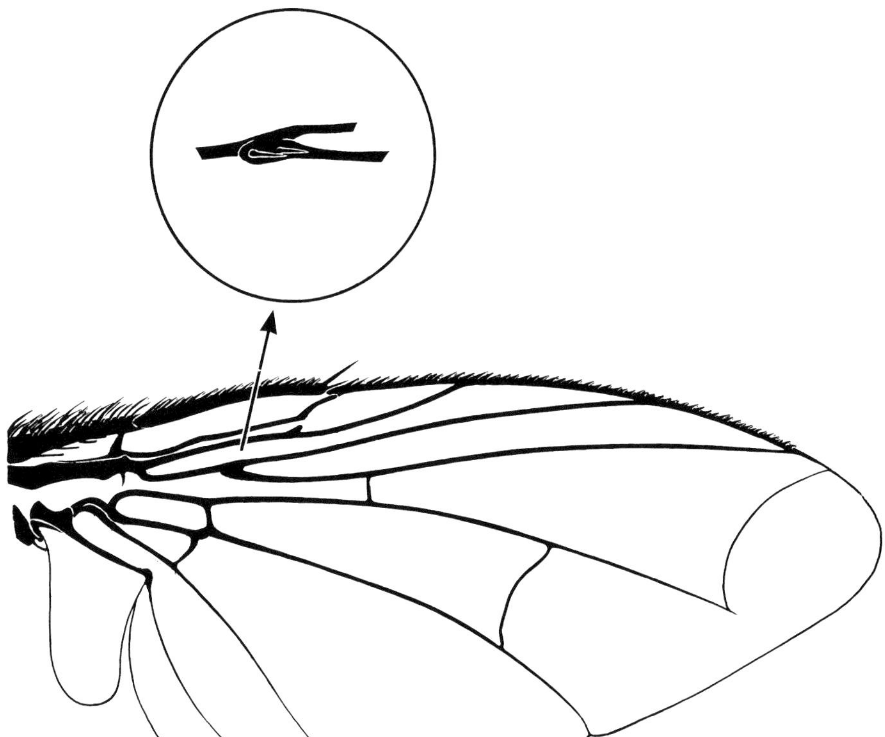

612

Figs 611, 612. Wings. (611) *Sturmiopsis parasitica*. (612) *Descampsina sesamiae*.

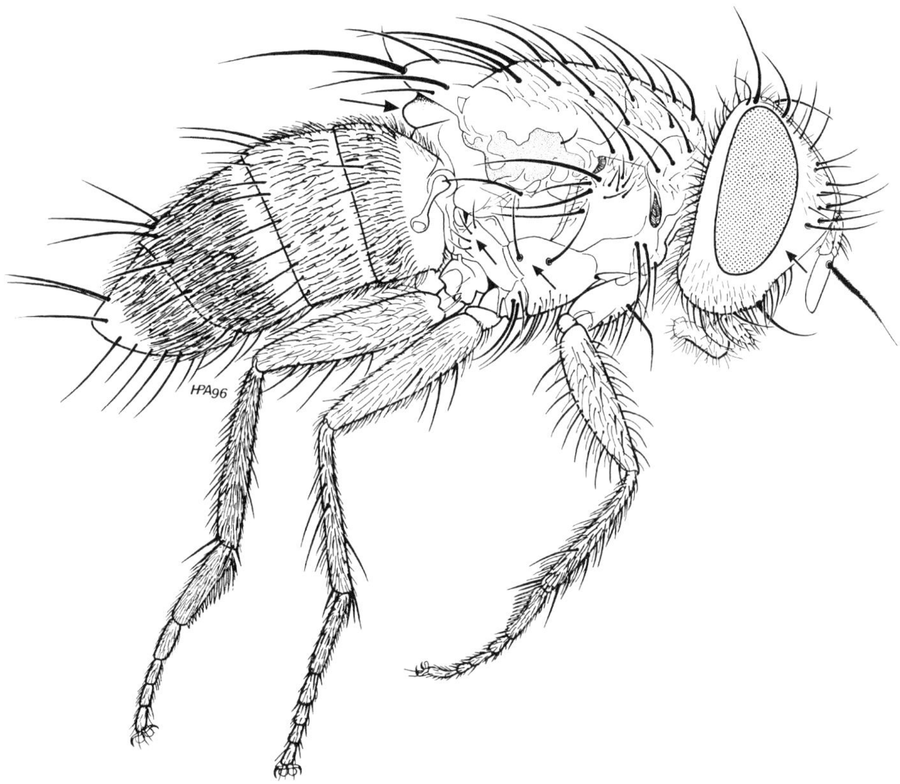

Fig. 613. *Descampsina sesamiae.*

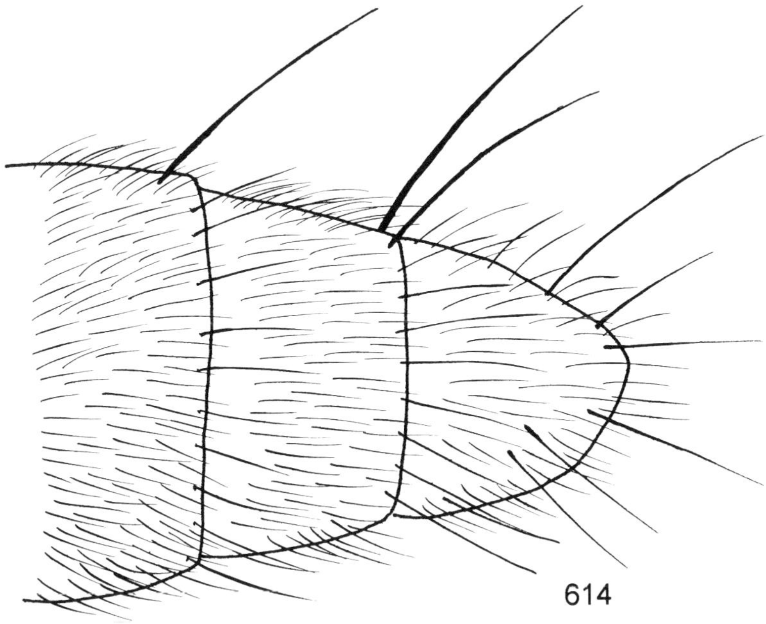

614

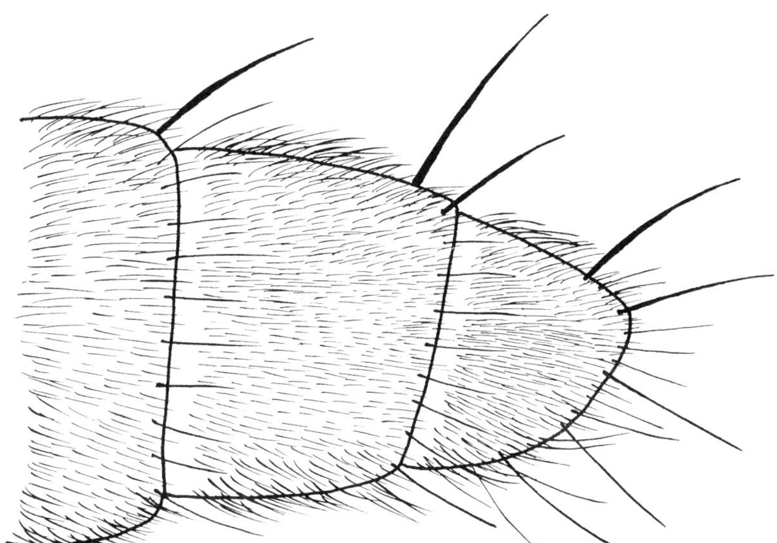

615

Figs 614, 615. Lateral view of male abdominal segments 3–5. (614) *Sturmiopsis parasitica.* (615) *Descampsina sesamiae.*

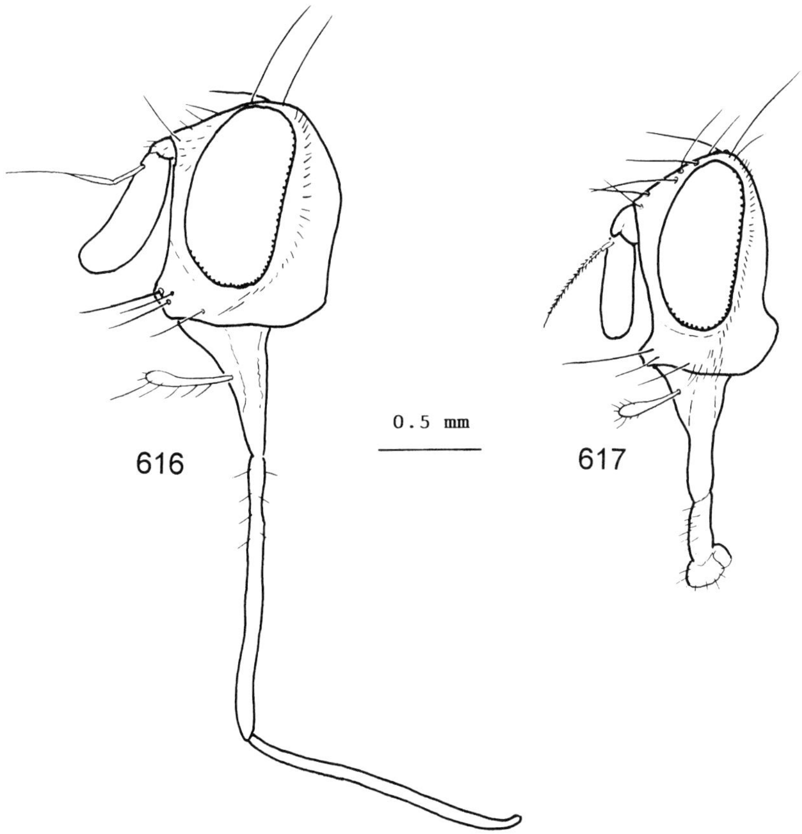

Figs 616, 617. Heads. (616) *Siphona murina*. (617) *Actia* sp.

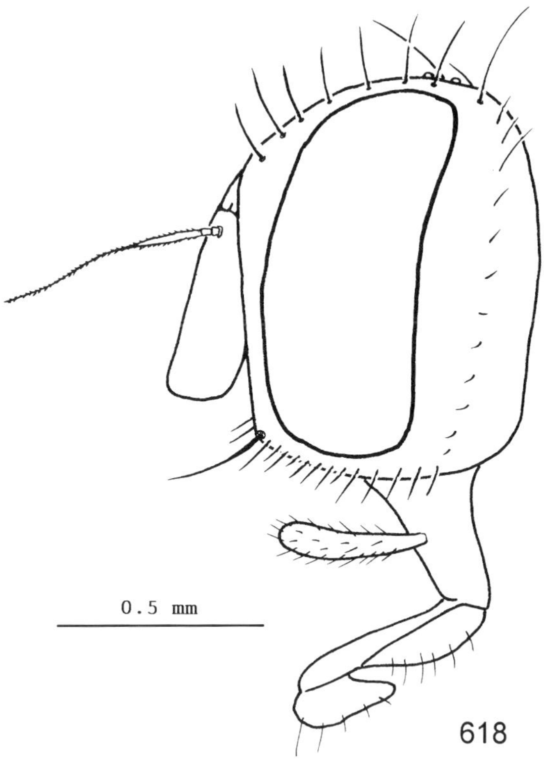

618

Fig. 618. *Atherigona*, head.

619

Fig. 619. Larval cephalopharyngeal skeleton of *Sturmiopsis parasitica*.

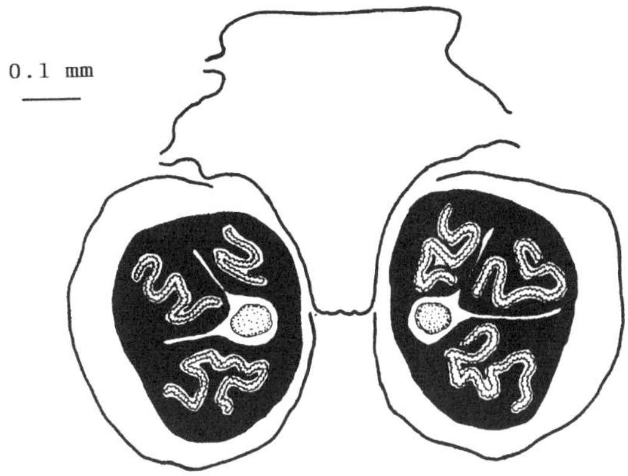

620

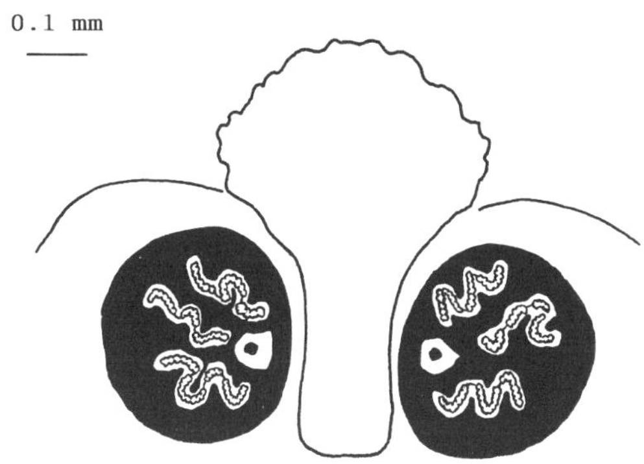

621

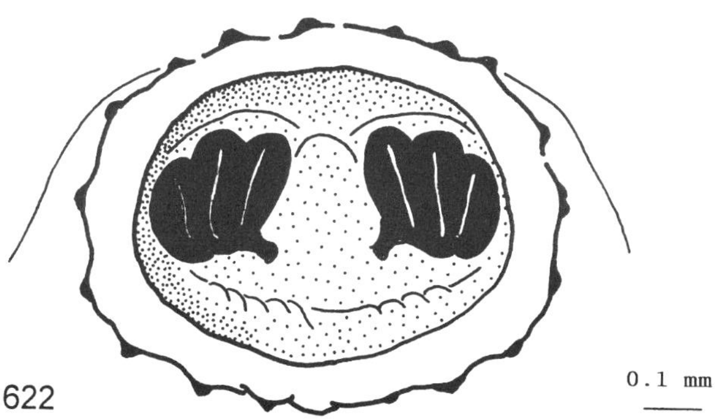

622

Figs 620–622. Posterior pupal spiracles. (620) *Sturmiopsis parasitica*. (621) *Descampsina sesamiae*. (622) Sarcophagidae (sunken).

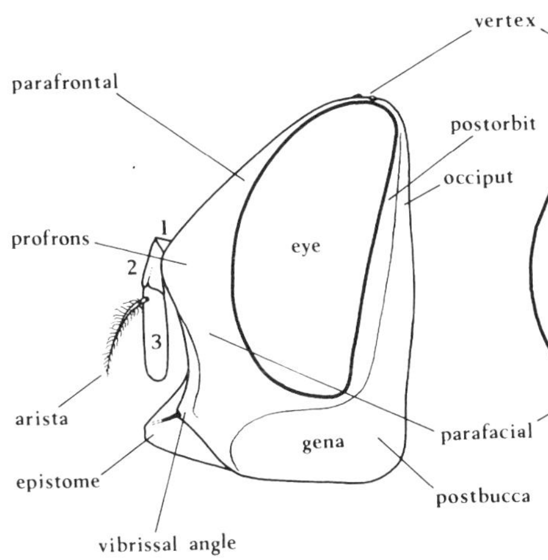

623

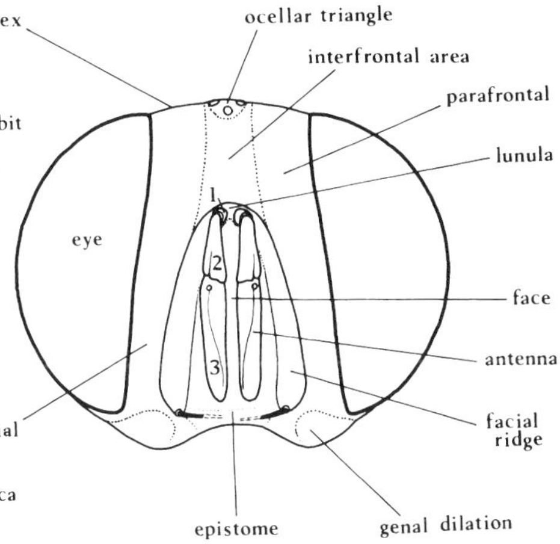

624

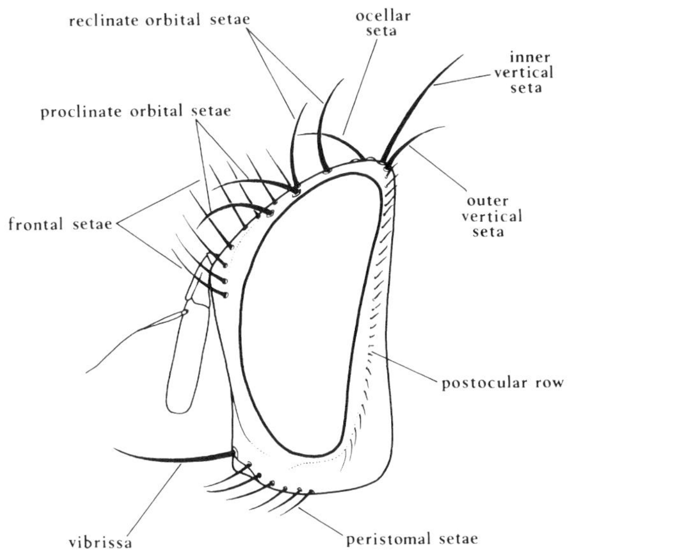

625

RWC

Figs 623–625. Terminology of tachinid head and main setae. (623) Left lateral view to show structure. (624) Facial view to show structure. (625) Left lateral view to show setae (Crosskey, 1984).

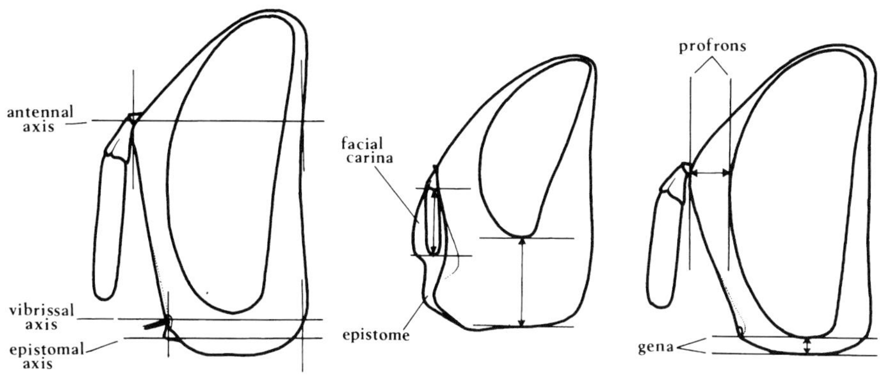

626 627 628

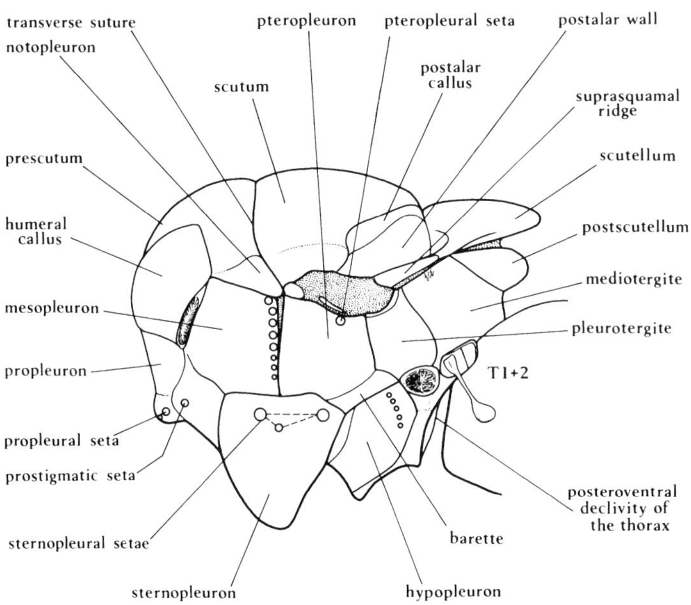

629 RWC

Figs 626–629. Terminology of tachinid head and thorax. (626–628) Left lateral view of some head shapes to show characters sometimes mentioned in keys. (629) Left lateral view of thorax (vestiture omitted but positions of main setae of pleural region indicated by open circles) (Crosskey, 1984).

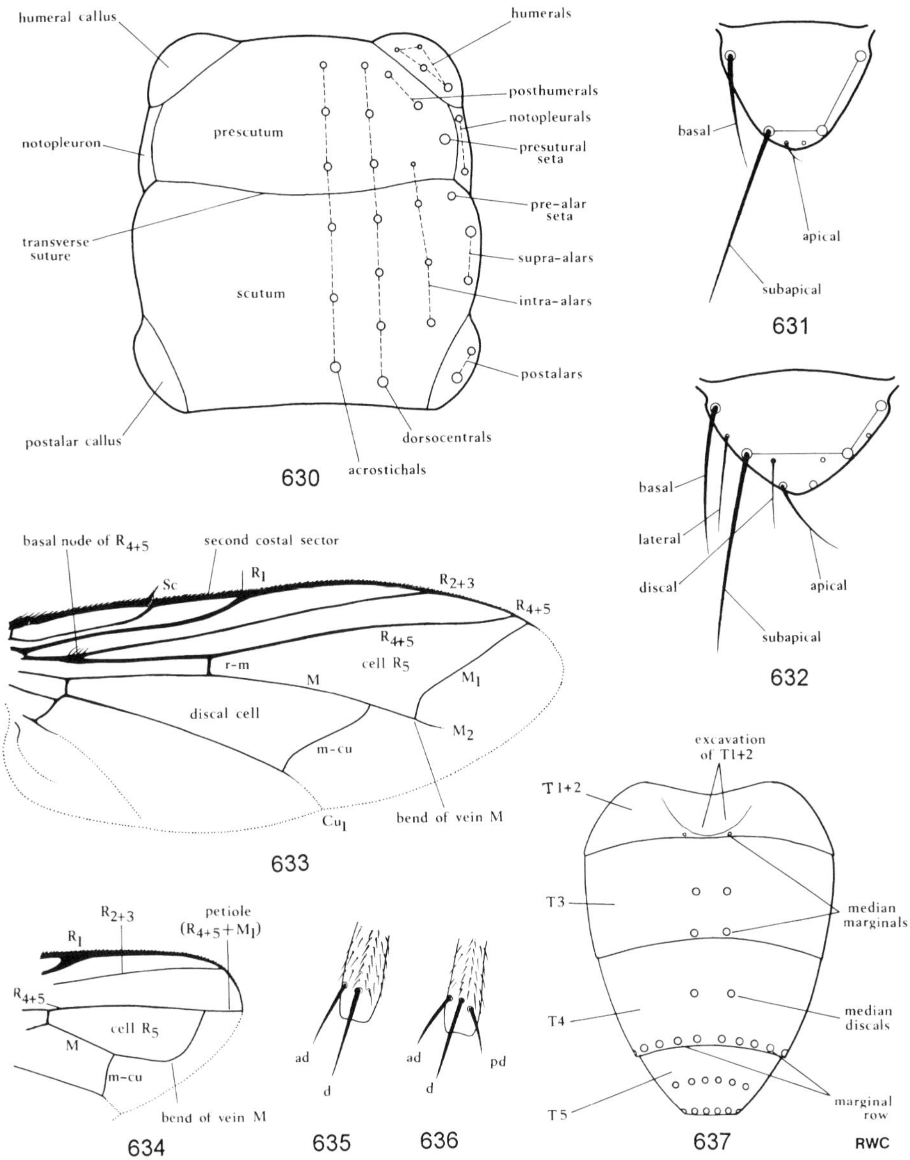

Figs 630–637. Terminology of some thoracic, appendage and abdominal features cited in keys. (630) Dorsal thoracic chaetotaxy (indicated schematically by pore positions of one side only). (631, 632) Scutellar setae (two typical shapes of scutellum, with fine lines connecting basal and subapical pores indicating important differences in proportions). (633) Wing venation of typical tachinid in which cell R_5 is open to wing edge. (634) Apical part of R_5 wing venation in a tachinid with closed cell and 'petiole' present in wing venation. (635, 636) Dorsal view of apex of hind tibia in tachinids with (635) and without (636) posterodorsal preapical setae (pd). (637) Dorsal view of abdomen to show terminology of tergites (T) and main setae (indicated by pore positions) (Crosskey, 1984).

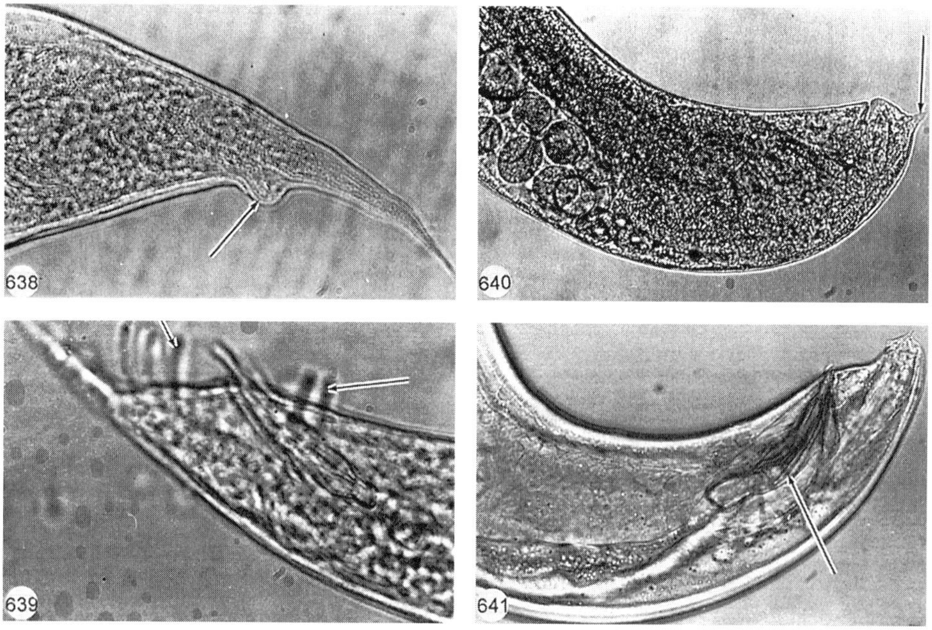

Figs. 638–641. Nematoda. (638) Female tail of *Heterorhabditis* sp. Note the anal swelling (arrow) and the attenuated tail. (639) Male tail of *Heterorhabditis* sp. Note the straight spicules and bursal rays (arrows). (640) Female tail of *Steinernema* sp. Note pointed tail projection (arrow). (641) Male tail of *Steinernema* sp. showing curved spicules (arrow) and absence of bursal rays.

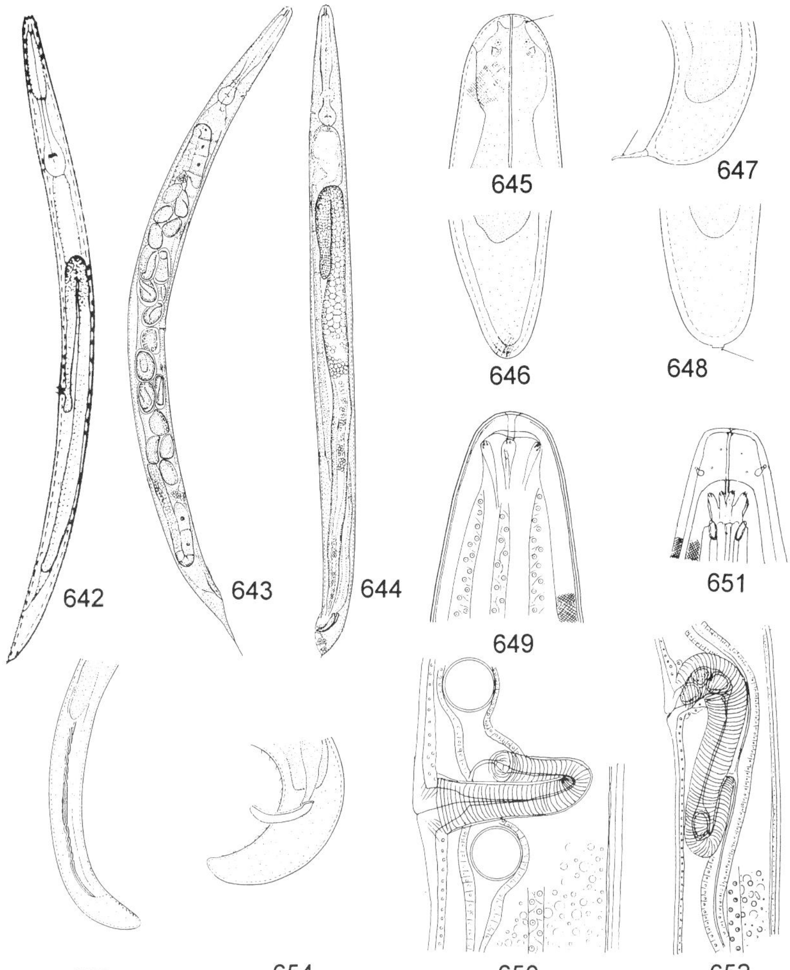

Figs 642–654. Nematoda. (642) Panagrolaimidae: note ovary reflexed past the vulva. (643) Rhabditidae: female. (644) Rhabditidae: male. (645) Ventral view of the head of *Mermis* sp. showing terminal mouth, paired lateral lip papillae (arrowed), four submedian head papillae, paired thread-like amphids and criss-cross fibres in the cuticle. (646) Tail of a juvenile *Mermis* sp. showing criss-cross fibres in the cuticle. (647) Tail of a juvenile mermithid with tail appendage. (648) Tail of a juvenile *Agamermis* showing the tail scar (arrowed). (649, 650) *Ovomermis* sp. head (649) and ovary (650). (651, 652) *Amphimermis* sp. head (651) and ovary (652). (653) Mermithidae, showing long, twisted spicules. (654) Mermithidae, showing short, paired, separate spicules. (649–652 after Rubtsov, 1976.)

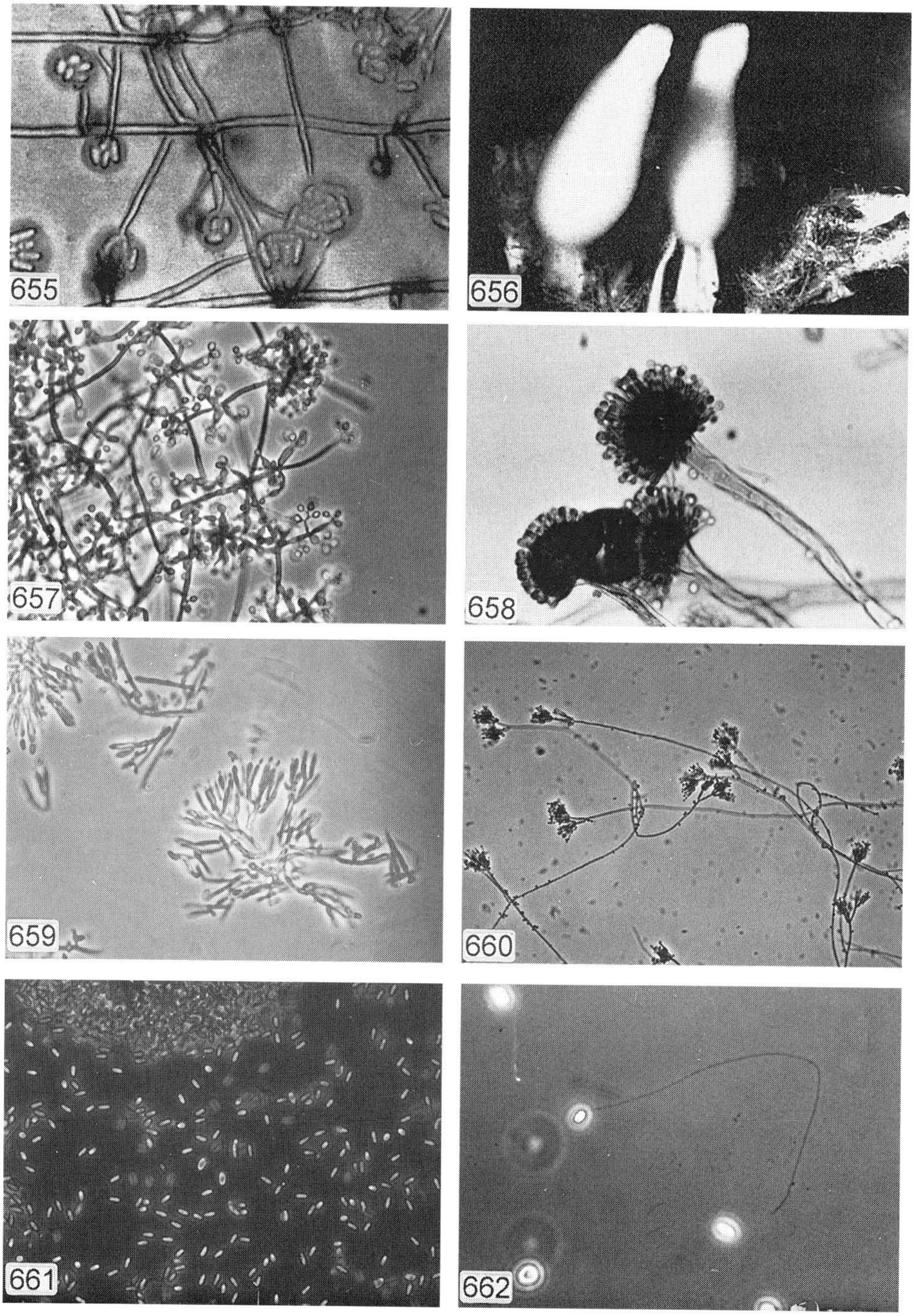

Figs 655–662. (655) *Cephalosporium* sp., conidia in gelatinous clusters. (656) Synnemata of *Isaria* sp. (657) Conidia and phialides of *Beauveria bassiana*. (658) Phialides and spores of *Aspergillus*. (659) Conidia and phialides of *Metarrhizium anisopliae*. (660) Conidia and phialides of *Paecilomyces farinosa*. (661) *Nosema* spores. (662) *Nosema* sp. with polar filament.

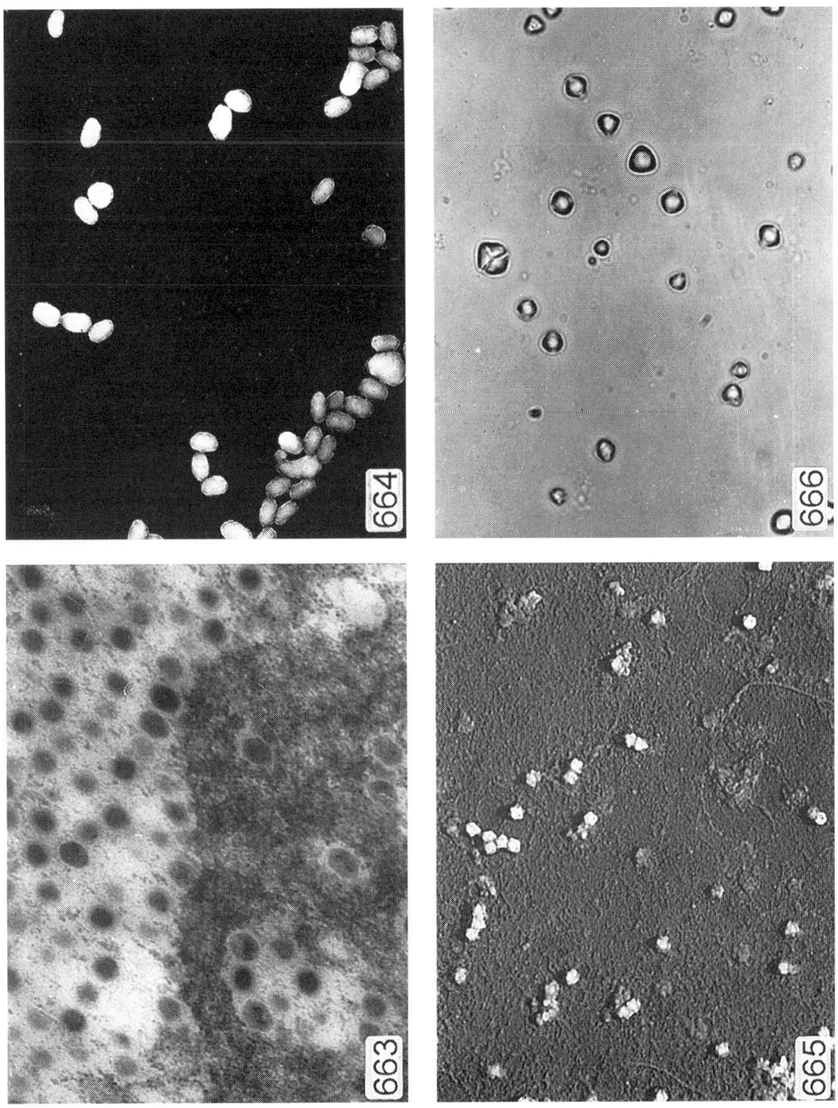

Figs 663–666. (663) Iridovirus. (664) Granulosis virus. (665) Cytoplasmic polyhedrosis virus. (666) Nuclear polyhedrosis virus.

Fig. 667. Improved results are obtained when insecticide spray is directed into whorls of plants.
Fig. 668. On the small scale, granular formulations of insecticides should be applied into whorls of plants.

669

670

Fig. 669. Sorghum plants stacked next to fields in Lesotho.
Fig. 670. Millet stems used for roofing in South Africa.

Index